Precision Agriculture and Crop Monitoring Based on Remote Sensing Methods

Precision Agriculture and Crop Monitoring Based on Remote Sensing Methods

Guest Editors

Renan Falcioni
Renato Herrig Furlanetto
Luis Crusiol

MDPI

Basel • Beijing • Wuhan • Barcelona • Belgrade • Novi Sad • Cluj • Manchester

Guest Editors
Renan Falcioni
Department of Biology
State University of Maringá
Maringá
Brazil

Renato Herrig Furlanetto
Institute of Food and
Agricultural Sciences
University of Florida
Wimauma, FL
USA

Luis Crusiol
Embrapa Soja, National
Soybean Research Center
Brazilian Agricultural
Research Corporation
Londrina
Brazil

Editorial Office
MDPI AG
Grosspeteranlage 5
4052 Basel, Switzerland

This is a reprint of the Special Issue, published open access by the journal *Remote Sensing* (ISSN 2072-4292), freely accessible at: https://www.mdpi.com/journal/remotesensing/special_issues/MFTY977N4C.

For citation purposes, cite each article independently as indicated on the article page online and as indicated below:

Lastname, A.A.; Lastname, B.B. Article Title. *Journal Name* **Year**, *Volume Number*, Page Range.

ISBN 978-3-7258-6610-6 (Hbk)
ISBN 978-3-7258-6611-3 (PDF)
https://doi.org/10.3390/books978-3-7258-6611-3

Cover image courtesy of Renan Falcioni
High-resolution aerial view of agricultural trial plots.

Contents

About the Editors

Renan Falcioni

Renan Falcioni (ORCID: 0000-0002-2343-5045) is a Professor at the State University of Maringá, Brazil, and contributes to graduate and undergraduate teaching and research in agronomy and plant sciences. His research integrates plant physiology with remote sensing, with a focus on hyperspectral and multispectral sensing, chlorophyll fluorescence, and multivariate and machine learning approaches for plant phenotyping, photosynthesis, and crop monitoring. He develops and applies sensing-based methods to quantify plant responses to abiotic and biotic stresses, linking optical signatures to biochemical, anatomical, and molecular traits, and works in collaboration with national and international research teams. He has published peer-reviewed studies across plant science and remote sensing and serves as Guest Editor of the *Remote Sensing* Special Issue "Precision Agriculture and Crop Monitoring Based on Remote Sensing Methods". He obtained his first undergraduate degree in Biological Sciences (BSc and BEd) and a second undergraduate degree in Agronomy (BSc), MSc in Agronomy (Crop Production), first PhD in Biological Sciences (Cell and Molecular Biology) and second PhD in Agronomy (Crop Production, with a focus on Plant Morphogenesis and Biochemical and Molecular Characterisation via Remote Sensing Sensors), and completed postdoctoral studies in Agronomy all at the State University of Maringá, Brazil.

Renato Herrig Furlanetto

Renato Herrig Furlanetto (PhD) (ORCID: 0000-0001-9586-5995) is a Postdoctoral Researcher at the University of Florida, USA, working at the interface of artificial intelligence, remote sensing, and precision agriculture. His research focuses on the development and application of AI-driven approaches—particularly deep learning and computer vision—for crop monitoring, weed detection, and site-specific management in specialty and row crop production systems. He integrates multispectral, hyperspectral, thermal, and LiDAR sensing from UAV, proximal, and ground-based platforms with geospatial analysis and machine learning to support variable-rate applications, smart sprayer technologies, and integrated pest and weed management strategies. His work emphasises translating sensing data into actionable agronomic decisions, improving input efficiency, and enhancing sustainability in agricultural systems. He collaborates with multidisciplinary national and international research teams and has authored numerous peer-reviewed publications across precision agriculture, remote sensing, soil and plant sciences, and artificial intelligence. He also serves as Guest Editor for several international journals, including *Remote Sensing* (MDPI) and *Frontiers in Plant Science*. Renato holds a BSc in Agronomy, an MSc in Agronomy (Soils and Plant Nutrition), and a PhD in Agronomy (Soils and Plant Nutrition) from the State University of Maringá, Brazil, and is currently engaged in postdoctoral research in AI-enabled precision agriculture at the University of Florida.

Luis Crusiol

Luis Crusiol (ORCID: 0000-0002-2387-964X) is a visiting Researcher at Embrapa Soja (National Soybean Research Center, Brazilian Agricultural Research Corporation) in Londrina, Brazil, working at the intersection of precision agriculture, spectroscopy, and remote sensing. His research focuses on multispectral, hyperspectral, and thermal data processing from UAV and satellite platforms, coupled with data-driven modelling, to monitor drought and nutrient stress, quantify crop water status, and predict soil and plant traits relevant to management decisions. He has contributed to international projects with the State University of Maringá, Embrapa Soja, and the Chinese Academy of Agricultural Sciences, including studies using high-spatial-resolution CubeSat NDVI time series for crop monitoring and leaf-based hyperspectral models for soybean water status and soil moisture prediction. He has authored peer-reviewed articles in journals such as *Remote Sensing*, *Agricultural Water Management*, and *International Journal of Remote Sensing*, and serves as a Guest Editor of the *Remote Sensing* Special Issue "Precision Agriculture and Crop Monitoring Based on Remote Sensing Methods (Second Edition)". He holds BSc and BEd degrees in Geography from the State University of Londrina and an MSc in Agronomy (Soils and Plant Nutrition), as well as a PhD in Agronomy (Soils and Plant Nutrition) from the State University of Maringá, Brazil. He has also carried out postdoctoral training at the Chinese Academy of Agricultural Sciences.

Preface

Reliable information is now as important to agriculture as water, seed, and soil. Climate variability, tightening input budgets, and the push for sustainability are all increasing the value of measurements that are timely, spatially explicit, and comparable across seasons. Remote sensing, whether from a handheld spectrometer, a drone, or a satellite, has moved from promise to practice. The remaining challenge is conversion: turning spectra, images, and point clouds into agronomic variables that are accurate enough to act on, robust to changing conditions, and efficient to deliver at scale.

This Special Issue brings together twelve papers that tackle this conversion problem from complementary angles. Several contributions focus on data acquisition and the practicalities that often decide whether an approach is usable outside of a research plot. The UAV LiDAR study in cotton demonstrates how flight settings and processing choices shape point-cloud quality and plant-height estimation, offering guidance for achieving centimetre-level performance without excessive processing. At the plant-counting end of the workflow, the corn detection study evaluates YOLOv9 under contrasting soil colours, growth stages, and flight heights, identifying conditions under which higher-altitude surveys can still deliver reliable detection. For 3D phenotyping and robotics, a new PP3D point-cloud dataset and a dual-representation network are introduced for panoptic plant recognition, addressing issues such as scale variability, occlusion, and class imbalance.

A second cluster of papers concentrates on retrieving biophysical and biochemical traits. Wheat canopy properties are estimated through radiative transfer model inversion supported by artificial neural networks, illustrating how sensor configuration and explicit treatment of soil contributions can improve the retrieval of plant area, nitrogen-related traits, and canopy water content. Above-ground biomass is tackled across the full season for spike-bearing crops using a growth-stage-aware strategy that combines canopy height with spectral information and adapts the model after heading, when traditional spectral relationships often saturate. Nitrogen monitoring is extended to agronomic outcomes in rice: UAV multispectral imagery and a squared chlorophyll-based index are used not only to estimate nitrogen uptake but also to relate canopy status to grain quality.

Stress detection forms a third theme. Hyperspectral imaging is used to map leaf water content in grapevine and to visualise how downy mildew changes water patterns during pathogenesis, including differences between compatible interactions and partial resistance responses. In soybean, hyperspectral spectroscopy combined with machine learning identifies early drought stress markers and shows that a reduced set of informative wavelengths can capture stress levels and biochemical changes with high accuracy, supporting the design of deployable sensing for precision irrigation.

The final theme is scale: moving from plot measurements to field and regional monitoring while keeping uncertainty visible. A physics-driven machine learning framework couples PROSAIL simulations with an optimised neural network to retrieve leaf area index from Sentinel-2 and to quantify uncertainty by separating physical and learning components. Maize mapping is addressed with TWDTW driven by optimised multi-source time-series features from Sentinel-1 and Sentinel-2, demonstrating that carefully selected observation windows can improve mapping accuracy and reduce computational cost. Rice field identification in mountainous terrain is strengthened through SAR-optical fusion, cloud removal, and a tailored graph convolutional network, achieving robust classification even with limited labelled data. Closing the volume, a review and comparative study of almond mapping examines the practical trade-offs between classical machine learning and deep learning segmentation and illustrates how model choice can support multi-year change analysis.

Together, these contributions show a discipline maturing towards operational, confidence-aware remote sensing for agriculture: methods that respect phenology, exploit complementary sensors, and report not only what is predicted but how certain we can be. We hope that this collection is useful for researchers building new algorithms and practitioners and decision-makers looking for dependable tools to support resilient and resource-efficient crop production.

We hope that you enjoy reading this Special Issue.

Renan Falcioni, Renato Herrig Furlanetto, and Luis Crusiol
Guest Editors

remote sensing

MDPI

Article

Optimizing Unmanned Aerial Vehicle LiDAR Data Collection in Cotton Through Flight Settings and Data Processing

Anish Bhattarai [1], Gonzalo J. Scarpin [1,2], Amrinder Jakhar [1], Wesley Porter [3], Lavesta C. Hand [3], John L. Snider [3] and Leonardo M. Bastos [1,*]

1 Department of Crop and Soil Sciences, University of Georgia, Miller Plant Sciences Building, 120 Carlton Street, Athens, GA 30602, USA; ab68010@uga.edu (A.B.)
2 Instituto Nacional de Tecnología Agropecuaria (INTA), Estación Experimental de Reconquista, Ruta Nacional N.° 11, km 773, Reconquista 3560, Santa Fe, Argentina
3 Department of Crop and Soil Sciences, University of Georgia, 2282 Rainwater Rd., Tifton, GA 31793, USA; jlsnider@uga.edu (J.L.S.)
* Correspondence: lmbastos@uga.edu

Abstract: Light Detection and Ranging (LiDAR) technology can be used to assess canopy height in cotton (*Gossypium hirsutum* L.), but standardized data acquisition and processing guidelines are lacking. Accurate canopy height estimation is crucial in cotton for optimizing growth regulator application and maximizing yield. The main goal of this study was to determine the optimal unmanned aerial vehicle flight settings—altitude and speed—and assess specific processing parameters' impact on data accuracy, processing time, and file size. Nine flight settings comprising three altitudes (12.2 m, 24.4 m, and 48.8 m) and three speeds (4.8 km/h, 9.6 km/h, and 14.4 km/h) were tested. LiDAR data were processed using DJI Terra software (v. 4.1.0), where two user-defined processing steps were examined: point-cloud thinning via grid size sub-sampling (0, 10, 20, 30, 40, and 50 cm) and slope classification (flat, gentle, and steep). The optimal flight altitude was 24.4 m, with no effect of flight speed. Grid sub-sampling up to 20 cm produced balanced accuracy, processing time, and file size. The choice of slope category had no significant effect on LiDAR-derived canopy height. These findings contribute to the development of standardized LiDAR data acquisition and processing guidelines for cotton to support crop management decision.

Keywords: point cloud; LAS; remote sensing; row crop; technology; agriculture

1. Introduction

Cotton (*Gossypium hirsutum* L.) is one of the most significant fiber crops globally. The U.S. produced around 14.4 million bales of cotton in the year 2023/24, ranking forth globally in cotton production after India, China, and Brazil [1]. Within the U.S., Texas (30%), Georgia (16%), and Arkansas (9%) are the largest cotton producing states, responsible for 55% of cotton lint production in the U.S. and 12% of world production [2].

Cotton is a perennial plant that is cultivated as an annual crop. This is achieved by using management techniques such as plant growth regulators, which help control growth rate and plant height, preventing the crop from expending too much energy on vegetative growth and directing it towards reproductive growth [3,4]. Balancing vegetative and reproductive sinks is important in cotton to optimize fiber yield and quality [5].

Different metrics exist to determine cotton growth rates and serve as a decision-making tools for deciding the timing and rate of synthetic plant growth regulator applications [6]. The most commonly used metrics are the height-to-node ratio (HNR) (ratio between plant height and total number of nodes) and length of the upper five internodes [7]. HNR values below growth-stage-specific thresholds indicate sub-optimal growth rates, whereas values above the threshold indicate excessive growth rates that need to be controlled using exogenously applied plant growth regulators like mepiquat chloride. For example, during the early bloom stage, if the HNR exceeds 6.4 cm/node, it is recommended to apply mepiquat chloride at a rate ranging from 0.5 to 2 L/ha at a standard concentration of 4.2% w/v [8]. The rate of application is determined based on multiple factors, including the cotton variety, the plant growth stage, the current rate of vegetative growth and weather conditions [8,9].

Traditionally, plant height is determined by visiting the field and collecting manual height measurements on plants across multiple areas in the field. This approach is time- and labor-intensive, and as cotton plants grow and form dense canopies, walking a field can potentially damage the crop. For large-scale commercial farming, such manual methods are impractical for achieving the required precision and speed in decision making. Additionally, capturing height variation accurately across large fields remains one of the biggest constraints, as it directly impacts yield estimation, canopy structure analysis, and variable rate input applications.

To overcome this issue, different technologies have been developed to create estimated continuous maps of plant height, including the unmanned aerial vehicle (UAV) imagery-based canopy height model (difference of the height at the top of the plant canopy and the elevation of the bare ground) [10,11]; tractor-mounted LiDAR (Light Detection and Ranging) [12]; and UAV-based LiDAR systems. The latter has emerged as a solution to bridge the gap between ground-based platforms, known for their high accuracy but low efficiency, and UAV imagery-based platforms, which offer extensive coverage but limited detail [13]. LiDAR directly measures crop height by emitting laser pulses, making it more accurate than UAV-mounted passive spectral imaging, which relies on ambient light and requires additional processing [14]. Height estimation from spectral data depends on photogrammetry techniques like structure-from-motion (SfM), which reconstructs 3D models based on overlapping images [15]. This approach requires ideal lighting, minimal occlusion, and accurate tie points, with additional steps to separate crop height from ground elevation. Errors in shadowed areas, vegetation movement, or point matching can reduce accuracy, making LiDAR the more reliable option for precise height measurements, especially in dense or complex canopies [16].

LiDAR application in row crops such as maize started around the mid-2010s, when it was used for evaluating crop structure [17] and predicting yield potential [18]. Studies on the use of UAV-mounted LiDAR to estimate plant height in cotton are scant [19]. This is a limitation especially because different flight settings like altitude and speed, and data processing steps like point cloud classification, grid size for height extraction and canopy modeling, can contribute to the accuracy of LiDAR-based height estimates, yet no studies exist that provide best-practice recommendations for the cotton crop.

For example, adjusting the point cloud density—the number of laser returns per square meter—is key [20] to generate accurate LiDAR-based height estimates. Achieving a high point cloud density enhances measurement accuracy but requires a UAV to fly at lower altitudes, slower speeds, or with greater overlap between flight paths, all of which extend flight time and increase the cost and complexity of data collection and data processing. A high point cloud density poses a challenge as it generates larger data files [21], which demand significant storage capacity and increase computing power requirements and/or processing times. Therefore, identifying flight characteristics and data processing steps that optimize crop height estimation accuracy while minimizing flight time and data storage and processing requirements is of utmost importance.

Another key processing step with LiDAR data is grid-based sub-sampling, where the raw LiDAR point cloud data are selectively thinned based on the highest-height point within each predefined grid cell [22]. By identifying a grid cell size that balances data retention and reduction, one can minimize file size, reduce computational requirements, and improve processing speed, all of which contribute to more efficient and cost-effective UAV-based LiDAR applications [20,23]. The decision on the size of the grid cell to be utilized at this step is left to the user, and no best-practice recommendations exist when using LiDAR in row crops in general, or cotton in specific.

Another important step when processing LiDAR data in DJI Terra (DJI, China), a LiDAR mapping and analysis software developed by DJI, is selecting the appropriate field slope type for the initial classification of point clouds. This choice helps the software distinguish between ground and non-ground points based on terrain variation. The classification process is influenced by the field slope characteristics, as variations in topography affect the differentiation between ground and non-ground points [24]. Like grid cell size selection, the decision on the slope choice is left to the user, and no best-practice recommendations exist.

The overall aim of this study was to establish best-practice guidelines for UAV LiDAR flight settings and data processing to ensure high-quality data acquisition for accurate cotton canopy height estimation. The first objective of this study was to determine and select the optimal UAV flight settings—altitude and speed—that provide the most accurate LiDAR-derived crop height estimates. Based on the best flight settings identified in the first objective, the second and third objectives focused on evaluating the impact of two user-dependent data processing steps on crop height estimation. Specifically, the second objective assessed how defining different grid cell sizes for sub-sampling point clouds affects crop height accuracy, processing time, and file size. The third objective examined the effect of selecting the field slope type on the accuracy of crop height estimation.

2. Materials and Methods

2.1. Experimental Site Description

This experiment was conducted in two different irrigated fields in Georgia, USA near Midville (32.875761, −82.222197) and Watkinsville (33.870858, −83.452716) for the year 2024. The field in Midville had a flat terrain predominately characterized by Tifton soil series and the Dothan soil series (fine-loamy, kaolinitic, thermic plinthic kandiudults), which are both deep, well-drained, and have a fine-loamy texture [25]. The field in Watkinsville has a somewhat flat terrain characterized by Cecil sandy loam soil with 2% to 6% slope [25]. Cotton was planted at a row spacing of 91 cm and seeding rate of 90,000 seeds/ha. The varieties planted were NexGen 3195 Bollgard® 3 XtendFlex® in Watkinsville and Dyna-Gro 3799 Bollgard® 3 XtendFlex® in Midville, and were selected according to local recommendations. Planting occurred on 26th of June in Midville and 2nd of July in Watkinsville.

Each trial was implemented as a nitrogen (N) fertilizer side-dress rate study, with N rate treatments of 0 + 0, 112 + 0, 40 + 26, 40 + 60, 40 + 94, and 40 + 128 kg/ha, denoted as preplant + side-dress application rates. The experiment followed a randomized complete block design with four blocks. In brief, all plots, except for the control (0 + 0) and the 112 + 0 treatment, received a baseline application of 40 kg N/ha at planting, applied using a tractor-mounted boom sprayer. The fertilizer source for the application was urea ammonium nitrate (UAN) 28% N, which is a liquid fertilizer. At the blooming crop stage (approximately 49–52 days after planting), side-dress N fertilizer treatments were applied with a variable rate applicator. For this manuscript, the treatment design of N rates was utilized only to generate variability in crop height, and therefore further statistical analysis did not incorporate N rate as an explanatory variable.

2.2. Aerial LiDAR Data Collection and Pre-Processing

The workflow for data acquisition, pre-processing, and analysis consisted of four main steps: (i) field characterization, (ii) data acquisition with different flight settings, (iii) data processing, and (iv) validation (Figure 1). Field characterization in relation to slope is an important initial step used to define processing settings in the DJI Terra (DJI, Shenzen, China) v. 4.1.0 software. DJI Terra provides the user with the option of identifying the field as flat (1–4 degrees slope), gentle (4–7 degrees slope), or steep (8–12 degrees slope) as one of the first steps in the processing pipeline. To conduct field characterization in our study, elevation data were retrieved for each field from the Google Earth Engine data catalogue of USGS 3 DEP 1 m National Map using the geemap library [26] in Python 3.9. Then, the elevation raster was imported into R software v. 2024.12.0 (R development team, 2024), where the terra package [27] v. 1.8-42 was used to calculate the slope of the field. The field was then classified into one of flat (1–4 degrees), gentle (4–7 degrees), or steep (8–12 degrees) slope to match the options provided by DJI Terra.

Data acquisition was conducted on August 10th in Midville (76 days after planting) and August 27th in Watkinsville (86 days after planting) in 2024. Cotton typically reaches its first flower at 50–60 days after planting, with peak bloom occurring 20–30 days later [28]. This period is critical for assessing management practices like plant growth regulator (PGR) application, making accurate crop height data essential for informed decision making. First, 10 plants from each of 28 plots had their height measured manually using a measuring tape, for a total of 280 plants measured in each field. On the same day, a DJI Matrice 350 (DJI, Shenzhen, China) UAV mounted with a Zenmuse L2 (DJI, Shenzhen, China) LiDAR sensor was used to collect the sensor data from the field from each combination of three different flight altitudes (12.2, 24.4, 48.8 m) and three different flight speeds (4.8, 9.6, 14.4 km/h), resulting in a total of nine unique flight settings per location. Each combination of flight altitude and speed was flown once per location, totaling 18 flights across the two locations. Flight altitudes of 12.2, 24.4, and 48.8 m and speeds of 4.8, 9.6, and 14.4 kmph were selected to balance data quality and operational efficiency in UAV LiDAR surveys. Lower altitudes (12.2 m) increase point density and detail, ideal for precise canopy measurements, while higher altitudes (48.8 m) enable faster, broader coverage with an acceptable resolution for field-scale assessment [29–31]. Similarly, slower speeds (4.8 kmph) improve point cloud density and accuracy by allowing for more LiDAR pulses per unit area, whereas faster speeds (14.4 kmph) prioritize survey efficiency at the cost of reduced detail [32]. For speed selection, 4.8 km/h was the default setting in the DJI flight planner application. To assess its impact on data acquisition and quality, additional speeds were chosen at double (9.6 km/h) and triple (14.4 km/h) the default value. Higher speeds were chosen to allow for a comparative analysis of speed's influence on data accuracy and efficiency.

The Zenmuse L2 integrates the Livox LiDAR module, a high-precision inertial moment unit (IMU), a mapping camera, and three-axis gimbals. The IMU includes a three-axis accelerometer, a three-axis angular velocity and a barometric altimeter, which is used to measure attitude. All the flights were conducted at a frontal overlap of 80% and side overlap of 10%. The scanner pulse repetition rate was set at 240 kHz, and the scanner angle to 90 degrees. The UAV was connected with a D-RTK 2 high precision GNSS mobile station (DJI, Shenzhen, China) that supports four global satellite navigation systems, BeiDou, GPS, Galileo, GLONASS, and 11-frequency satellite reception, providing real-time differential corrections to facilitate the aircraft in centimeter-level precision positioning and help to get the precise geolocation of each of the cloud points from the LiDAR.

Data were processed on a desktop computer with AMD Ryzen Threadripper PRO 5975WX 32, core 3.6 GHz processor, 128 GB RAM. The movements of the UAV were corrected using the DJI Terra software (DJI, Shenzhen, China) which combines the GNSS correction, and the movement data recorded by the IMU, with a manufacture-reported precision of ~4 cm for all flights. Data processing involved the steps of point cloud classification, DTM generation, normalization, DSM generation, canopy modeling, rasterization, and metric extraction using different software.

To produce a geo-referenced point cloud, the flight path and raw point cloud data were combined in DJI Terra software (DJI, Shenzhen, China). After uploading the data into the software, the user can initiate the subsampling method using a grid-based approach to systematically reduce point cloud density. This method involved partitioning the area into predefined grid sizes (00 cm or no sub-sampling, 10 cm, 20 cm, 30 cm, 40 cm, and 50 cm) and retaining only the highest point within each grid cell. The point clouds were also classified in the software as ground and non-ground points using a field slope-based approach. Previous studies have employed algorithms such as the cloth simulation filter [33] and progressive morphological filter [34], as well as software platforms including CloudCompare 2.10 [32] and ArcGIS 10.4.1 (ESRI, Redlands, CA, USA) [35] to perform this task. In this study, point cloud ground classification was conducted using DJI Terra software as it has been specifically developed by DJI to ensure seamless integration with their hardware systems, such as the Zenmuse L2 LiDAR sensor. In DJI Terra, at this stage, the user is required to select the field slope classification. For the fields in this study, the correct classification was "flat" based on open-source elevation data obtained pre-flight at the field characterization step. Because an improper selection of field slope can potentially impact the quality of the processed data, the impact of the slope type parameter selected during ground point classification in DJI Terra was evaluated by processing the LiDAR data using all the available slope settings: 'Flat', 'Gentle', and 'Steep'. These settings represent varying terrain slope conditions and influence how the software identifies and classifies ground points within the LiDAR dataset. The final co-registered and geo-referenced point cloud datasets were stored in LAS format (Figure 1).

The LAS file exported from DJI Terra was then further processed in R using the lidR package [36] v. 2.0.0 in five sub-steps: (i) digital terrain model generation; (ii) height normalization, (iii) noise filtering, (iv) digital surface model generation, and (v) canopy height model generation. The first sub-step in processing the LiDAR data in R involved creating a DTM from the LAS file obtained from DJI Terra. This was achieved using the triangular irregular network (TIN) [37] method, which leverages ground point data to model the DTM. TIN was selected as the triangulation method due to being straightforward and computationally efficient, as it does not require complex adjustments or additional parameters compared to other methods [24].

The second sub-step in processing was the height normalization which was performed based on the DTM generated on the previous sub-step. In this process, the terrain height values in the DTM were set as a baseline (considered zero). Also, any points that were within 6 cm height from the ground were considered as ground so to smooth out small bumps on the field, and the remaining points above this threshold were regarded as representing heights above the terrain, effectively isolating the plant heights. A similar approach was adopted by other studies [38,39], where a ground offset of 5 to 10 cm was applied to account for ground surface roughness. The third sub-step in processing was noise filtering which was implemented to improve data quality by flagging and filtering out noisy points in the point cloud. A noisy point was considered as any point that fell below the DTM (indicating negative heights). This filtering step ensured that only relevant data points were retained for accurate canopy height measurements.

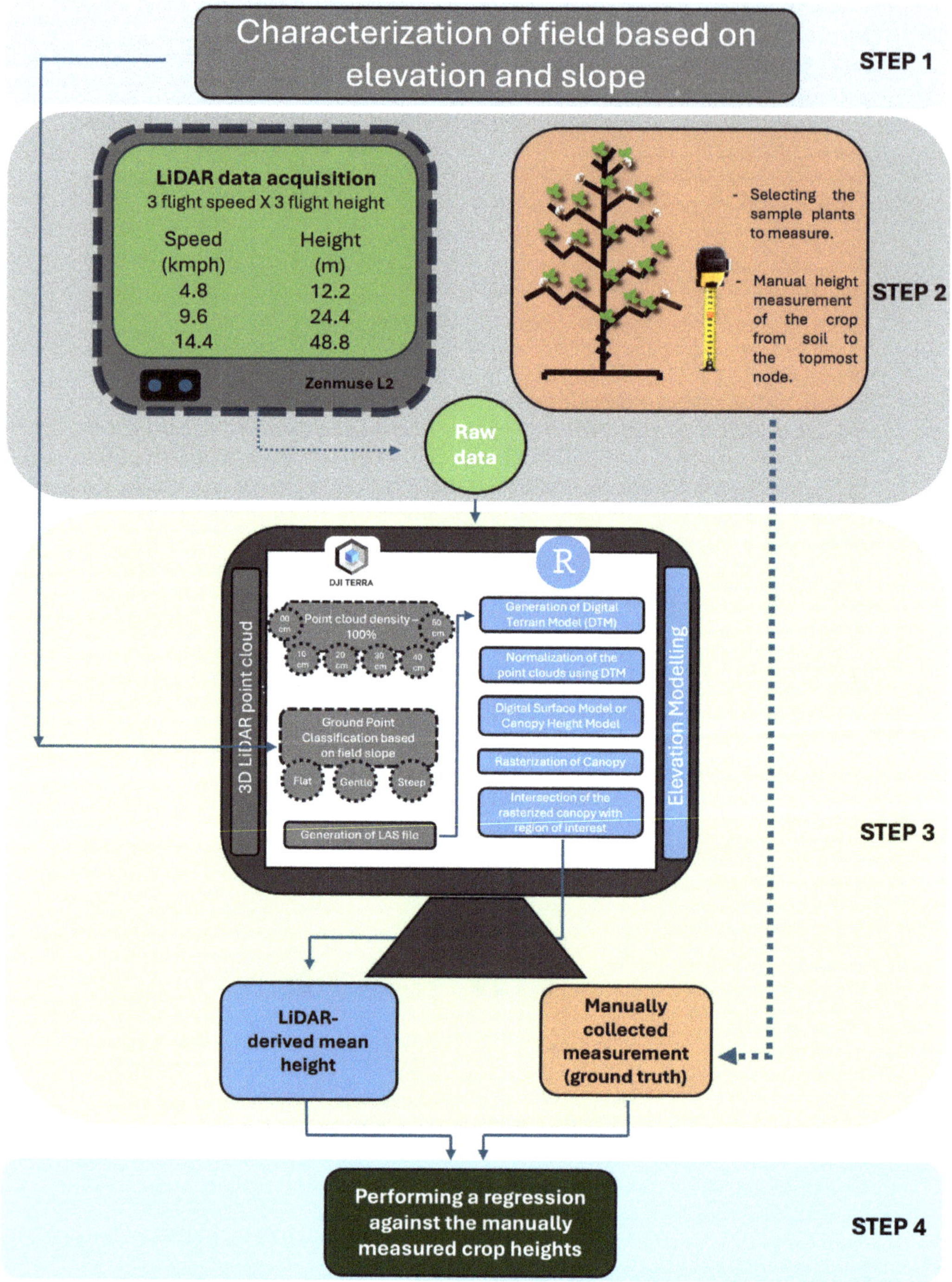

Figure 1. LiDAR-based crop height estimation workflow including step 1: field characterization (elevation and slope); step 2: data acquisition (varying flight settings and manual plant height measurements); step 3: data processing (canopy modeling and metric extraction); and step 4: validation (regression between LiDAR plant height estimate and ground truth plant height).

The fourth sub-step in processing was to generate a digital surface model (DSM) using a TIN and incorporating all points from the LiDAR point cloud, including those at higher elevations such as crop canopies, trees, or structures. Unlike a DTM, which uses only the lowest points to represent the bare ground, a DSM captures the highest elevation at each location. The TIN for a DSM is formed by connecting these points into an irregular mesh, creating a 3D representation of the entire surface, including all objects above the ground. Finally, the fifth sub-step in processing was to generate the canopy height model (CHM) which was calculated by subtracting the DSM and the DTM which is a commonly used method for canopy height estimation [34,40]. The CHM represents the canopy height, providing a measure of plant height above ground level. The resulting CHM was then rasterized to facilitate further spatial analysis.

To match the CHM with the manual measurements of plant height, the rasterized CHM was intersected with the bounding boxes around 10 marked plants per plot. The mean height for these 10 plants was calculated for both LiDAR and manual measurements. These averages were then compared using regression analysis.

2.3. Data Analysis and Validation

To assess the impact of flight altitude and speed, data from both locations were analyzed together to account for the potential site-specific variations as our objective was to determine the optimal flight setting for quality data retrieval. A regression model was used to examine the relationship between LiDAR-derived height and manual height measurements, with location incorporated as a random effect. However, for evaluating the optimal grid sub-sampling method, the data from each location were analyzed separately. This separate analysis was necessary because the two locations differed in field size, which directly influenced the amount of LiDAR data collected and the overall processing time. To assess grid sub-sampling performance, three metrics were evaluated: (i) file size (smaller file sizes improve storage efficiency), (ii) processing time (faster processing reduces computational burden), and (iii) height prediction quality (to ensure the subsampled data retained sufficient accuracy). To evaluate height prediction quality, a regression analysis comparing the LiDAR-derived heights obtained through different grid sub-sampling methods against manually measured height was conducted. The coefficient of determination (R^2, higher is better), root mean square error (RMSE, lower is better), and mean absolute error (MAE, lower is better) were used as the evaluation metrics. Furthermore, the intercept and slope of the equation were tested using the linearHypothesis function from the car package [41] v. 3.1.3 for significant differences from 0 and 1, respectively, with a significance level of 0.05.

Moreover, to evaluate the effect of the slope choice (flat, gentle and steep) for point cloud classification in DJI Terra software to derive the LiDAR height, the difference between manual and LiDAR-derived height was explained as a function of slope type, location, and their interaction as fixed effects, while the block nested within location was treated as random effect. Similarly, to further extend our understanding of the slope choice for point cloud classification the data from both locations were analyzed together. A regression model was used to examine the relationship between LiDAR-derived height and manual height measurements, with location incorporated as a random effect to account for variability between sites.

3. Results

The relationship between manual height measurements and LiDAR-derived height measurements for each combination of flight altitude and speed was analyzed combining the data from both locations (Figure 2). Among the tested settings, the UAV flight at a speed of 14.5 km/h and an altitude of 24.4 m had the best performance, achieving the highest R^2 value (0.97) and the lowest RMSE (3.65 cm) and MAE (3.09 cm). At higher altitudes (48.8 m), the accuracy decreased significantly as evidenced by slightly lower R^2 values (0.86–0.87), and increased RMSE (10.18–11.71 cm) and MAE (8.87–9.62 cm) values, indicating a weaker correlation between manual and LiDAR-derived height. In contrast, at the moderate altitude of 24.4 m, increasing speed improved accuracy, with the highest R^2 (0.97) and the lowest RMSE (3.65 cm) and MAE (3.09 cm) achieved at 14.5 km/h. However, at higher altitudes, faster speeds lead to reduced accuracy evidenced by increased RMSE. Overall, optimal accuracy is achieved at moderate altitudes (24.4 m) with higher flight speeds, while higher altitude compromised measurement precision, particularly at faster speeds. The intercept and slope were significantly different from 0 and 1, respectively, only for the highest flight altitude (48.8 m) at all three speed settings (Figure 2).

Using the results above, data from the UAV flight setting that optimized plant height prediction (i.e., flight altitude of 28.8 m and flight speed of 14.4 km/h) were selected and used in the downstream analysis of sub-sampling grid size for each location separately. For Watkinsville, as the grid sub-sampling size increased from 0 cm to 50 cm, there was a significant reduction in processing time and file size, demonstrating that larger grid sub-sampling improves computational efficiency and reduces storage requirements (Table 1). Point cloud density decreased drastically by 99.6%, from 5834 points/m^2 at 0 cm to just 21 points/m^2 at 50 cm. Total processing time followed a similar trend, reduced by 55.8% from 206 s at 0 cm to 91 s at 50 cm. File size decreased significantly, declining by 99.5% from 812 MB at 0 cm to just 4 MB at 50 cm (Table 1).

Table 1. Effect of grid sub-sampling size on LiDAR point cloud density, processing time and file size with data obtained from Watkinsville.

Grid Sub-Sampling (cm)	Point Cloud Density (points/m^2)	Benchmarking Time DJI Terra (3D LiDAR Point Cloud Modeling) (s)	Benchmarking Time R (Elevation Modeling) (s)	Total Time for Processing (s)	File Size (mb)
0	5834	143	63	206	812
10	539	101	55	156	149
20	110	83	37	120	35
30	41	82	21	103	13
40	20	80	19	99	6
50	21	77	14	91	4

An increase in sub-sampling grid size and concurrent decreases in processing time and file size concur in lower quality of LiDAR data by reducing the point cloud density (Table 1). In Watkinsville, the greatest agreement between manual height and LiDAR-derived height was obtained at no sub-sampling (i.e., sub-sampling grid size of 0 cm, R^2 = 0.88, RMSE = 2.86, MAE = 2.34 cm), with a 10 cm sub-sampling grid size producing acceptable agreement (R^2 = 0.85, RMSE = 4.29 cm, MAE = 3.56 cm, Figure 3). With the coarsest sub-sampling grid size, fit metrics were significantly compromised (R^2 = 0.38, RMSE = 10.78 cm, MAE = 7.97 cm). Both the intercept and slope were significantly different from 0 and 1, respectively, for the coarsest sub-sampling grid size (50 cm). Only the intercept showed a significant difference from 0 for the 30 cm and 40 cm sub-sampling grid sizes. In contrast, neither the intercept nor the slope differed significantly from 0 and 1, respectively, for grid sizes smaller than 30 cm (Figure 3).

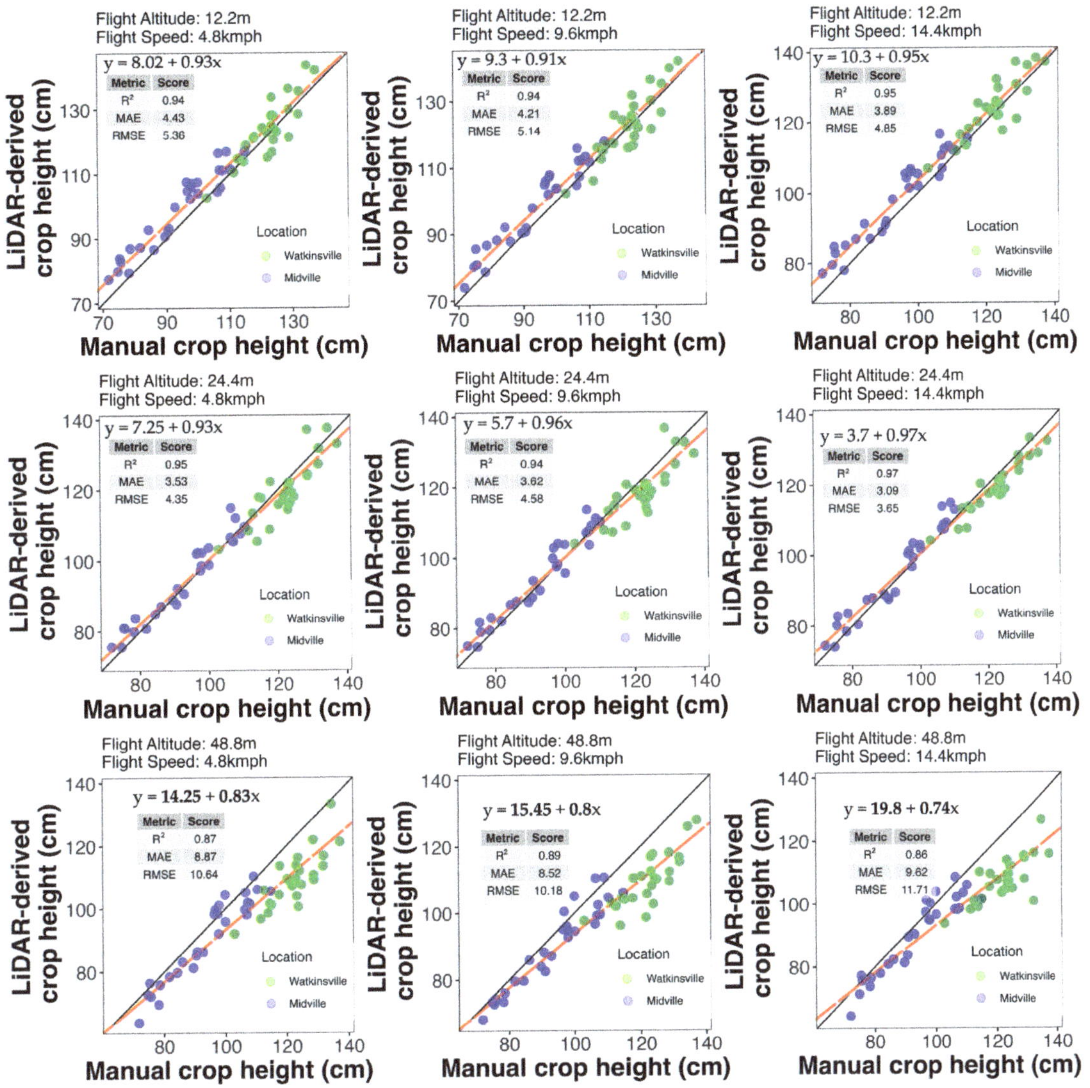

Figure 2. Scatterplots of manual measured plant height (cm) and LiDAR-derived plant height (cm) where points represent heights from two locations: Watkinsville (green) and Midville (blue). Each panel represents data from one unmanned aerial vehicle (UAV) flight altitude and speed combination. The solid black line represents the 1:1 line. The red line represents the regression line, with its equation shown inside each panel. The bolded intercept and slope values indicate a significant difference from 0 and 1, respectively, at $p < 0.05$; non-bold values are not significantly different. R^2 is the coefficient of determination. R^2 is the coefficient of determination. MAE is the mean absolute error. RMSE is the root mean squared error.

For Midville, as the grid sub-sampling size increased from 0 cm to 50 cm, there was a significant decline in all parameters, demonstrating the efficiency of larger grid sub-sampling (Table 2). As grid sub-sampling size increased from 0 cm to 50 cm, point cloud density decreased significantly, dropping by approximately 99.7% from 2637 points/m^2 to 9 points/m^2. Similarly, total processing time reduced by 57.4% from 183 s to 78 s, with benchmarking time in DJI Terra and R showing reductions of 47.2% and 80.4%, respectively. Additionally, file size decreased 99.6% from 718 MB to just 3 MB (Table 2).

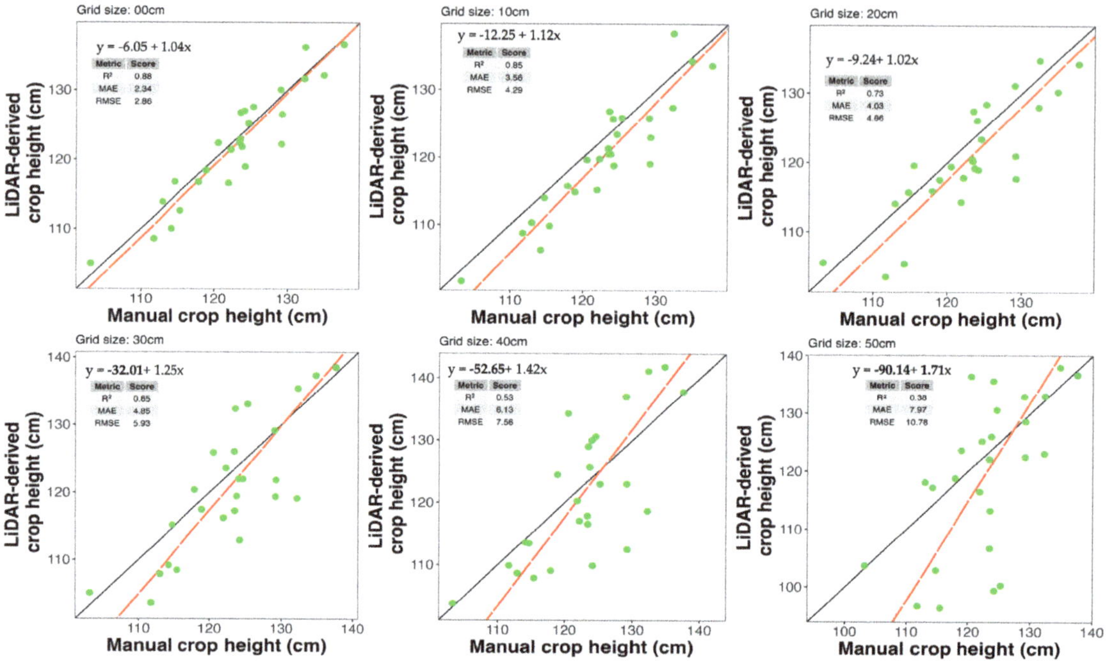

Figure 3. Scatterplots of manual height and LiDAR-derived height in Watkinsville, GA. Each panel represents data from different sub-sampling grid sizes in Watkinsville. The title on left top of each graph shows the grid size used for sub-sampling. The solid black line represents the 1:1 line. The red line represents the regression line, with its equation showed inside each panel. The bolded intercept and slope values indicate a significant difference from 0 and 1, respectively, at $p < 0.05$; non-bold values are not significantly different. R^2 is the coefficient of determination. MAE is the mean absolute error. RMSE is the root mean squared error. Grid size: 00cm denotes the original resolution layer, meaning the data is in its full detail with no subsampling performed.

Table 2. Effect of grid sub-sampling size on LiDAR point cloud density, processing time, and file size with data obtained from Midville.

Grid Sub-Sampling (cm)	Point Cloud Density (points/m^2)	Benchmarking Time DJI Terra (3D LiDAR Point Cloud Modeling) (s)	Benchmarking Time R (Elevation Modeling) (s)	Total Time for Processing (s)	File Size (mb)
0	2637	127	56	183	718
10	362	77	40	117	99
20	87	70	26	96	24
30	34	70	17	87	9
40	17	67	13	80	5
50	9	67	11	78	3

In Midville, the greatest agreement between manual height and LiDAR-derived height was obtained at no sub-sampling (i.e., sub-sampling grid size of 0 cm, $R^2 = 0.94$,

RMSE = 4.66 cm, MAE = 3.78 cm), with an up to 20 cm sub-sampling grid size still producing acceptable agreement (R^2 = 0.91, RMSE = 4.19 cm, Figure 4). With the coarsest sub-sampling grid size, fit metrics were significantly compromised (R^2 = 0.63, RMSE = 9.51 cm, MAE = 6.86 cm). Only the intercept showed a significant difference from 0 for coarsest and 40 cm sub-sampling grid sizes. In contrast, neither the intercept nor the slope differed significantly from 0 and 1, respectively, for grid sizes smaller than 40 cm (Figure 4).

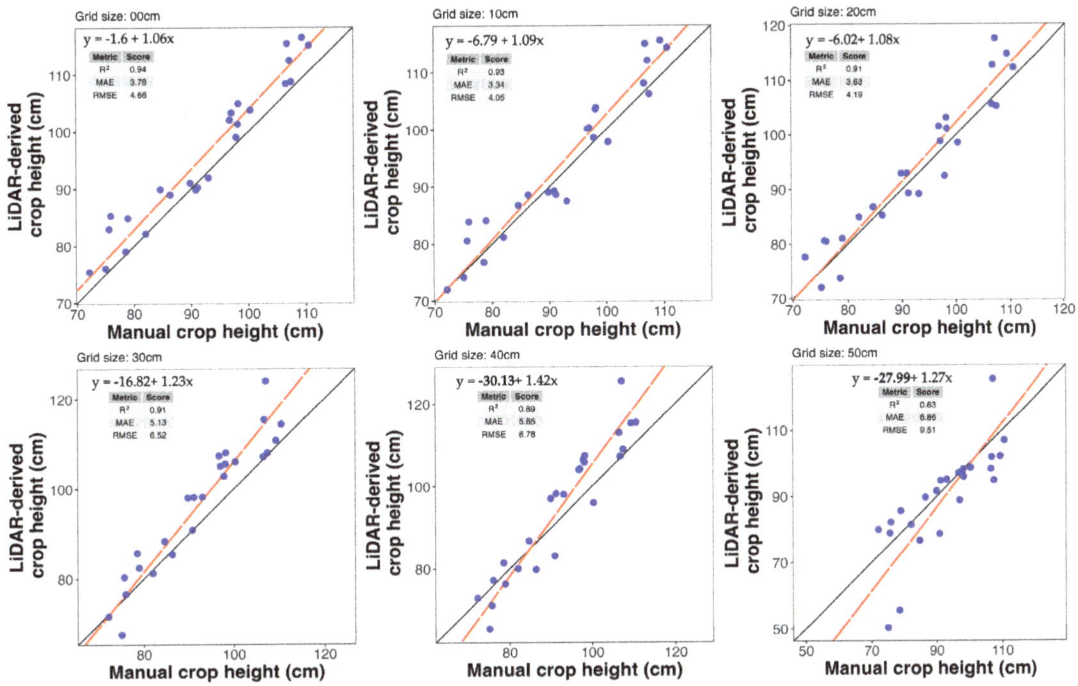

Figure 4. Scatterplots of manual height and LiDAR-derived height in Midville, GA. Each panel represents data from different sub-sampling grid sizes in Midville. The title on left top of each graph shows the grid size used for sub-sampling. The solid black line represents the 1:1 line. The red line represents the regression line, with its equation showed inside each panel. The bolded intercept and slope values indicate a significant difference from 0 and 1, respectively, at $p < 0.05$; non-bold values are not significantly different. R^2 is the coefficient of determination. MAE is the mean absolute error. RMSE is the root mean squared error. Grid size: 00cm denotes the original resolution layer, meaning the data is in its full detail with no subsampling performed.

Slope choice during the processing of the point clouds had no significant effect on the mean difference between manual and LiDAR-derived heights across two locations. The mean difference between manual and LiDAR-derived heights were −1.1 cm, −0.57 cm, and −0.24 cm for flat, gentle, and steep, respectively.

Similarly, the tested slope choices (flat, gentle and steep) in DJI Terra software for point cloud classification showed similar performance across both locations when compared to manually measured heights, as reflected by comparable R^2 (0.93–0.95), RMSE (4.42–4.77 cm) and MAE (3.41–3.6 cm, Figure 5). The intercept and slope were not significantly different from 0 and 1, respectively, for any of the three slope choices (Figure 5).

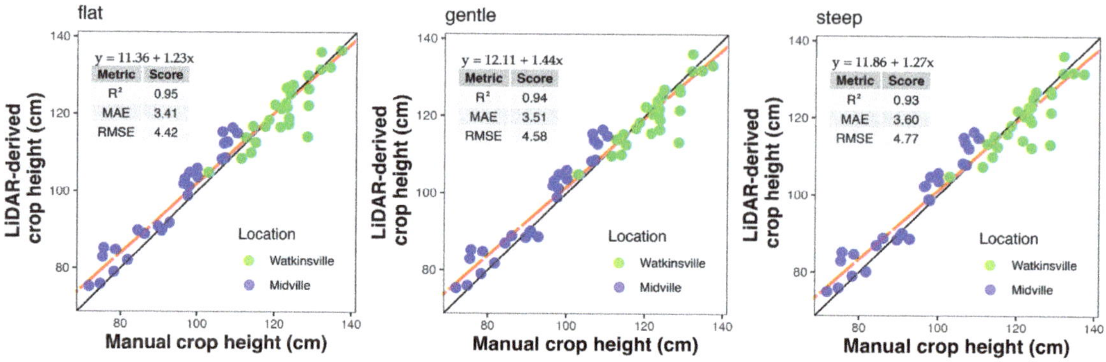

Figure 5. Scatterplots of manual height and LiDAR-derived height. Each panel represents data from different slope choices for point cloud classification for two locations, Watkinsville (green) and Midville (blue). The title on left top of each graph shows the slope choice made. Solid black line represents the 1:1 line. Red line represents the regression line, with its equation showed inside each panel. The bolded intercept and slope values indicate a significant difference from 0 and 1, respectively, at $p < 0.05$; non-bold values are not significantly different. MAE is the mean absolute error. RMSE is the root mean squared error.

4. Discussion

In this study, the effect of UAV flight parameters and data processing steps on cotton plant height estimation accuracy was investigated. This research establishes a framework for developing optimized flight plans and data processing protocols aimed at improving the accuracy, efficiency, and scalability of LiDAR-derived information. This work is unique and fills an important methodological gap in using UAV LiDAR for row crop height estimation. In specific, this research can serve as a foundational framework for selecting flight settings during the peak bloom stage of cotton that balance data quality with practical considerations, such as minimizing processing time and maintaining manageable file sizes. Additionally, it holds particular significance for cotton, where management practices like PGR application rely on accurate plant height measurements to optimize production [42]. Various studies have found that the application of PGR produced more flowers [4], reduced the spatial yield variability within a field [43], controlled excessive vegetative growth [44, 45], and, in some cases, increased cotton yield [46–48]. Additionally, the application of PGR has been shown to enhance cotton fiber quality by promoting earlier maturation of the cotton plant. In this context, [49] reported that delayed harvesting negatively affects the cotton fiber quality. Moreover, UAV-based LiDAR data can support variable rate application of PGR which has shown to reduce the PGR application rates ranging from 10 to 53% and PGR cost by 17% when compared to fixed-rate applications [50].

In this study, a UAV-mounted LiDAR sensor at different flight settings was tested as it often provides better height estimation compared to multispectral imagery or SfM technology. A study on wheat [51] reported that the multispectral imagery under-estimated the height and was less accurate compared to the LiDAR estimated height. Similarly, several studies reported that the correlation between digitally derived canopy height and ground measurements was found to range from 0.5 to 0.87 [52,53], which is low compared to the correlation obtained from LiDAR, ranging from 0.8 to 0.97 [12,54].

Our results showed that varying UAV flight speed had a minimal impact on the accuracy of plant height assessment compared to changes in flight altitude. This aligns with previous studies that have highlighted the significant influence of altitude on point cloud density and the subsequent accuracy of vegetation structure measurements [21,30].

Lower flight altitudes generally produce denser point clouds, which enhance measurement precision but may introduce more noise due to increased overlap and redundant data capture. Studies like [55] on forest canopy and [54] maize and soybean suggested that it is not necessary to have high LiDAR point cloud density to achieve the desired estimation precision of vegetation parameters. Conversely, higher altitudes result in sparser point clouds, potentially reducing measurement accuracy [31,33,56]. A study conducted in maize and soybeans [54] found that when a UAV-mounted LiDAR was flown at the height of 150 m above ground level, a weak correlation of R^2 (0.64 and 0.4, respectively, for each crop) was found. For instance, [57] demonstrated that a flight altitude of 85 m provided more accurate tree height estimations compared to 145 m, while ensuring sufficient coverage and sensor safety in complex terrains. These findings suggest that lower flight altitudes improve height estimation accuracy, likely due to higher point cloud density. However, there is a difference in LiDAR application between tree canopies and row crops such as cotton, stemming from their distinct structural characteristics. Unlike trees, which have loose and varied canopies, cotton canopies have dense, uniform foliage that heavily obscures the ground. This dense foliage complicates the accurate estimation of DTM, as also noted by [53,58,59]. In our study, lower flight altitudes were more effective in capturing detailed ground information under dense cotton canopies, demonstrating the importance of tailoring flight parameters to the specific structural characteristics of the vegetation being analyzed. [12] reported that a LiDAR sensor mounted on a tractor achieved a root mean square error (RMSE) of 6.5 cm in field-based cotton height measurements. In comparison, our study demonstrated that under optimal UAV flight settings, an RMSE of 3.65 cm could be achieved, indicating a higher level of accuracy. Furthermore, while both approaches provide valuable canopy height estimations, UAV-based LiDAR offers a significantly more efficient and rapid data acquisition method.

Our study further demonstrated the value of optimizing data processing techniques for LiDAR datasets. For instance, sub-sampling the point cloud to a grid size of up to 20 cm (86–90% reduction in point cloud density from the original data) maintained sufficient accuracy for height measurements while significantly reduced file size. This aligns with recent advancements in point cloud processing, such as shape-aware down-sampling techniques, which have been shown to preserve structural integrity while eliminating unnecessary data points, thereby improving the overall quality of the dataset [60]. Furthermore, reducing the point cloud density by 25% still resulted in accurate LiDAR-derived crop height estimations, maintaining an R^2 of 0.96 with the actual measured crop height [61]. Also, reducing the point cloud density to approximately 7.32 points/m^2 maintained a strong correlation between LiDAR-derived and manually measured canopy heights, with R^2 values ranging from 0.65 to 0.75. However, when the point cloud density was further reduced to 0.074 points/m^2, the accuracy declined significantly, with R^2 dropping to 0.3 [30]. This suggests that while some level of point cloud thinning is feasible without substantial accuracy loss, excessive reduction in point density compromises the reliability of canopy height estimations. The reduction in point cloud density also translated to smaller file sizes and shorter processing times, enhancing the practicality of working with LiDAR datasets in resource-limited environments. Larger file sizes tend to slow down analysis and increase computational overhead, which can be a significant bottleneck for large-scale studies.

Our study revealed that varying slope parameters during the classification of LiDAR point clouds into ground and non-ground points had no significant effect on the classification accuracy. This finding suggests that the underlying algorithms utilized by DJI Terra are robust and effectively handled variations in user-defined slope settings during the data processing workflow.

5. Conclusions

This study provided key recommendations for UAV LiDAR flight settings and data processing steps for cotton height estimation and serves as an initial step for new users of this technology. UAV flight speed had no significant effect on the accuracy of LiDAR data acquisition, whereas flight altitude significantly influenced data quality. As flight altitude increased, point cloud density decreased, which led to a gradual loss of structural integrity in the captured vegetation data, thereby affecting the accuracy of canopy height estimation. For any optimal flight height, to address challenges associated with longer data processing times and larger file sizes, sub-sampling the point cloud up to 20 cm grid size provided an effective balance between data reduction and information retention. Additionally, our findings indicate that the slope choices during point cloud classification did not affect the accuracy of LiDAR-derived measurements, further highlighting the effectiveness of DJI Terra's automated classification algorithm in ensuring reliable data outputs.

Although this work represents an important step in proposing recommendations for UAV-based LiDAR use in cotton, some limitations and gaps remain. The major limitation of this research was its focus on a single crop, cotton, which restricts the generalizability of findings to other crops with different canopy structures and growth patterns. Also, this study focused on a single growth stage when the canopy was fully developed and provided maximum coverage. While these results are highly relevant for critical cotton growth stages such as flowering and boll development—key periods for evaluating plant height for management practices—they may not be as applicable to the early growth stages when canopy cover is minimal. Early-stage flights would likely exhibit different LiDAR responses, including greater ground visibility and less point cloud interference from overlapping leaves. Moreover, other influential flight factors such as flight overlap, scan angles, and UAV trajectory planning were not addressed. Flight overlaps and scan angles, for example, play a critical role in ensuring data completeness and accuracy, particularly in terrains or fields with uneven surfaces [62].

Future studies should include UAV flights on various crops with differing canopy structures, such as maize, wheat, and soybean. Additionally, flights at different growth stages should be further tested to improve generalizability across the growing season. Research on varying flight overlap, scan angles, and UAV trajectories could further elucidate other aspects of flight settings to optimize data collection for different crops and field conditions, and merit attention in future studies.

Author Contributions: Conceptualization, A.B. and L.M.B.; Methodology, A.B. and L.M.B.; Software, A.B.; Validation, A.B.; Formal analysis, A.B.; Investigation, A.B. and L.M.B.; Resources, L.M.B.; Data curation, A.B.; Writing—original draft, A.B. and J.L.S.; Writing—review & editing, G.J.S., A.J., W.P., L.C.H. and L.M.B.; Visualization, A.B.; Supervision, L.M.B.; Project administration, L.M.B.; Funding acquisition, L.M.B. All authors have read and agreed to the published version of the manuscript.

Funding: This research was partially funded by the Institute of Integrative Precision Agriculture at the University of Georgia, and Deere & Company.

Data Availability Statement: The raw data supporting the conclusions of this article will be made available by the authors on request.

Conflicts of Interest: The authors declare no conflict of interest.

References

1. Meyer, L. *Cotton and Wool Outlook: October 2024*. 2024. Available online: https://ers.usda.gov/sites/default/files/_laserfiche/outlooks/110207/CWS-24j.pdf?v=42812 (accessed on 20 February 2025).

2. USDA. Cotton Explorer International Production Assessment Division, United States Department of Agriculture. 2024. Available online: https://ipad.fas.usda.gov/cropexplorer/cropview/commodityView.aspx?cropid=2631000 (accessed on 18 February 2025).
3. Cothren, J.T.; Oosterhuis, D.M. Use of Growth Regulators in Cotton Production. In *Physiology of Cotton*; Stewart, J.M., Oosterhuis, D.M., Heitholt, J.J., Mauney, J.R., Eds.; Springer: Dordrecht, The Netherlands, 2010; pp. 289–303. [CrossRef]
4. Pettigrew, W.T. Effects of Different Seeding Rates and Plant Growth Regulators on Early-planted Cotton. *J. Cotton Sci.* **2005**, *9*, 189–198.
5. Bradow, J.M.; Davidonis, G.H. Quantitation of Fiber Quality and the Cotton Production-Processing Interface: A Physiologist's Perspective. *J. Cotton Sci.* **2000**, *4*, 34–64.
6. Oosterhuis, D.M. Growth and Development of a Cotton Plant. In *Nitrogen Nutrition of Cotton: Practical Issues*; John Wiley & Sons, Ltd.: Hoboken, NJ, USA, 1990; pp. 1–24. [CrossRef]
7. Kerby, T.A.; Plant, R.E.; Horrocks, R.D. Height-to-Node Ratio as an Index of Early Season Cotton Growth. *J. Prod. Agric.* **1997**, *10*, 80–83. [CrossRef]
8. Hand, L.C.; Snider, J.; Roberts, P. *Cotton Growth Monitoring and PGR Management*; Circular 1244; University of Georgia: Athens, GA, USA, 2022.
9. Byrd, S. *Plant Growth Regulators in Cotton*; PSS-2189; Oklahoma Cooperative Extensive Service: Oklahoma City, OK, USA, 2019.
10. Kawamura, K.; Asai, H.; Yasuda, T.; Khanthavong, P.; Soisouvanh, P.; Phongchanmixay, S. Field phenotyping of plant height in an upland rice field in Laos using low-cost small unmanned aerial vehicles (UAVs). *Plant Prod. Sci.* **2020**, *23*, 452–465. [CrossRef]
11. Lu, W.; Okayama, T.; Komatsuzaki, M. Rice Height Monitoring between Different Estimation Models Using UAV Photogrammetry and Multispectral Technology. *Remote Sens.* **2022**, *14*, 78. [CrossRef]
12. Sun, S.; Li, C.; Paterson, A.H.; Jiang, Y.; Xu, R.; Robertson, J.S.; Snider, J.L.; Chee, P.W. In-field High Throughput Phenotyping and Cotton Plant Growth Analysis Using LiDAR. *Front. Plant Sci.* **2018**, *9*, 16. [CrossRef] [PubMed]
13. Watanabe, K.; Guo, W.; Arai, K.; Takanashi, H.; Kajiya-Kanegae, H.; Kobayashi, M.; Yano, K.; Tokunaga, T.; Fujiwara, T.; Tsutsumi, N.; et al. High-Throughput Phenotyping of Sorghum Plant Height Using an Unmanned Aerial Vehicle and Its Application to Genomic Prediction Modeling. *Front. Plant Sci.* **2017**, *8*, 421. [CrossRef]
14. Amann, M.-C.; Bosch, T.M.; Lescure, M.; Myllylae, R.A.; Rioux, M. Laser ranging: A critical review of unusual techniques for distance measurement. *Opt. Eng.* **2001**, *40*, 10–19. [CrossRef]
15. Snavely, N.; Seitz, S.M.; Szeliski, R. Modeling the World from Internet Photo Collections. *Int. J. Comput. Vis.* **2008**, *80*, 189–210. [CrossRef]
16. Tao, W.; Lei, Y.; Mooney, P. Dense point cloud extraction from UAV captured images in forest area. In Proceedings of the 2011 IEEE International Conference on Spatial Data Mining and Geographical Knowledge Services, Fuzhou, China, 29 June–1 July 2011; pp. 389–392. [CrossRef]
17. Garrido, M.; Paraforos, D.S.; Reiser, D.; Vázquez Arellano, M.; Griepentrog, H.W.; Valero, C. 3D Maize Plant Reconstruction Based on Georeferenced Overlapping LiDAR Point Clouds. *Remote Sens.* **2015**, *7*, 17077–17096. [CrossRef]
18. Anderson, S.L.; Murray, S.C.; Malambo, L.; Ratcliff, C.; Popescu, S.; Cope, D.; Chang, A.; Jung, J.; Thomasson, J.A. Prediction of Maize Grain Yield before Maturity Using Improved Temporal Height Estimates of Unmanned Aerial Systems. *Plant Phenome J.* **2019**, *2*, 190004. [CrossRef]
19. Xu, W.; Yang, W.; Wu, J.; Chen, P.; Lan, Y.; Zhang, L. Canopy Laser Interception Compensation Mechanism—UAV LiDAR Precise Monitoring Method for Cotton Height. *Agronomy* **2023**, *13*, 2584. [CrossRef]
20. Peng, X.; Zhao, A.; Chen, Y.; Chen, Q.; Liu, H. Tree Height Measurements in Degraded Tropical Forests Based on UAV-LiDAR Data of Different Point Cloud Densities: A Case Study on Dacrydium pierrei in China. *Forests* **2021**, *12*, 328. [CrossRef]
21. Singh, K.K.; Chen, G.; McCarter, J.B.; Meentemeyer, R.K. Effects of LiDAR point density and landscape context on estimates of urban forest biomass. *ISPRS J. Photogramm. Remote Sens.* **2015**, *101*, 310–322. [CrossRef]
22. Jakubowski, M.K.; Guo, Q.; Kelly, M. Tradeoffs between lidar pulse density and forest measurement accuracy. *Remote Sens. Environ.* **2013**, *130*, 245–253. [CrossRef]
23. Béjar-Martos, J.A.; Rueda-Ruiz, A.J.; Ogayar-Anguita, C.J.; Segura-Sánchez, R.J.; López-Ruiz, A. Strategies for the Storage of Large LiDAR Datasets—A Performance Comparison. *Remote Sens.* **2022**, *14*, 2623. [CrossRef]
24. Zhang, K.; Chen, S.-C.; Whitman, D.; Shyu, M.-L.; Yan, J.; Zhang, C. A progressive morphological filter for removing nonground measurements from airborne LIDAR data. *IEEE Trans. Geosci. Remote Sens.* **2003**, *41*, 872–882. [CrossRef]
25. Soil Survey Staff. Soil Survey Geographic Database (SSURGO) | Natural Resources Conservation Service. 2025. Available online: https://www.nrcs.usda.gov/resources/data-and-reports/soil-survey-geographic-database-ssurgo (accessed on 10 February 2025).
26. Wu, Q. geemap: A Python package for interactive mapping with Google Earth Engine. *J. Open Source Softw.* **2020**, *5*, 2305. [CrossRef]

27. Hijmans, R.J.; Barbosa, M.; Bivand, R.; Brown, A.; Chirico, M.; Cordano, E.; Dyba, K.; Pebesma, E.; Rowlingson, B.; Sumner, M.D. *terra: Spatial Data Analysis* (Version 1.8-15) [Computer software]. 2025. Available online: https://cran.rproject.org/web/packages/terra/index.html (accessed on 10 February 2025).
28. Mauney, J.R.; Stewart, J.M. *Cotton Physiology*. 1986. Available online: https://www.cotton.org/foundation/reference-books/cotton-physiology/upload/COTTON-PHYSIOLOGY.pdf (accessed on 10 February 2025).
29. Kuželka, K.; Slavík, M.; Surový, P. Very High Density Point Clouds from UAV Laser Scanning for Automatic Tree Stem Detection and Direct Diameter Measurement. *Remote Sens.* **2020**, *12*, 1236. [CrossRef]
30. Luo, S.; Chen, J.M.; Wang, C.; Xi, X.; Zeng, H.; Peng, D.; Li, D. Effects of LiDAR point density, sampling size and height threshold on estimation accuracy of crop biophysical parameters. *Opt. Express* **2016**, *24*, 11578–11593. [CrossRef]
31. Boucher, P.B.; Hockridge, E.G.; Singh, J.; Davies, A.B. Flying high: Sampling savanna vegetation with UAV-lidar. *Methods Ecol. Evol.* **2023**, *14*, 1668–1686. [CrossRef]
32. Brede, B.; Calders, K.; Lau, A.; Raumonen, P.; Bartholomeus, H.M.; Herold, M.; Kooistra, L. Non-destructive tree volume estimation through quantitative structure modelling: Comparing UAV laser scanning with terrestrial LIDAR. *Remote Sens. Environ.* **2019**, *233*, 111355. [CrossRef]
33. Zhang, Q.; Hu, M.; Zhou, Y.; Wan, B.; Jiang, L.; Zhang, Q.; Wang, D. Effects of UAV-LiDAR and Photogrammetric Point Density on Tea Plucking Area Identification. *Remote Sens.* **2022**, *14*, 1505. [CrossRef]
34. Navarro, A.; Young, M.; Allan, B.; Carnell, P.; Macreadie, P.; Ierodiaconou, D. The application of Unmanned Aerial Vehicles (UAVs) to estimate above-ground biomass of mangrove ecosystems. *Remote Sens. Environ.* **2020**, *242*, 111747. [CrossRef]
35. Moudrý, V.; Gdulová, K.; Fogl, M.; Klápště, P.; Urban, R.; Komárek, J.; Moudrá, L.; Štroner, M.; Barták, V.; Solský, M. Comparison of leaf-off and leaf-on combined UAV imagery and airborne LiDAR for assessment of a post-mining site terrain and vegetation structure: Prospects for monitoring hazards and restoration success. *Appl. Geogr.* **2019**, *104*, 32–41. [CrossRef]
36. Roussel, J.-R.; Auty, D.; Coops, N.C.; Tompalski, P.; Goodbody, T.R.H.; Meador, A.S.; Bourdon, J.-F.; de Boissieu, F.; Achim, A. lidR: An R package for analysis of Airborne Laser Scanning (ALS) data. *Remote Sens. Environ.* **2020**, *251*, 112061. [CrossRef]
37. Chang, Y.; Habib, A.; Lee, D.; Yom, J. Automatic classification of lidar data into ground and non-ground points. *Int. Arch. Photogramm. Remote Sens.* **2008**, *37*, 463–468.
38. Jimenez-Berni, J.A.; Deery, D.M.; Rozas-Larraondo, P.; Condon, A.G.; Rebetzke, G.J.; James, R.A.; Bovill, W.D.; Furbank, R.T.; Sirault, X.R.R. High Throughput Determination of Plant Height, Ground Cover, and Above-Ground Biomass in Wheat with LiDAR. *Front. Plant Sci.* **2018**, *9*, 237. [CrossRef]
39. Walter, J.D.C.; Edwards, J.; McDonald, G.; Kuchel, H. Estimating Biomass and Canopy Height With LiDAR for Field Crop Breeding. *Front. Plant Sci.* **2019**, *10*, 1145. [CrossRef]
40. Alexander, C.; Korstjens, A.H.; Hill, R.A. Influence of micro-topography and crown characteristics on tree height estimations in tropical forests based on LiDAR canopy height models. *Int. J. Appl. Earth Obs. Geoinf.* **2018**, *65*, 105–113. [CrossRef]
41. Fox, J.; Weisberg, S. *An R Companion to Applied Regression*, 3rd ed.; Sage: Washington, DC, USA, 2019; Available online: https://us.sagepub.com/en-us/nam/an-r-companion-to-applied-regression/book246125#resources (accessed on 20 January 2025).
42. Hand, L.C.; Stanley, C.; Jenna, V.; Glen, H.; Bob, K.; Liu, Y. 2024 Georgia Cotton Production Guide. *UGA Extension Cotton Team 2360 Rainwater Road Tifton, GA 31793*. 2024. Available online: https://secure.caes.uga.edu/extension/publications/files/pdf/AP%20124-4_2.PDF (accessed on 10 February 2025).
43. Vaz, C.M.P.; Franchini, J.C.; Speranza, E.A.; Inamasu, R.Y.; De CJorge, L.A.; Rabello, L.M.; De ONLopes, I.; Das Chagas, S.; De Souza, J.L.R.; De Souza, M.; et al. Zonal Application of Plant Growth Regulator in Cotton to Reduce Variability and Increase Yield in a Highly Variable Field. *J. Cotton Sci.* **2023**, *23*, 60–73. [CrossRef]
44. Fang, S.; Gao, K.; Hu, W.; Wang, S.; Chen, B.; Zhou, Z. Foliar and seed application of plant growth regulators affects cotton yield by altering leaf physiology and floral bud carbohydrate accumulation. *Field Crops Res.* **2019**, *231*, 105–114. [CrossRef]
45. Samples, C.; Dodds, D.M.; Catchot, A.L.; Golden, B.R.; Gore, J.; Varco, J.J. Determining optimum plant growth regulator application rates in response to fruiting structure and flower bud removal. *J. Cotton Sci.* **2015**, *19*, 359–367. [CrossRef]
46. Leal, A.J.F.; Piati, G.L.; Leite, R.C.; Zanella, M.S.; Osorio, C.R.W.S.; Lima, S.F. Nitrogen and mepiquat chloride can affect fiber quality and cotton yield. *Revista Brasileira de Engenharia Agrícola e Ambiental* **2020**, *24*, 238–243. [CrossRef]
47. Tung, S.A.; Huang, Y.; Hafeez, A.; Ali, S.; Liu, A.; Chattha, M.S.; Ahmad, S.; Yang, G. Morpho-physiological Effects and Molecular Mode of Action of Mepiquat Chloride Application in Cotton: A Review. *J. Soil Sci. Plant Nutr.* **2020**, *20*, 2073–2086. [CrossRef]
48. Sawana, Z.M.; Hafez, S.A.; Alkassas, A.R. Nitrogen, potassium and plant growth retardant effects on oil content and quality of cotton seed. *Grasas y Aceites* **2007**, *58*, 243–251. [CrossRef]
49. Scarpin, G.J.; Cereijo, A.E.; Dileo, P.N.; Winkler, H.H.M.; Muchut, R.J.; Lorenzini, F.G.; Roeschlin, R.A.; Paytas, M. Delayed harvest time affects strength and color parameters in cotton fibers. *Agron. J.* **2023**, *115*, 583–594. [CrossRef]
50. Trevisan, R.G.; Vilanova Júnior, N.S.; Eitelwein, M.T.; Molin, J.P. Management of Plant Growth Regulators in Cotton Using Active Crop Canopy Sensors. *Agriculture* **2018**, *8*, 101. [CrossRef]

51. Madec, S.; Baret, F.; De Solan, B.; Thomas, S.; Dutartre, D.; Jezequel, S.; Hemmerlé, M.; Colombeau, G.; Comar, A. High-throughput phenotyping of plant height: Comparing unmanned aerial vehicles and ground LiDAR estimates. *Front. Plant Sci.* **2017**, *8*, 2002. [CrossRef]
52. Wang, H.; Singh, K.; Poudel, H.; Ravichandran, P.; Natarajan, M.; Eisenreich, B. Estimation of Crop Height and Digital Biomass from UAV-Based Multispectral Imagery. In Proceedings of the 2023 13th Workshop on Hyperspectral Imaging and Signal Processing: Evolution in Remote Sensing (WHISPERS), Athens, Greece, 31 October–2 November 2023; pp. 1–4. [CrossRef]
53. Dandois, J.P.; Ellis, E.C. High spatial resolution three-dimensional mapping of vegetation spectral dynamics using computer vision. *Remote Sens. Environ.* **2013**, *136*, 259–276. [CrossRef]
54. Luo, S.; Liu, W.; Zhang, Y.; Wang, C.; Xi, X.; Nie, S.; Ma, D.; Lin, Y.; Zhou, G. Maize and soybean heights estimation from unmanned aerial vehicle (UAV) LiDAR data. *Comput. Electron. Agric.* **2021**, *182*, 106005. [CrossRef]
55. Roussel, J.-R.; Caspersen, J.; Béland, M.; Thomas, S.; Achim, A. Removing bias from LiDAR-based estimates of canopy height: Accounting for the effects of pulse density and footprint size. *Remote Sens. Environ.* **2017**, *198*, 1–16. [CrossRef]
56. Storch, M.; Jarmer, T.; Adam, M.; De Lange, N. Systematic Approach for Remote Sensing of Historical Conflict Landscapes with UAV-Based Laserscanning. *Sensors* **2021**, *22*, 217. [CrossRef]
57. Maguya, A.S.; Junttila, V.; Kauranne, T. Adaptive algorithm for large scale dtm interpolation from lidar data for forestry applications in steep forested terrain. *ISPRS J. Photogramm. Remote Sens.* **2013**, *85*, 74–83. [CrossRef]
58. Tsoulias, N.; Paraforos, D.S.; Fountas, S.; Zude-Sasse, M. Estimating Canopy Parameters Based on the Stem Position in Apple Trees Using a 2D LiDAR. *Agronomy* **2019**, *9*, 740. [CrossRef]
59. Stereńczak, K.; Ciesielski, M.; Balazy, R.; Zawiła-Niedźwiecki, T. Comparison of various algorithms for DTM interpolation from LIDAR data in dense mountain forests. *Eur. J. Remote Sens.* **2016**, *49*, 599–621. [CrossRef]
60. Li, D.; Zhou, Z.; Wei, Y. Unsupervised shape-aware SOM down-sampling for plant point clouds. *ISPRS J. Photogramm. Remote Sens.* **2024**, *211*, 172–207. [CrossRef]
61. Hämmerle, M.; Höfle, B. Effects of Reduced Terrestrial LiDAR Point Density on High-Resolution Grain Crop Surface Models in Precision Agriculture. *Sensors* **2014**, *14*, 24212–24230. [CrossRef]
62. Ding, Q.; Chen, W.; King, B.; Liu, Y.; Liu, G. Combination of overlap-driven adjustment and Phong model for LiDAR intensity correction. *ISPRS J. Photogramm. Remote Sens.* **2013**, *75*, 40–47. [CrossRef]

remote sensing

MDPI

Article

Imaging of Leaf Water Patterns of *Vitis vinifera* Genotypes Infected by *Plasmopara viticola*

Erich-Christian Oerke * and Ulrike Steiner

INRES–Plant Pathology, Rheinische Friedrich-Wilhelms-Universitaet Bonn, Nussallee 9, D-53115 Bonn, Germany; u-steiner@uni-bonn.de
* Correspondence: ec-oerke@uni-bonn.de

Abstract: The water status of plants is affected by abiotic and biotic environmental factors and influences the growth and yield formation of crops. Assessment of the leaf water content (LWC) of grapevine using hyperspectral imaging (1000–2500 nm) was investigated under controlled conditions for its potential to study the effects of the downy mildew pathogen *Plasmopara viticola* on LWC of host tissue in compatible and incompatible interactions. A calibration curve was established for the relationship between LWC and the Normalized Difference Leaf Water Index ($NDLWI_{1937}$) that uses spectral information from the water absorption band and NIR for normalization. LWC was significantly lower for abaxial than for adaxial leaf sides, irrespective of grapevine genotype and health status. Reflecting details of leaf anatomy, vascular tissue exhibited effects reverse to intercostal areas. Effects of *P. viticola* on LWC coincided with the appearance of first sporangia on the abaxial side and increased during further pathogenesis. Continuous water loss ultimately resulted in tissue death, which progressed from the margins into central leaf areas. Tiny spots of brown leaf tissue related to the reaction of partial resistant cultivars could be monitored only at the sensor's highest spatial resolution. Proximal sensing enabled an unprecedented spatial resolution of leaf water content in host–pathogen interactions and confirmed that resistance reactions may produce a combination of dead and still-living cells that enable the development of biotrophic *P. viticola*.

Keywords: downy mildew; hyperspectral imaging; leaf anatomy; leaf morphology; water balance

1. Introduction

Grapevine (*Vitis vinifera* L.) was grown for the production of table grapes and wine, in 2023, on about 6.6 million ha worldwide (https://www.fao.org/faostat/en/#data, accessed on 3 January 2025). As many vineyards are in regions where water is a limiting resource and drought stress may reduce the quality and quantity of grapevine production, monitoring of the crop water status is very important in viticulture and has become a major research area [1].

Stomatal conductance, water potential, canopy temperature, and vegetation indices are widely used to characterize the water status of grapevine and other crop species [1–4]. Measurements of stomatal conductance and water potential of leaves, stems, or plants result in a mean value per measuring object, deliver no spatial information, and require contact between plant tissue and sensor. In contrast, thermal and spectral sensing and imaging provide various vegetation indices, are contactless and may be applied on various platforms and scales; e.g., ground-based spectrometry and thermography [5], aerial thermography [6,7], aerial multispectral and thermal imagery [8–10], and ground-based hyperspectral imaging [11].

The effects of water content on leaf reflectance are greatest in spectral bands centered at 1450, 1940, and 2500 nm. Secondary effects occur at 980 nm and 1240 nm [12–14]. In satellite and airborne imagery, the SWIR wavebands are usually saturated with atmospheric water vapor, which has a higher concentration over the full atmospheric column between sensor and vegetation than the liquid water of the plant tissue [15]. As the SWIR water absorption bands are not suitable for remote sensing of the water content of vegetation, wavebands from the NIR range are used instead. Spectral band ratios developed to estimate the water content of vegetation from remote sensing include the normalized difference water index (NDWI; [16]) and the plant water index (PWI; [17]). These spectral ratio indices use the (broad) wavebands provided by satellite-borne or airborne sensors.

Wavebands around 1200 nm and 1400 nm have been used to predict the water status of grapevines, also from aerial SWIR data [18,19]. The best models used SWIR data (750–1700 nm) from UAV and allowed predicting stem water potential with R^2 of 0.54. Random Forest and Support Vector Machine models used wavebands located around 1420 nm and 1460 nm, and 1050 nm and 1200 nm, respectively [20].

In proximal sensing, the content of atmospheric water vapor of the column between leaf tissue and sensor is considerably lower than the content of liquid water in living plant tissue. Reflection from ground-based hyperspectral imaging at 1410 nm and 1520 nm was correlated with the water status of tomato leaves [21]. For the period 2019 to 2022, Sanaeifar et al. [22] listed 41 studies in which proximal hyperspectral sensors were used for drought detection of various plant species on the leaf and canopy scale; 20 non-imaging sensors, 18 in the VISNIR range, and 2 using SWIR data, in grapevine [23] and in oak phenotyping [24], respectively. Oerke & Steiner [25] successfully used a normalized difference water index $NDWI_{1937}$ for characterizing the water status of apple leaves infected by *Venturia inaequalis*.

Knowledge of the effect of leaf diseases on the water status of crop plants is limited (see review by Grimmer et al. [26]). Leaf diseases such as downy mildew, powdery mildew, and various leaf spot diseases are described to affect yield formation of grapevine and to aggravate the effects of drought conditions on host physiology without knowing details on their effect on leaf water status [27,28].

Downy mildew of grapevine is caused by the obligate biotrophic oomycete *Plasmopara viticola* (Berk. & M.A. Curtis) Berl & De Toni and is one of the most devastating diseases worldwide [29]. Despite frequent fungicide use, it causes considerable production losses in grapevine production. As the completion of its life cycle depends on host plant resistance and environmental conditions, the use of (partially) resistant cultivars and the use of fungicides according to forecast models, depending on leaf wetness duration, are essential tools in downy mildew control [30–33].

Plant-pathogen interactions may be classified as compatible (successful infection leading to disease and pathogen reproduction) or incompatible (successful plant defense, no disease and no pathogen survival) [34]. In more detail, however, there is a continuum in the level of compatibility of interactions. The interaction between the susceptible genotype Mueller-Thurgau, a typical representative of European grapevine *Vitis vinifera*, and *P. viticola* is fully compatible and results in the abundant formation of abaxial sporangia. Partial, i.e., incomplete resistance of grapevine cultivars to *P. viticola* depends on genes (*Rpv*) introgressed from *Vitis* species from the Americas and Asia [35,36]. *Rpv3* (derived from *V. rupestris*) and *Rpv10* (from *V. amurensis*) confer a high level of resistance of cultivars Regent (carrying *Rpv3-1*) and Solaris (*Rpv3-3, Rpv10*), respectively, to *P. viticola*. Resistance reactions of Regent and Solaris include the formation of spots of brown tissue with some necrotic (dead) cells. Earlier studies, however, revealed that *P. viticola* can survive and even sporulate on this tissue, which includes living cells containing haustoria of the pathogen [37].

The influence of downy mildew on leaf transpiration has been investigated using thermography in cucumber [38,39] and grapevine [40]. Hyperspectral imaging of the leaf water content of apple leaves infested by *Venturia inaequalis* demonstrated that increased transpiration does not inevitably reduce the water content of colonized leaf tissue, as the plant may compensate for the increased water loss through the cuticle provided water supply is not limiting [25]. Only in advanced stages of pathogenesis, the imaging sensor system revealed highly localized water gradients around infections. The study also revealed that the adaxial sides of non-infected apple leaves had higher leaf water content than the abaxial side. However, the relationship between leaf water content (LWC) and Normalized Difference Leaf Water Index was calibrated only for adaxial apple leaf tissue. It was not clear whether the difference in $NDLWI_{1937}$ values between leaf sides resulted from an effect on the SWIR water absorption bands, on the tissue structure-characterizing NIR waveband at 1047 nm, or on both leaf characteristics. In dicotyledonous plants, the arrangement of leaf tissues is strongly asymmetric, with the chloroplast-rich palisade parenchyma near the upper surface to intercept incoming light and the spongy mesophyll and large air spaces on the abaxial side of the leaf near the lower surface where the stomata are located [15]. This bifacial arrangement is likely to cause differences in the spectral characteristics between adaxial and abaxial leaf surfaces. The water-transporting vascular bundle is oriented more to the abaxial side of leaves, with xylem and phloem often located underneath the palisade parenchyma.

Imaging sensors enable the detection, mapping, and modelling of the biochemical and physiological processes in compatible and incompatible host–pathogen interactions [37,41–43]. The objective of the study reported herein was to improve our understanding of the effect of *P. viticola* on the leaf water status of grapevine leaves, depending on the compatibility of the host–pathogen interaction, by applying an innovative method of LWC mapping on the tissue scale. Four objectives were addressed; (I) to establish the relationship between LWC measured gravimetrically and a normalized difference vegetation index based on the SWIR water absorption band(s) for adaxial and abaxial sides of grapevine leaves; (II) to check the validity of this relationship under various environmental conditions, e.g., background and grapevine genotype; (III) to map the effect of downy mildew on the leaf water status in compatible and incompatible interactions; and (IV) to track the changes in LWC during pathogenesis in longitudinal studies.

2. Materials and Methods

2.1. Plant Growth

Grapevine (*Vitis vinifera* L.) plants of cultivars Mueller-Thurgau, Regent, and Solaris were grown from green cuttings in commercial substrate ED 73 (Balster Einheitserdewerk, Froendenberg, Germany) in plastic pots (10 × 10 cm, 0.9 L volume) in a greenhouse at 23/20 °C (day/night), 60% relative humidity (RH), and a photoperiod of 16 h (>300 $\mu mol\ m^{-2}\ s^{-1}$). Plants were watered as necessary and fertilized before the experiment once with 100 mL of a 0.2% solution of Poly Crescal (Aglukon, Düsseldorf, Germany). The cultivars differ in field susceptibility to downy mildew: cv. Mueller-Thurgau (white wine, rating 7), cv. Regent (red wine, 3), cv. Solaris (white wine, 3) (scale 1 = resistant, to 9 = susceptible; [44]).

2.2. Inoculum Preparation and Inoculation

Sporangia of *Plasmopara viticola* produced on the abaxial side of grapevine leaves of the susceptible cv. Mueller-Thurgau were stored at −18 °C. Sporangia were washed off the leaves with water containing 0.1% Tween-20, adjusted to a concentration of 5×10^4 mL^{-1}, and sprayed onto the abaxial side of the upper four leaves (1.2 mL per leaf) of grapevines by turning the plants upside down during spraying.

For each cultivar, 4–6 plants were inoculated with the pathogen when at least six leaves were fully developed. Inoculated plants were incubated for 24 h at 100% RH, 23/20 °C, 16 h photoperiod; subsequently, they were grown at 60% RH. Non-inoculated control plants were not subjected to spells of 100% RH.

2.3. Disease Assessment

The severity of *P. viticola* colonization of grapevine leaves was assessed visually starting 3 days post-inoculation (d p.i.). To induce sporulation of *P. viticola*, infected plants were incubated for 14 h (19.00 to 09.00) at 100% RH. For cv. Mueller-Thurgau, the percent (abaxial) leaf area covered by whitish sporangia was estimated using the classes 0, 1, 2, 5, 10, 20, 30, . . ., 90, 100%. For cvs. Regent and Solaris, the percentage leaf area covered by tiny brownish leaf spots was estimated using the same classes.

On images of leaf tissue, the size of individual leaf spots was assessed by counting the number of pixels with $NDLWI_{1937}$ values < 0.55.

2.4. Leaf Thickness

The thickness of grapevine leaves was assessed using a digital sliding caliper (Workzone DMV-SL05, Dario GmbH & Co KG, Hamburg, Germany). The thickness of intercostal areas and second-order leaf veins of ten mature, fully developed leaves and 10 young leaves (leaf level 2 and 3 from top) per grapevine genotype was measured.

2.5. Gravimetric Assessment of Leaf Water Content

The leaf water content (LWC) of grapevine leaves was measured gravimetrically as LWC = (fresh mass − dry mass)/fresh mass × 100. Per treatment, the fresh mass of 8–10 leaves was weighed directly after leaf detachment, and the dry mass after drying at 72 °C for at least 12 h. The LWC value is in contrast to the relative water content, as it does not consider the maximum water content of plant tissue. The experiment was carried out twice.

2.6. Hyperspectral Imaging (HSI)

2.6.1. Measuring System

Hyperspectral images in the shortwave infrared (SWIR) range (940 to 2544 nm) were recorded using a camera with the ImSpector N25E spectrograph (Spectral Imaging Ltd., Oulu, Finland) and an objective with a focal length of 30 mm and a field-of-view of 18°. An effective sensor slit length of 9.6 mm and 320 pixels per line result in a sensor pixel size of 30 µm. Spectral resolution was 6.3 nm (256 bands). Three ASD-Pro-Lamps (Analytical Spectral Devices Inc., Boulder, CO, USA) on each side of the camera provided a near-solar light spectrum and gave even illumination of the scene. Camera and lamps were arranged on a motorized line stage (SP-X-Stage-Dual2, Spectral Imaging Ltd.), which moved the measuring system above the plants. Camera and line stage were controlled using the SpectralCube software (Spectral Imaging Ltd.). Imaging data were recorded in a darkened room for optimal and reproducible measuring conditions.

During HSI measurements, the leaves attached to grapevine stems were fixed between two grids (mesh size 30 × 40 mm) of black fibers in a frame 0.35 m above a black polypropylene background to support the leaves from below and to smooth them from above. The potted grapevine plants and the supporting system were placed on a motorized table in order to vertically adjust the leaves to a working distance of 0.3 m (for maximum spatial resolution) to 0.6 m (for maximal spatial coverage). A set of four reflection images per object were recorded in order to allow the calculation of reflectance values: (I) white reference bar (Spectral Imaging Ltd.) with horizontal size and vertical level similar to the object area; (II) dark current image; (III) image of the object area of interest with optimized exposure time; and (IV) dark current image of the object area.

Hyperspectral images were recorded from leaves without visible water droplets on the surfaces. Leaf surfaces of plants incubated under 100% RH for 14 h overnight in order to induce *P. viticola* sporulation had to be dried under ambient conditions for >6 h before measurements.

2.6.2. Processing of Hyperspectral Data

The raw image data (reflection values, 240 wavebands in the range from 997 to 2500 nm) were converted into hyperspectral reflectance images using the software ENVI + IDL 5.0 (EXELIS Visual Information Solutions, Boulder, CO, USA). Data from the 4 images of an object were imported and normalized by a customized IDL tool. Spectral vegetation indices—Normalized Difference Leaf Water Index (NDLWI) values—were calculated using the values of the water absorption bands at 1453 nm (band #82) and 1937 nm (band #159), respectively, and 1124 nm (near infrared, band #30) for normalization: $NDLWI_{1453} = (R_{1124} - R_{1453})/(R_{1124} + R_{1453})$ and $NDLWI_{1937} = (R_{1124} - R_{1937})/(R_{1124} + R_{1937})$, respectively.

Graphic illustrations of x-profiles and y-profiles along transects through leaf areas were prepared after data transfer from ENVI + IDL to Excel.

2.7. Calibration of Normalized Difference Leaf Water Indices

Non-inoculated young and mature leaves of grapevine (cv. Mueller-Thurgau) attached to the plants—with optimal leaf water content—and after detachment and drying for 2, 5, and 8 h, respectively, were recorded with the hyperspectral SWIR camera using a working distance of 60 cm. Young leaves (n = 10) dried under ambient air conditions (20–22 °C, 50% RH), mature leaves (n = 10) were dried at 40 °C in a drying chamber at maximum ventilation. After spectral recording, the weight of leaves was measured gravimetrically. After the fourth hyperspectral measurement, all leaves were desiccated to dryness at 70 °C for 14 h, and the dry matter of leaves was measured after equilibration with ambient air.

Marking the perimeter of the leaves recorded, averaging the spectral information of these regions of interest (ROIs), and extracting the reflectance values at 1124 nm, 1453 nm, and 1937 nm yielded the spectral information for the calibration curves $NDLWI_{1453}$ vs. leaf water content and $NDLWI_{1937}$ vs. leaf water content, respectively. Due to the venation of the grapevine leaf, ROIs included varying percentages of leaf veins and intercostal areas, respectively.

2.8. Statistical Analysis

For statistical analysis, individual grapevine leaves were used as the experimental unit. Statistical analyses were conducted using the software SPSS Statistics vers. 26.0 (IBM Germany GmbH, Ehningen, Germany). Data were tested for normal distribution and equality of variances using the Shapiro–Wilk test and Levene's test, respectively. For normally distributed data, a standard ANOVA was performed. For significant F-values, mean comparisons were performed using the Tukey–Kramer test at a significance level of $p \leq 0.05$. *T*-test statistics (significance level $p \leq 0.05$) were applied to detect significant differences between healthy and infected leaves as well as adaxial and abaxial leaf sides. Data series were related to each other by the Pearson correlation coefficient (r). All experiments were conducted in biologically independent repetitions at least twice.

3. Results

3.1. Relationship Between Leaf Water Content and Normalized Difference Leaf Water Index

Spectral information of non-inoculated grapevine leaves (cv. Mueller-Thurgau) recorded before detachment and after different times of drying, respectively, and gravimetric leaf water content (LWC) values of these leaves were combined to generate calibration curves for the relationship between the Normalized Difference Leaf Water Index (NDLWI). Reflectance at 1453 nm (R_{1453}) and R_{1937}, representing water absorption bands, were used, as well as R_{1124}—the maximum reflectance of grapevine leaves in the NIR range –for normalization. For calibration, young and mature grapevine leaves were used, and reflectance was recorded for the adaxial and the abaxial leaf side, respectively, in separate experiments (Figure S1). As leaf age had no significant effect, data for young and mature leaves were combined into general calibration curves, which had high coefficients of determination (R^2) for both indices (Figure 1). With a stronger slope than $NDLWI_{1453}$, $NDLWI_{1937}$ was used for all following experiments. The difference in the calibration curves between adaxial and abaxial leaf sides was consistent, but statistically not significant.

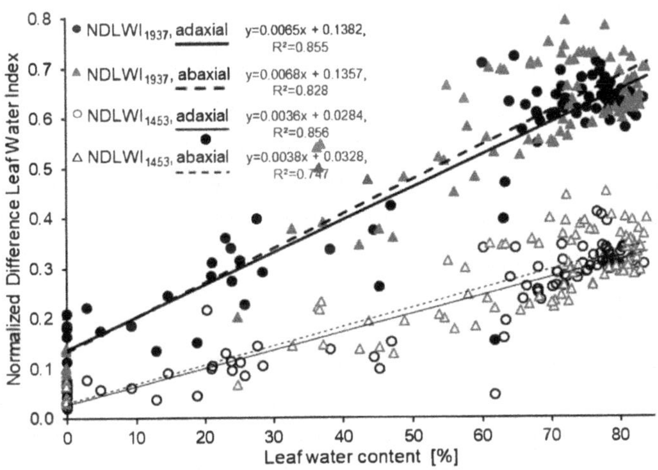

Figure 1. Relationship between water content of grapevine (cv. Mueller-Thurgau) leaves and Normalized Difference Leaf Water Index calculated using the water absorption bands at 1453 nm and 1937 nm, respectively. Hyperspectral imaging of adaxial and abaxial leaf sides in two independent experiments (n = 88).

As the regression between LWC and $NDLWI_{1937}$ was calculated from mean values (of both parameters) from leaves with varying portions of leaf veins and intercostal areas differing in tissue thickness and water content, spectral index values were used instead of derived LWC values to investigate the effect of *P. viticola* infections on the spatial patterns of leaf water content heterogeneity of grapevine leaves. Only for average values per leaf, LWC was calculated from $NDLWI_{1937}$. For images with high spatial resolution, LWC values were not used because of the heterogeneity in the thickness of grapevine leaf tissue varying among and within leaves (Figure S2).

In order to characterize the origin of the spectral leaf water signal, effects of the background (color) of the recorded leaves and of the stacking of leaves differing significantly in LWC were examined (Figure 2). Black and white paperboard below the recorded leaf differed in their effects on leaf reflectance in the SWIR range, which, in response, had a strong effect on the NDLWI values derived. However, a black background resulted in reflectance values very similar to those of leaf areas recorded "without" background (black background 0.35 m below the rack carrying the leaves) represented by the rectangle in the center of leaves, a white background substantially increased reflectance values as well as NDLWI values by about 0.2 units. Therefore, all hyperspectral images were recorded using a stage of fine black fibers (3 × 4 cm spacing) supporting the grapevine leaves to be imaged. The stacking of dried and naturally hydrated leaves in both combinations indicated that the origin of the SWIR signal (and NDLWI value) was the top leaf layer, provided the time of contact between tissues differing in hydration was short, and that the contribution of lower layers to the signal was neglectable (Figure 2b).

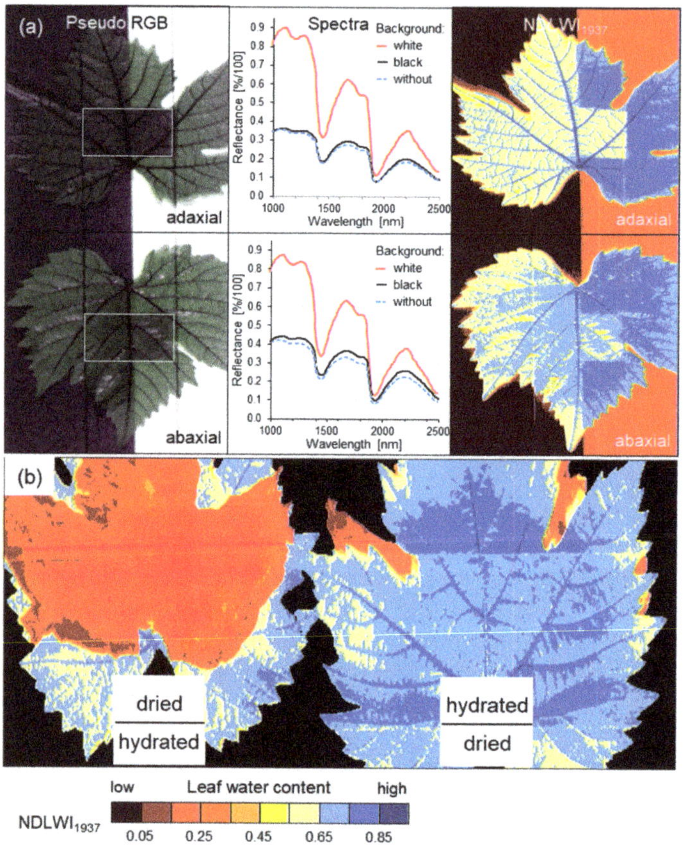

Figure 2. Effects of background (color) and stacking of dried and naturally hydrated grapevine (cv. Mueller-Thurgau) leaves on the assessment of leaf water content by $NDLWI_{1937}$ (NADIR images). (**a**) Grapevine leaves with adaxial and abaxial side up were recorded with black and white paperboard underneath—the center of leaves (marked by white outline on the left) was without paper background on the stage of black fibers; (**b**) effect of stacking of naturally hydrated (LWC 78%) and dried grapevine leaves on $NDLWI_{1937}$ values from hyperspectral imaging.

3.2. *Effect of Plasmopara viticola Development on Spatial Pattern of Grapevine Leaf Water Content in Compatible Interaction*

Depending on inoculum density and greenhouse temperature, *Plasmopara viticola* caused typical downy mildew symptoms—formation of whitish sporangia on the abaxial leaf side, sometimes in combination with oil flecks on the adaxial side—5 to 8 days post-inoculation (d p.i.) on cv. Mueller-Thurgau. These symptoms occurred only after induction of sporulation by spells of high RH (100% RH for 14 h overnight). Without sporangia induction, intensive tissue colonization by *P. viticola* caused spots with reduced chlorophyll content, slight wilting symptoms of leaves, and tissue damage in later stages (11–18 d p.i.).

The effect of downy mildew development on the leaf water status of cv. Mueller-Thurgau leaves was investigated in a series of hyperspectral images with the first images recorded 5 d p.i. (Figure 3; all sub-figures in Figure S3a–e). The water content of non-diseased leaf blades was characterized by high values of main veins, lower values of veins of higher order, and tapering off to intercostal areas and the leaf margin. Focusing on the abaxial leaf side where sporangia—when induced—protrude through stomata, the longitudinal analysis revealed severe effects on leaf water patterns at 9 d p.i., with $NDLWI_{1937}$ dropping <0.55 for large areas of infected leaves. In earlier stages, the latent colonization had no effect on leaf water status. Inoculated leaves without induction of sporangia formation caused an increased heterogeneity of intercostal $NDLWI_{1937}$ values at 9 d p.i., which developed into necrotisation of leaf margins characterized by $NDLWI_{1937}$ values < 0.3 in later stages. This phenomenon was much stronger for leaves with sporulation, even after mechanical removal of sporangia. This treatment indicated that the sporangia themselves did not contribute to the low NDLW values.

Figure 3. Effect of *P. viticola* development on the spatial pattern of leaf water content of grapevine (cv. Mueller-Thurgau) leaves as visualized by Normalized Difference Leaf Water Index ($NDLWI_{1937}$) for the period 5 to 14 d p.i. Assessment of abaxial $NDLWI_{1937}$ of non-inoculated leaves (**left**) and inoculated leaves without induction of sporulation (**second left**), with sporangia (**second right**), and with sporangia mechanically removed (**right**).

At higher magnification and with smaller steps of $NDLWI_{1937}$ levels in water content visualization, first modifications in the leaf water pattern due to *P. viticola* were observed 7 d p.i. when first sporangia appeared on the abaxial leaf side (Figure 4a). Large parts of the leaf were still free from sporangia and had a rather homogenous $NDLWI_{1937}$ pattern, especially when visualized at full range (0.05–0.85) of $NDLWI_{1937}$ values. With narrow steps in the range 0.45 to 0.85, the leaf map indicated several spots with $NDLWI_{1937}$ values < 0.55. The comparison of SWIR information with RGB images confirmed the spatial coincidence of reduced leaf water content and sites of sporangia formation. The permanent overlay of grapevine leaf parts resulted in a small area of increased $NDLWI_{1937}$ values (Figure 4a, bottom right). Even higher spatial resolution enabled the precise assignment of $NDLWI_{1937}$ values to infection sites. X- and Y-profiles of $NDLWI_{1937}$ values along transects through sites of sporulation demonstrated the highly localized—4 and 8 pixels (=0.13 and 0.25 mm), respectively—reduction of leaf water content (Figure 4b). The effect on $NDLWI_{1937}$ intensified with the size of the sporulating area from 0.45–0.50 to 0.40–0.45.

Figure 4. Effect of *P. viticola* on leaf water content of grapevine (cv. Mueller-Thurgau) leaves, 7 d p.i. (**a**) RGB images (center) and $NDLWI_{1937}$ images of adaxial (**top**) and abaxial (**middle**) sides of leaf after induction of sporangia formation by 100% RH for 14 h. $NDLWI_{1937}$ levels displayed for the range 0.05–0.85 (**left**) and 0.45–0.85 (**right**), respectively. Bottom row with details of sporangia formation and the highly localized decrease in leaf water content at the sites of sporangia formation: Note higher NDLWI values at the site of tissue overlap; (**b**) $NDLWI_{1937}$ profiles (range 0.45–0.85) of two infection sites with sporangia formation.

In case the formation of abaxial sporangia was not induced, progress in downy mildew pathogenesis resulted in decreased chlorophyll content and heterogenous leaf water content 9 d p.i. (Figure 5a,c). $NDLWI_{1937}$ values exhibited higher within-leaf variation, but not a modified level—except for some spots with reduced LWC at the margin. Non-inoculated control leaves had the typical pattern of veins high in LWC, intercostal areas with values 0.1 to 0.15 units lower, and reduced water content values tapering off to the leaf margin. Leaf colonization associated with strong abaxial sporangia formation by *P. viticola* caused a large-scale decrease in the water content of intercostal areas and veins (of higher order) by 0.1–0.2 $NDLWI_{1937}$ units and desiccation of marginal leaf tissue ($NDLWI_{1937}$ values < 0.3; Figure 5d,e). The water content of non-diseased tissue of infected leaves was very similar to that of non-infected leaves. Mechanical removal of the desiccated sporangia caused a slight increase in abaxial $NDLWI_{1937}$ values.

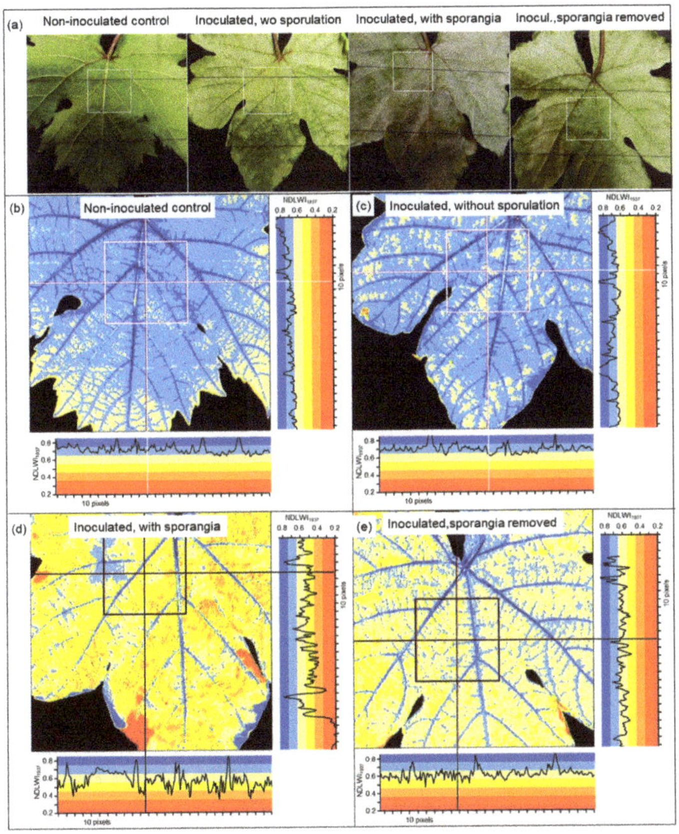

Figure 5. Effect of downy mildew on spatial pattern of leaf water content of grapevine (cv. Mueller-Thurgau) leaves, 9 d p.i. (**a**) RGB images; (**b–e**) $NDLWI_{1937}$ images of abaxial leaf side of (**b**) non-inoculated control; (**c**) inoculated leaf without induction of sporangia formation; (**d**) leaf with sporangia formation induced by 100% RH for 14 h; (**e**) leaf after mechanical removal of sporangia.

The negative effect of *P. viticola* on the water status of compatible grapevine leaves depended on the severity of downy mildew symptoms per leaf and was even promoted by spells of 100% RH, inducing abundant sporangia formation on abaxial leaf sides (Figure 6). Although more pronounced on the abaxial side, the effect of downy mildew on the pattern of leaf water status could also be measured on the adaxial side, even at 7 d p.i. when downy mildew reduced $NDLWI_{1937}$ values by 0.2 units or less. The desiccation of leaf margins and tissue spots in later stages caused identical $NDLWI_{1937}$ values < 0.3 on both leaf sides. As $NDLWI_{1937}$ of adaxial sides was always about 0.05 units higher than that of abaxial sides—irrespective whether infected or not—the contrast in leaf water contact between affected and non-affected areas was stronger for the adaxial side (Figure S3). LWC of veins of strongly damaged grapevine tissue indicated a still functional system of water transport.

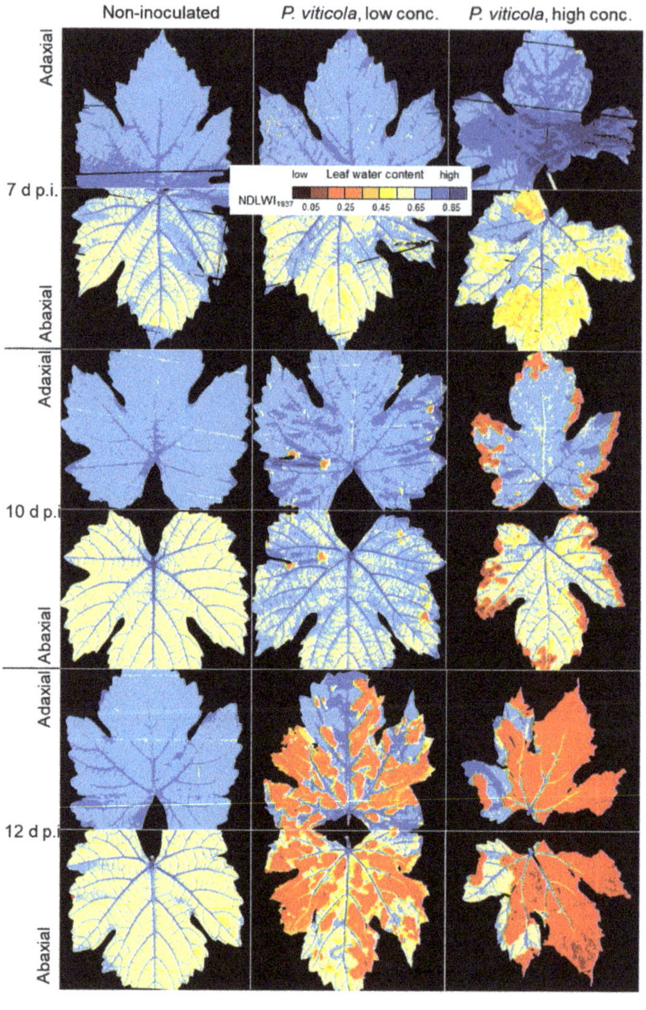

Figure 6. Effect of downy mildew disease severity and time on spatial patterns of water content of grapevine (cv. Mueller-Thurgau) leaves. $NDLWI_{1937}$ images of adaxial and abaxial leaf side of non-inoculated control (**left**), leaves inoculated with low (**center**) and high (**right**) *P. viticola* inoculum density. Sporangia formation on abaxial side of leaves was induced before each measurement time by 100% RH for 14 h.

3.3. *Effect of Plasmopara viticola Development on Spatial Patterns of Water Content of Grapevine Leaves in Interactions with Reduced Compatibility*

The partially resistant grapevine cultivars Regent and Solaris responded to *P. viticola* infection with the formation of tiny spots of brownish tissue, which were visible on both leaf sides irrespective of RH during plant growth after inoculation. These symptoms of resistance appeared 7 to 11 d p.i. and progressed with time in a cultivar-dependent manner.

First experiments on the influence of *P. viticola* infection of cvs. Regent and Solaris with partial downy mildew resistance were not successful, as the spatial resolution of the sensor system was not sufficient when the working distance was 0.6 m (0.67 mm per pixel;

Figure 7a). The tiny brownish tissue spots (Ø 1–2 mm) on leaves were clearly visible in RGB images. In corresponding $NDLWI_{1937}$ images, however, they were hardly detected or not at all. Imaging data with the best spatial resolution available—0.34 mm per pixel, working distance 0.3 m—were suitable to detect and to characterize the spots of resistance reactions for both host genotypes (Figure 7b–e). $NDLWI_{1937}$ profiles of the affected leaf tissue of both *V. vinifera* cultivars demonstrated a sharp and strong decline in water content with $NDLWI_{1937}$ values as low as 0.3. The spatial dimension of this effect was strictly limited—2 to 6 pixels—and the water content of neighboring tissue was not affected.

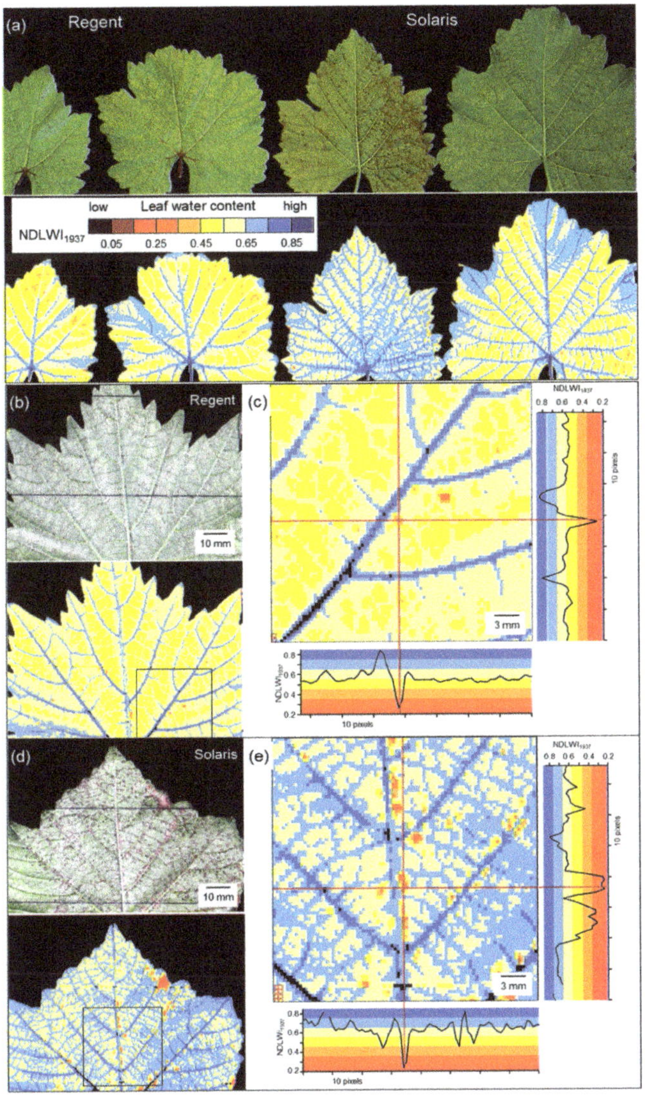

Figure 7. Detection of the effect of resistance reactions of grapevine cvs. Regent and Solaris to infection by *P. viticola*—i.e., spots of brownish tissue—on the water status of the abaxial leaf side 9 d p.i. (**a**) Non-successful attempt because of insufficient spatial sensor resolution; RGB images clearly showed brown spots; (**b–e**) detection and characterization of the effect on water status of leaves of cv. Regent (**b**,**c**) and Solaris (**d**,**e**) with improved spatial resolution.

Longitudinal studies covering the development of tissue reaction of both partially resistant cultivars for the period 5 to 18 d p.i. revealed no effects of *P. viticola* infection before spots of brown tissue appeared. The number and size of brownish spots of leaf tissue increased with time (Figure S4). Detailed analysis of $NDLWI_{1937}$ images of cv. Regent revealed first tissue reactions 11 d p.i. and an increase in the number and size of spots, which were sharply confined and strongly reduced in water content (Figure 8). The development over time was very similar on both leaf sides and was in agreement with observations of symptoms in RGB images. The small step size in the visualization of $NDLWI_{1937}$ values (0.05 units per color) confirmed the local character of the infection effect. The size of one spot increased from 1 pixel 11 d p.i. to 5 and 7 pixels 14 and 18 d p.i., respectively, another spot grew from 3 to 6 pixels within 4 days after first appearance.

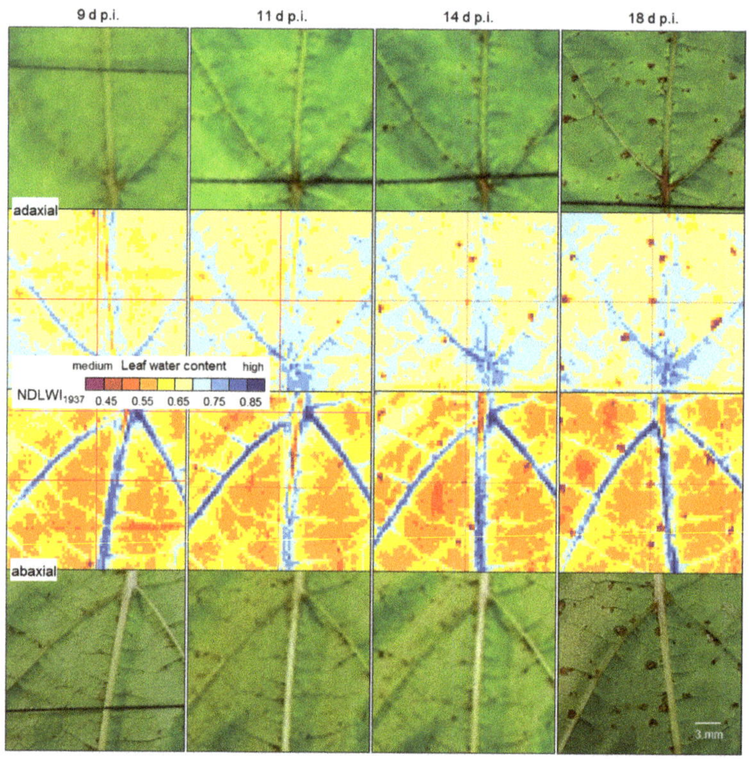

Figure 8. Longitudinal study of the effect of the development of brownish spots due to the resistance reaction of grapevine cv. Regent to infection by *P. viticola* on leaf water status of adaxial (**top** rows) and abaxial (**bottom** rows) leaf sides. The number and size of spots visible in RGB and $NDLWI_{1937}$ images increased with time.

Tracking of individual spots on leaves for the period 7 to 18 d p.i. demonstrated different progress in the effect of resistance reaction to *P. viticola* infection on the water status of spots of cv. Solaris (Figure 9). Although the reduction in water content of colonized tissue was detectable on both leaf sides already 9 d p.i., the area and the degree of water content reduction varied with the leaf side. On the adaxial side, brownish tissue started with a large area of $NDLWI_{1937}$ values < 0.55 (e.g., 26 pixels), but this area decreased to 6 pixels within the next 7 d. On abaxial side, the spot size of $NDLWI_{1937}$ values < 0.55 was smaller (10 pixels) 9 d p.i. and increased to 21 18 d p.i. RGB images gave no evidence for changes in the size of brown spots; however, some spots revealed the transient deposition of amorphous material on the abaxial side 11 d p.i.

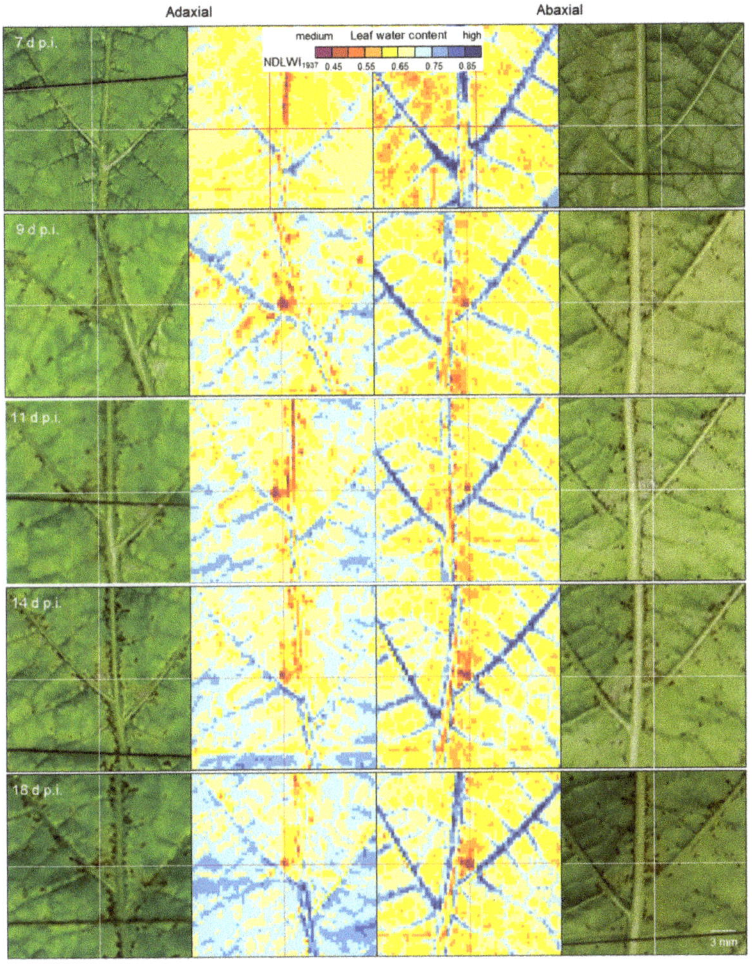

Figure 9. Longitudinal study of the effect of the development of brownish spots due to the resistance reaction of grapevine cv. Solaris to *P. viticola* infection on leaf water status of adaxial (**left**) and abaxial (**right**) leaf sides. The number of spots visible in RGB and $NDLWI_{1937}$ images increased with time. The size and effect on leaf water content, however, varied between leaf sides—a decrease and an increase for the adaxial and abaxial sides, respectively.

Profiles of $NDLWI_{1937}$ values along transects through individual leaf spots caused by *P. viticola* infection indicated a rapid decrease of leaf water content from 0.65–0.70 to 0.45–0.55 and 0.30–0.35 at 14 and 18 d p.i., respectively (Figure 10). The size of the affected tissue area increased with the level of $NDLWI_{1937}$ reduction. Brown spots of cv. Solaris had a strong reduction in leaf water content in the early stage, a transient resurge 11 d p.i. (when deposits on the stomata reduce stomatal transpiration) and a subsequent aggravation and attenuation—in both the level of reduction and the area affected—of the effect on water content for the abaxial and adaxial leaf sides, respectively.

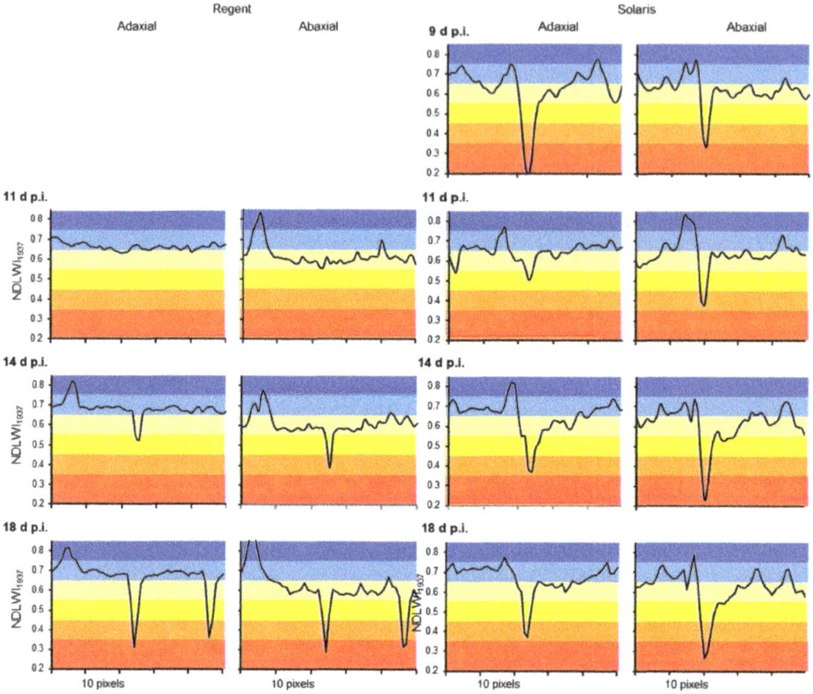

Figure 10. Effect of individual *P. viticola* infections on the leaf water content of grapevine cultivars Regent (**left**) and Solaris (**right**) during the appearance and development of brown tissue spots. Profiles of $NDLWI_{1937}$ values of transects through brown leaf spots for adaxial and abaxial leaf sides 9, 11, 14, and 18 d p.i., respectively.

3.4. Reassessment of Water Content Differences Between Adaxial and Abaxial Leaf Sides

Consistently lower $NDLWI_{1937}$ values of abaxial leaf sides prompted the question whether this difference really indicated significant differences in the water status across small distances between palisade parenchyma and spongy parenchyma (leaf thickness 0.2–0.3 mm) or if it was due to changes in reflectance characteristics of plant tissue in the NIR range. Spectra of leaf tissue from grapevine cultivars Mueller-Thurgau, Regent, and Solaris, non-inoculated and 8 d p.i. *P. viticola*-infected, respectively, were analyzed in more detail (Figure 11). Averages of four ROIs with >6500 pixels per treatment (cultivar × leaf side × infection) revealed that the spectral difference between the sides of leaves at 1124 nm was larger than the differences among grapevine genotypes. Differences in leaf thickness among cultivars could not be retrieved in the spectra. The water absorption band at 1937 nm exhibited higher values (=lower water content) for the abaxial than for the adaxial leaf side (Figure 11a). Substantial sporangia formation on the abaxial side of Mueller-Thurgau leaves increased adaxial R_{1124}, had no effect on abaxial value, and increased both R_{1937} values. Therefore, the effect of downy mildew on mean $NDLWI_{1937}$ per leaf was not significant (Figure 11b). Little formation of tiny brown spots on the leaves of cv. Regent had no significant effect on adaxial and abaxial leaf spectrum and $NDLWI_{1937}$, whereas substantial formation of brown spots of cv. Solaris reduced both R_{1124} values and R_{1937} values and resulted in $NDLWI_{1937}$ values significantly different among treatments (Figure 11d).

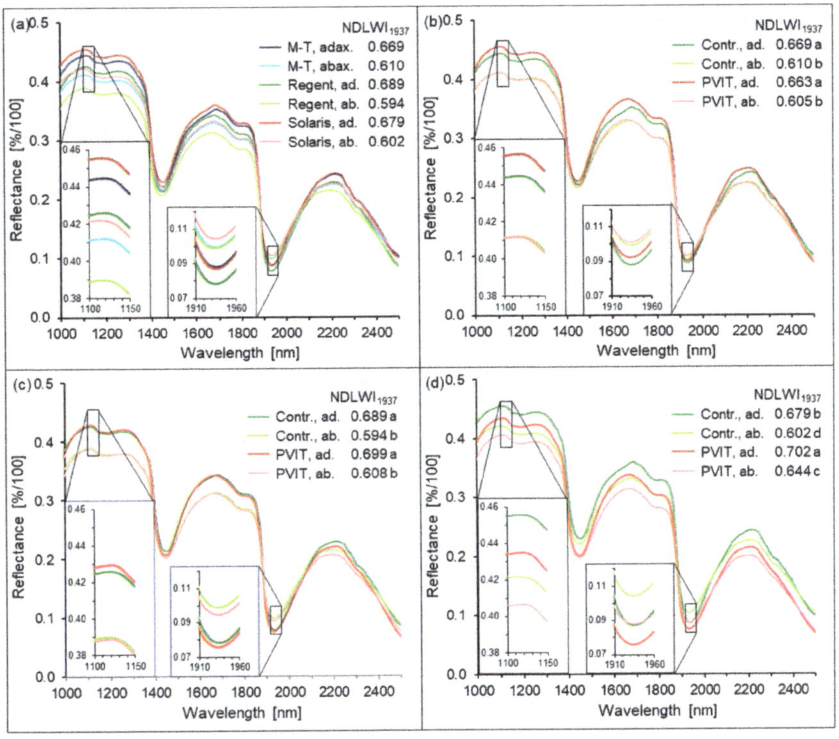

Figure 11. Effects of cultivar, leaf side, and infection by *P. viticola* on spectral characteristics of grapevine leaves, highlighting the spectral ranges used to calculate the $NDLWI_{1937}$ (inserts). (**a**) Effect of grapevine genotype and leaf side; (**b**) effect of leaf side and downy mildew on reflectance of cv. Mueller-Thurgau, 8 d p.i.; (**c**) effect of leaf side and tiny spots of brown tissue on reflectance of cv. Regent, 8 d p.i.; (**d**) effect of leaf side and tiny spots of brown tissue on reflectance of cv. Solaris, 8 d p.i. Mean spectra of >6500 pixels per leaf (n = 4). ab(ax)., abaxial leaf side; ad(ax)., adaxial leaf side; $NDLWI_{1937}$ values followed by the same letter were not significantly different (Tukey–Kramer test, $p \leq 0.05$).

Separate assessment of $NDLWI_{1937}$ for leaf veins (first and second order) and intercostal areas, respectively, for adaxial and abaxial leaf surfaces of the three grapevine cultivars confirmed that intercostal areas had abaxial $NDLWI_{1937}$ values significantly ($p < 0.01$) lower than the corresponding adaxial areas (Figure 12). The veins of these leaves showed the opposite effect with higher abaxial $NDLWI_{1937}$ values. Assuming an area ratio of 0.8:0.2 between intercostal area and veins for grapevine leaves, the differences between adaxial and abaxial $NDLWI_{1937}$ values per leaf were +0.034, +0.059, and +0.038 for cvs. Mueller-Thurgau, Regent, and Solaris, respectively. This effect explains the higher contrast of abaxial leaf sides in $NDLWI_{1937}$ visualizations, which was consistent for the three genotypes during these studies.

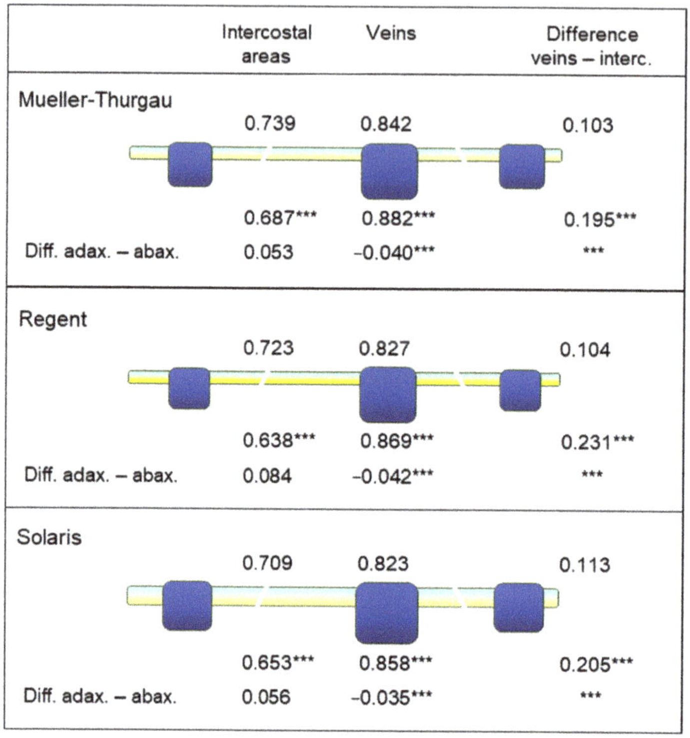

Figure 12. Effect of leaf side and tissue type—veins and intercostal areas, respectively—on $NDLWI_{1937}$ values of three grapevine genotypes under greenhouse conditions. For all genotypes, leaf water values were significantly higher for the adaxial than for the abaxial leaf side (n = 6). Leaf veins had higher leaf water content than intercostal areas, and the difference between adaxial and abaxial leaf sides was positive and negative for intercostal areas and leaf veins, respectively (*** highly significant according to *t*-test). Thickness of intercostal areas and leaf veins (2nd order) of grapevine leaves differed among genotypes as indicated by the size of these structures.

4. Discussion

4.1. Relationship Between Leaf Water Content and $NDLWI_{1937}$

The Normalized Difference Leaf Water Index for grapevine uses the water absorption band at 1940 nm and the maximum NIR reflectance at 1124 nm, and is very similar to the NDLWI established for apple leaves [25]. Proximal sensing enables the use of SWIR water absorption bands without the risk of corruption of the water signal on its path from leaf to sensor. Normalization by a NIR waveband proved to be very useful when studying the variation of this spectral information depending on grapevine genotype and leaf tissue variability. The greater y-axis intercept of the $NDLWI_{1937}$ as compared to apple leaves is likely to result from the inclusion of young grapevine leaves, which had higher reflectance of dried leaves than older ones. The higher tissue variability of grapevine leaves covering intercostal areas and veins of several orders also contributed to a lower coefficient of determination of the regression line for grapevine leaves (0.855 compared to 0.918 for apple leaves; Figure S2). As downy mildew pathogens primarily produce abaxial disease symptoms, a calibration curve was also established for the lower leaf side of grapevine leaves in a separate experiment, which was almost identical to the relationship for the adaxial side.

Grapevine genotypes significantly differed in the thickness of leaf tissue and veins, and intercostal areas caused considerable variability of tissue thickness within leaves. Although the normalization of the leaf water index using the sum of both wavebands proved to increase its robustness to tissue variability and illumination conditions, $NDLWI_{1937}$ was used to visualize patterns of the leaf water status of infected leaves instead of LWC values calculated from the calibration curve. The variability in the relationship between spectral information and leaf water content resulted from using mean values for both the water content of whole leaves and averages of spectral information. Although spectral information may be sub-classified easily to veins (of various orders) and intercostal areas, it is hard to obtain the ground truth of the water content of specific leaf parts. The use of LWC values calculated from the calibration curve for maps of the leaf water status would neglect the effect of within-leaf variability. Moreover, the relative differences displayed in $NDLWI_{1937}$ maps were used for leaves of grapevine cultivars differing in leaf thickness.

Interestingly, experiments with different backgrounds and with the stacking of leaves varying in leaf water content demonstrated that the reflection of the normally transmitted part of the irradiation by a white and dry background significantly increased SWIR reflectance and $NDLWI_{1937}$ values by 0.15 to 0.20 units. Using a non-reflecting background about 0.35 m below the measured leaf, grapevine tissue with a leaf water content strongly differing from that of the leaf of interest above, in contrast, had hardly any effect on the spectral information of the incoming signal. This lack of influence suggested that back-scattering SWIR originates from the upper part of irradiated tissue. The idea was supported by slightly increased $NDLWI_{1937}$ values for permanently overlapping parts of grapevine leaves where the lower tissue impairs the water loss through transpiration of the upper tissue.

4.2. Effects of Leaf Side

Adaxial leaf sides of grapevine leaves had significantly higher mean $NDLWI_{1937}$ values than corresponding abaxial sides, irrespective of grapevine genotype and health status of living plant tissue. This result confirmed earlier observations on apple leaves [25]. As valid for all cultivars, this observation was independent of leaf thickness. Interestingly, the difference between adaxial and abaxial tissue water content was observed only for living intercostal tissue. Necrotic tissue did not exhibit this difference.

A more detailed data analysis confirmed this effect for the intercostal areas, whereas $NDLWI_{1937}$ values of leaf veins were higher for the abaxial leaf side. Reflectance spectra revealed that grapevine genotype, leaf side, and host tissue reaction to *P. viticola* infection affected reflectance in the NIR range (represented by R_{1124}) as well as at the water absorption bands. By calculating $NDLWI_{1937}$ as a proxy for the leaf water status on the tissue scale, the normalized difference index reliably eliminated spectral differences from tissue as influenced by the volume and structure of plant cells and the intercellular space typical for the bifacial leaves of dicotyledonous plant species [15]. The higher water content of adaxial leaf sides is likely to result from the combination of the compact cell layer of the palisade parenchyma, rich in cytoplasm and metabolically highly active, and the effectiveness of an intact cuticle as a water barrier. The spongy parenchyma of the adaxial leaf side, in contrast, comprises a high portion of intercellular spaces responsible for the allocation of CO_2 and O_2 within tissue and is the site where constantly liquid water is transformed into water vapor, which escapes the leaf through the stomata of the abaxial epidermal layer.

Vascular strands are located below the palisade parenchyma and protrude the abaxial surface of intercostal areas, significantly in the case of lower-order veins (Figure S2). Localization, specific morphology (pear-shaped, prominent abaxial side considerably wider than adaxial side), and histological organization (less pronounced bifacial character, lack of stomata, large parenchymatic cells below the vascular bundle) of veins within the leaf plain result in a higher abaxial water content of veins. The higher contrast between veins and intercostal areas for the abaxial side confirms the fine differences in water content on a small scale. The ability to measure LWC differences between both sides of the same leaf can be explained only by the fact that reflectance from tissue near the surface directed to the sensor has a significantly higher contribution to the overall signal than lower tissue.

4.3. Spatial Resolution of Spectral Information

Compared to RGB cameras and hyperspectral sensor systems working in the VISNIR range, the spatial resolution of SWIR cameras is often considerably lower. This discrepancy may result in uncertainty whether no differences in SWIR data really indicate no effect on leaf water status or whether the effect is too small to be detected because of technical limitations. Spatial resolution limits the sensing system's capability to detect small deviations from the normal status (healthy), i.e., in the case of plant pathology, to detect tiny primary disease symptoms with a diameter often <1 mm. Spectral anomaly may be small in spatial dimension as well as in spectral reflectance: The important role of spatial resolution in the detection and monitoring of brown leaf spots, i.e., the resistance reaction in interactions with reduced compatibility to *P. viticola* infection, highlighted that it also limits sensor sensitivity. The spatial resolution of the sensor (with squared sensing units) has to be high; the pixel size of reflectance images has to be smaller than a third of the diameter of the usually circular symptoms of pathogen infection. Only when spatial resolution was adequate the full dimension of leaf water reduction due to the tiny spots of brown leaf tissue could be assessed (Figure 13). For large-scale symptoms (of downy mildew sporangia or disease symptoms from other pathogens) or factors like nutrient supply, which affect all plant parts similarly, spatial resolution is not critical, and effects on plant water status may be detectable using non-imaging devices.

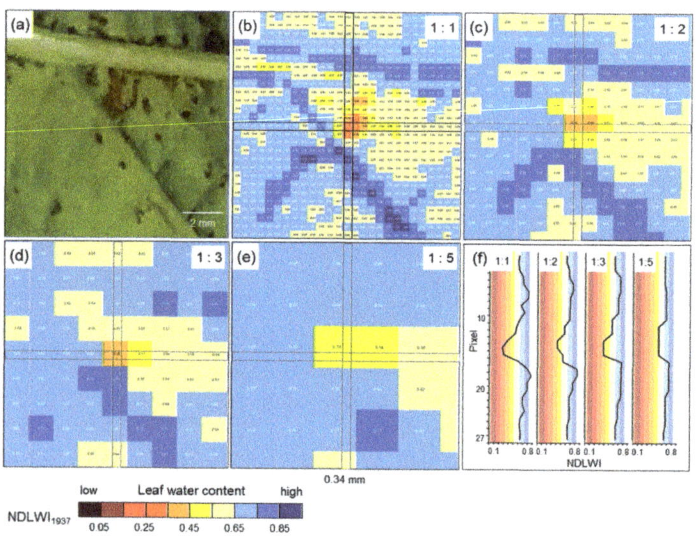

Figure 13. Effect of spatial resolution on the sensitivity of hyperspectral assessment of LWC of grapevine (cv. Solaris) 9 d p.i. with spots of brown issue in response to *P. viticola* infection by applying $NDLWI_{1937}$. (**a**) RGB image; (**b**) original spatial resolution of hyperspectral sensor with 0.34 mm pixel size (27 × 27 pixels); (**c**) spatial resolution reduced by 1:2; (**d**) spatial resolution reduced by 1:3; (**e**) spatial resolution reduced by 1:5; (**f**) $NDLWI_{1937}$ profile of column #14 at different spatial resolutions demonstrating the reduction in sensitivity.

The highest spatial resolution of the SWIR camera was not only suitable to detect the tiny brown tissue spots in response to *P. viticola* infection, but also to quantify and differentiate between response levels. Brown tissue spots had $NDLWI_{1937}$ values lower than grapevine tissue with abaxial downy mildew symptoms; however, levels were often higher than those of leaf necroses and water status of leaf spots varied with the time of pathogenesis, sometimes in both directions. This information is in agreement with observations of the little, but consistent development of biotrophic *P. viticola* in these brown spots of cvs. Regent and Solaris, which also results in pathogen sporulation and spread [37]. The transient resurge in the water content of cv. Solaris spots also indicated that brownish tissue differs from necroses and may represent a combination of necrotic and living cells. The area of brown spots visible in the VIS range differed from the area with modified tissue water content, stressing the difference between visible symptom and the effect on the water status as assessed from SWIR information.

4.4. Leaf Water Response to *P. viticola* Infection

As expected, the leaf water status of grapevine leaves infected by the downy mildew pathogen *P. viticola* depended on the compatibility of the interaction. For partial resistant grapevine genotypes, infections resulted in the formation of tiny spots of brownish tissue associated with highly localized considerable LWC reduction; however, as the affected area generally remained small and LWC of non-infected tissue was hardly affected, the leaves maintained greenness and functionality. Disease resistance, though only partial, successfully protected leaves from severe damage and safeguarded their contribution to overall plant development.

The *Rpv3-1*-mediated resistance of cv. "Regent" is associated with a defense mechanism that triggers the synthesis of fungi-toxic stilbenes and programmed cell death [45]. Cultivar "Solaris" confers resistance to *P. viticola* associated with necrosis formation, callose deposition, and stilbene accumulation [46]. In the early stages of pathogenesis in cv. Solaris, the collapse of infected palisade cells largely restricts the expansion of *P. viticola* to the spongy parenchyma, where tissue colonization ultimately may result in the formation of sporangiophores with sporangia on the abaxial leaf surface [37,47]. The formation of callose and amorphous material on stomata has been reported to impede *P. viticola* sporulation on cv. Solaris [47,48]. Reduced or delayed sporulation may also (transiently) limit the effect on leaf water status of brownish leaf spots as observed 11 d p.i. in this study.

In the compatible interaction, the first effects of *P. viticola* on the LWC pattern of leaves coincided with first sporangia formation or, without the induction of sporulation by high RH, with strong colonization (and production of pathogen biomass) of leaf tissue. The effect on leaf water status, initially limited to sites of primary sporangia, increased with the density of sporulation and caused large-scale LWC reduction of colonized leaf tissue and ultimately resulting in tissue collapse. Although necrotic tissue had NDLWI values of 0.2 and lower, xylem and phloem retained their functionality as long as possible.

The reduced LWC of leaf tissue with abundant sporangia formation revealed a rather uniform and large-scale effect of *P. viticola* on the water status of the susceptible cv. Mueller-Thurgau. Although non-diseased intercostal areas of infected leaves were not affected, the high number of sporangiophores with sporangia protruding through stomata had a wicking activity on apoplastic leaf water and resulted in LWC values 10 to 20% lower than those of non-diseased leaves. As sporangiophores prevented diurnal closure of stomata, the slow but continuing loss of water led to a dehydration of intercostal areas, which started at leaf margins and expanded to central leaf parts in later stages of pathogenesis; leaf veins remained functional for longer periods of time. The pathogen biomass withdraws water (and nutrients) from the apoplast and (via haustoria) from the plant symplast and transfers it to ambient air, to the plant's disadvantage. Without another spell of high RH of ambient air, *P. viticola* sporangia dried within 24 h, but preserved pathogenicity for several days. The large surface of sporangia and sporangiophores protruding into the relatively dry ambient air explains the severe effect of downy mildew on leaf vitality observed in vineyards, which may be even aggravated by insufficient water supply from the soil.

5. Conclusions

Non-destructive and contactless measurement of the leaf water status of grapevine leaves infected by *P. viticola* demonstrated a high sensitivity of spectral imaging of SWIR water absorption bands to small-scale differences and enabled longitudinal assessment of water status patterns related to downy mildew symptoms. Difficulties in the assessment of tiny spots of brownish tissue associated with resistance reactions in incompatible interactions disclosed that better spatial resolution is desirable for SWIR applications in investigations of early effects of pathogens on the leaf water status of plants.

Proximal sensing offers an unprecedented spatial resolution of leaf water content assessment, enabling studies to track effects of pathogens on the plant water balance in host–pathogen interactions. Beyond proximal sensing, $NDLWI_{1937}$ may be applied in phenotyping for genotypes resilient to drought stress caused by biotic and abiotic factors (climate change). The use of hyperspectral SWIR imaging in the field is limited by the interferences of water vapor in the air and crop geometry.

Supplementary Materials: The following supporting information can be downloaded at: https://www.mdpi.com/article/10.3390/rs17101788/s1, Figure S1: Calibration curves for the relationship between water content of grapevine leaves and Normalized Difference Leaf Water Indices—effect of leaf side and ontogenetic stage of leaves; Figure S2: Variability of tissue thickness within grapevine leaves; Figure S3a–e: Effect of *P. viticola* on the water status of leaves of compatible grapevine (cv. Mueller-Thurgau) over time, 5, 7, 9, 12 and 14 d p.i., respectively; Figure S4: Effect of resistance reactions of grapevine cultivars Regent and Solaris to infection by *P. viticola* on leaf water status of leaves.

Author Contributions: Conceptualization, E.-C.O. and U.S.; methodology, E.-C.O. and U.S.; investigation, E.-C.O. and U.S.; validation, E.-C.O. and U.S.; data curation, E.-C.O.; writing—original draft preparation, E.-C.O. and U.S.; writing—review and editing, E.-C.O. and U.S. All authors have read and agreed to the published version of the manuscript.

Funding: This research received no external funding.

Data Availability Statement: The raw data supporting the conclusions of this article will be made available by the authors on request.

Acknowledgments: The authors thank Alaa Al Hlwany for experimental support.

Conflicts of Interest: The authors declare no conflicts of interest.

References

1. Salgado-Pirata, M.; Marques da Silva, J.R. An overview on vine water status assessment. *Ciência E Técnica Vitivinic.* **2024**, *39*, 93–102. [CrossRef]
2. Gutiérrez, S.; Diago, M.P.; Fernández-Novales, J.; Tardaguila, J. Vineyard water status assessment using on-the-go thermal imaging and machine learning. *PLoS ONE* **2018**, *13*, e0192037. [CrossRef]
3. Giovos, R.; Tassopoulos, D.; Kalivas, D.; Lougkos, N.; Priovolou, A. Remote sensing vegetation indices in viticulture: A critical review. *Agriculture* **2021**, *11*, 457. [CrossRef]
4. Mirás-Avalos, J.M.; Araujo, E.S. Optimization of vineyard water management: Challenges, strategies, and perspectives. *Water* **2021**, *13*, 746. [CrossRef]
5. Cogato, A.; Jewan, S.Y.Y.; Wu, L.; Marinello, F.; Meggio, F.; Sivilotti, P.; Sozzi, M.; Pagay, V. Water stress impacts on grapevines (*Vitis vinifera* L.) in hot environments: Physiological and spectral responses. *Agronomy* **2022**, *12*, 1819. [CrossRef]

6. Araújo-Paredes, C.; Portela, F.; Mendes, S.; Valín, M.I. Using aerial thermal imagery to evaluate water status in *Vitis vinifera* cv. Loureiro. *Sensors* **2022**, *22*, 8056. [CrossRef]
7. Atencia Payares, L.K.; Gomez-del-Campo, M.; Tarquis, A.M.; García, M. Thermal imaging from UAS for estimating crop water status in a Merlot vineyard in semi-arid conditions. *Irrig. Sci.* **2025**, *43*, 87–103. [CrossRef]
8. Buunk, T.; Vélez, S.; Ariza-Sentís, M.; Valente, J. Comparing nadir and oblique thermal imagery in UAV-based 3D crop water stress index applications for precision viticulture with LiDAR validation. *Sensors* **2023**, *23*, 8625. [CrossRef]
9. Mertens, S.; Verbraeken, L.; Sprenger, H.; De Meyer, S.; Demuynck, K.; Cannoot, B.; Merchie, J.; De Block, J.; Vogel, J.T.; Bruce, W.; et al. Monitoring of drought stress and transpiration rate using proximal thermal and hyperspectral imaging in an indoor automated plant phenotyping platform. *Plant Methods* **2023**, *19*, 132. [CrossRef]
10. Nowack, J.C.; Atencia-Payares, L.K.; Tarquis, A.M.; Gomez-del-Campo, M. Application of unmanned aerial vehicle (UAV) sensing for water status estimation in vineyards under different pruning strategies. *Plants* **2024**, *13*, 1350. [CrossRef]
11. Kang, C.C.; Diverres, G.; Paudel, A.; Karkee, M.; Zhang, Q.; Keller, M. Estimating soil and grapevine water status using ground based hyperspectral imaging under diffused lighting conditions: Addressing the effect of lighting variability in vineyards. *Comput. Electron. Agr.* **2023**, *212*, 108175. [CrossRef]
12. Carter, G.A. Primary and secondary effects of water content of the spectral reflectance of leaves. *Amer. J. Bot.* **1991**, *78*, 916–924. [CrossRef]
13. Filella, I.; Peñuelas, J. The red edge position and shape as indicators of plant chlorophyll content, biomass and hydric status. *Int. J. Remote Sens.* **1994**, *15*, 1459–1470. [CrossRef]
14. Ustin, S.L.; Roberts, D.A.; Gamon, J.A.; Asner, G.P.; Green, R.O. Using imaging spectroscopy to study ecosystem processes and properties. *BioScience* **2004**, *54*, 523–553. [CrossRef]
15. Ustin, S.L.; Jacquemoud, S. How the optical properties of leaves modify the absorption and scattering of energy and enhance leaf functionality. In *Remote Sensing of Plant Diversity*; Cavender-Bares, J., Gamon., J.A., Townsend, P.A., Eds.; Springer Open: Cham, Switzerland, 2020; pp. 349–384.
16. Gao, B. NDWI-a normalized difference water index for remote sensing of vegetation liquid water from space. *Remote Sens. Environ.* **1996**, *58*, 257–266. [CrossRef]
17. Peñuelas, J.; Piñol, J.; Ogaya, R.; Filella, I. Estimation of plant water concentration by the reflectance Water Index WI (R900/R970). *Int. J. Remote Sens.* **1997**, *18*, 2869–2875. [CrossRef]
18. Kandylakis, Z.; Falagas, A.; Karakizi, C.; Karantzalos, K. Water stress estimation in vineyards from aerial SWIR and multispectral UAV data. *Remote Sens.* **2020**, *12*, 2499. [CrossRef]
19. Laroche-Pinel, E.; Duthoit, S.; Albughdadi, M.; Costard, A.D.; Rousseau, J.; Chéret, V.; Clenet, H. Towards vine water status monitoring on a large-scale using Sentinel-2 images. *Remote Sens.* **2021**, *13*, 1837. [CrossRef]
20. Laroche-Pinel, E.; Vasquez, K.R.; Brillante, L. Assessing grapevine water status in a variably irrigated vineyard with NIR/SWIR hyperspectral imaging from UAV. *Prec. Agric.* **2024**, *25*, 2356–2374. [CrossRef]
21. Zhao, T.J.; Nakano, A.; Iwaski, Y.; Umeda, H. Application of hyperspectral imaging for assessment of tomato leaf water status in plant factories. *Appl. Sci.* **2020**, *10*, 4665. [CrossRef]
22. Sanaeifar, A.; Yang, C.; de la Guardia, M.; Zhang, W.K.; Li, X.; He, Y. Proximal hyperspectral sensing of abiotic stresses in plants. *Sci. Total Environ.* **2023**, *861*, 160652. [CrossRef] [PubMed]
23. Zovko, M.; Žibrat, U.; Knapič, M.; Bubalo Kovačić, M.; Romić, D. Hyperspectral remote sensing of grapevine drought stress. *Prec. Agric.* **2019**, *20*, 335–347. [CrossRef]
24. Mazis, A.; Choudhury, S.D.; Morgan, P.B.; Stoerger, V.; Hiller, J.; Ged, Y.; Awada, T. Application of high-throughput plant phenotyping for assessing biophysical traits and drought response in two oak species under controlled environment. *For. Ecol. Manag.* **2020**, *465*, 118101. [CrossRef]
25. Oerke, E.C.; Steiner, U. Hyperspectral imaging reveals small-scale water gradients in apple leaves due to minimal cuticle perforation by *Venturia inaequalis* conidiophores. *J. Exp. Bot.* **2024**, *75*, 3125–3140. [CrossRef]
26. Grimmer, M.K.; Foulkes, M.J.; Pavely, N.D. Foliar pathogenesis and plant water relations: A review. *J. Exp. Bot.* **2012**, *63*, 4321–4331. [CrossRef]
27. Ayres, P.G. Growth responses induced by pathogens and other stresses. In *Response of Plants to Multiple Stresses*; Mooney, J.A., Winner, W.E., Pell, E.J., Eds.; Academic Press: London, UK, 1991; pp. 227–249.
28. Wilcox, W.F.; Gubler, W.D.; Uyemoto, J.K. *Compendium of Grape Diseases, Disorders, and Pests*, 2nd ed.; APS Press: St. Paul, MN, USA, 2015.
29. Toffolatti, S.L.; Russo, G.; Campia, P.; Bianco, P.A.; Borsa, P.; Coatti, M.; Torriani, S.F.F.; Sierotzki, H. A time-course investigation of resistance to the carboxylic acid amide mandipropamid in field populations of *Plasmopara viticola* treated with anti-resistance strategies. *Pest. Manag. Sci.* **2018**, *74*, 2822–2834. [CrossRef]
30. Campbell, S.E.; Brannen, P.M.; Scherm, H.; Eason, N.; MacAllister, C. Efficacy of fungicide treatments for *Plasmopara viticola* control and occurrence of strobilurin field resistance in vineyards in Georgia, USA. *Crop Prot.* **2021**, *139*, 105371. [CrossRef]

31. Massi, F.; Torriani, S.F.F.; Borghi, L.; Toffolatti, S.L. Fungicide resistance evolution and detection in plant pathogens: *Plasmopara viticola* as a case study. *Microorganisms* **2021**, *9*, 119. [CrossRef]
32. Possamai, T.; Wiedemann-Merdinoglu, S. Phenotyping for QTL identification: A case study of resistance to *Plasmopara viticola* and *Erysiphe necator* in grapevine. *Front. Plant Sci.* **2022**, *13*, 930954. [CrossRef]
33. Velasquez-Camacho, L.; Otero, M.; Basile, B.; Pijuan, J.; Corrado, G. Current trends and perspectives on predictive models for mildew diseases in vineyards. *Microorganisms* **2023**, *11*, 73. [CrossRef]
34. Glazebrook, J. Contrasting mechanisms of defense against biotrophic and necrotrophic pathogens. *Annu. Rev. Phytopathol.* **2005**, *43*, 205–227. [CrossRef] [PubMed]
35. Boso, S.; Kassemeyer, H.H. Different susceptibility of European grapevine cultivars for downy mildew. *Vitis* **2008**, *47*, 39–49.
36. Boso, S.; Alonso-Villaverde, V.; Gago, P.; Santiago, J.L.; Martínez, M.C. Susceptibility to downy mildew (*Plasmopara viticola*) of different *Vitis* varieties. *Crop Prot.* **2014**, *63*, 26–35. [CrossRef]
37. Oerke, E.C.; Juraschek, L.; Steiner, U. Hyperspectral mapping of the response of grapevine cultivars to *Plasmopara viticola* infection at the tissue scale. *J. Exp. Bot.* **2023**, *74*, 377–395. [CrossRef]
38. Lindenthal, M.; Steiner, U.; Dehne, H.W.; Oerke, E.C. Effect of downy mildew development on transpiration of cucumber leaves visualized by digital infrared thermography. *Phytopathology* **2005**, *95*, 233–240. [CrossRef]
39. Oerke, E.C.; Steiner, U.; Dehne, H.W.; Lindenthal, M. Thermal imaging of cucumber leaves affected by downy mildew and environmental conditions. *J. Exp. Bot.* **2006**, *57*, 2121–2132. [CrossRef]
40. Stoll, M.; Schultz, H.R.; Berkelmann-Loehnertz, B. Exploring the sensitivity of thermal imaging for *Plasmopara viticola* pathogen detection in grapevines under different water status. *Funct. Plant Biol.* **2008**, *35*, 281–288. [CrossRef]
41. Kuska, M.; Wahabzada, M.; Leucker, M.; Dehne, H.W.; Kersting, K.; Oerke, E.C.; Steiner, U.; Mahlein, A.K. Hyperspectral phenotyping on the microscopic scale: Towards automated characterization of plant-pathogen interactions. *Plant Methods* **2015**, *11*, 28. [CrossRef]
42. Gold, K.M.; Townsend, P.A.; Herrmann, I.; Gevens, A.J. Investigating potato late blight physiological differences across potato cultivars with spectroscopy and machine learning. *Plant Sci.* **2020**, *295*, 110316. [CrossRef]
43. Maina, A.W.; Oerke, E.C. Characterization of rice-*Magnaporthe oryzae* interactions by hyperspectral imaging. *Plant Dis.* **2023**, *107*, 3139–3147. [CrossRef]
44. Anonymous. *Beschreibende Sortenliste. Reben [Descriptive List of Cultivars. Grapevine]*; Bundessortenamt: Hannover, Germany, 2015.
45. Eisenmann, B.; Czemmel, S.; Ziegler, T.; Buchholz, G.; Kortekamp, A.; Trapp, O.; Rausch, T.; Dry, I.; Bogs, J. *Rpv*-mediated resistance to grapevine downy mildew is associated with specific host transcriptional responses and the accumulation of stilbenes. *Plant Biol.* **2019**, *19*, 343.
46. Juerges, G.; Kassemeyer, H.H.; Durrenberger, M.; Duggelin, M.; Nick, P. The mode of interaction between *Vitis* and *Plasmopara viticola* Berk. and Curt. Ex de Bary depends on the host species. *Plant Biol.* **2009**, *11*, 886–898. [CrossRef]
47. Juraschek, L.; Matera, C.; Steiner, U.; Oerke, E.C. Pathogenesis of *Plasmopara viticola* depending on resistance mediated by *Rpv3_1* and *Rpv10* and *Rpv3_3* and by the vitality of leaf tissue. *Phytopathology* **2022**, *112*, 1486–1499. [CrossRef]
48. Gindro, K.; Pezet, R.; Viret, O. Histological study of the responses of two *Vitis vinifera* cultivars (resistant and susceptible) to *Plasmopara viticola* infections. *Plant Physiol. Biochem.* **2003**, *41*, 846–853. [CrossRef]

remote sensing

MDPI

Article

Estimating Wheat Traits Using Artificial Neural Network-Based Radiative Transfer Model Inversion

Lukas J. Koppensteiner [1], Hans-Peter Kaul [1,*], Sebastian Raubitzek [2], Philipp Weihs [3], Pia Euteneuer [4], Jaroslav Bernas [5], Gerhard Moitzi [4], Thomas Neubauer [2,6], Agnieszka Klimek-Kopyra [7], Norbert Barta [8] and Reinhard W. Neugschwandtner [1]

1 Institute of Agronomy, Department of Agricultural Sciences, University of Natural Resources and Life Sciences, Vienna, Konrad-Lorenz-Straße 24, 3430 Tulln, Austria; lukas.koppensteiner@boku.ac.at (L.J.K.); reinhard.neugschwandtner@boku.ac.at (R.W.N.)
2 Institute of Information Systems Engineering—Data Science, University of Technology Vienna, Favoritenstraße 9-11, 1040 Vienna, Austria; sraubitzek2@sba-research.org (S.R.); tneubauer@sba-research.org (T.N.)
3 Institute of Meteorology and Climatology, Department of Ecosystem Management, Climate and Biodiversity, University of Natural Resources and Life Sciences, Vienna, Gregor-Mendel-Straße 33, 1180 Vienna, Austria; philipp.weihs@boku.ac.at
4 Experimental Farm Groß-Enzersdorf, Department of Agricultural Sciences, University of Natural Resources and Life Sciences, Vienna, Schloßhofer Straße 31, 2301 Groß-Enzersdorf, Austria; pia.euteneuer@boku.ac.at (P.E.); gerhard.moitzi@boku.ac.at (G.M.)
5 Department of Agroecosystems, Faculty of Agriculture and Technology, University of South Bohemia in Ceske Budejovice, Studentska 1668, 370 05 Ceske Budejovice, Czech Republic; bernas@fzt.jcu.cz
6 FFoQSi—Austrian Competence Centre for Feed and Food Quality, Safety and Innovation, 3430 Tulln, Austria
7 Department of Agroecology and Plant Production, University of Agriculture in Kraków, Al. Mickiewicza 21, 31-120 Kraków, Poland; agnieszka.klimek@urk.edu.pl
8 Institute of Agricultural Engineering, Department of Agricultural Sciences, University of Natural Resources and Life Sciences, Vienna, Peter-Jordan-Straße 82, 1190 Vienna, Austria; norbert.barta@boku.ac.at
* Correspondence: hans-peter.kaul@boku.ac.at

Abstract: Estimating wheat traits based on spectral reflectance measurements and machine learning remains challenging due to the large datasets required for model training and testing. To overcome this limitation, a simulated dataset was generated using the radiative transfer model (RTM) PROSAIL and inverted based on an artificial neural network (ANN). Field experiments were conducted in Eastern Austria to measure spectral reflectance and destructively sample plants to measure the wheat traits plant area index (PAI), nitrogen yield (NY), canopy water content (CWC), and above-ground dry matter (AGDM). Four ANN-based RTM inversion models were setup, which varied in their spectral resolution, hyperspectral or multispectral, and the inclusion or exclusion of background soil spectra correction. The models were also compared to a simple vegetation index approach using Normalized Difference Vegetation Index (NDVI) and Normalized Difference Red-Edge (NDRE). The RTM inversion model with hyperspectral input data and background soil spectra correction was the best among all tested models for estimating wheat traits during the vegetative developmental stages (PAI: R^2 = 0.930, RRMSE = 17.9%; NY: R^2 = 0.908, RRMSE = 14.4%; CWC: R^2 = 0.967, RRMSE = 17.0%) as well as throughout the whole growing season (PAI: R^2 = 0.845, RRMSE = 27.7%; CWC: R^2 = 0.884, RRMSE = 20.0%; AGDM: R^2 = 0.960, RRMSE = 13.7%). Many models presented in this study provided suitable estimations of the relevant wheat traits PAI, NY, CWC, and AGDM for application in agronomy, breeding, and crop sciences in general.

Keywords: phenotyping; machine learning; simulation learning; vegetation indices; background soil spectra

https://doi.org/10.3390/rs17111904

1. Introduction

Traditionally, measurements of wheat traits (*Triticum aestivum* L.) are often destructive and time-consuming. An alternative is the estimation of traits using remote sensing, i.e., the measurement of the reflected or emitted electromagnetic radiation of objects, which is non-destructive, faster, and less labor-intensive than destructive sampling [1–3].

A common approach to estimate plant traits based on remote sensing is the application of vegetation indices, which are calculations based on spectral reflectance data of vegetation. This technique is simple and generally provides good results [4–8]. The calculation of vegetation indices, however, is usually based on only two or three spectral bands, i.e., regions of the electromagnetic spectrum defined by their wavelength in nm. Therefore, not all collected spectral data are used [9,10].

An alternative approach is the use of machine learning algorithms [11,12], i.e., the development of model algorithms including training and testing. This technique can utilize all available spectral data, but usually requires large datasets for model training [13]. To overcome this limitation of machine learning, Reinosch et al. [14] proposed to train models with empirically collected data augmented by simulated data, which are generated with existing mechanistic models. In the context of estimating plant traits based on spectral data, radiative transfer models (RTMs) can be utilized as the mechanistic models, as suggested by Reinosch et al. [14].

In general, RTMs simulate the spectral reflectance of leaves and canopies (RTM output), which requires information on various leaf and canopy traits (RTM input). To estimate plant traits based on spectral data, however, the inputs and outputs of RTMs need to be inverted [9,15,16]. This can be achieved by, e.g., look-up tables [17,18] or machine learning algorithms [19–21]. With this approach, all available spectral data can be used for estimating plant traits [16], similar to classic machine learning. Additionally, the limitation by scarce empirical data in traditional machine learning is offset by the inclusion of simulated data based on mechanistic models, e.g., inverted RTMs [14]. Furthermore, the use of mechanistic models allows for a wider generalization of the results and, thus, reduces the need for recalibration for different cultivars or environments [16].

The RTM inversion approach results in complex models that require specific training for the individual spectral resolution of a given sensor, e.g., the position, width, and number of spectral bands [20]. The simulation of canopy reflectance spectra is also substantially affected by the soil background reflectance [22]. This effect is especially prevalent when the canopy cover is low, e.g., at early and late developmental stages [16]. Furthermore, the soil background reflectance varies greatly across and within seasons and locations [23,24]. Therefore, wheat traits were estimated in this study using the spectral reflectance of both canopies and underlying soil combined to adequately account for soil background effects.

In this study, the traits plant area index (PAI, $m^2\ m^{-2}$), nitrogen yield (NY, $g\ m^{-2}$), canopy water content (CWC, $g\ m^{-2}$), and above-ground dry matter (AGDM, $g\ m^{-2}$) were estimated in wheat using an artificial neural network (ANN)-based inversion of the RTM PROSAIL [25]. The models were further trained and tested with data on destructively sampled plants as well as spectral reflectance, which were collected from wheat field experiments in Eastern Austria for two seasons. Different aspects of these field experimental data were investigated in previous studies. Koppensteiner et al. [26] presented a detailed yield and yield component analysis. Bernas et al. [27] conducted a life cycle assessment to find the environmentally optimal N fertilization rate in wheat. Furthermore, Moitzi et al. [28] studied the effects of sowing time and nitrogen fertilizer rate on energy efficiency.

The objectives of this study were to conduct (1) the ANN-based RTM inversion and (2) estimate wheat traits based on the RTM inversion models to present the feasibility of this approach and provide a guide for setting up similar models also for different crops or other spectral sensors. Furthermore, this study evaluated certain aspects of model performances: the differences between varying spectral resolutions, i.e., hyperspectral and multispectral, the effects of including or excluding the background soil spectra correction, as well as the comparison of performances between complex RTM inversion models and simple vegetation index models.

2. Materials and Methods

2.1. Overview

In this section, an overview of the main steps conducted in this study is presented. In the following sections, the individual steps are presented in detail.

Firstly, field experimental data (Section 2.2) were collected in two seasons, which consists of reflectance measurements and destructive plant samples. Secondly, the RTM PROSAIL (Section 2.3) was used to create a simulated dataset (Section 2.4). Thirdly, an ANN (Section 2.5) was trained on the simulated dataset to achieve the RTM inversion. The model input was reflectance data. The model output consisted of the leaf and canopy traits of the RTM PROSAIL. Fourthly, the experimentally collected reflectance data were processed (Section 2.5) and then used by the RTM inversion model to calculate wheat trait estimates. Furthermore, the same experimental reflectance data were applied to calculate the vegetation indices NDVI and NDRE.

Finally, simple regression models were created for model evaluation (Section 2.6). The model inputs were the wheat trait estimates by the RTM inversion models or vegetation indices. The model outputs were the target wheat traits PAI, NY, CWC, and AGDM, which were measured destructively in the field experiments. The regression models were trained on one season of the field experimental data and tested on the field experimental data of the other season.

2.2. The Field Experimental Dataset

Field experiments were performed in the growing seasons of 2019/20 and 2020/21 at the Experimental Farm Groß-Enzersdorf of the University of Natural Resources and Life Sciences, Vienna (48° 11′N, 16° 33′E) in Eastern Austria. The soil at the experimental site was classified as chernozem of alluvial origin, which is rich in calcareous sediments and has a pH_{CaCl2} of 7.6 and 2.3% soil organic carbon [29].

The mean annual temperature and the mean annual precipitation sum from 1980 to 2018 were 10.9 °C and 546 mm, respectively. The two experimental seasons showed highly varying environmental conditions with low amounts of precipitation in March and April for both seasons and in June 2021, high temperature in March 2020, as well as low temperature in April and May 2021. Further detailed information on the temperature and precipitation during the growing periods of the conducted field experiments can be obtained from Koppensteiner et al. [26].

In both experimental seasons, the preceding crop was winter barley. A tine cultivator was used for primary soil tillage to 20 cm depth. Seedbed preparation was conducted using a power harrow at a shallow tilling depth.

The field experiments were established in a split-plot design. The experimental factors were the sowing time with two factor levels, autumn- and spring-sowing, as well as nitrogen fertilization consisting of five levels of 0, 5, 10, 15, and 20 g N m^{-2}. The main plot factor sowing time was set up in complete blocks, while the sub-plot factor nitrogen fertilization was randomized among sub-plots within the main plots. The field experiments were conducted in three replications.

The facultative wheat cultivar Lennox was sown with 300 germinable seeds m^{-2} on 17 October 2019 and 10 March 2020 as well as 27 October 2020 and 8 March 2021 using an

Oyjard plot drill seeder at 4 cm depth. Plot dimensions were 12 × 3 m (36 m^2). Row spacing was 12.5 cm. Nitrogen fertilizer was applied as stabilized urea (46% nitrogen, ALZON 46®, SKW Piesteritz, Lutherstadt Wittenberg, Germany) on 10 March 2020 and 10 March 2021. Weeds, pests, and diseases were controlled according to good agricultural practice. Details on plant protection measures were presented in Koppensteiner et al. [26].

In both seasons, data on crop development, plant area index (PAI, m^2 m^{-2}), above-ground dry matter (AGDM, g m^{-2}), nitrogen yield (NY, g m^{-2}), and canopy water content (CWC, g m^{-2}) were collected at approximately 14-day intervals from March until harvest in July.

Crop development was recorded according to the BBCH scale [30,31]. Plants were sampled by manually cutting 0.6 m^2 per plot at soil surface during the growing season except for the final harvest date, when 1.2 m^2 per plot were cut. The plant area index (m^2 m^{-2}) of fresh plant material was measured using the LI-3100C Area Meter (LI-COR Inc., Lincoln, OR, USA).

Plant samples were weighed fresh and after drying at 105 °C for 24 h (UNE 600, Memmert GmbH & Co. KG, Schwabach, Germany). Weight before drying corresponds to above-ground fresh matter (AGFM, g m^{-2}). Weight after drying is referred to as AGDM gravimetric canopy water content (CWC, g m^{-2}) and is calculated as:

$$\text{CWC}\left(\text{g m}^{-2}\right) = \text{AGFM}\left(\text{g m}^{-2}\right) - \text{AGDM}\left(\text{g m}^{-2}\right) \quad (1)$$

Plant material was milled to pass through a 1 mm sieve and dried at 105 °C for 24 h. Milled plant samples were analyzed for nitrogen concentration by the Dumas combustion method [32] using an element analyzer (vario MAX cube CNS, Elementar Analysesysteme, Langenselbold, Germany). The resulting nitrogen concentration values (NC, %) were multiplied by AGDM to calculate the nitrogen yield (NY, g m^{-2}):

$$\text{NY}\left(\text{g m}^{-2}\right) = \text{AGDM}\left(\text{g m}^{-2}\right) \times \text{NC}(\%) \times 10^{-2} \quad (2)$$

Autumn-sown wheat was harvested on 6 July 2020 and 13 July 2021. Spring-sown wheat was harvested on 20 July 2020 and 26 July 2021.

Furthermore, the destructively measured PAI values were linearly interpolated to achieve a time series of daily PAI values for each plot from the first sampling date until harvest for both seasons. The daily PAI values were then integrated over time to calculate the plant area duration (m^2 m^{-2} d):

$$\text{PAD}\left(\text{m}^2\ \text{m}^{-2}\ \text{d}\right) = \int_a^b \text{PAI}(t)\text{d}t \quad (3)$$

where a is the first day of sampling in a given season, b is the last day of sampling in the same season, and t is the integration variable time in days.

Spectral reflectance measurements on wheat canopy were conducted using the spectroradiometer FieldSpec Handheld 2 (ASD Inc., Lincoln, OR, USA). This sensor is passive and non-imaging, and it provides hyperspectral reflectance data from 325 to 1075 nm with a wavelength accuracy of ±1 nm, a spectral resolution of <3 nm at 700 nm, a sampling interval of 1.5 nm, and a field of view of 25 degrees. Spectral reflectance was measured on sampling dates at solar noon under clear sky conditions in the nadir position. The sensor was mounted on a tripod and positioned 1 m above the canopy. The diameter of the resulting measurement area was 44.3 cm. A 99% Spectralon calibration panel (Labsphere Inc., North Sutton, NH, USA) was used to convert measured data into spectral reflectance values. Per plot and sampling date, six reflectance measurements were conducted across the whole plot area. Then, the average spectral reflectance per plot was calculated. The measured spectral reflectance was interpolated to achieve 1 nm increments using ViewSpecPro software version 6.0 (ASD Inc., Lincoln, OR, USA).

2.3. The Radiative Transfer Model PROSAIL

The RTM PROSAIL consists of the two RTMs: PROSPECT and SAIL.

The RTM PROSPECT simulates the spectral reflectance of leaves from 400 to 2500 nm in 1 nm increments using information on leaf characteristics. It requires the following input parameters: leaf structure index (N, unitless), chlorophyll a + b content (C_{ab}, μg cm^{-2}), carotenoid content (C_{cx}, μg cm^{-2}), brown pigment content (C_{bp}, unitless), leaf mass per area (LMA, g cm^{-2}), and equivalent water thickness (EWT, g cm^{-2}) [16,33,34].

The RTM SAIL simulates the spectral reflectance of vegetation canopy, which requires information on leaf reflectance and transmittance, leaf area index (LAI, m^2 m^{-2}), average leaf inclination angle (ALIA, degrees), hot-spot effect (Hot, m m^{-1}), the fraction of diffuse illumination (skyl, %), as well as viewing geometry [25]. The Hot parameter refers to the hot-spot effect, which is explained in detail by Kuusk [35]. In the RTM SAIL, the Hot parameter is described by the ratio of leaf width to canopy height. The fraction of diffuse illumination is set to 23% for standard clear sky conditions [36]. For more information, Spitters et al. [36] also provides equations and detailed analysis on the estimation of direct and diffuse components of irradiance. Viewing geometry includes information on the sun and observer zenith angles (SZA and OZA, degrees) as well as the relative azimuth angle between sun and sensor (rAA, degrees). The spectral reflectance of background soil (ρ_{soil}, % reflectance) can be adapted in the RTM SAIL. Furthermore, a soil reflectance factor (α_{soil}, unitless) is used to simulate effects of varying soil moisture on soil reflectance [37].

The necessary data on leaf reflectance and transmittance for the RTM SAIL were provided by the previously mentioned RTM PROSPECT. The combination of the RTMs PROSPECT and SAIL is generally referred to as the RTM PROSAIL. The RTM PROSAIL simulates the spectral reflectance of vegetation canopy at a spectral resolution of 400 to 2500 nm at 1 nm increments [16,25]. Table 1 provides an overview of PROSAIL input parameters, including respective ranges specific to wheat, from the literature [24,36,38]. This study utilized the RTM PROSAIL (version 5B) in the package hsdar (version 1.0.3) for the R programming language (version 4.1.1).

2.4. The Simulated Dataset

A simulated dataset consisting of 100,000 observations was created for model training and testing. Each observation included a set of PROSAIL input parameter values, which were randomly selected from uniform distributions within the wheat-specific ranges of the PROSAIL parameters N, C_{ab}, C_{cx}, C_{bp}, LMA, EWT, LAI, ALIA, Hot, and SZA (Table 1). Spectral measurements in field experiments were conducted in the nadir, i.e., downward vertical, position. Thus, OZA and rAA were set to 0 degrees, while SZA was randomly selected from a uniform distribution between 0 and 90 degrees. Furthermore, spectral reflectance for background soil was varied among observation in the simulated dataset. Thus, the available data on soil reflectance by the ICRAF-ISRIC Soil MIR Spectral Library of the International Soil Reference and Information Centre (ISRIC) were used. Measurements on soil reflectance were conducted at the World Agroforestry Centre (ICRAF) Soil and Plant Spectral Diagnostic Laboratory [39]. It was downloaded using the Open Soil Spectral Library developed by the project Soil Spectroscopy 4 Global Good (Falmouth, MA, USA) [40]. This soil reflectance dataset includes 3717 soil reflectance spectra in the visible and near infrared range from 350 to 2500 nm in 2 nm increments. The spectral resolution was reduced to the range from 400 to 1000 nm and interpolated linearly to match the spectral resolution of the RTM PROSAIL ranging from 400 to 2500 nm with 1 nm increments.

For each observation in the simulated dataset, 1 of the 3717 soil reflectance spectra was randomly selected as spectral reflectance of background soil. The soil brightness factor was set to 0. The hyperspectral canopy reflectance was simulated using the RTM PROSAIL for each observation in the simulated dataset based on the randomly selected PROSAIL input parameter set and the randomly selected background soil reflectance.

Table 1. Input parameters for the radiative transfer model PROSAIL (version 5B), which consists of the two radiative transfer models PROSPECT and SAIL. Names, symbols, units, and typical variable ranges for wheat published by Danner et al. [24], Spitters et al. [36] and Kong et al. [38] are presented.

Radiative Transfer Model PROSPECT			
Parameter	**Symbol**	**Unit**	**Wheat-Specific Ranges**
Leaf structure index	N	unitless	1.0 to 2.5
Chlorophyll a + b content	C_{ab}	$\mu g\ cm^{-2}$	0 to 80
Carotenoid content	C_{cx}	$\mu g\ cm^{-2}$	1 to 24
Brown pigment content	C_{bp}	unitless	0 to 1
Leaf mass per area	LMA	$g\ cm^{-2}$	0.001 to 0.02
Equivalent water thickness	EWT	$g\ cm^{-2}$	0.001 to 0.05
Radiative transfer model SAIL			
Parameter	Symbol	Unit	Wheat-specific ranges
Leaf area index	LAI	$m^2\ m^{-2}$	0 to 8
Average leaf inclination angle	ALIA	degrees	20 to 90
Hot-spot parameter	Hot	$m\ m^{-1}$	0.01 to 0.5
Fraction of diffuse illumination	skyl	%	23
Sun zenith angle	SZA	degrees	0 to 90
Observer zenith angle	OZA	degrees	0 to 90
Relative azimuth angle	rAA	degrees	0 to 90
Soil reflectance	ρ_{soil}	%	*
Soil brightness factor	α_{soil}	unitless	0 to 1

* Soil reflectance refers to the spectral reflectance of the underlying soil.

In summary, each of the 100,000 observations in the simulated dataset consisted of a set of randomly selected PROSAIL input parameters, a randomly selected background soil reflectance, and the simulated canopy reflectance based on the RTM PROSAIL.

Additionally, a second simulated dataset was set up. This second simulated dataset was identical to the first one; however, it excluded the previously mentioned background soil spectra. Instead, each observation of this second simulated dataset included a standard soil spectrum and a value for the soil brightness factor, α_{soil}, which was randomly selected from a uniform distribution ranging from 0 to 1 (Table 1). This second dataset was used to evaluate the relevance of background soil spectra correction for estimating wheat traits.

Both simulated datasets, including and excluding background soil spectra, were divided into a train and a test set in a 9 to 1 ratio for model training (90,000 observations) and testing (10,000 observations).

2.5. *The Artificial Neural Network and Spectral Data Processing*

The approach proposed in this study was based on an artificial neural network (ANN). General information on ANNs presented in this section and further detailed information are presented by Stanford University [41]. This ANN consists of multiple layers. The first layer provides the input data. The final layer produces an output. In between, there are multiple layers comprising artificial neurons. Each individual artificial neuron is provided with input data from the previous layer, processes this information using an activation function, and forwards the resulting output to the next layer. An activation function is required to decide whether a neuron should be activated or not, i.e., to introduce non-linearity to the model. A neural network without the activation function is more similar to a linear regression model. There are many different types of activation functions. In general, they follow this formula:

$$y = \text{Activation}\left(\sum_{i=1}^{n} (x_i w_i) + b\right) \quad (4)$$

where y is the neuron output, x_i is the ith input to the neuron from the previous layer, w_i is the weight parameter for the ith input to the neuron, and b is the bias parameter of the neuron. In this study, the widely used ReLU activation function was applied. Furthermore, all layers were fully connected, i.e., all input values or neuron outputs from the previous layer were used as input values for all neurons in the subsequent layer.

To determine the performance of this model, a cost function is required. Many different cost functions exist. In this study, the mean squared error was used:

$$\text{MSE} = \frac{1}{n}\sum_{i=1}^{n} \left(y_i - \hat{y}_i\right)^2 \quad (5)$$

where n is the sample size, y_i is the measured value for the ith observation, and $\hat{y}_i$ is the estimated value for the ith observation.

Finally, to optimize the cost function, an optimization algorithm is required. In this paper, the adaptive optimizer algorithm Adam was applied. To avoid overfitting, the number of neuron parameter training epochs was set to a maximum of 500 with early stopping at 20. To set up the ANN, Google Colaboratory (Google LLC, Mountain View, CA, USA), an available Keras implementation in Python (version 3.8), was used. All parameter settings not mentioned above, e.g., learning rates, regularization, and cross validation, were set to default values.

In this study, an ANN was set up to achieve the inversion of the RTM PROSAIL. Firstly, this ANN was trained and tested using the simulated dataset. The model inputs were SZA as well as background soil reflectance and canopy reflectance, both from 400 to 1000 nm in 1 nm increments. The model outputs were the PROSAIL parameters N, C_{ab}, C_{cx}, C_{bp}, EWT, LMA, LAI, ALIA, and Hot. For the second simulated dataset excluding background soil spectra, the PROSAIL parameter α_{soil} was an additional model output.

Spectral measurements in field experiments were conducted in the nadir position. Therefore, OZA and rAA were set to 0 degrees. The spectral resolutions of background soil reflectance, simulated canopy reflectance, and spectral measurements conducted with the spectroradiometer FieldSpec Handheld 2 in the field experiments were matched. This resulted in a spectral resolution from 400 to 1000 nm in 1 nm increments. All experimentally measured hyperspectral reflectance data of both canopy and background soil were smoothed using the Savitzky–Golay filter in the R package hsdar (version 1.0.3). The filter length was set to 25, and a third-order polynomial was applied.

Additionally, all hyperspectral data of canopy and background soil spectra, both simulated and measured experimentally, were transformed to multispectral data in accordance with the Sentinel-2A spectral response functions for the bands from 1 to 9 including band 8A [42]. As a result, spectral data were available at two spectral resolutions, hyperspectral (400 to 1000 nm in 1 nm increments) and multispectral (according to Sentinel-2A bands 1 to 9 including 8A).

The PROSAIL parameters, i.e., the ANN outputs, showed highly varying ranges (Table 1). Therefore, the PROSAIL parameter values were normalized before the training of the ANN using minimum–maximum normalization [43]. After ANN training, all PROSAIL parameter data were transformed back to their original range.

The ANN consisted of three fully connected layers with 150 neurons each for hyperspectral input data and two fully connected layers with 25 neurons each for multispectral input data. The number of layers and neurons were determined through grid search within feasible layer and neuron number ranges.

Estimations of LAI, C_{ab}, and EWT using ANN-based inversions of the RTM PROSAIL are presented. Experimentally measured PAI, NY, and CWC were estimated using LAI, LAI $\times$ C_{ab}, and LAI $\times$ EWT estimations. Four different RTM inversions were used varying in the spectral resolution of input data, i.e., hyper- and multispectral, as well as including or excluding background soil spectra correction. This was conducted to simulate the performance of sensors with different spectral resolution, i.e., multi- and hyperspectral, as well as models of varying complexity, i.e., including or excluding background soil spectra correction.

To compare complex RTM inversion models with standard vegetation index models, the normalized difference vegetation index (NDVI) and the normalized difference red-edge (NDRE) were calculated. This was performed based on the measured canopy reflectance spectra with multispectral resolution:

$$\mathrm{NDVI} = \frac{\mathrm{NIR} - \mathrm{RED}}{\mathrm{NIR} + \mathrm{RED}} \tag{6}$$

$$\mathrm{NDRE} = \frac{\mathrm{NIR} - \mathrm{RE}}{\mathrm{NIR} + \mathrm{RE}} \tag{7}$$

where NIR is near-infrared reflectance (Sentinel-2A band 8, central wavelength 842 nm), RED is red reflectance (Sentinel-2 band 4A, central wavelength 665 nm), and RE (Sentinel-2A band 7, central wavelength 783 nm) is red-edge reflectance.

2.6. Model Training and Testing

Six different models were tested for estimating PAI, NY, CWC, and AGDM. These models used varying inputs: NDVI, NDRE, or trait estimates of four different RTM inversions, which differed in their spectral resolution of input data, i.e., hyper- and multispectral, and the inclusion or exclusion of background soil spectra correction. The RTM inversion outputs for estimating PAI, NY, and CWC were LAI, LAI $\times$ C_{ab}, and LAI $\times$ EWT, respectively.

In accordance with previous studies, the nitrogen content and water content of plants were calculated as ground area-based traits, i.e., NY ($\mathrm{g\,m^{-2}}$) and CWC ($\mathrm{g\,m^{-2}}$) [22,44].

The field experimental data of the season 2020/21 were applied for model training. The resulting estimations on PAI, NY, and CWC using NDVI, NDRE, as well as four different RTM inversions were then tested with the field experimental data of the 2019/20 season. The 2020/21 season featured a higher range in PAI, NY, and CWC values compared to the 2019/20 season. Therefore, it was more suitable to use the 2020/21 season for model training and the 2019/20 season for model testing to avoid extrapolation in the model testing step.

For model training, the values of NDVI, NDRE, and RTM inversion estimates were fitted to measured PAI, NY, and CWC using linear regression (8), quadratic regression (9), or exponential function (10). Based on the visual assessment of the model fit, the best regression type for model training was selected.

$$y = a + bx \tag{8}$$

$$y = a + bx + cx^2 \tag{9}$$

$$y = ae^{zx} \tag{10}$$

where x is the predictor variable, y is the response variable, a is the intercept, b is the first order regression coefficient, c is the second order regression coefficient, and z is the exponential coefficient.

To estimate AGDM, the traits PAD, NDVI integral, and leaf area duration (LAD, $m^2\ m^{-2}$ d) were applied. In Section 2.2, the calculation of PAD based on destructively measured PAI values was presented. Accordingly, NDVI values as well as RTM inversion-based LAI values were linearly interpolated to achieve a time series of daily NDVI and LAI values for each plot from the first sampling date until harvest for both seasons. Daily NDVI values as well as daily RTM inversion-based LAI values were then integrated over time to calculate LAD (11) and NDVI integral (12), respectively:

$$\text{NDVI integral} = \int_a^b \text{NDVI}(t)\text{d}t \tag{11}$$

$$\text{LAD}\left(\text{m}^2\ \text{m}^{-2}\ \text{d}\right) = \int_a^b \text{LAI}(t)\text{d}t \tag{12}$$

where a is the first measurement day in a given season, b is the last measurement day in the same season, and t is the integration variable time in days. Model training and testing to estimate AGDM were conducted analogous to the estimations of PAI, NY, and CWC.

Regression functions were calculated based on the estimated marginal means of model inputs, i.e., NDVI, NDRE, and RTM inversion estimates, and model outputs, i.e., measured values of PAI, NY, CWC, and AGDM. Thus, all model inputs and outputs were fitted using a linear mixed model based on the lme function of the R package nlme (version 3.1-164). As mentioned in Section 2.2, the field experiments were set up in a split-plot design. The main plot factor sowing time (ST) was established in complete blocks (BLOCK). The sub-plot factor nitrogen fertilization (NF) was completely randomized among sub-plots (SUB) within the main plots (MAIN). A mixed model analysis was conducted to adequately account for several random sources of variation in the split-plot design [45]. The factors ST and NF were implemented in the statistical model as fixed factors. The following mixed model specification was applied:

SD + NF + SD.NF: BLOCK.MAIN + BLOCK.MAIN.SUB (13)

The colon separates fixed and random effects. The residual is underlined. Then, estimated marginal means for each nitrogen fertilization level of autumn- and spring-sown wheat in both seasons were calculated using the emmeans functions of the R package emmeans (version 1.10.0).

Finally, model performance was evaluated using coefficients of determination (R^2, 14) as well as relative root mean square error (RRMSE, 16), which is based on the root mean square error (RMSE, 15):

$$\text{R}^2 = \frac{1}{n}\frac{\sum_{i=1}^{n}(y_i - \hat{y}_i)^2}{\sum_{i=1}^{n}(y_i - \overline{y})^2} \tag{14}$$

$$\text{RMSE} = \sqrt{\frac{1}{n}\sum_{i=1}^{n}(y_i - \hat{y}_i)^2} \tag{15}$$

$$\text{RRMSE}(\%) = \frac{\text{RMSE}}{\overline{y}} \times 10^{-2} \tag{16}$$

where n is the sample size, y_i is the measured value for the ith observation, $\hat{y}_i$ is the estimated value for the ith observation, and $\overline{y}$ is the mean value. Model performance was evaluated based on the testing dataset (2019/20 season).

Calculations and regression models were conducted in Microsoft Excel (version 16.0). Graphs were created using SigmaPlot (version 14.5).

3. Results

3.1. Radiative Transfer Model Inversion

Table 2 presents the coefficients of determination (R^2) for PROSAIL parameter estimations based on the simulated test dataset using four different RTM inversions. The models differed in the spectral resolution of input data, i.e., hyper- and multispectral, as well as the inclusion or exclusion of background soil spectra correction.

Table 2. Coefficients of determination for estimating PROSAIL parameters using the simulated test dataset. Estimations were compiled by the artificial neural network-based inversions of the radiative transfer model PROSAIL. Four different models were tested. They differed in the spectral resolution of input data, i.e., hyper- and multispectral, as well as the inclusion or exclusion of background soil spectra correction. Investigated PROSAIL parameters were leaf structure index (N, untiless), chlorophyll content (C_{ab}, μg cm^{-2}), carotenoid content (C_{cx}, μg cm^{-2}), brown pigments (C_{bp}, unitless), equivalent water thickness (EWT, g cm^{-2}), leaf mass per area (LMA, g cm^{-2}), soil reflectance (α_{soil}, %), leaf area index (LAI, m^2 m^{-2}), average leaf inclination angle (ALIA, °), and hot-spot parameter (Hot, m m^{-1}).

Spectral Resolution	Background Correction	N	C_{ab}	C_{cx}	C_{bp}	EWT	LMA	α_{soil}	LAI	ALIA	Hot
Hyperspectral	Included	0.546	0.947	0.885	0.922	0.906	0.770	–	0.885	0.892	0.108
Hyperspectral	Excluded	0.605	0.953	0.936	0.936	0.881	0.590	0.320	0.872	0.928	0.215
Multispectral	Included	0.227	0.891	0.607	0.806	0.699	0.556	–	0.848	0.847	0.030
Multispectral	Excluded	0.305	0.900	0.740	0.839	0.777	0.439	0.260	0.808	0.834	0.022

The following estimations showed R^2 values above 0.8: C_{ab} (all), C_{cx} (hyperspectral input data), C_{bp} (all), EWT (hyperspectral input data), LAI (all), and ALIA (all). Furthermore, R^2 was close to or above 0.7 for estimating C_{cx} (multispectral input data and no background soil spectra correction), EWT (multispectral input data), and LMA (hyperspectral input data and background soil spectra correction).

In general, the R^2 values of RTM inversions using hyperspectral input data were higher compared to those of models using multispectral input data. The difference between the R^2 values of RTM inversions with hyper- and multispectral input data varied substantially among PROSAIL parameters. The R^2 values of C_{ab} estimations, for example, showed low variation between hyper- (0.947 and 0.953) and multispectral input data (0.891 and 0.900), while EWT estimations differed substantially in R^2 depending on the spectral resolution (hyperspectral: 0.906 and 0.881; multispectral: 0.699 and 0.777).

Furthermore, RTM inversion without background soil spectra correction showed higher R^2 values compared to models with background soil spectra correction for many PROSAIL parameters. Exceptions were, for example, EWT (hyperspectral input data) as well as LMA and LAI (hyper- and multispectral input data). The differences in R^2 between RTM inversions with and without background soil spectra correction, however, were generally low. An exception was the highly varying R^2 of LMA estimations using the RTM inversion with hyperspectral input data and including background soil spectra correction (0.770) compared to the model with hyperspectral input data and excluding background soil spectra correction (0.590).

The PROSAIL parameter α_{soil} could only be estimated by models excluding background soil spectra correction.

Figure 1 presents the estimations of LAI, C_{ab}, and EWT using the RTM inversion with hyperspectral input data and including background soil spectra correction. These estimations, based on the simulated test dataset, resulted in RRMSE values of 18.5% for LAI (RMSE = 0.743 m^2 m^{-2}, mean = 4.008 m^2 m^{-2}), 14.5% for C_{ab} (RMSE = 5.81 μg cm^{-2}, mean = 40.15 μg cm^{-2}), and 13.8% for EWT (RMSE = 0.00352 g cm^{-2}, mean = 0.02560 g cm^{-2}).

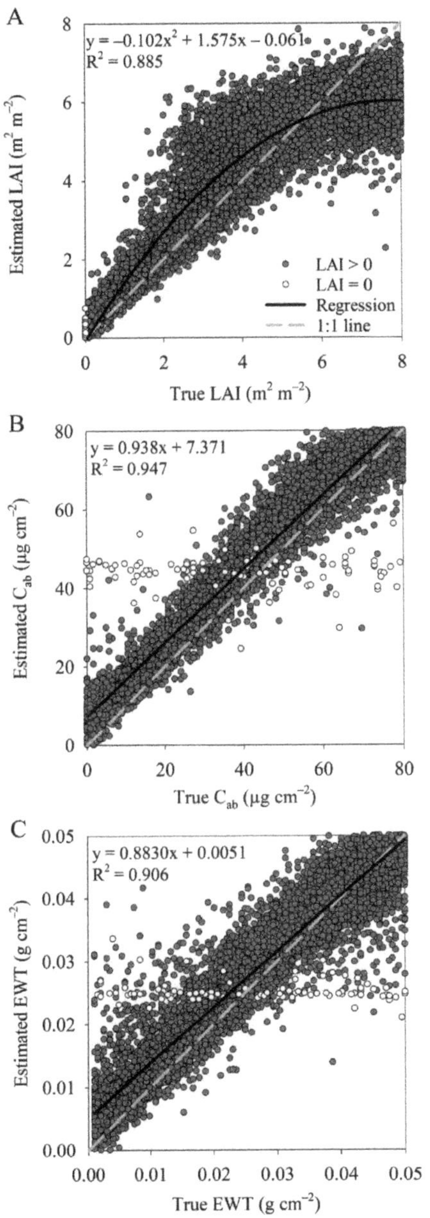

Figure 1. True leaf area index (LAI, m^2 m^{-2}, (**A**)), chlorophyll content (C_{ab}, μg cm^{-2}, (**B**)), and equivalent water thickness (EWT, g cm^{-2}, (**C**)) versus estimated values using the simulated test dataset. Estimations were compiled by an artificial neural network-based inversion of the radiative transfer model PROSAIL. Hyperspectral input data were used, and background soil spectra correction was included. Coefficients of linear or quadratic regression (*a*, *b* and *c*) as well as respective coefficients of determination (R^2) are presented.

When LAI was 0, the estimations of C_{ab} and EWT in Figure 1B,C showed random scattering around the respective mean values.

3.2. Estimating Wheat Traits

Table 3 provides an overview of data collected in field experiments on individual sampling dates in the 2019/2020 and 2020/21 seasons. It presents developmental stages as well as estimated marginal means and standard deviation values of PAI, NY, CWC, and AGDM for autumn- and spring-sown wheat across five nitrogen fertilization levels from 0 to 20 g N m^{-2}.

The mean PAI of the sampling dates ranged from 0.037 m^2 m^{-2} (2019/20, spring-sowing, BBCH 11) to 3.751 m^2 m^{-2} (2020/21, autumn-sowing, BBCH 57). The average AGDM of the sampling dates was the lowest in 2020/21 for spring-sown wheat at BBCH 10 (2.4 g m^{-2}) and the highest also in 2020/21 for autumn-sown wheat at harvest, i.e., BBCH 89 (1151.9 g m^{-2}). The mean NY of the sampling dates ranged from 0.13 g m^{-2} (2020/21, spring-sowing, BBCH 10) to 14.21 g m^{-2} (2020/21, autumn-sowing, BBCH 89). The lowest average CWC of the sampling dates was 7.4 g m^{-2} (2019/20, spring-sowing, BBCH 11), while the highest CWC was 2099.5 g m^{-2} (2020/21, autumn-sowing, BBCH 57).

Table 4 presents the results of estimating PAI, NY, and CWC using NDVI, NDRE, and four different RTM inversions during vegetative developmental stages from leaf development (BBCH 10) until the start of anthesis (BBCH 60). The models differed in the spectral resolution of input data and the inclusion or exclusion of background soil spectra correction. The traits PAI, NY, and CWC were estimated by the RTM inversion outputs LAI, LAI × C_{ab}, and LAI × EWT, respectively. For model training based on the field experimental data of the 2020/21 season, regression types (linear, quadratic, and exponential), regression coefficients (a, b, c, and z) as well as R^2 values are presented in Table 4. Model testing was based on the field experimental data of the 2019/20 season and included linear regression coefficients (a and b), R^2 values, and RRMSE values. In all regression models, the estimated marginal means were used for model inputs and outputs.

Regarding PAI, all models showed high R^2 for model training ranging from 0.869 (RTM inversion with multispectral input data and no background soil spectra correction) to 0.964 (NDRE). In model testing, R^2 values were lower, but still close to or above 0.8 for all models except for RTM inversion with multispectral input data and no background soil spectra correction (R^2 = 0.521). The testing RRMSE was the lowest for RTM inversion with hyperspectral input data and background soil spectra correction (RRMSE = 17.9%, RMSE = 0.215 m^2 m^{-2}, mean = 1.200 m^2 m^{-2}). Other models that showed good testing PAI estimationss were NDVI (R^2 = 0.893, RRMSE = 22.4%, RMSE = 0.269 m^2 m^{-2}) and RTM inversion with multispectral input data and background soil spectra correction (R^2 = 0.792, RRMSE = 24.4%, RMSE = 0.293 m^2 m^{-2}).

The results on NY show high training R^2 from 0.911 (RTM inversion with hyperspectral input data and no background soil spectra correction) to 0.970 (NDRE). In model testing, however, there were only two models with high R^2 and low RRMSE. These were the RTM inversion with hyperspectral input data and background soil spectra correction (R^2 = 0.908; RRMSE = 14.4%, RMSE = 0.608 g m^{-2}, mean = 4.223 g m^{-2}) and NDRE (R^2 = 0.904; RRMSE = 23.0%, RMSE = 0.971 g m^{-2}).

Lastly, in Table 4, CWC estimations also resulted in high training R^2 for all models ranging from 0.826 (NDVI) to 0.944 (RTM inversion with hyperspectral input data and background soil spectra correction). In model testing, however, only the RTM inversion with hyperspectral input data and background soil spectra correction showed both high R^2 (0.967) and low RRMSE (17.0%, RMSE = 82.2 g m^{-2}, mean = 483.0 g m^{-2}).

Figure 2 graphically presents the model training and testing of the RTM inversion with hyperspectral input data and background soil spectra correction for the estimation of PAI, NY, and CWC during the vegetative developmental stages.

Table 3. Developmental stage (BBCH) as well as estimated marginal means (Mean) and standard deviation (SD) of plant area index (PAI, $m^2\ m^{-2}$), nitrogen yield (NY, $g\ m^{-2}$), canopy water content (CWC, $g\ m^{-2}$), and above-ground dry matter (AGDM, $g\ m^{-2}$) for autumn- and spring-sown wheat across five nitrogen fertilization levels from 0 to 20 g N m^{-2} on individual sampling dates in 2019/20 and 2020/21.

Date	Autumn-Sown Wheat, 2019/20 Season									Spring-Sown Wheat, 2019/20 Season								
	BBCH	PAI		NY		CWC		AGDM		BBCH	PAI		NY		CWC		AGDM	
		Mean	SD	Mean	SD	Mean	SD	Mean	SD		Mean	SD	Mean	SD	Mean	SD	Mean	SD
9 March 2020	23	0.641	0.051	2.20	0.10	124.9	14.1	52.5	1.5									
23 March 2020	24	1.050	0.261	3.31	0.59	265.6	64.9	85.6	19.3									
6 April 2020	30	1.318	0.236	3.95	1.02	349.6	82.1	141.6	24.5	11	0.037	0.006	0.19	0.03	7.4	1.8	4.0	0.6
20 April 2020	32	2.447	0.439	6.58	1.70	801.6	165.0	285.1	43.0	13	0.110	0.023	0.55	0.11	32.1	8.3	12.8	2.6
4 May 2020	45	2.354	0.463	7.65	2.08	1112.0	217.7	503.1	72.3	30	0.342	0.101	1.38	0.35	87.8	30.0	33.1	8.4
17 May 2020	59	2.298	0.550	9.42	3.15	1198.4	260.6	691.1	82.2	37	0.809	0.189	3.62	0.71	272.7	72.3	107.0	20.8
1 June 2020	77	2.011	0.534	9.14	3.98	1224.0	256.1	794.3	123.5	51	1.349	0.238	5.97	1.29	774.3	138.5	257.3	37.6
15 June 2020	85	1.986	0.405	11.65	3.78	1296.5	276.5	1050.7	178.9	71	1.490	0.333	8.88	2.33	986.6	228.4	466.0	67.0
6 July 2020	89	1.535	0.283	10.74	3.70	106.7	19.2	1048.8	184.6	85	1.033	0.169	10.93	1.94	688.3	155.8	685.8	79.1
20 July 2020										89	1.034	0.160	9.94	2.05	73.1	7.8	725.1	88.9
Date	**Autumn-sown wheat, 2020/21 Season**									**Spring-sown wheat, 2020/21 Season**								
	BBCH	**PAI**		**NY**		**CWC**		**AGDM**		**BBCH**	**PAI**		**NY**		**CWC**		**AGDM**	
		Mean	**SD**	**Mean**	**SD**	**Mean**	**SD**	**Mean**	**SD**		**Mean**	**SD**	**Mean**	**SD**	**Mean**	**SD**	**Mean**	**SD**
9 March 2021	21	0.162	0.031	0.67	0.12	33.9	5.7	14.9	2.5									
23 March 2021	22	0.224	0.035	0.90	0.14	40.2	7.0	21.1	2.9									
6 April 2021	23	0.508	0.106	1.64	0.33	121.4	28.1	39.4	7.6	10	0.043	0.010	0.13	0.03	9.2	2.1	2.4	0.5
20 April 2021	30	0.795	0.106	2.76	0.42	198.4	31.0	67.5	8.9	12	0.074	0.016	0.31	0.05	14.8	4.1	6.4	1.2
2 May 2021	31	1.596	0.350	5.15	1.39	498.8	102.0	136.8	22.3	21	0.262	0.081	0.99	0.28	51.0	19.3	18.1	4.5
18 May 2021	41	3.116	0.598	9.59	2.75	1205.0	254.1	343.1	39.8	31	1.154	0.242	3.76	0.84	253.5	80.4	72.4	14.6
31 May 2021	57	3.751	0.962	11.42	3.69	2099.5	524.3	600.3	94.3	33	2.325	0.556	7.98	2.15	928.2	242.6	206.5	41.4
15 June 2021	75	3.204	1.003	13.82	4.47	1752.0	408.6	928.9	146.7	56	3.583	0.837	10.32	2.77	1608.2	331.8	497.2	79.4
27 June 2021	85	2.284	0.632	12.69	4.61	839.3	190.6	1144.2	242.5	75	2.460	0.510	11.07	2.77	1099.9	245.4	790.8	111.1
13 July 2021	89	2.187	0.568	14.21	4.53	141.9	26.8	1151.9	213.6	85	2.520	0.467	11.73	2.67	311.9	91.4	931.2	131.4
26 July 2021										89	2.609	0.336	11.95	2.48	106.0	11.6	887.4	92.3

Table 4. Estimating plant area index (m^2 m^{-2}), nitrogen yield (g m^{-2}) and canopy water content (g m^{-2}) using normalized difference vegetation index (NDVI), normalized difference red-edge (NDRE), and artificial neural network-based radiative transfer model inversion (RTMI) during the vegetative developmental stages from leaf development (BBCH 10) to the start of anthesis (BBCH 60). The four RTMIs differed in the spectral resolution of input data (hyper- or multispectral) and the inclusion or exclusion of background soil spectra correction. The following RTMI outputs were used for respective trait estimations: leaf area index (LAI, m^2 m^{-2}), LAI multiplied by chlorophyll content (LAI × C_{ab}, g m^{-2}), and LAI multiplied by equivalent water thickness (LAI × EWT, g m^{-2}). Vegetation index values and RTMI outputs were trained using the experimental data of the 2020/21 season. For training, regression type (linear, quadratic, or exponential), regression coefficients (*a*, *b*, *c*, and *z*), as well as coefficient of determination (R^2) are presented. Estimations of PAI, NY, and CWC based on the trained models using vegetation indices and RTMI outputs were tested with the experimental data of the 2019/20 season. For model testing, linear regression coefficients (*a* and *b*), coefficients of determination, and relative root mean square error (RRMSE) are provided.

Model	Spectral Resolution	Background Correction	Regression Type	Model Training (2020/21 Season)					Model Testing (2019/20 Season)			
				a	*b*	*c*	*z*	R^2	*a*	*b*	R^2	RRMSE
Plant area index (m^2 m^{-2}), before anthesis												
NDVI	–	–	Exponential	0.0318			5.07	0.915	−0.049	1.112	0.893	22.4
NDRE	–	–	Linear	−1.0276	27.1640			0.964	−0.204	1.309	0.903	31.2
LAI (RTMI)	Hyperspectral	Included	Quadratic	0.0595	0.3723	0.0881		0.955	0.108	1.032	0.930	17.9
LAI (RTMI)	Hyperspectral	Excluded	Quadratic	−0.3313	0.2390	0.0917		0.898	0.270	1.231	0.852	46.9
LAI (RTMI)	Multispectral	Included	Quadratic	0.4115	0.4479	0.0305		0.922	0.134	0.826	0.792	24.4
LAI (RTMI)	Multispectral	Excluded	Quadratic	−0.0034	0.1778	0.1028		0.869	0.534	0.783	0.521	53.1
Nitrogen yield (g m^{-2}), before anthesis												
NDVI	–	–	Exponential	0.127			4.791	0.920	0.356	0.889	0.747	31.5
NDRE	–	–	Linear	−3.076	83.888			0.970	−0.513	1.144	0.904	23.0
LAI × C_{ab} (RTMI)	Hyperspectral	Included	Quadratic	0.306	2.292	0.132		0.961	0.524	0.871	0.908	14.4
LAI × C_{ab} (RTMI)	Hyperspectral	Excluded	Linear	−0.118	3.061			0.911	−0.776	0.916	0.788	35.0
LAI × C_{ab} (RTMI)	Multispectral	Included	Quadratic	1.440	2.404	0.186		0.950	0.553	0.756	0.740	28.2
LAI × C_{ab} (RTMI)	Multispectral	Excluded	Linear	−0.143	3.635			0.943	0.096	0.992	0.766	29.8
Canopy water content (g m^{-2}), before anthesis												
NDVI	–	–	Exponential	4.578			6.17	0.826	76.30	0.78	0.664	40.8
NDRE	–	–	Quadratic	−69.214	−152.890	61,573.509962		0.929	−22.07	1.12	0.782	32.9
LAI × EWT (RTMI)	Hyperspectral	Included	Quadratic	−50.943	0.544	0.000219		0.944	−40.93	1.15	0.967	17.0
LAI × EWT (RTMI)	Hyperspectral	Excluded	Linear	−167.070	1.341			0.920	300.04	1.00	0.825	62.6
LAI × EWT (RTMI)	Multispectral	Included	Quadratic	62.502	0.404	0.000123		0.901	0.72	−2.58	0.818	31.5
LAI × EWT (RTMI)	Multispectral	Excluded	Quadratic	−33.379	0.316	0.000178		0.850	208.78	0.72	0.738	40.9

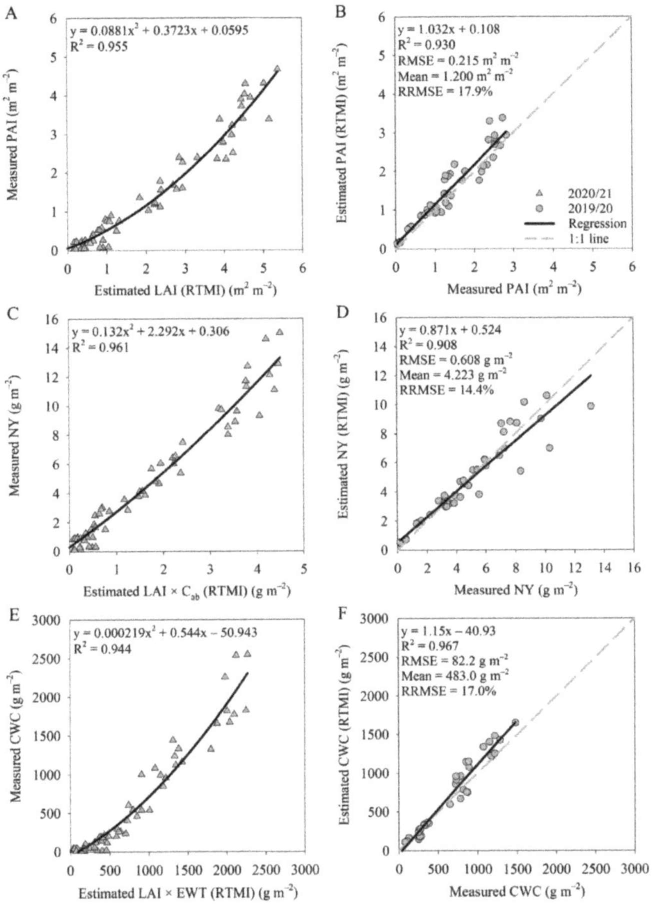

Figure 2. Estimating plant area index (PAI, m^2 m^{-2}, (**A**,**B**)), nitrogen yield (NY, g m^{-2}, (**C**,**D**)), and canopy water content (CWC, g m^{-2}, (**E**,**F**)) using artificial neural network-based radiative transfer model inversion (RTMI) during the vegetative stages from leaf development (BBCH 10) to the start of anthesis (BBCH 60). This RTMI used hyperspectral input data and included background soil spectra correction. The outputs of RTMI were leaf area index (LAI, m^2 m^{-2}, (**A**)), LAI multiplied by chlorophyll content (LAI × C_{ab}, g m^{-2}, (**C**)), and LAI multiplied by equivalent water thickness (LAI × EWT, g m^{-2}, (**E**)). For training, the field experimental data of the 2020/21 season was used (**A**,**C**,**E**). The model testing was performed with the field experimental data of the 2019/20 season (**B**,**D**,**F**).

Table 5 presents the results of PAI and CWC estimations throughout the growing season from leaf development (BBCH 10) until harvest (BBCH 89) using NDVI, NDRE, and the four different RTM inversions. For both PAI and CWC, one model featured the highest training and testing R^2 as well as the lowest testing RRMSE. This was the RTM inversion with hyperspectral input data and background soil spectra correction. Estimations of PAI using the RTM inversion with hyperspectral input data and background soil spectra correction resulted in a training R^2 of 0.922, a testing R^2 of 0.845, and a testing RRMSE of 27.7% (RMSE = 0.364 m^2 m^{-2}, mean = 1.317 m^2 m^{-2}), while CWC estimations showed a training R^2 of 0.927, a testing R^2 of 0.884, and a testing RRMSE of 20.0% (RMSE = 115 g m^{-2}, mean = 574 g m^{-2}). All other models for PAI and CWC featured lower training and testing R^2 values as well as higher testing RRMSE values. Figure 3 visually presents the training and testing of PAI and CWC estimations throughout the growing season based on the RTM inversion with hyperspectral input data and background soil spectra correction. Severe deviations from the 1:1 line in model testing for PAI and CWC estimations in Figure 3B,D occurred partially for data collected at harvest (BBCH 89).

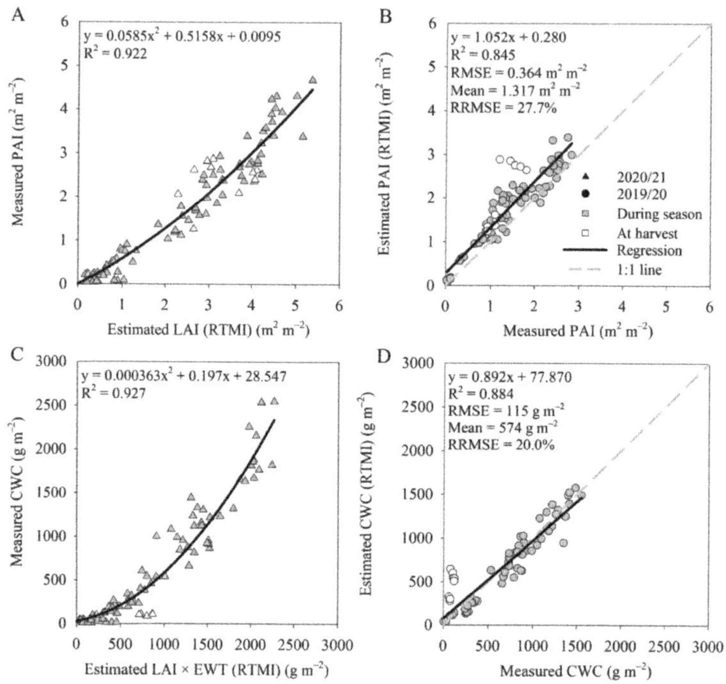

Figure 3. Estimating plant area index (PAI, $m^2\ m^{-2}$, (**A**,**B**)) and canopy water content (CWC, $g\ m^{-2}$, (**C**,**D**)) using artificial neural network-based radiative transfer model inversion (RTMI) throughout the growing season from leaf development (BBCH 10) to harvest (BBCH 89). This RTMI used hyperspectral input data and included background soil spectra correction. Outputs of RTMI were leaf area index (LAI, $m^2\ m^{-2}$, (**A**)) and LAI multiplied by equivalent water thickness (LAI × EWT, $g\ m^{-2}$, (**C**)). For training, the field experimental data of the 2020/21 season was used (**A**,**C**). The model testing was performed with the field experimental data of the 2019/20 season (**B**,**D**).

Estimating NY from leaf development (BBCH 10) to harvest (BBCH 89) was tested; however, R^2 values were low and RRMSE values were high for all models.

Finally, AGDM estimations throughout the growing season from leaf development (BBCH 10) to harvest (BBCH 89) were conducted based on PAD, NDVI integral, as well as LAD (Table 6). PAD was calculated using destructively measured PAI. LAD was calculated using LAI estimations based on the RTM inversion with hyperspectral input data and background soil spectra correction. Accordingly, the integral over time was also calculated for NDVI. Models were trained using the field experimental data of the 2020/21 season and tested using the field experimental data of the 2019/20 season. All three models, PAD, NDVI integral, and LAD, showed high training and testing R^2 as well as low testing RRMSE. Training R^2 values were similar ranging from 0.965 (RTM inversion-based LAD) to 0.972 (measured PAD). Testing R^2 was the lowest for measured PAD (0.914) and highest for RTM inversion-based LAD (0.960). Testing RRMSE was 18.1% (RMSE = 77.0 $g\ m^{-2}$, mean = 426.0 $g\ m^{-2}$) for the measured PAD, 24.3% (RMSE = 103.3 $g\ m^{-2}$) for NDVI integral, and 13.7% (RMSE = 58.6 $g\ m^{-2}$) for RTM inversion-based LAD.

Table 5. Estimating plant area index ($m^2\ m^{-2}$) and canopy water content ($g\ m^{-2}$) using normalized difference vegetation index (NDVI), normalized difference red-edge (NDRE), and artificial neural network-based radiative transfer model inversion (RTMI) throughout the growing season from leaf development (BBCH 10) to harvest (BBCH 89). The four RTMIs differed in the spectral resolution of input data (hyper- or multispectral) and the inclusion or exclusion of background soil spectra correction. The following RTMI outputs were used for respective trait estimations: leaf area index (LAI, $m^2\ m^{-2}$) and LAI multiplied by equivalent water thickness (LAI × EWT, $g\ m^{-2}$). Vegetation index values and RTMI outputs were trained using the field experimental data of the 2020/21 season. For model training, regression type, regression coefficients (*a*, *b*, *c*, and *z*), as well as coefficient of determination (R^2) are presented. Model testing was performed with the field experimental data of the 2019/20 season. For model testing, linear regression coefficients (*a* and *b*), coefficients of determination, and relative root mean square error (RRMSE) are provided.

Model	Spectral Resolution	Background Correction	Regression Type	Model Training (2020/21 Season)					Model Testing (2019/20 Season)			
				a	*b*	*c*	*z*	R^2	*a*	*b*	R^2	RRMSE
Plant area index ($m^2\ m^{-2}$), all developmental stages												
NDVI	–	–	Exponential	0.1629			3.1927	0.485	0.177	0.936	0.664	31.2
NDRE	–	–	Quadratic	−0.4831	28.2480	−32.7020		0.670	0.318	1.046	0.704	41.3
LAI (RTMI)	Hyperspectral	Included	Quadratic	0.0095	0.5158	0.0585		0.922	0.280	1.052	0.845	27.7
LAI (RTMI)	Hyperspectral	Excluded	Linear	−0.7270	0.7114			0.778	0.677	0.926	0.695	46.1
LAI (RTMI)	Multispectral	Included	Quadratic	0.6204	0.8408	−0.0508		0.830	0.559	0.898	0.723	39.9
LAI (RTMI)	Multispectral	Excluded	Quadratic	−0.0237	0.2212	0.0879		0.867	0.616	0.695	0.496	40.1
Canopy water content ($g\ m^{-2}$), all developmental stages												
NDVI	–	–	Exponential	13.855			5.053	0.756	117.54	0.67	0.566	42.7
NDRE	–	–	Linear	−526.450	12,335.294			0.891	14.27	0.96	0.710	39.0
LAI × EWT (RTMI)	Hyperspectral	Included	Quadratic	28.547	0.197	0.000363		0.927	77.87	0.89	0.884	20.0
LAI × EWT (RTMI)	Hyperspectral	Excluded	Linear	−168.710	1.189			0.785	340.33	0.80	0.739	44.9
LAI × EWT (RTMI)	Multispectral	Included	Linear	35.600	0.731			0.813	132.28	0.65	0.637	35.7
LAI × EWT (RTMI)	Multispectral	Excluded	Quadratic	29.498	0.017	0.000293		0.842	121.52	0.78	0.806	33.3

Table 6. Estimating above-ground dry matter (AGDM, g m^{-2}) throughout the growing season from leaf development (BBCH 10) to harvest (BBCH 89) using plant area duration (PAD, m^2 m^{-2} d), integral of normalized difference vegetation index (NDVI) over time, as well as leaf area duration (LAD, m^2 m^{-2}). The latter was calculated with an artificial neural network-based radiative transfer model inversion (RTMI). This RTMI applied hyperspectral input data and background soil spectra correction. The RTMI output was leaf area index (LAI, m^2 m^{-2}). The destructively measured plant area index (PAI, m^2 m^{-2}) was used to calculate PAD. The integral over time was calculated for measured PAI, NDVI, as well as RTMI-based LAI. The integrals of PAI and LAI over time are PAD and LAD, respectively. Values of PAD, NDVI integral, and LAD were trained using the experimental data of the 2020/21 season. For training, regression type, regression coefficients (*a*, *b*, and *c*), as well as coefficient of determination (R^2) are presented. Model testing was performed with the field experimental data of the 2019/20 season. For model testing, linear regression coefficients (*a* and *b*), coefficients of determination, and relative root mean square error (RRMSE) are provided.

Model	Model Training (2020/21 Season)					Model Testing (2019/20 Season)			
	Regression Type	*a*	*b*	*c*	R^2	*a*	*b*	R^2	RRMSE
PAD (measured)	Linear	11.80	5.24		0.972	−19.60	0.92	0.914	18.1
NDVI integral	Quadratic	−26.87	6.23	0.13	0.969	−77.38	1.27	0.927	24.3
LAD (RTMI)	Linear	−16.78	3.88		0.965	−21.29	1.12	0.960	13.7

The training and testing of the RTM inversion-based LAD model with hyperspectral input data and background soil spectra correction are presented visually in Figure 4.

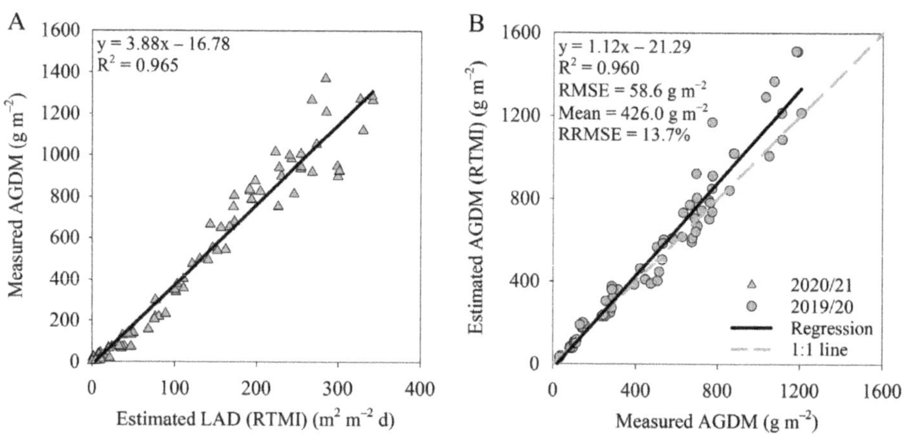

Figure 4. Estimating above-ground dry matter (AGDM, g m^{-2}) throughout the growing season from leaf development (BBCH 10) to harvest (BBCH 89) using artificial neural network-based radiative transfer model inversion (RTMI). This RTMI used hyperspectral input data and included background soil spectra correction. Leaf area duration (LAD, m^2 m^{-2} d) is the integral of leaf area index (LAI, m^2 m^{-2}) over time. Values of LAI were estimated by the RTMI. For training, the field experimental data of the 2020/21 season was used (**A**). Model testing was performed with the field experimental data of the 2019/20 season (**B**).

4. Discussion

4.1. Radiative Transfer Model Inversion

The RTM inversion was successful for the relevant traits in this study, i.e., LAI, C_{ab}, and EWT, as shown by high R^2 (Table 2) and low RRMSE for the PROSAIL parameter estimations based on the simulated test dataset (Section 3.1.). Furthermore, the estimations of other PROSAIL parameters also featured high R^2 in the simulated test dataset, e.g., C_{cx},

C_{bp}, ALIA, and LMA. These additional traits, however, could not be analyzed further, because no field experimental data on these traits were available.

The estimations of LAI based on a simulated test dataset were similar or superior to the results reported by Atzberger [19] (R^2 = 0.85, RMSE = 0.59 m^2 m^{-2}), Weiss and Baret [22] (RMSE = 0.89 m^2 m^{-2}), Bacour et al. [46] (R^2 = 0.80, RMSE = 1.15 m^2 m^{-2}), and Verger et al. [47] (RMSE = 0.91 m^2 m^{-2}). Furthermore, Atzberger [19] showed similar performance for the estimations of C_{ab} (R^2 = 0.83, RMSE = 7.8 µg cm^{-2}) and EWT (R^2 = 0.92, RMSE = 0.003 g cm^{-2}) based on the simulated test data. These studies, however, simulated varying spectral resolutions and did not limit their dataset to a specific crop. In this paper, the spectral resolution was limited to the spectral range from 400 to 1000 nm in 1 nm increments, i.e., hyperspectral, or Sentinel-2A bands 1 to 9, including 8A, i.e., multispectral.

The use of hyperspectral input data resulted in superior trait estimations compared to multispectral input data (Table 2). This is in accordance with Baret and Buis [9], Atzberger et al. [10] and Weiss et al. [48], who reason that an increase in the number of spectral bands can improve trait estimations in terms of noise sensitivity, robustness, generalization, and reproducibility.

Furthermore, RTM inversions excluding the background soil spectra correction featured a superior R^2 compared to RTM inversions including the background soil spectra correction (Table 2). The two simulated datasets, i.e., one including and one excluding background soil spectra correction, drastically differed in their variability in background soil spectra. The standard soil spectrum in the simulated dataset excluding background correction was generated using a single default soil spectrum for all observations and a randomly selected soil brightness factor α_{soil}. In comparison, the soil spectrum in the simulated dataset including background soil spectra correction was randomly selected from the 3717 highly varying soil spectra of the ICRAF-ISRIC Soil MIR Spectral Library [40] for each observation. Therefore, the variability in soil spectra was substantially higher in the simulated dataset including background soil spectra correction. The ANN models that were trained on the two variants of the simulated dataset were of identical structure, i.e., number of layers and neurons. Therefore, the model tested on the simulated dataset of lower variability, i.e., excluding background correction, showed superior performance compared to the model tested on the highly variable simulated dataset, i.e., including background correction.

Furthermore, C_{ab} and EWT could not be estimated when LAI was 0 in the simulated test dataset. In these cases, the reflectance spectra were only dependent on the underlying soil spectra. Thus, the resulting estimations for C_{ab} and EWT defaulted to their respective mean values in the simulated dataset (Figure 1).

4.2. Estimating Wheat Traits

The empirically collected data showed high variation in PAI, NY, CWC, and AGDM (Table 3) due to two distinct sowing times, autumn- and spring-sowing, two seasons with substantially differing weather conditions, 2019/20 and 2020/21, as well as five nitrogen fertilization rates from 0 to 20 g N m^{-2} and a high number of sampling dates within seasons. Therefore, the available field experimental data were a suitable, highly diverse empirical dataset for model training and testing.

The RTM inversion with hyperspectral input data and background soil spectra correction was the best model for the estimation of PAI, NY, and CWC during the vegetative stage from BBCH 10 to BBCH 60: PAI: train R^2 = 0.955, test R^2 = 0.930, RRMSE = 17.9%, RMSE = 0.215 m^2 m^{-2}, mean = 1.200 m^2 m^{-2}; NY: train R^2 = 0.961, test R^2 = 0.908, RRMSE: 14.4%, RMSE = 0.608 g m^{-2}, mean = 4.223 g m^{-2}; CWC: train R^2 = 0.944, test R^2 = 0.967, RRMSE: 17.0%, RMSE = 82.2 g m^{-2}, and mean = 483.0 g m^{-2} (Table 4). These trait estimation results were similar or superior compared to the literature. Herrmann et al. [49], for example, estimated LAI in wheat during the vegetative stage using Partial Least Squares Regression based on Sentinel-2 bands with an R^2 of 0.903 and a RMSE of 0.62 m^2 m^{-2}. Estimations of NY in wheat during the vegetative stage were conducted by Fitzgerald et al. [44] (R^2 = 0.97, RMSE = 0.65 g m^{-2}) and Palka et al. [8] (R^2 = 0.9, RMSE = 1.84 g m^{-2}) using the CCCI-CNI approach. Zhang et al. [50] reported CWC estimations in wheat throughout the growing season using the vegetation index NDWI (Normalized Difference Water Index) with an R^2 of 0.68 and an RMSE of 148 g m^{-2}.

Other RTM inversions in this study showed worse estimations of PAI, NY, and CWC for the empirical test dataset. For RTM inversions with multispectral input data, the decrease in model performance can be attributed to the reduction in spectral resolution and thus lower noise sensitivity, robustness, generalization, and reproducibility [9,10,48]. The inversions of the RTM without background soil spectra correction were not able to handle the highly varying soil spectra among sampling dates in an adequate way, which resulted in worse trait estimations. Weiss and Baret [22] as well as Danner et al. [24] also highlighted the importance of accounting for background soil spectra variation in RTM inversion. Nevertheless, PAI estimations using the RTM inversion with multispectral data and background soil spectra correction were adequate (train R^2 = 0.922, test R^2 = 0.792, RRMSE = 24.4%, RMSE = 0.293 m^2 m^{-2}, mean = 1.200 m^2 m^{-2}) (Table 4).

The simple approach using vegetation indices showed promising trait estimations during the vegetative stage based on the empirical test dataset, e.g., PAI estimations using NDVI (train R^2 = 0.915, test R^2 = 0.893, RRMSE = 22.4%, RMSE = 0.269 m^2 m^{-2}, mean = 1.200 m^2 m^{-2}) and NY estimations using NDRE (train R^2 = 0.970, test R^2 = 0.904, RRMSE = 23.0%, RMSE = 0.971 g m^{-2}, mean = 4.223 g m^{-2}). Estimations of CWC using the vegetation index approach, however, showed poor results.

The best model to estimate PAI and CWC from BBCH 10 to harvest (BBCH 89) was also the RTM inversion with hyperspectral input data and soil correction (PAI: train R^2 = 0.922, test R^2 = 0.845, RRMSE = 27.7%, RMSE = 0.364 m^2 m^{-2}, mean = 1.317 m^2 m^{-2}; CWC: train R^2 = 0.927, test R^2 = 0.884, RRMSE = 20.0%, RMSE = 115 g m^{-2}, mean = 574 g m^{-2}) (Table 5). Model performance, however, was generally worse compared to estimations limited to the vegetative stage (BBCH 10 to 60). Nevertheless, the performance of this model, i.e., hyperspectral input data and inclusion of the background soil spectra correction, was substantially better than all other models tested in this study. Other models based on vegetation indices and RTM inversions using multispectral input data or excluding background soil spectra correction featured a lower performance, especially in model testing. In Figure 3B,D, estimations at harvest (BBCH 89) showed severe deviations from the 1:1 line in model testing for PAI and CWC.

The presented estimations on PAI and CWC throughout the season were similar or superior to the results reported in the literature: Atzberger et al. [51], for example, conducted an ANN-based RTM inversion using hyperspectral input from 450 to 2500 nm data to estimate LAI (R^2 = 0.86%, RMSE = 0.83 m^2 m^{-2}) in wheat throughout the growing season. Vuolo et al. [52] reported LAI estimations using RTM inversion based on Sentinel-2 bands across different crops with an R^2 of 0.83 and an RMSE of 0.32 m^2 m^{-2}. Pan et al. [53] conducted trait estimations in wheat throughout the season using a look-up table-based RTM inversion approach on Sentinel-2 bands resulting in an RRMSE of 11% for LAI and 32% for CWC. Wu et al. [54] estimated CWC in wheat throughout the season using vegetation indices based on multispectral data. The vegetation index $NDWI_{1940}$ resulted, for example, in CWC estimations with an R^2 of 0.83.

The estimation of NY throughout the growing season (BBCH 10 to 89) was evaluated; however, the results were generally poor for all models. This can be explained by the change in the allocation of nitrogen from the vegetative to the generative stage. From germination to flowering, the available nitrogen is generally used to produce chlorophyll, which can be detected spectrally. From the start of the generative stage onwards, however, the nitrogen in chlorophyll is translocated to the grains for storage proteins. Thus, NY can be estimated by the chlorophyll content during the vegetative stage; however, the chlorophyll content is not a good predictor of NY during the generative stage [55].

Finally, there were three models tested for estimating AGDM: measured PAD, NDVI integral, and RTM inversion-based LAD. All three models showed a suitable performance; however, the results of the RTM inversion with hyperspectral input data and background soil spectra correction (RTM inversion-based LAD) were superior to the other models tested in this study (train R^2 = 0.965, test R^2 = 0.960, RRMSE = 13.7%, RMSE = 58.6 g m^{-2}, mean = 426.0 g m^{-2}). This is especially noteworthy, since one of the alternative models was based on destructively measured PAI values (train R^2 = 0.972, test R^2 = 0.914, RRMSE = 18.1%). The comparatively simple NDVI integral approach also showed good results for AGDM estimations (train R^2 = 0.969, test R^2 = 0.927, RRMSE = 24.3%). In comparison, Tucker et al. [56] were among the first to correlate spectral data with AGDM in wheat and achieved an R^2 of 0.79. Mistele and Schmidhalter [57] determined dry matter yields in maize using curvilinear models between spectral indices and AGDM with R^2 ranging from 0.67 to 0.91 in three seasons. Fabbri et al. [58] tested various vegetation indices to estimate AGDM in durum wheat. The best performing index for AGDM was EVI2 (R^2 = 0.951, RRMSE = 17.9%) during the vegetative stage.

The presented model performances are promising; nevertheless, it needs to be mentioned that these models also feature certain limitations. The selected RTM PROSAIL is a basic RTM that features a simplified description of canopy architecture compared to more complex RTMs [25]. Furthermore, the field experimental dataset is limited in the following aspects: limited data variability, i.e., solely one crop (wheat) was investigated in one region (Eastern Austria) in two seasons; limited dataset size, i.e., a modest number of data points; and limitations of the applied spectral sensor, i.e., no spatial variation and limited spectral resolution from 400 to 1000 nm. In the context of CWC estimations, for example, the spectral reflectance above 1000 nm is of substantial importance [54]. To offset the limitation of the small field experimental dataset, a simulated dataset based on an RTM inversion was introduced in this study. The extensive use of simulated data as well as the previously mentioned aspects on the limitations of the RTM PROSAIL and the field experimental dataset, however, limit the generalizability of the proposed models.

Although the proposed models show certain limitations, they nevertheless resulted in suitable model performances for estimating PAI, CWC, and AGDM throughout the growing season as well as NY from leaf development (BBCH 10) to anthesis (BBCH 60) for wheat in Eastern Austria. The presented models can be used in future studies for quick and non-destructive measurements of PAI, CWC, NY, and AGDM based on spectral measurements of wheat in Eastern Austria during the aforementioned developmental stages. This can reduce the required workload drastically. Furthermore, when applied jointly, the proposed models for estimating PAI, CWC, NY, and AGDM can improve the evaluation of stress conditions in wheat canopies, e.g., nitrogen deficiencies and drought. Additionally, the models and methods proposed in this study can be used as templates for setting up models for additional wheat cultivars, crops, regions, or target traits in the future. Furthermore, the approach presented in this study can be adapted to establish models for different spectral resolutions and spatial resolutions of other sensors that can be mounted on a tripod or a tractor for proximal measurements, on a drone, an airplane, or also satellites. The proposed methodology can also be advanced by applying more complex RTMs, e.g., 4SAIL2 [59,60] and SCOPE [61], or different machine learning algorithms, e.g., random forests, support vector machines, more complex ANNs, and gradient boosting or convolution neural networks for object detection. More complex methodologies can additionally introduce a comparison of performances across different models or investigate the model interpretability, e.g., explainable artificial intelligence [62].

5. Conclusions

The results indicate that both differences in spectral resolution, i.e., hyperspectral versus multispectral, as well as the inclusion versus exclusion of background soil spectra correction substantially affected the model estimations of the simulated dataset as well as the field experimental dataset.

Many RTM inversion-based models presented in this study resulted in suitable estimations of wheat traits. Thus, the approach proposed by Reinosch et al. [14], which combines empirically collected data with mechanistic models, e.g., RTMs, was successful in building models based on machine learning with a limited experimental dataset. The RTM inversion with hyperspectral input data and background soil spectra correction was the best among all tested models to estimate PAI, NY, and CWC during the vegetative stages (BBCH 10 to 60) as well as PAI and CWC throughout the growing season (BBCH 10 to 89). Additionally, this model was also the best for estimating AGDM based on the integral of estimated LAI values over days, i.e., the LAD. Furthermore, this model was equivalent or superior compared to similar trait estimation results reported in the literature.

With this study, a promising new model to estimate PAI, NY, CWC, and AGDM was established. This model was specifically designed for the available spectroradiometer with a spectral resolution from 400 to 1000 nm in 1 nm increments. Going forward, this combination of sensor and model can now be applied in wheat field experiments to reduce labor and costs for analyzing PAI, NY, CWC, and AGDM. Nevertheless, the simple approach using vegetation indices also showed good results, especially for the estimation of PAI and NY during vegetative stages (BBCH 10 to 60) as well as the estimation of AGDM using NDVI integral. The estimations of PAI throughout the growing season and CWC in general, however, were substantially worse using vegetation indices compared to the RTM inversion with hyperspectral input data and including background soil spectra correction.

Many models presented in this study, both vegetation index as well as RTM inversion-based, provided suitable estimations of the relevant wheat traits PAI, NY, CWC, and AGDM for various applications in, for example, agronomy, breeding, and crop sciences in general.

Author Contributions: Conceptualization, L.J.K., H.-P.K., S.R., P.W., P.E., T.N., N.B. and R.W.N.; methodology, L.J.K., H.-P.K., S.R., P.W. and R.W.N.; software, L.J.K., H.-P.K., S.R., P.W., N.B. and R.W.N.; validation, L.J.K., H.-P.K. and R.W.N.; formal analysis, L.J.K., H.-P.K., S.R., P.W. and R.W.N.; investigation, L.J.K., H.-P.K., A.K.-K. and R.W.N.; resources, L.J.K., H.-P.K., P.E., T.N., N.B. and R.W.N.; data curation, L.J.K., H.-P.K., S.R., P.W., J.B., G.M., G.M. and R.W.N.; writing—original draft preparation, L.J.K.; writing—review and editing, L.J.K., H.-P.K., S.R., P.W., P.E., J.B., G.M., T.N., A.K.-K., N.B. and R.W.N.; visualization, L.J.K., H.-P.K. and R.W.N.; supervision, H.-P.K., T.N., A.K.-K. and R.W.N.; project administration, L.J.K., H.-P.K., T.N., N.B. and R.W.N.; funding acquisition, H.-P.K., T.N., N.B. and R.W.N. All authors have read and agreed to the published version of the manuscript.

Funding: This research was funded by the project "DiLaAg—Digitalization and Innovation Laboratory in Agricultural Sciences" of the private foundation "Forum Morgen, Austria" and the Federal State of Lower Austria.

Data Availability Statement: The raw data supporting the conclusions of this article will be made available by the authors on request.

Acknowledgments: We thank Ivica Simonovic and Johannes Kemetter for managing the field experiments; our student assistants Anna Hofer, Caroline Huber, Laura Sturm, and Vincent Aubry for their assistance in the field and laboratory; as well as Craig Jackson for his field assistance and for conducting laboratory analyses.

Conflicts of Interest: The authors declare no conflicts of interest.

References

1. Myers, V.I.; Allen, W.A. Electrooptical Remote Sensing Methods as Nondestructive Testing and Measuring Techniques in Agriculture. *Appl. Opt.* **1968**, *7*, 1819–1838. [CrossRef] [PubMed]
2. Wiegand, C.L.; Richardson, A.J.; Kanemasu, E.T. Leaf Area Index Estimates for Wheat from LANDSAT and Their Implications for Evapotranspiration and Crop Modeling. *Agron. J.* **1979**, *71*, 336–342. [CrossRef]
3. Khanal, S.; Kushal, K.C.; Fulton, J.P.; Shearer, S.; Ozkan, E. Remote Sensing in Agriculture—Accomplishments, Limitations, and Opportunities. *Remote Sens.* **2020**, *12*, 3783. [CrossRef]
4. Baret, F.; Guyot, G. Potentials and limits of vegetation indices for LAI and APAR assessment. *Remote Sens. Environ.* **1991**, *35*, 161–173. [CrossRef]
5. Haboudane, D.; Miller, J.R.; Pattey, E.; Zarco-Tejada, P.J.; Strachan, I.B. Hyperspectral vegetation indices and novel algorithms for predicting green LAI of crop canopies: Modeling and validation in the context of precision agriculture. *Remote Sens. Environ.* **2004**, *90*, 337–352. [CrossRef]
6. Marti, J.; Bort, J.; Slafer, G.A.; Araus, J.L. Can wheat yield be assessed by early measurements of Normalized Difference Vegetation Index? *Ann. Appl. Biol.* **2007**, *150*, 253–257. [CrossRef]
7. Mulla, D.J. Twenty-five years of remote sensing in precision agriculture: Key advances and remaining knowledge gaps. *Biosyst. Eng.* **2013**, *114*, 358–371. [CrossRef]
8. Palka, M.; Manschadi, A.M.; Koppensteiner, L.J.; Neubauer, T.; Fitzgerald, G.J. Evaluating the performance of the CCCI-CNI index for estimating N status of winter wheat. *Eur. J. Agron.* **2021**, *130*, 126346. [CrossRef]
9. Baret, F.; Buis, S. Estimating canopy characteristics from remote sensing observations: Review of methods and associated problems. In *Advances in Land Remote Sensing*; Springer: Dordrecht, The Netherlands, 2008; pp. 173–201.
10. Atzberger, C.; Darvishzadeh, R.; Immitzer, M.; Schlerf, M.; Skidmore, A.; le Maire, G. Comparative analysis of different retrieval methods for mapping grassland leaf area index using airborne imaging spectroscopy. *Int. J. Appl. Earth Obs. Geoinf.* **2015**, *43*, 19–31. [CrossRef]
11. Bian, C.; Shi, H.; Wu, S.; Zhang, K.; Wei, M.; Zhao, Y.; Sun, Y.; Zhuang, H.; Zhang, X.; Chen, S. Prediction of Field-Scale Wheat Yield Using Machine Learning Method and Multi-Spectral UAV Data. *Remote Sens.* **2022**, *14*, 1474. [CrossRef]
12. Cheng, E.; Zhang, B.; Peng, D.; Zhong, L.; Yu, L.; Liu, Y.; Xiao, C.; Li, C.; Li, X.; Chen, Y.; et al. Wheat yield estimation using remote sensing data based on machine learning approaches. *Front. Plant Sci.* **2022**, *13*, 1090970. [CrossRef] [PubMed]
13. van Klompenburg, T.; Kassahun, A.; Catal, C. Crop yield prediction using machine learning: A systematic literature review. *Comput. Electron. Agric.* **2020**, *177*, 105709. [CrossRef]
14. Reinosch, N.; Münzberg, A.; Martini, D.; Niehus, A.; Seuring, L.; Troost, C.; Kumar Srivastava, R.; Berger, T.; Streck, T.; Bernardi, A. SIMLEARN—Ontologiegestützte Integration von Simulationsmodellen, Systemen für maschinelles Lernen und Planungsdaten. In *43. GIL-Jahrestagung, Resiliente Agri-Food-Systeme*; Gesellschaft für Informatik e.V.: Bonn, Germany, 2023; pp. 477–482, ISBN 978-3-88579-724-1.
15. Monteith, J.L. Light Distribution and Photosynthesis in Field Crops. *Ann. Bot.* **1965**, *29*, 17–37. Available online: https://www.jstor.org/stable/42908627 (accessed on 22 August 2022). [CrossRef]
16. Berger, K.; Atzberger, C.; Danner, M.; D'Urso, G.; Mauser, W.; Vuolo, F.; Hank, T. Evaluation of the PROSAIL Model Capabilities for Future Hyperspectral Model Environments: A Review Study. *Remote Sens.* **2018**, *10*, 85. [CrossRef]
17. Richter, K.; Atzberger, C.; Vuolo, F.; D'Urso, G. Evaluation of Sentinel-2 Spectral Sampling for Radiative Transfer Model Based LAI Estimation of Wheat, Sugar Beet, and Maize. *IEEE J. Sel. Top. Appl. Earth Obs. Remote Sens.* **2011**, *4*, 458–464. [CrossRef]
18. Locherer, M.; Hank, T.; Danner, M.; Mauser, W. Retrieval of Seasonal Leaf Area Index from Simulated EnMAP Data through Optimized LUT-Based Inversion of the PROSAIL Model. *Remote Sens.* **2015**, *7*, 10321–10346. [CrossRef]
19. Atzberger, C. Object-based retrieval of biophysical canopy variables using artificial neural nets and radiative transfer models. *Remote Sens. Environ.* **2004**, *93*, 53–67. [CrossRef]
20. Duveiller, G.; Weiss, M.; Baret, F.; Defourny, P. Retrieving wheat Green Area Index during the growing season from optical time series measurements based on neural network radiative transfer inversion. *Remote Sens. Environ.* **2011**, *115*, 887–896. [CrossRef]
21. Zhu, J.; Lu, J.; Li, W.; Wang, Y.; Jiang, J.; Cheng, T.; Zhu, Y.; Cao, W.; Yao, X. Estimation of canopy water content for wheat through combining radiative transfer model and machine learning. *Field Crops Res.* **2023**, *302*, 109077. [CrossRef]
22. Weiss, M.; Baret, F.; Jay, S. *S2ToolBox Level 2 Products: LAI, FAPAR, FCOVER*; INRAE: Paris, France, 2016.
23. Huete, A.R. A soil-adjusted vegetation index (SAVI). *Remote Sens. Environ.* **1988**, *25*, 295–309. [CrossRef]

24. Danner, M.; Berger, K.; Wocher, M.; Mauser, W.; Hank, T. Retrieval of Biophysical Crop Variables from Multi-Angular Canopy Spectroscopy. *Remote Sens.* **2017**, *9*, 726. [CrossRef]
25. Jacquemoud, S.; Verhoef, W.; Baret, F.; Bacour, C.; Zarco-Tejada, P.J.; Asner, G.P.; François, C.; Ustin, S.L. PROSPECT + SAIL models: A review of use for vegetation characterization. *Remote Sens. Environ.* **2009**, *113*, 56–66. [CrossRef]
26. Koppensteiner, L.J.; Kaul, H.-P.; Piepho, H.-P.; Barta, N.; Euteneuer, P.; Bernas, J.; Klimek-Kopyra, A.; Gronauer, A.; Neugschwandtner, R.W. Yield and yield components of facultative wheat are affected by sowing time, nitrogen fertilization and environment. *Eur. J. Agron.* **2022**, *140*, 126591. [CrossRef]
27. Bernas, J.; Koppensteiner, L.J.; Tichá, M.; Kaul, H.-P.; Klimek-Kopyra, A.; Euteneuer, P.; Moitzi, G.; Neugschwandtner, R.W. Optimal environmental design of nitrogen application rate for facultative wheat using life cycle assessment. *Eur. J. Agron.* **2023**, *146*, 126813. [CrossRef]
28. Moitzi, G.; Koppensteiner, L.J.; Klimek-Kopyra, A.; Bernas, J.; Kaul, H.-P.; Wagentristl, H.; Euteneuer, P.; Neugschwandtner, R.W. Effects of sowing date and nitrogen applications on the energy efficiency of facultative wheat (*Triticum aestivum* L.) in a Pannonian environment. *Heliyon* **2024**, *10*, e37923. [CrossRef]
29. Neugschwandtner, R.W.; Liebhard, P.; Kaul, H.-P.; Wagentristl, H. Soil chemical properties as affected by tillage and crop rotation in a long-term field experiment. *Plant Soil Environ.* **2014**, *60*, 57–62. [CrossRef]
30. Witzenberger, A.; Hack, H.; van den Boom, T. Erläuterungen zum BBCH-Dezimal-Code für die Entwicklungsstadien des Getreides—Mit Abbildungen [Commentary to the BBCH-decimal-code for growth stages of cereals—With images]. *Gesunde Pflanz.* **1989**, *41*, 384–388. (In German)
31. Meier, U.; Bleiholder, H.; Buhr, L.; Feller, C.; Hack, H.; Hess, M.; Lancashire, P.D.; Schnock, U.; Stauss, R.; van den Boom, T.; et al. The BBCH system to coding the phenological growth stages of plants—History and publications. *J. Kult.* **2009**, *61*, 41–52. [CrossRef]
32. Winkler, R.; Botterbrodt, S.; Rabe, E.; Lindhauer, M.G. Stickstoff-/Proteinbestimmung mit der Dumas-Methode in Getreide und Getreideprodukten [Nitrogen and protein determination using the Dumas method in cereal and cereal products]. *Getreide Mehl Brot* **2000**, *54*, 86–91. (In German)
33. Jacquemoud, S.; Baret, F. PROSPECT: A model of leaf optical properties spectra. *Remote Sens. Environ.* **1990**, *34*, 75–91. [CrossRef]
34. Feret, J.-B.; François, C.; Asner, G.P.; Gitelson, A.A.; Martin, R.E.; Bidel, L.P.R.; Ustin, S.L.; le Maire, G.; Jacquemoud, S. PROSPECT-4 and 5: Advances in the leaf optical properties model separating photosynthetic pigments. *Remote Sens. Environ.* **2008**, *112*, 3030–3043. [CrossRef]
35. Kuusk, A. The Hot Spot Effect in Plant Canopy Reflectance. In *Photon-Vegetation Interactions: Applications in Optical Remote Sensing and Plant Ecology*; Myneni, R.B., Ross, J., Eds.; Springer: Berlin/Heidelberg, Germany, 1991; pp. 139–159.
36. Spitters, C.J.T.; Toussaint, H.A.J.M.; Goudriaan, J. Separating the diffuse and direct component of global radiation and its implications for modeling canopy photosynthesis Part I. Components of incoming radiation. *Agric. For. Meteorol.* **1986**, *38*, 217–229. [CrossRef]
37. Richter, K.; Vuolo, F.; D'Urso, G.; Palladino, M. Evaluation of near-surface soil water status through the inversion of soil-canopy radiative transfer models in the reflective optical domain. *Int. J. Remote Sens.* **2012**, *33*, 5473–5491. [CrossRef]
38. Kong, W.P.; Huang, W.J.; Zhou, X.F.; Song, X.Y.; Casa, R. Estimation of carotenoid content at the canopy scale using the carotenoid triangle ratio index from in situ and simulated hyperspectral data. *J. Appl. Remote Sens.* **2016**, *10*, 026035. [CrossRef]
39. Van Reeuwjik, L.P. *Procedure for Soil Analysis*, 6th ed.; ISRIC: Wageningen, The Netherlands, 2002.
40. SoilSpec4GG Open Soil Spectral Library Explorer. Soil Spectroscopy 4 Global Good. 2022. Available online: https://explorer.soilspectroscopy.org/ (accessed on 7 July 2022).
41. Stanford University. *CS231n Deep Learning for Computer Vision*; Stanford University: Stanford, CA, USA, 2024. Available online: https://cs231n.github.io/neural-networks-1/ (accessed on 3 January 2025).
42. ESA (European Space Agency). Sentinel-2 Spectral Response Functions (S2-SRF), Version: 3.2. 10 October 2023. Available online: https://sentinels.copernicus.eu/documents/d/sentinel/s2-srf_cope-gseg-eopg-tn-15-0007_3-2 (accessed on 7 January 2024).
43. Akanbi, O.; Amiri, I.; Fazeldehkordi, E. Chapter 4—Feature Extraction. In *A Machine-Learning Approach to Phishing Detection and Defense*; Akanbi, O., Amiri, I., Fazeldehkordi, E., Eds.; Syngress: Rockland, MA, USA, 2015; pp. 45–54. [CrossRef]
44. Fitzgerald, G.; Rodriguez, D.; O'Leary, G. Measuring and predicting canopy nitrogen nutrition in wheat using a spectral index —The canopy chlorophyll content index (CCCI). *Field Crops Res.* **2010**, *116*, 318–324. [CrossRef]
45. Piepho, H.P.; Büchse, A.; Emrich, K. A Hitchhiker's Guide to Mixed Models for Randomized Experiments. *J. Agron. Crop Sci.* **2003**, *189*, 310–322. [CrossRef]
46. Bacour, C.; Baret, F.; Béal, D.; Weiss, M.; Pavageau, K. Neural network estimation of LAI, fAPAR, fCover and LAI$\times C_{ab}$, from top of canopy MERIS reflectance data: Principles and validation. *Remote Sens. Environ.* **2006**, *105*, 313–325. [CrossRef]
47. Verger, A.; Baret, F.; Camacho, F. Optimal modalities for radiative transfer-neural network estimation of canopy biophysical characteristics: Evaluation over an agricultural area with CHRIS/PROBA observation. *Remote Sens. Environ.* **2011**, *115*, 415–426. [CrossRef]

48. Weiss, M.; Baret, F.; Myneni, R.B.; Pragnere, A.; Knyazikhin, Y. Investigation of a model inversion technique to estimate canopy biophysical variables from spectral and directional reflectance data. *Agronomie* **2000**, *20*, 3–22. [CrossRef]
49. Herrmann, I.; Pimstein, A.; Karnieli, A.; Cohen, Y.; Alchanatis, V.; Bonfil, D.J. LAI assessment of wheat and potato crops by Venμs and Sentinel-2 bands. *Remote Sens. Environ.* **2011**, *115*, 2141–2151. [CrossRef]
50. Zhang, C.; Pattey, E.; Liu, J.; Cai, H.; Shang, J.; Dong, T. Retrieving leaf and canopy water content of winter wheat using vegetation water indices. *IEEE J. Sel. Top. Appl. Earth Obs. Remote Sens.* **2017**, *11*, 112–126. [CrossRef]
51. Atzberger, C.; Jarmer, T.; Schlerf, M.; Kötz, B.; Werner, W. Retrieval of wheat bio-physical attributes from hyperspectral data and SAILH+PROSPECT radiative transfer model. In Proceedings of the 3rd EARSeL Workshop Imaging Spectroscopy, Oberpfaffenhof, Germany, 13–16 May 2003.
52. Vuolo, F.; Żółtak, M.; Pipitone, C.; Zappa, L.; Wenng, H.; Immitzer, M.; Weiss, M.; Baret, F.; Atzberger, C. Data service platform for Sentinel-2 surface reflectance and value-added products: System use and examples. *Remote Sens.* **2016**, *8*, 938. [CrossRef]
53. Pan, H.; Chen, Z.; Ren, J.; Li, H.; Wu, S. Modeling winter wheat leaf area index and canopy water content with three different approaches using sentinel-2 multispectral instrument data. *IEEE J. Sel. Top. Appl. Earth Obs. Remote Sens.* **2018**, *12*, 482–492. [CrossRef]
54. Wu, C.; Niu, Z.; Tang, Q. Predicting vegetation water content in wheat using normalized difference water indices derived from ground measurements. *J. Plant Res.* **2009**, *122*, 317–326. [CrossRef]
55. Simpson, R.J.; Lambers, H.; Dalling, M.J. Nitrogen Redistribution during Grain Growth in Wheat (*Triticum aestivum* L.): IV. Development of a Quantitative Model of the Translocation of Nitrogen to the Grain. *Plant Physiol.* **1983**, *71*, 7–14. [CrossRef] [PubMed]
56. Tucker, C.J.; Holben, B.N.; Elgin, J.H.; McMurtrey, J.E. Remote sensing of total dry-matter accumulation in winter wheat. *Remote Sens. Environ.* **1981**, *11*, 171–189. [CrossRef]
57. Mistele, B.; Schmidhalter, U. Spectral measurements of the total aerial N and biomass dry weight in maize using a quadrilateral-view optic. *Field Crops Res.* **2008**, *106*, 94–103. [CrossRef]
58. Fabbri, C.; Mancini, M.; dalla Marta, A.; Orlandini, S.; Napoli, M. Integrating satellite data with a nitrogen nutrition curve for precision top-dress fertilization of durum wheat. *Eur. J. Agron.* **2020**, *120*, 126148. [CrossRef]
59. Verhoef, W.; Bach, H. Coupled soil–leaf-canopy and atmosphere radiative transfer modeling to simulate hyperspectral multi-angular surface reflectance and TOA radiance data. *Remote Sens. Environ.* **2007**, *109*, 166–182. [CrossRef]
60. Verhoef, W.; Jia, L.; Xiao, Q.; Su, Z. Unified Optical-Thermal Four-Stream Radiative Transfer Theory for Homogeneous Vegetation Canopies. *IEEE Trans. Geosci. Remote Sens.* **2007**, *45*, 1808–1822. [CrossRef]
61. van der Tol, C.; Verhoef, W.; Timmermans, J.; Verhoef, A.; Su, Z. An integrated model of soil-canopy spectral radiances, photosynthesis, fluorescence, temperature and energy balance. *Biogeosciences* **2009**, *6*, 3109–3129. [CrossRef]
62. Holzinger, A. Explainable AI (ex-AI). *Inform. Spektrum* **2018**, *41*, 138–143. [CrossRef]

remote sensing

MDPI

Article

Remote Estimation of Above-Ground Biomass Throughout the Entire Growth Period for Crops with Conspicuous Spikes

Qiaoling Zhang [1,2], Yan Gong [3,4], Yubin Chen [5], Yalan Huang [1,2], Tingfan Wang [3], Siyu Zhang [1,2], Minzi Wang [6], Yi Peng [3,4], Feng Jiang [1,2], Fan Yang [1,2] and Xingqi Wang [3,*]

[1] Kweichow Moutai Co., Ltd., Zunyi 564500, China; 1500030@moutai.com.cn (Q.Z.); 20244514@moutai.com.cn (Y.H.); zsy@mouai.com.cn (S.Z.); 1400134@moutai.com.cn (F.J.); yf@moutai.com.cn (F.Y.)

[2] Chishui River Middle Basin, Watershed Ecosystem, Observation and Research Station of Guizhou Province, Zunyi 564500, China

[3] School of Remote Sensing and Information Engineering, Wuhan University, Wuhan 430079, China; gongyan@whu.edu.cn (Y.G.); wangtingfan@whu.edu.cn (T.W.); ypeng@whu.edu.cn (Y.P.)

[4] Lab of Remote Sensing for Precision Phenomics of Hybrid Rice, Wuhan University, Wuhan 430079, China

[5] Kweichow Moutai Distillery (Group) Hongyingzi Agricultural Science and Technology Development Co., Ltd., Zunyi 564500, China; 1700497@moutai.com.cn

[6] Department of Resource and Environment, Moutai Institute, Zunyi 564507, China; wangminzi@mtxy.edu.cn

* Correspondence: xingqiwang@whu.edu.cn; Tel.: +86-18130271297

Abstract: Above-ground biomass (AGB) is an important factor in crop yield. However, most AGB estimation methods for crops with conspicuous spikes, such as rice and sorghum, can achieve high accuracy during the vegetative stage but low accuracy during the reproductive stage. In this study, we explored an AGB estimation model throughout the entire growth period. Firstly, we divided the growth period of crops into two stages—before heading and after heading—and adopted different strategies according to the characteristics of the different stages. Before heading, we estimated AGB by multiplying the multi-spectral vegetation index (VI) and the crop canopy height (H) square. After heading, we added spectral absorption characteristic parameters to characterize spike biomass and used a multiple linear regression model. This model can accurately estimate AGB in both rice and sorghum throughout the entire growth period, which has a coefficient of determination (R^2) above 0.88 and the relative root mean square error (rRMSE) below 20.13% in both crops. Compared with the direct estimation of AGB throughout the entire growth period using $H^2 \times$ VI, this model effectively improved the accuracy of AGB estimation for crops with conspicuous spikes in the reproductive stage, which can provide reliable information for evaluating crop growth at plot scale.

Keywords: crop; above-ground biomass (AGB); heading; spectral absorption characteristic parameters; vegetation index (VI); canopy height

1. Introduction

AGB is a key growth parameter for evaluating the crop growth situation, which reflects the weight of organic matter accumulated by the crop after photosynthesis, and can be used to evaluate the net productivity of the crop [1]. The traditional method of acquiring AGB is destructive field measurement, which mainly obtains AGB through manual destructive sampling, drying processes, and integration with manual measurements. This method is time-consuming and labor-intensive, inefficient, and will damage crop plants, so it is impossible to continuously observe the same crop plant. Crop growth is a dynamic and continuous process, and it is necessary to continuously analyze the growth parameters of the crop at different growth stages to monitor crop growth, so the traditional method of obtaining growth parameters obviously cannot meet this requirement. Compared with time-consuming and labor-intensive destructive manual observation, remote sensing technology is mainly based on the spectral characteristics of crop plants, which offers advantages such as fast, non-destructive, and continuous observation [2].

There are various remote sensing estimation methods for crop AGB, including multispectral data [3,4], hyperspectral data [5,6], and lidar data [7,8]. According to the estimation principle, these can be divided into mechanism models [9] and empirical models [10].

Mechanism models include the crop growth model [11], the light energy utilization model [12], and the canopy radiation transfer model (RTM) [13]. The crop growth model simulates the growth process of crops through environmental information such as meteorology, soil, planting density, and planting methods, as well as biological information such as crop variety characteristics, then simulates the changes in volume and mass of crops during the growth process to estimate crop AGB [14]. Common mechanism models include the decision support system for agrotechnology (DSSAT) model [15–19], the World Food Study (WOFOST) model [20–22] and the agricultural product system simulator (APSIM) model [23–26]. The light energy utilization model uses the light energy utilization rate as the efficiency of crop dry matter accumulation during a certain period of time [27]. Common light energy utilization models include the Carnegie–Ames–Stanford approach (CASA) model [28–30] and the global production efficiency model (GLO-PEM) [31]. RTM is a physical remote sensing method, based on physical principles and mathematical equations, that is used to explain the canopy's radiation transfer process [32]. The PROSAIL model, one of the major RTMs, has been widely applied to the inversion of biochemical variables and biophysical parameters of vegetation, such as the leaf area index (LAI) [33] and AGB [34]. However, the mechanism model is a model based on relatively complex biological theories, which requires a large amount of environmental information, accurate crop information, and complex biological theories as support, and the modeling process is rather complicated.

Empirical models generally do not pay attention to the internal mechanism of the crop growth process, but directly establish the relationship between crop surface reflectance, vegetation index (VI), or geometric parameters and crop growth parameters according to the spectral characteristics of crops, and estimate crop growth parameters based on statistical models. In this study, empirical models are divided into AGB estimation models based on multispectral data [35], estimation models based on hyperspectral data [36], estimation models based on texture information [37], and estimation models integrating multiple data sources according to different feature parameters used in the models [38,39].

VI calculated from multispectral data is a common method to estimate AGB, which mainly carries out statistical analysis on crop AGB and corresponding VI of different growth periods or entire growth periods to obtain the best regression model [40–42]. The hyperspectral data have higher spectral resolution, providing more accurate crop reflectance and also allowing the spectral curve of the crop to be obtained, which can also allow the crop to be observed in a detailed and comprehensive way [43,44]. Feng et al. [45] applied hyperspectral data to deep learning and proposed a VGC-AGB model to improve the accuracy of potato AGB estimation. Texture information is the embodiment of the change rule of image on the gray scale, which can well reflect the structural characteristics of crop canopy in the horizontal direction [46–49]. Fu et al. [50,51] found that the introduction of Gabor texture into the spectral index model could improve the estimation accuracy of winter wheat nitrogen content. Liu et al. [52] improved the estimation accuracy of rice AGB by changing the window size and direction of the grey level co-occurrence matrix GLCM, with R^2 reaching 0.84.

Although different types of data have been introduced for estimating crop AGB, when only a single type of data is used to estimate crop AGB, various problems often arise. For example, when crop AGB is estimated by using texture information, most texture indexes are weakly correlated with the AGB in the early period of crop growth [46]. Therefore, many studies have combined different types of data to estimate crop AGB, such as VI, texture index, canopy height, and vegetation coverage. Jimenez-Berni et al. [53] used a LiDAR sensor to extract three-dimensional geometric information of wheat, then obtained wheat plant height and vegetation coverage. They combined plant height with vegetation coverage to estimate wheat AGB, with R^2 reaching 0.92. Tilly et al. [54] used a ground-based laser scanner to obtain wheat plant height and combined it with ground-measured spectrum, finding that the combination of the two can effectively improve the accuracy of estimating AGB. Yue et al. [55] found that the correlation between VI multiplied by height and wheat AGB was higher than that between single VI or height and AGB. Navarro et al. [56] used the RGB image of an unmanned aerial vehicle (UAV) to extract a three-dimensional point cloud, then combined it with Sentinel-1 SAR and Sentinel-2 optical images, and found that the estimation model established by the combination of the three had the highest accuracy. Shu et al. [57] used UAV digital surface model (DSM) images to extract maize plant height, then multiplied the VI and plant height to characterize LAI. Finally, they multiplied LAI and plant height to build an AGB cube model, with R^2 reaching 0.79 and the relative root mean square error (rRMSE) reaching 29.19%. Liu et al. [58,59] combined UAV RGB data with hyperspectral data to estimate potato AGB, effectively reducing the root mean square error (RMSE). Liu et al. [60] combined remote sensing features with structural features to estimate wheat AGB, with R^2 reaching 0.88.

However, despite fusing different types of data, the effects of most crop AGB estimation methods are still relatively poor during the heading stage and subsequent growth stages, especially for crops with conspicuous spikes. Miao et al. [61] divided the growth period of rice into four stages from the tillering stage to the heading stage, and used the optimal multi-narrowband reflectance (OMNBR) model based on the original reflectance of hyperspectral data, R^2 reached 0.77 at the tillering stage and 0.50 at the heading stage. Xu et al. [62] used VI combined with the corresponding-band texture (VI-CBT) to explore the effect of the fusion characteristics of VI and texture in estimating the AGB of rice at different growth stages. The effect was good at the tillering and jointing stages, but began to deteriorate after the booting stage. As a crop with conspicuous spikes, the presence of spikes in rice during the booting stage and subsequent growth stages can affect the canopy spectrum. Moreover, most of the existing crop AGB estimation methods are based on leaf biomass and rarely take spike biomass into account, which leads to a significant decline in the AGB estimation effect starting from the booting stage, especially at the heading stage. There is a lack of effective methods for estimating crop AGB in the heading stage and subsequent growth stages, especially for crops with conspicuous spikes. Therefore, a crop AGB estimation method is needed that takes into account both the composition of the crop canopy structure and the spectral changes of the canopy before and after heading. The main objectives of the study are to (1) evaluate the necessity of introducing hyperspectral features to improve the estimation of AGB of crops with conspicuous spike after heading; (2) find a spectral feature that can better characterize the spike biomass to improve the estimation accuracy of AGB; and (3) compare the effects of different methods in estimating the AGB of crops with conspicuous spikes after heading and propose a crop AGB estimation method based on the entire growth period.

2. Materials and Methods

2.1. Study Area

There were two rice study areas in this study: the hybrid rice experiment base of Wuhan University in Lingshui, Hainan province, China (18°31′N, 110°3′E), and that in Wuhan, Hubei province, China (30°33′N, 114°32′E). 1. The two rice study areas are in different climatic zones, which have disparate meteorological conditions. Hainan is located in the tropics and has a tropical monsoon climate with high temperatures throughout the year, while Hubei belongs to the subtropics and has a subtropical monsoon climate with an annual average temperature of 15.8 °C. In Hubei, rice plants were usually sown in mid-May, transplanted in early June, and harvested in October, with a growth period of around 4 months. By contrast, rice plants in Hainan were sown in mid-December of the previous year, transplanted in early January, and harvested in May, with a growth period of around 5 months. At the same time, there is also a sorghum study area in Zunyi, Guizhou province, China (27°42′N, 106°22′E). Guizhou province has a subtropical humid monsoon climate with moderate rainfall and an average annual temperature of 15.8 °C. The planting time of sorghum varies due to the significant difference in altitude. In this study, sorghum was planted in late March and harvested in late August.

As shown in Figure 1, there were 42 cultivars in 2018 Hainan, 27 cultivars in 2022 Hubei, 48 cultivars in 2023 Hubei, and 54 cultivars in 2024 Guizhou. Each cultivar had a separate field plot, and the white mark placed at the edge of the plots was used to distinguish different cultivars.

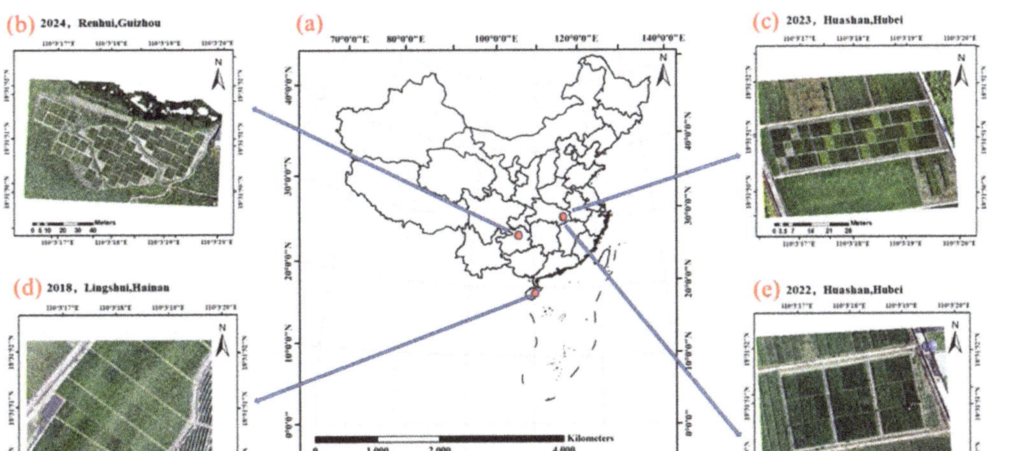

Figure 1. Study area: (**a**) geographical location; (**b**) 2024, Zunyi, Guizhou; (**c**) 2023, Wuhan, Hubei; (**d**) 2018, Lingshui, Hainan; (**e**) 2022, Wuhan, Hubei.

2.2. AGB Measurements

The collection of crop AGB began at the initial stage of tillering and continued until the wax ripening stage. During each sampling, three representative crop plants were randomly selected from each experimental plot, then the crop was shoveled out and put into the sampling bag. After the sampling bags were numbered, a small amount of field water was placed in the sampling bag to ensure the survival of the crop. The leaves, stems, and spikes of the crop were numbered and sealed in separate paper bags, then dried in an oven at 105 °C for 30 min, which was then adjusted to 80 °C for about two days at a constant temperature until the weight remained the same. At the end of drying, each sample was weighed separately and recorded. The leaf, stem, and spike biomass in the same plot were added to obtain three crop plants in the field total AGB (g). The AGB conversion formula for this cell is shown in Equation (1), and the unit of AGB is g/m^2.

$$\mathrm{AGB} = \frac{\mathrm{W}}{3} \times \mathrm{d} \tag{1}$$

where W is the total three crop plants biomass of each plot (g) and d is the planting density of each plot (strains/ m^2).

2.3. Manual Determination of Heading Date in Crop

The heading date in crops is generally defined as the time when approximately 50% of the spikes have exserted, which was determined by manual visual observations in the field. The heading date of the studied different crop cultivars varied between 59 and 73 Days After Transplanting (DAT) in rice. In this study, the growing season of each crop cultivar can be roughly divided based on heading date into before-heading and after-heading stages.

2.4. Canopy Hyperspectral Reflectance Retrieved from ASD

The ground hyperspectral data were collected using an Analytical Spectral Device (ASD). The ASD FieldSpec4 (Analytical Spectral Devices, Inc., West Gosford NSW, Malvern Panalytical, Worcs, UK) spectrometer has a spectral resolution of 350 nm~2500 nm and a spectral resolution of 3 nm in the range of 350 nm~1000 nm. The spectral resolution in the 1000 nm~2500 nm wave range is 8 nm. The instrument can be used to measure a variety of parameters such as reflectivity, transmittance, radiant brightness, and radiance.

2.5. UAV-Based Data Collection

Before the UAV imaging, the ground control points (GCPs) were evenly arranged in the field using black and white calibration plates measuring 40 cm in length and width. Intelligent real-time kinematic (RTK) was used to obtain the accurate geographical positions of each control point. UAV images were collected throughout the entire growth period in 2018 Hainan, 2022 Hubei, 2023 Hubei, and 2024 Guizhou. The 12-band multispectral images were obtained for the study site on the first flight, and the RGB images were then taken on the other flight.

The UAV data used in this study include RGB and multispectral images. The RGB images were used to extract canopy height, and the multispectral images were used to extract VI.

2.5.1. RGB Image

The RGB images of the study site were taken with a DJI FC 6310R camera mounted on a DJI Phantom 4 RTK UAV (SZ DJI Technology Co., Ltd., Shenzhen, China). The UAV is equipped with an adjustable cloud terrace (pitch angle ranges from +30° to −90°). The horizontal field of view of the camera was 84°, the vertical field of view of the camera was ±10°, and the ground sampling distance (GSD) was 0.8 cm. The flight altitude was 30 m above ground level. The forward overlap was set to 85%, and the lateral overlap was set to 80%.

2.5.2. Multispectral Image

The 12-band images of the study site were obtained by the Mini-MCA camera system (Tetracam, Inc., Chatsworth, CA, USA) mounted on an M8 UAV (Beijing TT Aviation Technology Co., Ltd, Beijing, China). This array of 12 cameras was equipped with customer-specified band pass filters centered at a wavelength of 490 nm, 520 nm, 550 nm, 570 nm,

670 nm, 680 nm, 700 nm, 720 nm, 800 nm, 850 nm, 900 nm, and 950 nm, respectively. The 12 camera lenses were co-registered in the laboratory prior to the flight so that corresponding pixels of each lens were spatially overlapping in the same focal plane. The flights were taken in sunny and cloudless weather between 10:00 AM and 2:00 PM at an altitude of 100 m and with an image spatial resolution of 5.5 cm.

For image radiometric calibration, a linear relationship was assumed between surface reflectance (ρ) and image digital numbers (DNs):

$$\rho_\lambda = DN_\lambda \times Gain_\lambda + Offset_\lambda \quad (2)$$

where ρ_λ and DN_λ were the surface reflectance and corresponding image digital numbers at wavelength λ. Eight near-Lambertian calibration canvases, at the constant reflectance of 0.03, 0.06, 0.12, 0.24, 0.36, 0.48, 0.56, and 0.80, were placed in the camera's field of view to solve $Gain_\lambda$ and $Offset_\lambda$ value in different bands using the least square method for image radiometric calibration. In this case, the canopy reflectance at 12 bands can be calculated based on Equation (2).

2.6. Features Derived from Remote Sensing Data

For each crop plot, a rectangular region of interest (ROI) of the same size was defined that maximally fit the plot. The ROI included three rows of crops, corresponding to around 800 pixels in the 12-band image and 2000 pixels in the RGB image. The actual area of each area is 2.42 m^2, accounting for approximately 70% of the area of a single plot. The average reflectance and the average height of all pixels within the ROI were taken as the plot-level canopy reflectance and canopy height, respectively, which correspond to the AGB at the field level.

2.6.1. VI

The plot-level VI was calculated from plot-level canopy reflectance. Four VIs, which are widely used for AGB estimation in many studies and can be easily applied in current satellite sensors, were used in this study for AGB estimation (Table 1).

Table 1. VI calculated from UAV images in this study.

Features	Description	Equation
NDVI [63]	Normalized difference vegetation index	$(\rho_{NIR} - \rho_{Red})/(\rho_{NIR} + \rho_{Red})$
EVI2 [64]	Two-band enhanced vegetation index	$2.5 \times \frac{\rho_{NIR} - \rho_{Red}}{\rho_{NIR} + 2.4 \times \rho_{Red} + 1}$
NDRE [65]	Normalized difference red-edge vegetation index	$\left(\rho_{NIR} - \rho_{RedEdge}\right)/\left(\rho_{NIR} + \rho_{RedEdge}\right)$
OSAVI [66]	Optimized soil adjusted vegetation index	$(1 + 0.16) \times \frac{\rho_{NIR} - \rho_{Red}}{\rho_{NIR} + \rho_{Red} + 0.16}$

2.6.2. Spectral Absorption Characteristics Parameter

The spectral curve of the crop canopy collected by the ASD FieldSpec4 spectrometer can effectively reflect the absorption and reflection characteristics of the ground object. In the visible light band, due to the absorption effect of chlorophyll, there are two absorption peaks in the blue band and the red band. In the near-infrared band, due to the scattering effect of the canopy, a reflection platform with a relatively high reflectance is formed. The measurement is generally carried out from 10:00 to 14:00. Three positions are selected for each plot for measurement, and the measurement is repeated three times for each plot. The height of the spectrometer probe vertically downward from the top of the canopy is 0.7 to 1.0 m. To fully exploit the spectral information of the crop contained in the spectral curve, this study introduces the parameterization of spectral absorption characteristics into the estimation of crop growth parameters, fully considering the reflectance of multiple bands of the crop canopy and the waveform characteristics of the spectral curve.

Spectral absorption characteristics parameter quantifies the absorption peaks of the spectral curve into multiple spectral absorption characteristic parameters [67,68]. These spectral absorption characteristic parameters are used to locate the spectral absorption positions and quantify the shapes of absorption valleys. They include six parameters: absorption position (AP), absorption depth (AD), absorption width (AW), absorption area (AA), spectral absorption index (SAI), and absorption symmetry (AS). To extract the spectral absorption characteristic parameters, the spectral curve needs to be subjected to envelope line elimination first. The specific process of the envelope line algorithm is as follows: the first step is to find all the reflection rate maximum points of the spectral curve, and the second step is to find the envelope line points from the maximum point to the left and right, respectively. The envelope line points should satisfy the condition that the line segment formed by the previous envelope line point and the next envelope line point, and its subsequent extension line, are all above all the points after the previous envelope line point. Similarly, if the envelope line points are sought from the front, it is necessary to ensure that the line segment formed by the previous envelope line point and the next envelope line point and its subsequent extension line are all above all the points before the previous envelope line point.

AP: As shown in Figure 2a, the absorption position is the wavelength corresponding to the deepest absorption of the absorption valley after the envelope of the spectral curve is eliminated.

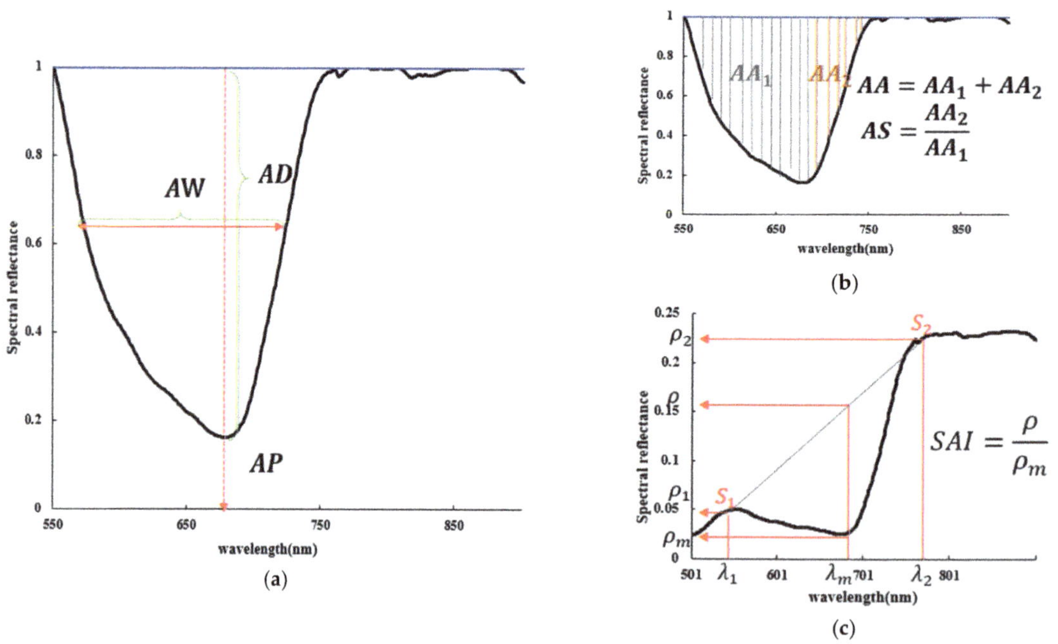

Figure 2. Quantitative extraction of spectral absorption characteristics parameter: (**a**) AW, AD, and AP; (**b**) AA and AS; (**c**) SAI.

AD: As shown in Figure 2a, the absorption depth is the deepest absorption depth of the absorption valley after the envelope of the spectral curve is eliminated.

AW: As shown in Figure 2a, the absorption width is the spectral width at half of the maximum absorption depth in the absorption valley resulting from the elimination of the envelope of the spectral curve.

AA: As shown in Figure 2b, the spectral absorption area is the integral of the absorption depth of the absorption valley over the wavelength. If the spectrum is continuous, the absorption area equation is Equation (3).

$$AA = \int_{\lambda_1}^{\lambda_2} (1 - \rho_\lambda) d\lambda \tag{3}$$

If the spectrum is discrete, the absorption area equation is Equation (4).

$$AA = \sum_{i=m}^{n} (1 - \rho_i) \tag{4}$$

AS: As shown in Figure 2b, the vertical line of the absorption position is the boundary, and the ratio of the area of the right region to the area of the left region.

SAI: As shown in Figure 2c, the two shoulders S_1 and S_2 of the absorption valley of a spectral curve, and the lines between S_1 and S_2 are called the non-absorption baseline. The reflectivity ρ_m, ρ is obtained from the spectral curve and non-absorption baselines S_1 and S_2, respectively. Equation (5) for calculating the spectral absorption index is as follows:

$$SAI = \frac{\rho}{\rho_m} = \frac{d \cdot \rho_2 + (1 - d) \cdot \rho_1}{\rho_m} \tag{5}$$

where d is the proportional parameter $d = (\lambda_m - \lambda_1)/(\lambda_2 - \lambda_1)$.

The absorption of crops is mainly concentrated in the visible light band, so this experiment mainly extracted the absorption characteristic parameters of the visible light band absorption valley of the crop canopy. After envelope elimination in the crop canopy spectrum, there is a large absorption valley in the visible light band, which has two absorption bands, namely the blue band and the red band, and the absorption depth of the red band is usually greater than that of the blue band. If the two absorption bands of red and blue are taken as one absorption valley to extract the absorption feature, the absorption position of the absorption valley will be shifted to the red band, while the absorption position of the blue wave segment will be ignored. Moreover, the large absorption area of the blue wave segment will also affect the absorption area, absorption width, and absorption symmetry of the red band. In order to more accurately investigate the spectral absorption characteristic parameters of the two absorption bands, this study divided the large absorption valley in the visible light band into two absorption bands, red and blue, and extracted the absorption characteristic parameters of the two absorption bands, respectively, with the green band (550 nm) as the limit.

In the process of crop growth, the absorption of the blue band and red band will change to some extent. The absorption characteristic parameters of the two absorption bands can be used as the ratio or difference to reflect the absorption difference between the red and blue bands. Therefore, on the basis of extracting the absorption characteristics of the two absorption bands, the absorption characteristics of the two absorption bands are differentiated or compared, and the ratio of absorption characteristics parameters is proposed. The absorption characteristic parameters extracted from the hyperspectral data of crop canopy in this study are summarized in Table 2.

Table 2. Spectral absorption characteristics parameter in this study.

Features	Description	Equation
RAA	Ratio absorption area	AA_{Red}/AA_{Blue}
RAW	Ratio absorption width	AW_{Red}/AW_{Blue}
RAD	Ratio absorption depth	AD_{Red}/AD_{Blue}
DSAI	Difference spectral absorption index	$(SAI_{Red} - SAI_{Blue})$
NDSAI	Normalized difference spectral absorption index	$(SAI_{Red} - SAI_{Blue})/(SAI_{Red} + SAI_{Blue})$
RSAI	Ratio spectral absorption index	SAI_{Red}/SAI_{Blue}

2.6.3. Canopy Height

The RGB images of the study were used for canopy 3-D reconstruction. Using Pix4Dmapper, the canopy DSM was generated. The canopy height (H) was calculated as:

$$H = DSM - DEM \tag{6}$$

where the DEM is equal to the DSM on the date before crop transplanting. This approach is widely applied to retrieve canopy height, with an accuracy of around 3 cm [69].

2.7. Analysis Methods

In this study, a tridimensional conceptual model of estimating maize AGB was proposed by combining the canopy height, VI, and spectral absorption characteristics parameters extracted from geometric data and multisource spectral data. Before crop heading, a linear function was constructed using the product of NDVI and the square of canopy height to estimate crop AGB. Due to the influence of spike dry weight on the estimation accuracy of crop AGB, the RSAI was introduced to characterize spike biomass on the basis of the original function, and the multiple linear regression model was used to estimate the total AGB. The accuracy of the AGB estimation model was evaluated by using the K-fold cross-validation method, where K was taken as 10. The main process is shown in Figure 3.

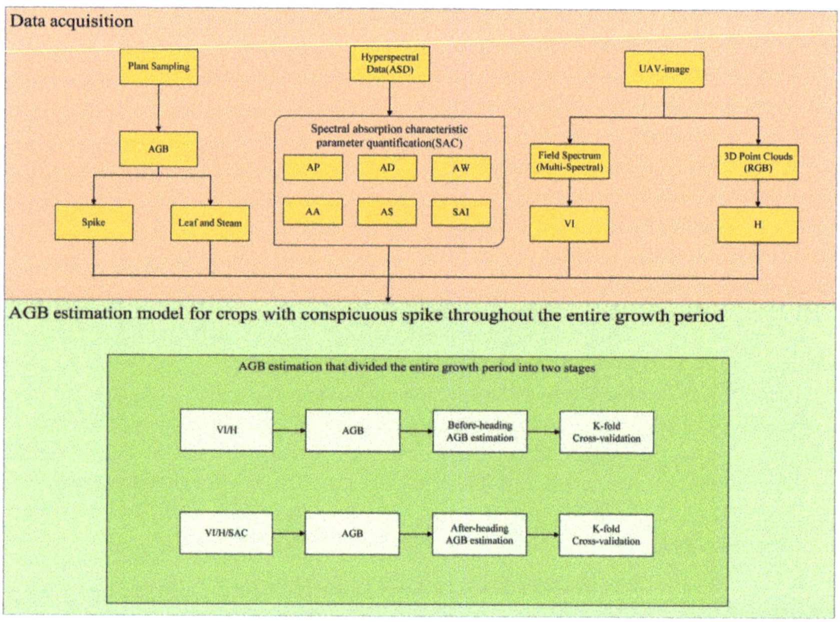

Figure 3. The main process of this study.

2.7.1. AGB Estimation Based on VI and Canopy Height

Many studies have shown that estimating AGB solely based on spectral data may lack the description of crop canopy structure [70,71], and combining spectral and geometric data can significantly improve the accuracy of AGB estimation [54–56]. In previous studies, some scholars used H × VI to estimate crop LAI and achieved good results [72], as shown in Equation (7). Some scholars also proposed a three-dimensional cube model to estimate maize AGB, in which AGB is the volume, LAI is the base area, canopy height is the height, and H × LAI is used to estimate crop AGB [57], as shown in Equation (8). Combined with the above studies, this study proposed $H^2 \times VI$ to estimate crop AGB, as shown in Equation (9).

$$LAI \propto (VI \times H) \tag{7}$$

$$AGB \propto (LAI \times H) \tag{8}$$

$$AGB \propto \left(VI \times H^2\right) \tag{9}$$

2.7.2. Multiple Linear Regression

Multiple linear regression can fully consider a variety of features to participate in AGB estimation, and combine different features to build a regression model. Stepwise multiple linear regression on the basis of multiple linear regression, variables are introduced into the multiple linear regression model one by one. For each new variable introduced, the old variables of the selected model are tested one by one, and the variables that are considered meaningless and can be replaced are deleted. This method can effectively screen multiple features and build a linear regression model that integrates multiple features.

2.7.3. Random Forest Regression

The random forest regression algorithm is an ensemble learning algorithm based on decision trees. It combines the idea of random selection of set features and the ensemble idea, and is a combined classifier based on decision trees. Ensemble learning can incorporate multiple algorithms of different types or the same type. Random forest combines multiple algorithms of the same type to form a decision tree and establish a more powerful prediction model. In this study, due to the small number of features and sufficient sample size, and in order to avoid the model being overly complex, the max_depth was 15 and the n_estimators were 250. To avoid overfitting, the max_samples were 0.8, and regularization techniques were introduced during model training, with other model parameters set to default values.

2.7.4. Support Vector Regression

Support vector regression (abbreviated as SVR) is a regression analysis method based on support vector machine (SVM), and is widely used in the fields of prediction and pattern recognition. SVR achieves the prediction of continuous variables by finding an optimal hyperplane in a high-dimensional space to maximize the interval between data points and the hyperplane. This article will deeply explore the theoretical basis, mathematical principles, model construction, parameter selection, training and optimization, application scenarios of SVR, as well as its advantages and limitations. In this study, due to the small number of features and sufficient sample size, the kernel was 'rbf'. Meanwhile, to avoid overfitting, the gamma was 1 and the C was 50, and other model parameters were set to default values.

3. Results

3.1. AGB Estimation Performance in Two Stages

In this study, the $H^2 \times$ VI model was used to estimate crop AGB. NDVI, NDRE, EVI2, and OSAVI were selected by VI, and the model was divided into two parts: before heading and after heading. The results are shown in Tables 3 and 4 and Figure 4.

Table 3. Rice AGB estimation results based on $H^2 \times$ VI before heading and after heading.

Model	Stage	R^2	RMSE (g/m^2)	rRMSE (%)
$H^2 \times$ NDVI	Before heading	0.86	150.87	30.17
	After heading	0.26	390.11	54.13
$H^2 \times$ EVI2	Before heading	0.82	177.71	35.54
	After heading	0.32	388.82	53.98
$H^2 \times$ NDRE	Before heading	0.83	187.40	37.48
	After heading	0.15	467.73	62.90
$H^2 \times$ OSAVI	Before heading	0.85	155.47	31.10
	After heading	0.21	407.65	56.12

Table 4. Sorghum AGB estimation results based on $H^2 \times$ VI before heading and after heading.

Model	Stage	R^2	RMSE (g/m^2)	rRMSE(%)
$H^2 \times$ NDVI	Before heading	0.93	35.05	18.31
	After heading	0.01	329.18	36.45
$H^2 \times$ EVI2	Before heading	0.93	75.86	39.64
	After heading	0.07	320.02	35.57
$H^2 \times$ NDRE	Before heading	0.92	37.37	19.53
	After heading	0.09	315.93	34.17
$H^2 \times$ OSAVI	Before heading	0.92	37.29	19.49
	After heading	0.11	313.15	33.89

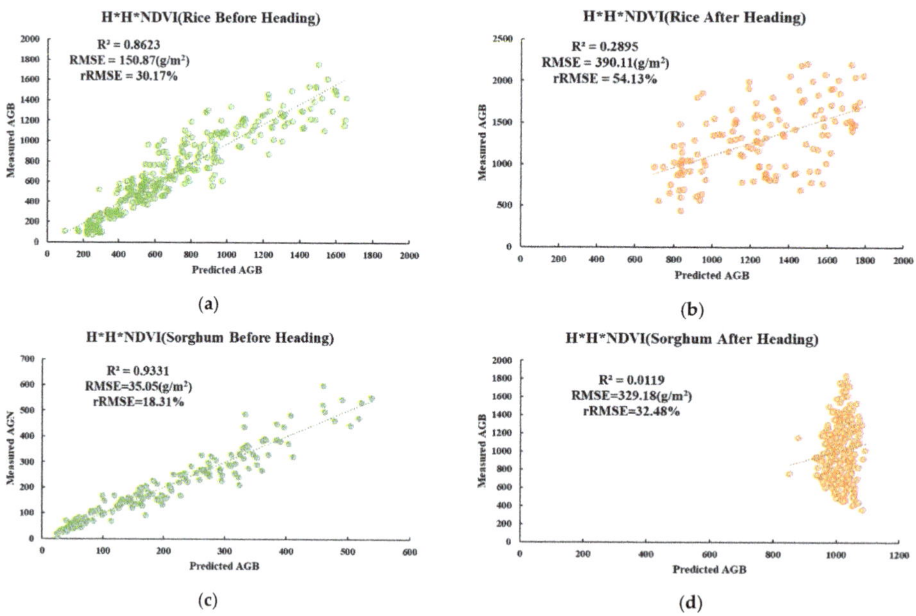

Figure 4. AGB estimation based on $H^2 \times$ NDVI in two stages: (**a**) rice before heading; (**b**) rice after heading; (**c**) sorghum before heading; (**d**) sorghum after heading.

According to the above results, the model has a good effect on estimating crop AGB before heading, but a poor effect after heading, especially for sorghum. Therefore, it is necessary to improve the after-heading model to improve the accuracy of AGB estimation.

From the perspective of crop canopy spectrum, the canopy spectrum before heading was composed of the interaction between stem spectrum and leaf spectrum, and the effect of spike spectrum was increased after heading, as shown in Figure 5. From the perspective of crop growth law, the AGB before heading only consisted of stem and leaf biomass, while the spike biomass increased after heading, as shown in Figure 6 for rice.

Figure 5. Comparison of crop canopy before and after heading: (**a**) rice before heading, (**b**) rice after heading, (**c**) sorghum before heading, (**d**) sorghum after heading.

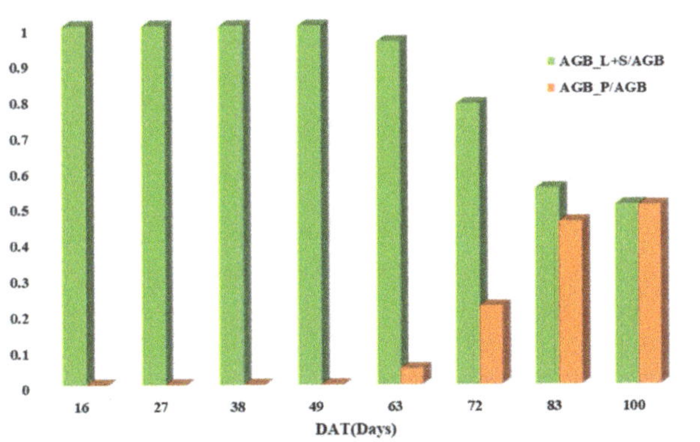

Figure 6. Trend of the proportion of rice AGB in different parts of total AGB (AGB_L+S: leaf biomass + stem biomass; AGB_P: spike biomass).

We speculated that the model's accuracy was reduced due to the influence of spike biomass, so we separately analyzed the correlation between $H^2 \times$ VI and stem and leaf biomass and spike biomass after heading. The results are shown in Table 5.

Table 5. R^2 between $H^2 \times$ VI AGB in different parts after heading.

Features	R^2 (AGB_Leaf+Stem)	R^2 (AGB_Spike)
$H^2 \times$ NDVI	0.5587	0.0034
$H^2 \times$ EVI2	0.5663	0.0151
$H^2 \times$ NDRE	0.5517	0.0749
$H^2 \times$ OSAVI	0.5078	0.0019

It can be seen from the results of Table 4 that $H^2 \times$ VI has a strong correlation with stem and leaf biomass after heading, but has almost no correlation with spike biomass. Therefore, to estimate the crop AGB model, it is necessary to introduce features that have a strong correlation with spike biomass to characterize spike biomass after heading, so as to improve the estimation accuracy of the model.

3.2. Adding Spectral Absorption Characteristic Parameter After Heading

In order to find the features related to spike biomass, we separately measured the end element spectra of leaves and spikes, and found that the reflectance of spikes gradually increased with time in the red band and blue wave segment, as shown in Figure 7a. Compared with the leaf spectrum, the absorption valley in the red band was more different in different stages after heading, which suggested that the absorption characteristics of the spectrum in the red band had a certain correlation with spike biomass.

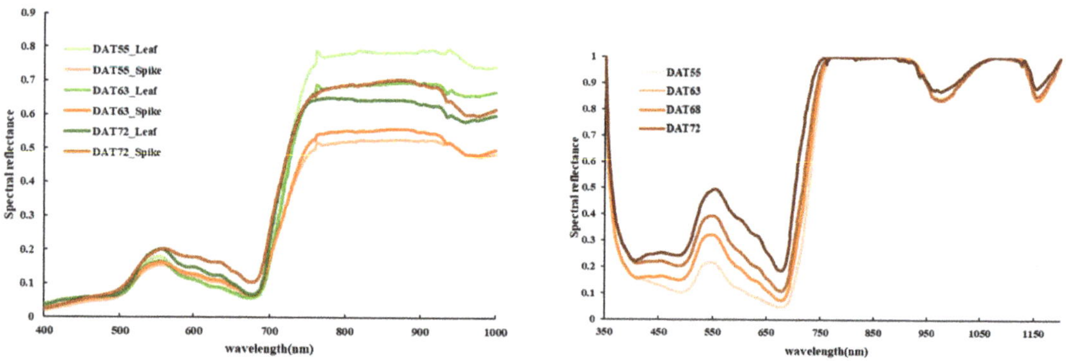

(a) (b)

Figure 7. Spectral reflectance plot: (**a**) comparison of leaf end-member spectra and spike end-member spectra after heading; (**b**) canopy spectra after heading (after envelope removing).

After heading, the canopy reflectance of the crop in the visible band increased slowly with the growth of the crop. There was an absorption valley in the red and blue bands, and a reflection peak in the green band. Reflection decreased first and then increased in the near infrared band, which was inconsistent with the change law of spike biomass. Moreover, the reflection peak in the green band was mainly caused by leaf reflection. Therefore, the reflectance of red and blue wave segments is selected as the research object, and the difference degree of features in different stages is increased by envelope removal. Finally, the characteristics of the absorption valley of red and blue wave segments are extracted by spectral feature quantization.

After envelope elimination, the reflectance of the canopy spectral curve after crop heading had a large difference in the red band and a small difference in the blue band, as shown in Figure 7b. The spectral absorption features of the two bands were extracted, and the differences in the characteristic values in different stages were increased by ratio, difference, and normalization operations.

Correlation analysis was conducted between the new characteristics obtained by the above calculation and spike biomass. The results are shown in Table 6, in which the RSAI had the highest correlation with spike biomass and a strong correlation. Therefore, we used the RSAI to characterize spike biomass.

Table 6. R^2 between spectral absorption characteristic parameter and spike AGB after heading.

Features	R^2 (AGB_Spike)
RAA	0.2059
RAW	0.1688
RAD	0.2649
DSAI	0.0624
NDSAI	0.5363
RSAI	0.6152

$H^2 \times$ NDVI, which had the highest correlation with stem and leaf biomass, and RSAI, which had the highest correlation with spike biomass, were selected as the input features of the model to construct the after-heading AGB estimation model, and the regression effects of random forest regression, support vector machine regression, and multiple linear regression models were tested. Finally, the K-fold crossover model was used to verify the results to ensure they were statistically significant. The results are shown in Table 7.

Table 7. R^2 and RMSE of AGB estimation based on $H^2 \times$ VI and spectral absorption characteristic parameter after heading.

Model	R^2 (Training Set)	RMSE (g/m^2) (Training Set)	R^2 (Test Set)	RMSE (g/m^2) (Test Set)
Random Forest Regression	0.93	201.34	0.82	269.56
Support Vector Regression	0.78	392.91	0.78	394.70
Multiple Linear Regression	0.89	260.58	0.89	262.86

Among the three models, we selected the multiple linear regression model with the highest test set accuracy to estimate the rice AGB after heading, and we used the linear regression model of $H^2 \times$ NDVI before heading. The model equations are shown in Table 8, and the final result is shown in Figure 8.

Table 8. AGB estimation model expressions.

Crop	Stage	Model
Rice	Before heading	AGB = 1412.1 × H^2 × NDVI + 75.93
Rice	After heading	AGB = 931.26 × H^2 × NDVI − 595.08 × RSAI + 1845.64
Sorghum	Before heading	AGB = 113.46 × H^2 × NDVI + 9.01
Sorghum	After heading	AGB = 59.921 × H^2 × NDVI − 343.5 × RSAI + 1488.08

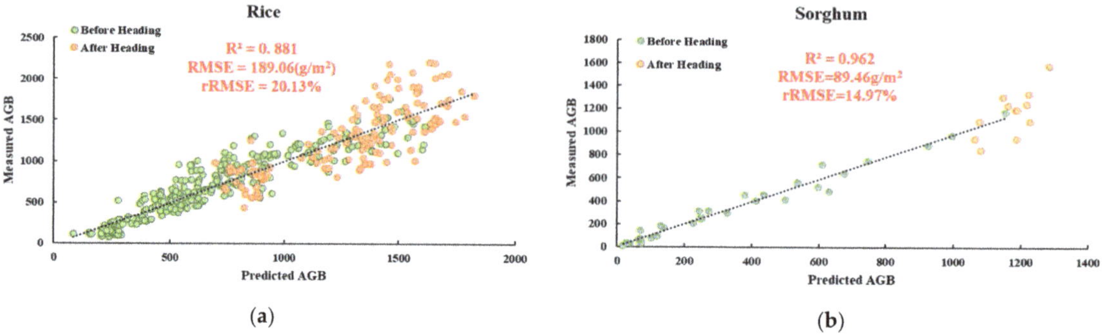

Figure 8. AGB estimation based on $H^2 \times$ NDVI and spectral absorption characteristic parameter that divides the entire growth period into before-heading stage and after-heading stage: (**a**) rice; (**b**) sorghum.

4. Discussion

Before heading, the vegetative growth of the crop was the main factor, and the height of the canopy gradually increased. At the same time, the AGB of the crop was mainly increased by stem and leaf biomass, and the increase in stem and leaf biomass was mainly related to chlorophyll. With the increase in chlorophyll content, the reflectance of the canopy gradually increased in the near-infrared band, while the reflectance in the green band gradually decreased. At the same time, NDVI and other VIs can well reflect the change of chlorophyll content, so the product of VI and the square of canopy height can be used to estimate crop AGB, as shown in Tables 3 and 4 and Figure 4.

However, after crop heading, the crop mainly grows by reproduction, and the AGB is composed of stem, leaf, and spike biomass, and the canopy height basically does not change. As can be seen from Figures 5 and 6, the increase in AGB was dominated by spike biomass, while the stem and leaf biomass increased slowly, and spike biomass accounted for an increasing proportion of AGB. The increase in spike biomass was mainly related to carotenoids, so the effect of estimating AGB by $H^2 \times$ VI decreased significantly, as shown in Tables 3 and 4 and Figure 4.

As can be seen from Figure 7, with the decrease in carotenoid content, the absorption of the red and blue bands gradually decreases, and the canopy reflectance in these two bands gradually increases. Then, the difference of canopy reflectance in different stages is eliminated and increased through the envelope. Because the increase in reflectance in the red band is smaller than that in the blue band, the change in the RSAI is exactly the same as that of carotenoid. Therefore, the introduction of the RSAI on the basis of the product of the original VI and the square of the canopy height can effectively estimate crop AGB after heading, as shown in Table 6.

As can be seen from Table 7, the support vector machine regression model has the worst effect. Although the random forest model has a good effect on the training set, its effect on the test set is not as good as that of the multiple linear regression model. Moreover, the multiple linear regression model has a simple structure and few parameters, so we adopt the multiple linear regression model.

The effect of the constructed multiple linear regression model ($H^2 \times$ NDVI, RSAI) in estimating AGB after heading was significantly improved compared with NDVI, $H^2 \times$ NDVI, and RSAI, as shown in Table 9.

Table 9. R^2 and RMSE of AGB estimation based on different characteristics after heading.

Model	R^2 (Training Set)	RMSE (g/m^2) (Training Set)	R^2 (Test Set)	RMSE (g/m^2) (Test Set)
NDVI	0.08	455.80	0.04	461.47
$H^2 \times$ NDVI	0.53	390.11	0.51	391.91
RSAI	0.72	307.89	0.70	309.76
Multiple Linear Regression ($H^2 \times$ NDVI, RSAI)	0.89	260.58	0.89	262.86

The linear regression model of $H^2 \times$ NDVI was used before heading, and the multiple linear regression models of $H^2 \times$ NDVI and RSAI were used to estimate AGB in different stages after heading. Compared with the model that directly estimates AGB throughout the entire growth period using $H^2 \times$ VI, the effect of our model was significantly improved. The test set results of K-fold cross-validation are shown in Tables 10 and 11.

Table 10. Rice AGB estimation results (the test set results of K-fold cross-validation) based on different characteristics throughout the entire growth period.

Model	R^2	RMSE (g/m^2)	rRMSE (%)
$H^2 \times$ NDVI	0.65	294.68	35.32
$H^2 \times$ EVI2	0.64	298.60	36.67
$H^2 \times$ NDRE	0.63	306.76	37.48
$H^2 \times$ OSAVI	0.63	298.21	36.62
f($H^2 \times$ NDVI, RSAI)	0.88	189.06	20.13

Table 11. Sorghum AGB estimation results (the test set results of K-fold cross-validation) based on different characteristics throughout the entire growth period.

Model	R^2	RMSE (g/m^2)	rRMSE (%)
$H^2 \times$ NDVI	0.63	296.49	42.68
$H^2 \times$ EVI2	0.48	357.28	51.43
$H^2 \times$ NDRE	0.44	367.89	52.95
$H^2 \times$ OSAVI	0.41	379.10	54.57
f($H^2 \times$ NDVI, RSAI)	0.96	89.46	14.97

Xu et al. constructed a multiple linear regression model using the texture features and spectral features in the hyperspectral images of UAV to estimate rice AGB, with R^2 = 0.76, and rRMSE = 27.10% [1]. Tunca et al. used high-resolution UAV data and machine learning (ML) to estimate sorghum AGB, with R^2 = 0.80, and RMSE = 290 g/m^2 [73]. Compared with previous studies, our method has achieved partial improvement in estimating rice AGB and significant improvement in estimating sorghum AGB.

As can be seen from Figure 8, although the estimation effect of crop AGB after heading is significantly improved, the accuracy of estimating rice AGB in the later stage still decreases. This is due to the influence of spike weight, which leads to a reduction in canopy height and thus a decline in the accuracy of estimating stem and leaf biomass by $H^2 \times$ VI. One of the reasons why the AGB estimation of sorghum is significantly better than that of rice is due to the small amount of sorghum data, and another reason is that the spike of sorghum changes from green to red, and the absorption in the red band changes more obviously than that of rice, so RSAI is more sensitive to the change of spike biomass. In subsequent studies, we will consider replacing the canopy height with the variation of the canopy height in two adjacent periods to improve the accuracy of AGB estimation.

Meanwhile, our method may also be extended and applied to satellite remote sensing in the future to improve the accuracy of large-scale crop AGB estimation.

5. Conclusions

In this study, aiming to address the problem that the accuracy of $H^2 \times VI$ estimation of crop AGB after heading is greatly reduced, we introduced the spectral absorption characteristic parameter RSAI, which is consistent with the change of spike biomass, from the perspective of the spectral variation rule of crop canopy after heading, and used it to characterize spike biomass. Together with $H^2 \times VI$, we constructed a multiple linear regression model to estimate crop AGB after heading. Compared with only using $H^2 \times VI$, the accuracy is effectively improved, the R^2 of the entire growth period was increased to 0.88, and the rRMSE was reduced by 15% in rice; the R^2 of the entire growth period was increased to 0.96, and the rRMSE was reduced by 27% in sorghum.

This study provided an accurate AGB estimation model throughout the entire growth period for crops with conspicuous spikes, which effectively improved the estimation accuracy in the reproductive stage and provided reliable information for evaluating crop growth at the plot scale.

Author Contributions: Conceptualization: T.W.; methodology: Y.G.; validation: Y.P.; formal analysis: Y.H.; investigation: S.Z.; resources: M.W.; data curation: Y.C.; writing—original draft preparation: Q.Z. and X.W.; supervision: F.J.; funding acquisition: F.Y. All authors have read and agreed to the published version of the manuscript.

Funding: This research was funded by Kweichow Moutai Co., Ltd.

Data Availability Statement: As the data involves privacy, the relevant data can be obtained by contacting the author.

Acknowledgments: We sincerely thank Kweichow Moutai Co., Ltd. for their experimental guidance and data provision. We also acknowledge Kweichow Moutai Distillery (Group) and Hongyingzi Agricultural Science and Technology Development Co., Ltd. for providing great experimental conditions.

Conflicts of Interest: Qiaoling Zhang, Yalan Huang, Siyu Zhang, Feng Jiang, and Fan Yang were employed by Kweichow Moutai Co. Yubin Chen was employed by Kweichow Moutai Distillery (Group) Hongyingzi Agricultural Science and Technology Development Co. The remaining authors declare that the research was conducted in the absence of any commercial or financial relationships that could be construed as a potential conflict of interest.

References

1. Xu, T.Y.; Wang, F.M.; Xie, L.L.; Yao, X.P.; Zheng, J.Y.; Li, J.L.; Chen, S.T. Integrating the Textural and Spectral Information of UAV Hyperspectral Images for the Improved Estimation of Rice Aboveground Biomass. *Remote Sens.* **2022**, *14*, 2534. [CrossRef]
2. Chen, Z.X.; Ren, J.Q.; Tang, H.J.; Shi, Y.; Len, P.; Liu, J.; Wang, L.M.; Wu, W.B.; Yao, Y.M.; Hasituya. Progress and prospect of agricultural remote sensing research and application. *J. Remote Sens.* **2016**, *20*, 748–767.
3. Yang, S.X.; Feng, Q.S.; Liang, T.G.; Liu, B.K.; Zhang, W.J.; Xie, H.J. Modeling grassland above-ground biomass based on artificial neural network and remote sensing in the Three-River Headwaters Region. *Remote Sens. Environ.* **2018**, *204*, 448–455. [CrossRef]
4. Yu, Y.; Pan, Y.; Yang, X.; Fan, W. Spatial Scale Effect and Correction of Forest Aboveground Biomass Estimation Using Remote Sensing. *Remote Sens.* **2022**, *14*, 2828. [CrossRef]
5. Liu, Y.; Feng, H.; Yue, J.; Li, Z.; Jin, X.; Fan, Y.; Feng, Z.; Yang, G. Estimation of Aboveground Biomass of Potatoes Based on Characteristic Variables Extracted from UAV Hyperspectral Imagery. *Remote Sens.* **2022**, *14*, 5121. [CrossRef]
6. Wocher, M.; Berger, K.; Verrelst, J.; Hank, T. Retrieval of carbon content and biomass from hyperspectral imagery over cultivated areas. *ISPRS J. Photogramm. Remote Sens.* **2022**, *193*, 104–114. [CrossRef] [PubMed]
7. Musthafa, M.; Singh, G. Improving Forest Above-Ground Biomass Retrieval Using Multi-Sensor L- and C- Band SAR Data and Multi-Temporal Spaceborne LiDAR Data. *Front. For. Glob. Chang.* **2022**, *5*, 822704. [CrossRef]

8. Santoro, M.; Cartus, O.; Wegmüller, U.; Besnard, S.; Carvalhais, N.; Araza, A.; Herold, M.; Liang, J.; Cavlovic, J.; Engdahl, M.E. Global estimation of above-ground biomass from spaceborne C-band scatterometer observations aided by LiDAR metrics of vegetation structure. *Remote Sens. Environ.* **2022**, *279*, 113114. [CrossRef]
9. Quan, X.W.; He, B.B.; Yebra, M.; Yin, C.M.; Liao, Z.M.; Zhang, X.T.; Li, X. A radiative transfer model-based method for the estimation of grassland aboveground biomass. *Int. J. Appl. Earth Obs. Geoinf.* **2017**, *54*, 159–168. [CrossRef]
10. Rotter, R.P.; Palosuo, T.; Kersebaum, K.C.; Angulo, C.; Bindi, M.; Ewert, F.; Ferrise, R.; Hlavinka, P.; Moriondo, M.; Nendel, C.; et al. Simulation of spring barley yield in different climatic zones of Northern and Central Europe: A comparison of nine crop models. *Field Crops Res.* **2012**, *133*, 23–36. [CrossRef]
11. Kaur, S.; Singh, M. Modeling the crop growth—A review. *Mausam* **2020**, *71*, 103–114.
12. Yuan, W.P.; Cai, W.W.; Xia, J.Z.; Chen, J.Q.; Liu, S.G.; Dong, W.J.; Merbold, L.; Law, B.; Arain, A.; Beringer, J.; et al. Global comparison of light use efficiency models for simulating terrestrial vegetation gross primary production based on the La Thuile database. *Agric. For. Meteorol.* **2014**, *192*, 108–120. [CrossRef]
13. Zhang, L.; Gao, H.; Zhang, X. Combining Radiative Transfer Model and Regression Algorithms for Estimating Aboveground Biomass of Grassland in West Ujimqin, China. *Remote Sens.* **2023**, *15*, 2918. [CrossRef]
14. Cheng, Z.Q.; Meng, J.H.; Jiang, J.F.; Wang, Y.; Fang, H.T.; Yu, L.H. Estimation of above-ground biomass of maize in late growing period based on WOFOST model and UAV data. *J. Remote. Sens.* **2020**, *24*, 1403–1418.
15. Jones, J.W.; Hoogenboom, G.; Porter, C.H.; Boote, K.J.; Batchelor, W.D.; Hunt, L.A.; Wilkens, P.W.; Singh, U.; Gijsman, A.J.; Ritchie, J.T. The DSSAT cropping system model. *Eur. J. Agron.* **2003**, *18*, 235–265. [CrossRef]
16. Lara, M.A.S.; Pedreira, C.G.S.; Boote, K.J.; Pedreira, B.C.; Moreno, L.S.B.; Alderman, P.D. Predicting Growth of Panicum maximum: An Adaptation of the CROPGRO-Perennial Forage Model. *Agron. J.* **2012**, *104*, 600–611. [CrossRef]
17. Pedreira, B.C.; Pedreira, C.G.S.; Boote, K.J.; Lara, M.A.S.; Alderman, P.D. Adapting the CROPGRO perennial forage model to predict growth of Brachiaria brizantha. *Field Crops Res.* **2011**, *120*, 370–379. [CrossRef]
18. Pequeno, D.N.L.; Pedreira, C.G.S.; Boote, K.J. Simulating forage production of Marandu palisade grass (*Brachiaria brizantha*) with the CROPGRO-Perennial Forage model. *Crop Pasture Sci.* **2014**, *65*, 1335–1348. [CrossRef]
19. Sau, F.; Boote, K.J.; Bostick, W.M.; Jones, J.W.; Minguez, M.I. Testing and improving evapotranspiration and soil water balance of the DSSAT crop models. *Agron. J.* **2004**, *96*, 1243–1257. [CrossRef]
20. Chen, J.S.; Huang, J.X.; Lin, H.; Pei, Z.Y. Rice yield estimation method based on remote sensing information and crop growth model assimilation. *Sci. China Inf. Sci.* **2010**, *40*, 173–183.
21. de Wit, A.; Boogaard, H.; Fumagalli, D.; Janssen, S.; Knapen, R.; van Kraalingen, D.; Supit, I.; van der Wijngaart, R.; van Diepen, K. 25 years of the WOFOST cropping systems model. *Agric. Syst.* **2019**, *168*, 154–167. [CrossRef]
22. Zhuo, W.; Fang, S.; Gao, X.; Wang, L.; Wu, D.; Fu, S.; Wu, Q.; Huang, J. Crop yield prediction using MODIS LAI, TIGGE weather forecasts and WOFOST model: A case study for winter wheat in Hebei, China during 2009–2013. *Int. J. Appl. Earth Obs. Geoinf.* **2022**, *106*, 102668. [CrossRef]
23. Holzworth, D.P.; Huth, N.I.; Devoil, P.G.; Zurcher, E.J.; Herrmann, N.I.; McLean, G.; Chenu, K.; van Oosterom, E.J.; Snow, V.; Murphy, C.; et al. APSIM—Evolution towards a new generation of agricultural systems simulation. *Environ. Model. Softw.* **2014**, *62*, 327–350. [CrossRef]
24. Asseng, S.; Keating, B.A.; Fillery, I.R.P.; Gregory, P.J.; Bowden, J.W.; Turner, N.C.; Palta, J.A.; Abrecht, D.G. Performance of the APSIM-wheat model in Western Australia. *Field Crops Res.* **1998**, *57*, 163–179. [CrossRef]
25. Asseng, S.; van Keulen, H.; Stol, W. Performance and application of the APSIM Nwheat model in the Netherlands. *Eur. J. Agron.* **2000**, *12*, 37–54. [CrossRef]
26. Keating, B.A.; Carberry, P.S.; Hammer, G.L.; Probert, M.E.; Robertson, M.J.; Holzworth, D.; Huth, N.I.; Hargreaves, J.N.G.; Meinke, H.; Hochman, Z.; et al. An overview of APSIM, a model designed for farming systems simulation. *Eur. J. Agron.* **2003**, *18*, 267–288. [CrossRef]
27. Potter, C.S.; Randerson, J.T.; Field, C.B.; Matson, P.A.; Klooster, S.A. Terrestrial Ecosystem Production: A Process Model Based on Global Satellite and Surface Data. *Glob. Biogeochem. Cycles* **1993**, *7*, 811–841. [CrossRef]
28. Raza, S.M.H.; Mahmood, S.A. Estimation of Net Rice Production through Improved CASA Model by Addition of Soil Suitability Constant ((h)over-bar alpha). *Sustainability* **2018**, *10*, 1788. [CrossRef]
29. Liu, Z.Z.; Zhang, X.W.; Chen, Y.S.; Zhang, C.C.; Qin, F.; Zeng, H.W. Remote sensing estimation of regional winter wheat biomass based on CASA mode. *Trans. Agric. Eng.* **2017**, *33*, 225–233+315–316.
30. Zhao, Y.; Mao, D.; Zhang, D.; Wang, Z.; Du, B.; Yan, H.; Qiu, Z.; Feng, K.; Wang, J.; Jia, M. Mapping Phragmites australis Aboveground Biomass in the Momoge Wetland Ramsar Site Based on Sentinel-1/2 Images. *Remote Sens.* **2022**, *14*, 694. [CrossRef]
31. Goetz, S.J.; Prince, S.D.; Goward, S.N.; Thawley, M.M.; Small, J. Satellite remote sensing of primary production: An improved production efficiency modeling approach. *Ecol. Model.* **1999**, *122*, 239–255. [CrossRef]
32. Darvishzadeh, R.; Skidmore, A.; Schlerf, M.; Atzberger, C. Inversion of a radiative transfer model for estimating vegetation LAI and chlorophyll in a heterogeneous grassland. *Remote Sens. Environ.* **2008**, *112*, 2592–2604. [CrossRef]

33. Adeluyi, O.; Harris, A.; Verrelst, J.; Foster, T.; Claya, G.D. Estimating the phenological dynamics of irrigated rice leaf area index using the combination of PROSAIL and Gaussian Process Regression. *Int. J. Appl. Earth Obs. Geoinf.* **2021**, *102*, 102454. [CrossRef] [PubMed]
34. Wang, Z.; He, L.; He, Z.; Wang, X.; Li, L.; Kang, G.; Bai, W.; Chen, X.; Zhao, Y.; Xiao, Y. Integrating the PROSAIL and SVR Models to Facilitate the Inversion of Grassland Aboveground Biomass: A Case Study of Zoigê Plateau, China. *Remote Sens.* **2024**, *16*, 1117. [CrossRef]
35. Elshikha, D.E.M.; Hunsaker, D.J.; Waller, P.M.; Thorp, K.R.; Dierig, D.; Wang, G.; Cruz, V.M.V.; Katterman, M.E.; Bronson, K.F.; Wall, G.W.; et al. Estimation of direct-seeded guayule cover, crop coefficient, and yield using UAS-based multispectral and RGB data. *Agric. Water Manag.* **2022**, *265*, 107540. [CrossRef]
36. Li, B.; Xu, X.; Zhang, L.; Han, J.; Bian, C.; Li, G.; Liu, J.; Jin, L. Above-ground biomass estimation and yield prediction in potato by using UAV-based RGB and hyperspectral imaging. *ISPRS J. Photogramm. Remote Sens.* **2020**, *162*, 161–172. [CrossRef]
37. Yue, J.; Yang, G.; Tian, Q.; Feng, H.; Xu, K.; Zhou, C. Estimate of winter-wheat above-ground biomass based on UAV ultrahigh-ground-resolution image textures and vegetation indices. *ISPRS J. Photogramm. Remote Sens.* **2019**, *150*, 226–244. [CrossRef]
38. Luo, S.; Jiang, X.; He, Y.; Li, J.; Jiao, W.; Zhang, S.; Xu, F.; Han, Z.; Sun, J.; Yang, J.; et al. Multi-dimensional variables and feature parameter selection for aboveground biomass estimation of potato based on UAV multispectral imagery. *Front. Plant Sci.* **2022**, *13*, 948249. [CrossRef]
39. Xu, L.; Zhou, L.; Meng, R.; Zhao, F.; Lv, Z.; Xu, B.; Zeng, L.; Yu, X.; Peng, S. An improved approach to estimate ratoon rice aboveground biomass by integrating UAV-based spectral, textural and structural features. *Precis. Agric.* **2022**, *23*, 1276–1301. [CrossRef]
40. Chen, P.F.; Wang, J.L.; Liao, X.Y.; Yin, F.; Chen, B.R.; Liu, R. Inversion of above-ground biomass of grassland in Hulunbuir based on remote sensing data from environmental disaster reduction satellites. *J. Nat. Resour.* **2010**, *25*, 1122–1131.
41. Bao, Y.; Wei, G.; Gao, Z. Estimation of winter wheat biomass based on remote sensing data at various spatial and spectral resolutions. *Front. Earth Sci. China* **2009**, *3*, 118–128. [CrossRef]
42. Mutanga, O.; Skidmore, A.K. Narrow band vegetation indices overcome the saturation problem in biomass estimation. *Int. J. Remote Sens.* **2004**, *25*, 3999–4014. [CrossRef]
43. Fu, Y.Y.; Yang, G.J.; Wang, J.H.; Song, X.Y.; Feng, H.K. Winter wheat biomass estimation based on spectral indices, band depth analysis and partial least squares regression using hyperspectral measurements. *Comput. Electron. Agric.* **2014**, *100*, 51–59. [CrossRef]
44. Thenkabail, P.S.; Smith, R.B.; De Pauw, E. Hyperspectral vegetation indices and their relationships with agricultural crop characteristics. *Remote Sens. Environ.* **2000**, *71*, 158–182. [CrossRef]
45. Feng, H.; Fan, Y.; Yue, J.; Bian, M.; Liu, Y.; Chen, R.; Ma, Y.; Fan, J.; Yang, G.; Zhao, C. Estimation of potato above-ground biomass based on the VGC-AGB model and deep learning. *Comput. Electron. Agric.* **2025**, *232*, 110122. [CrossRef]
46. Duan, B.; Liu, Y.; Gong, Y.; Peng, Y.; Wu, X.; Zhu, R.; Fang, S. Remote estimation of rice LAI based on Fourier spectrum texture from UAV image. *Plant Methods* **2019**, *15*, 124. [CrossRef] [PubMed]
47. Freitas, R.G.; Pereira, F.R.S.; Dos Reis, A.A.; Magalhães, P.S.G.; Figueiredo, G.K.D.A.; do Amaral, L.R. Estimating pasture aboveground biomass under an integrated crop-livestock system based on spectral and texture measures derived from UAV images. *Comput. Electron. Agric.* **2022**, *198*, 107122. [CrossRef]
48. Lu, D.; Batistella, M. Exploring TM image texture and its relationships with biomass estimation in Rondônia, Brazilian Amazon. *Acta Amaz.* **2005**, *35*, 249–257. [CrossRef]
49. Nichol, J.E.; Sarker, M.L.R. Improved Biomass Estimation Using the Texture Parameters of Two High-Resolution Optical Sensors. *IEEE Trans. Geosci. Remote Sens.* **2011**, *49*, 930–948. [CrossRef]
50. Fu, Y.Y.; Yang, G.J.; Li, Z.H.; Song, X.Y.; Li, Z.H.; Xu, X.G.; Wang, P.; Zhao, C.J. Winter Wheat Nitrogen Status Estimation Using UAV-Based RGB Imagery and Gaussian Processes Regression. *Remote Sens.* **2020**, *12*, 3778. [CrossRef]
51. Fu, Y.Y.; Yang, G.J.; Song, X.Y.; Li, Z.H.; Xu, X.G.; Feng, H.K.; Zhao, C.J. Improved Estimation of Winter Wheat Aboveground Biomass Using Multiscale Textures Extracted from UAV-Based Digital Images and Hyperspectral Feature Analysis. *Remote Sens.* **2021**, *13*, 581. [CrossRef]
52. Liu, J.; Zhu, Y.; Song, L.; Su, X.; Li, J.; Zheng, J.; Zhu, X.; Ren, L.; Wang, W.; Li, X. Optimizing window size and directional parameters of GLCM texture features for estimating rice AGB based on UAVs multispectral imagery. *Front. Plant Sci.* **2023**, *14*, 1284235. [CrossRef]
53. Jimenez-Berni, J.A.; Deery, D.M.; Rozas-Larraondo, P.; Condon, A.G.; Rebetzke, G.J.; James, R.A.; Bovill, W.D.; Furbank, R.T.; Sirault, X.R.R. High Throughput Determination of Plant Height, Ground Cover, and Above-Ground Biomass in Wheat with LiDAR. *Front. Plant Sci.* **2018**, *9*, 237. [CrossRef] [PubMed]
54. Tilly, N.; Aasen, H.; Bareth, G. Fusion of Plant Height and Vegetation Indices for the Estimation of Barley Biomass. *Remote Sens.* **2015**, *7*, 11449–11480. [CrossRef]

55. Yue, J.B.; Feng, H.K.; Jin, X.L.; Yuan, H.H.; Li, Z.H.; Zhou, C.Q.; Yang, G.J.; Tian, Q.J. A Comparison of Crop Parameters Estimation Using Images from UAV-Mounted Snapshot Hyperspectral Sensor and High-Definition Digital Camera. *Remote Sens.* **2018**, *10*, 1138. [CrossRef]
56. Navarro, J.A.; Algeet, N.; Fernandez-Landa, A.; Esteban, J.; Rodriguez-Noriega, P.; Guillen-Climent, M.L. Integration of UAV, Sentinel-1, and Sentinel-2 Data for Mangrove Plantation Aboveground Biomass Monitoring in Senegal. *Remote Sens.* **2019**, *11*, 77. [CrossRef]
57. Meiyan, S.; Mengyuan, S.; Qizhou, D.; Xiaohong, Y.; Baoguo, L.; Yuntao, M. Estimating the maize above-ground biomass by constructing the tridimensional concept model based on UAV-based digital and multi-spectral images. *Field Crops Res.* **2022**, *282*, 108491. [CrossRef]
58. Liu, Y.; Feng, H.; Yue, J.; Jin, X.; Fan, Y.; Chen, R.; Bian, M.; Ma, Y.; Li, J.; Xu, B.; et al. Improving potato AGB estimation to mitigate phenological stage impacts through depth features from hyperspectral data. *Comput. Electron. Agric.* **2024**, *219*, 108808. [CrossRef]
59. Liu, Y.; Feng, H.; Yue, J.; Jin, X.; Fan, Y.; Chen, R.; Bian, M.; Ma, Y.; Song, X.; Yang, G. Improved potato AGB estimates based on UAV RGB and hyperspectral images. *Comput. Electron. Agric.* **2023**, *214*, 108260. [CrossRef]
60. Liu, T.; Wang, J.; Wang, J.; Zhao, Y.; Wang, H.; Zhang, W.; Yao, Z.; Liu, S.; Zhong, X.; Sun, C. Research on the estimation of wheat AGB at the entire growth stage based on improved convolutional features. *J. Integr. Agric.* **2025**, *24*, 1403–1423. [CrossRef]
61. Gnyp, M.L.; Miao, Y.; Yuan, F.; Ustin, S.L.; Yu, K.; Yao, Y.; Huang, S.; Bareth, G. Hyperspectral canopy sensing of paddy rice aboveground biomass at different growth stages. *Field Crops Res.* **2014**, *155*, 42–55. [CrossRef]
62. Xu, T.; Wang, F.; Shi, Z.; Xie, L.; Yao, X. Dynamic estimation of rice aboveground biomass based on spectral and spatial information extracted from hyperspectral remote sensing images at different combinations of growth stages. *ISPRS J. Photogramm. Remote Sens.* **2023**, *202*, 169–183. [CrossRef]
63. Rouse, J.W.; Haas, R.H.; Schell, J.A.; Deering, D.W. Monitoring Vegetation Systems in the Great Plains with Erts. *NASA Spec. Publ.* **1974**, *351*.
64. Jiang, Z.Y.; Huete, A.R.; Didan, K.; Miura, T. Development of a two-band enhanced vegetation index without a blue band. *Remote Sens. Environ.* **2008**, *112*, 3833–3845. [CrossRef]
65. Fitzgerald, G.J.; Rodriguez, D.; Christensen, L.K.; Belford, R.; Sadras, V.O.; Clarke, T.R. Spectral and thermal sensing for nitrogen and water status in rainfed and irrigated wheat environments. *Precis. Agric.* **2006**, *7*, 233–248. [CrossRef]
66. Rondeaux, G.; Steven, M.; Baret, F. Optimization of soil-adjusted vegetation indices. *Remote Sens. Environ.* **1996**, *55*, 95–107. [CrossRef]
67. Clark, R.N.; Roush, T.L. Reflectance spectroscopy: Quantitative analysis techniques for remote sensing applications. *J. Geophys. Res.* **1984**, *89*, 6329–6340. [CrossRef]
68. Kokaly, R.F.; Clark, R.N. Spectroscopic Determination of Leaf Biochemistry Using Band-Depth Analysis of Absorption Features and Stepwise Multiple Linear Regression. *Remote Sens. Environ.* **1999**, *67*, 267–287. [CrossRef]
69. Jiang, Q.; Fang, S.; Peng, Y.; Gong, Y.; Zhu, R.; Wu, X.; Ma, Y.; Duan, B.; Liu, J. UAV-Based Biomass Estimation for Rice-Combining Spectral, TIN-Based Structural and Meteorological Features. *Remote Sens.* **2019**, *11*, 890. [CrossRef]
70. Wang, D.; Fahad, S.; Saud, S.; Kamran, M.; Khan, A.; Khan, M.N.; Hammad, H.M.; Nasim, W. Morphological acclimation to agronomic manipulation in leaf dispersion and orientation to promote "Ideotype" breeding: Evidence from 3D visual modeling of "super" rice (*Oryza sativa* L.). *Plant Physiol. Biochem.* **2019**, *135*, 499–510. [CrossRef]
71. Zeng, Y.; Hao, D.; Huete, A.; Dechant, B.; Berry, J.; Chen, J.M.; Joiner, J.; Frankenberg, C.; Bond-Lamberty, B.; Ryu, Y.; et al. Optical vegetation indices for monitoring terrestrial ecosystems globally. *Nat. Rev. Earth Environ.* **2022**, *3*, 477–493. [CrossRef]
72. Yang, K.; Mo, J.; Luo, S.; Peng, Y.; Fang, S.; Wu, X.; Zhu, R.; Li, Y.; Yuan, N.; Zhou, C.; et al. Estimation of Rice Aboveground Biomass by UAV Imagery with Photosynthetic Accumulation Models. *Plant Phenomics* **2023**, *5*, 56. [CrossRef] [PubMed]
73. Tunca, E.; Köksal, E.S.; Akay, H.; Öztürk, E.; Taner, S.Ç. Novel machine learning framework for high-resolution sorghum biomass estimation using multi-temporal UAV imagery. *Int. J. Environ. Sci. Technol.* **2025**, *22*, 1–16. [CrossRef]

remote sensing

MDPI

Article

Monitoring Nitrogen Uptake and Grain Quality in Ponded and Aerobic Rice with the Squared Simplified Canopy Chlorophyll Content Index

Gonzalo Carracelas [1,2,*], John Hornbuckle [1] and Carlos Ballester [1]

1 Centre for Regional and Rural Futures, Faculty of Science Engineering & Built Environment, Deakin University, Hanwood, NSW 2680, Australia; j.hornbuckle@deakin.edu.au (J.H.); carlos.ballesterlurbe@deakin.edu.au (C.B.)

2 National Institute for Agricultural Research (INIA), Tacuarembó 45000, Uruguay

* Correspondence: gcarracelas@inia.org.uy

Abstract

Remote sensing tools have been proposed to assist with rice crop monitoring but have been developed and validated on ponded rice. This two-year study was conducted on a commercial rice farm with irrigation automation technology aimed to (i) understand how canopy reflectance differs between high-yielding ponded and aerobic rice, (ii) validate the feasibility of using the squared simplified canopy chlorophyll content index ($SCCCI^2$) for N uptake estimates, and (iii) explore the $SCCCI^2$ and similar chlorophyll-sensitive indices for grain quality monitoring. Multispectral images were collected from an unmanned aerial vehicle during both rice-growing seasons. Above-ground biomass and nitrogen (N) uptake were measured at panicle initiation (PI). The performance of single-vegetation-index models in estimating rice N uptake, as previously published, was assessed. Yield and grain quality were determined at harvest. Results showed that canopy reflectance in the visible and near-infrared regions differed between aerobic and ponded rice early in the growing season. Chlorophyll-sensitive indices showed lower values in aerobic rice than in the ponded rice at PI, despite having similar yields at harvest. The $SCCCI^2$ model (RMSE = 20.52, Bias = -6.21 Kg N ha^{-1}, and MAPE = 11.95%) outperformed other models assessed. The $SCCCI^2$, squared normalized difference red edge index, and chlorophyll green index correlated at PI with the percentage of cracked grain, immature grain, and quality score, suggesting that grain milling quality parameters could be associated with N uptake at PI. This study highlights canopy reflectance differences between high-yielding aerobic (averaging 15 Mg ha^{-1}) and ponded rice at key phenological stages and confirms the validity of a single-vegetation-index model based on the $SCCCI^2$ for N uptake estimates in ponded and non-ponded rice crops.

Keywords: vegetation indices; remote sensing; drones; irrigation; non-ponded rice; water-saving strategies; N fertilization management

1. Introduction

The traditional water management in rice farms, in which the crop is ponded for most of the growing season, is threatening the sustainability of the rice industry due to high water use, particularly in arid and semi-arid regions [1–4]. Aiming to increase rice water use productivity, irrigation strategies that shorten to a greater or lesser extent the period the crop remains flooded have been suggested as alternative techniques to traditional management [5–9]. Aerobic rice production is one of the strategies that has been shown to increase water productivity in comparison with ponded rice [10–13].

In aerobic rice production, the crop is watered when the soil is in an unsaturated state and drained later so permanent water ponding is not maintained at any time during the growing season. The effect on grain yield for aerobic rice, and water-saving strategies that allow soil to dry to a certain level, is influenced by environmental and management factors, with soil water potential being the main driver of yield variation [14]. The average soil water potential threshold where rice grain yield is not reduced by applying water-saving management strategies has been estimated in a recent meta-analysis as -19 kPa, with variations among rice growing areas depending mainly on climate and soil type [14].

In temperate areas of Australia, aerobic rice production is being considered as a water management strategy of interest for the industry due to the recurrent annual fluctuations in irrigation water availability [15]. The main barriers to the adoption of this irrigation strategy by farmers at commercial scale are the high labor required for water management compared with ponded rice, and famers' perception of a high crop susceptibility to water stress when the soil is not saturated and/or to cold stress at the microspore stage [16,17], both associated with yield reductions. However, achieving high yields under aerobic rice management at commercial scale is possible with adequate water and nitrogen (N) management. Automation technology for gravity surface irrigation systems is key for reducing labor requirements and enabling timely watering of crops [18–20].

Over recent years, a variety of remote sensing tools have been proposed to assist in the monitoring of rice crops for optimizing different aspects of crop management. As the main irrigation water strategy employed for rice production, the majority of these tools have been developed and validated on ponded rice crops.

Due to the differences generally observed in canopy development between aerobic and ponded rice, with the former taking longer to reach full canopy cover [5] and in image background (differences in canopy light extinction coefficient [21] and presence or absence of surface water [22,23]), remote sensing tools developed for ponded rice (e.g., prediction of panicle initiation, nitrogen uptake, or grain yield) may not be appropriate for rice grown aerobically. Carracelas et al. [24] indicated that common chlorophyll-sensitive vegetation indices reported as optimum for N uptake estimates were significantly influenced by the level of surface water in bays at the moment of image collection. These authors observed that among the vegetation indices they assessed the squared simplified chlorophyll canopy content index ($SCCCI^2$) was the least affected by the level of surface water and observation time, suggesting it could be used as a general index for N uptake estimates regardless of the irrigation strategy undertaken. Validation of the model reported by Carracelas et al. [24] in different conditions to those in which it was developed (site, growing season, and variety) would demonstrate the $SCCCI^2$ validity for N uptake estimates and provide the industry with a valuable tool for optimizing N fertilizer management in rice farming.

Apart from being essential for maximizing rice grain yield, optimum N management is important for producing high grain quality [25–28]. Plant N availability, along with other factors, affect grain physical qualities [29]. It is hypothesized that vegetation indices that are sensitive to chlorophyll content could explain crop variability in grain quality. However, there are few studies dealing with rice grain quality monitoring from remote sensing measurements. In a study with two cultivars and a range of N application rates, Tsukaguchi et al. [30] reported a good correlation between the chlorophyll index green (Clg) obtained from drone imagery around heading and the percentage of imperfect grains. Hou et al. [31] reported a good correlation between spectral measurements and brown rice rate, moisture content, and taste value. In Australia, price penalties are imposed to crops with a percentage of cracked and immature grain exceeding established thresholds as both are the main parameters used to estimate the overall quality score. Thus, monitoring grain quality variability within bays could be of interest for growers in order to reduce within crop variability.

The objective of this study conducted at a commercial rice farm was to understand how canopy reflectance can differ between high-yielding ponded and aerobically grown rice crops over the growing season, validating the feasibility of using the SCCCI^2 for N uptake estimation regardless of the water management strategy undertaken, and exploring the potential use of the SCCCI^2 and similar chlorophyll-sensitive indices for grain quality monitoring.

2. Materials and Methods

2.1. Study Site Description and Field Management

The study was conducted in two bays (7.4 ha and 5.4 ha) of a commercial rice farm located in the Murrumbidgee rice growing region in New South Wales, Australia (Latitude: 340 21′ 21.9″S, Longitude: 1470 10′ 14.8″E) with different water management (aerobic and traditional) during two consecutive rice growing seasons (2022/2023 and 2023/2024) (Figure 1).

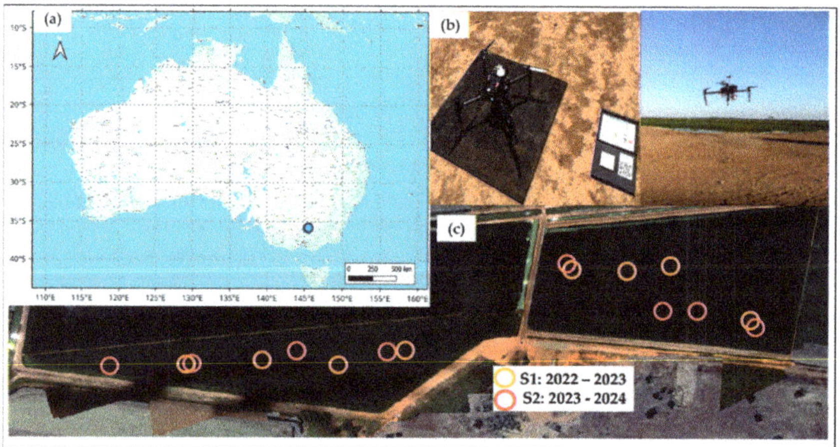

Figure 1. (**a**) Location of the commercial rice farm in the Murrumbidgee region of New South Wales, Australia (indicated with a blue dot). (**b**) Image of the unmanned-aerial-vehicle-based multispectral system and calibration panel used in the study. (**c**) Monitored bays and areas of interest for the 2022/2023 (orange circles) and 2023/2024 (red circles) growing seasons.

The climate in the Murrumbidgee rice growing region is temperate with hot dry summers and low humidity. A description of selected weather parameters obtained from the nearest weather station (~15 Km) for the period from 1 October to 15 April in each rice growing season is reported in Table 1. Total evapotranspiration was 164.3 mm lower in the first growing season than in the second one. The first growing season had 13 fewer days with maximum temperatures above 35 °C, and 13 more days with minimum temperatures below 15 °C from PI to flowering than the second growing season (Table 1).

A soil corer sample machine (Christie Engineering, Delaware Road, Horsley Park, NSW, Australia) was used prior to sowing to collect a total of 16 soil samples (0–30 cm depth) to determine soil chemical and physical properties in both bays. Soil samples were analyzed at the professional laboratory (EAL Environmental Analysis Laboratory, Lismore, NSW, Australia) (Table 2). The soil type was brownish clay in both bays, and was classified as Transitional red brown earth [32], commonly used for rice production in the area. Relative to the indicative guidelines for a heavy clay soil [33,34], the soil organic matter percentage, phosphorus and nitrate nitrogen on average were low. Potassium levels were higher in relation to the indicative guidelines. Most micronutrients (Zn, Mg, Cu) were on average below the indicative guidelines for a heavy clay soil type.

Table 1. Description of weather parameters obtained from the nearest meteorological station to the rice farm located at Griffith Airport, NSW, Australia (Bureau of Meteorology, Australian Government, www.bom.gov.au/watl/eto/, accessed on 10 May 2025) from the 1 October to the 15 April in each growing season.

Weather Parameters	Rice Growing Season	
	2022–2023	2023–2024
Total Evapotranspiration (mm)	1098.9	1263.2
Daily Evapotranspiration (mm d^{-1})	5.4	6.2
Total Rain (mm)	373.0	332.4
Maximum Temperature (°C)	27.7	30.0
Days with Max. temp. above 35 °C	29.0	42.0
Minimum Temperature (°C)	12.8	14.3
Days with Min. temp. below 15 °C	121.0	94.0
Days with Min. temp. below 15 °C (from PI to Flowering)	24.0	11.0
Maximum Relative Humidity (%)	88.2	83.4
Minimum Relative Humidity (%)	35.2	28.7
Wind-speed at 10 m (m/s)	4.1	4.0
Solar Radiation (MJ m^{-2} d^{-1})	21.8	22.4

Table 2. Selected soil properties determined prior to sowing at the Environmental Analysis Laboratory—EAL (www.scu.edu.au/eal, accessed on 1 December 2024). The data shown is the average of the two rice growing seasons and monitored bays.

Soil Parameters	Value	Soil Parameters	Value
pH (water)	7.6	Zn-Zinc (mg/kg)	0.9
pH (CaCl2)	7.0	Mn-Manganese (mg/kg)	20.5
Estimated Organic Matter (% OM)	1.5	Fe-Iron (mg/kg)	78.8
Phosphorus (mg/kg P) (Colwell)	41.4	Cu-Copper (mg/kg)	3.2
Nitrate Nitrogen (mg/kg N)	4.7	B-Boron (mg/kg)	1.1
Ammonium Nitrogen (mg/kg N)	3.4	Si-Silicon (mg/kg Si)	44.0
Sulfur (mg/kg S)	27.6	Total Carbon (%)	0.9
Electrical Conductivity (dS/m)	0.1	Total Nitrogen (%)	0.1
Ca—Exchangeable Calcium (mg/kg)	2.6	Carbon/Nitrogen Ratio	8.8
Mg—Exchangeable Magnesium (mg/kg)	1.2	Texture (ISSS classification)	Clay
K-Exchangeable Potassium (mg/kg)	309.7	Sand > 20 µm	38.3%
Na-Exchangeable Sodium (mg/kg)	102.1	Silt (2–20 µm)	8.9%
Al-Exchangeable Aluminum (mg/kg)	1.9	Clay (< 2 µm)	52.8%
ECEC—Effective Cation Exchange Capacity ($cmol^{+}$/kg)	25.3	Gravel > 2 mm Basic Color	0.2% Brownish

2.2. Field Management

The study was conducted during the second and third year of consecutively growing rice crops at the farm. Minimum tillage was performed on leveled bays. All tillage operations, herbicides applications and the first N fertilizer applications were undertaken on dry soil before there was any irrigation event.

In the first year of study, due to a wet start of the growing season (204 mm recorded in October, 55% of the total rainfall received throughout the entire crop cycle) the variety V071 (medium grain, Japonica) was broadcasted with pregerminated seed on the 19 October 2022 with an average seeding rate of 158 kg ha^{-1} (Table 1). Bays were irrigated the following day after sowing and remained flooded for 15 days to promote plant establishment. During the second rice growing season, the same variety was direct drilled (row spacing of 0.20 m) on the 2 October 2023 with an average seeding rate of 142 kg ha^{-1}. The first irrigation event occurred on the 10 October 2023, and it was followed by two more irrigation events to promote emergence. On average for both rice growing seasons, plant establishment was 169 plants per m^2, ranging from 40 to 305 plants per m^2 (n = 41).

The fertilization management included a basal starter (custom blend fertilizer—rice bouncer: N8%-P17%-S7% Ca8% Zn0.6%) application of ~220 kg ha^{-1} in 2022 and 150 kg ha^{-1} in 2023. This was followed by urea applications in late November and early in January. In the 2022/23 rice growing season, urea was broadcast at 355 kg ha^{-1} and 145 kg ha^{-1} in the first and second applications at both bays (500 kg ha^{-1} in total). In the 2023/24 rice growing season, 275 and 350 kg ha^{-1} of urea were applied in the aerobic and traditionally managed bays, respectively, in late November, and 125 kg ha^{-1} were applied in both bays early in January. To level the total amount of urea applied to the aerobic and traditional bays, an additional 100 kg ha^{-1} of urea were applied to the aerobic bay. In total in this second season, the aerobic and ponded rice received 500 and 475 kg ha^{-1} of urea.

2.3. Irrigation Management

The traditional water management consisted in maintaining the soil under anaerobic conditions. This was achieved by maintaining a layer of surface water that fluctuated over the growing seasons as reported in Figure 2. At microsporogenesis, as the standard practice followed in the rice growing region, the water layer was increased up to 0.20–0.25 m to protect the crop against the incidence of low temperatures (Figure 2b). The bay was drained in both growing seasons late November early December for the first urea application.

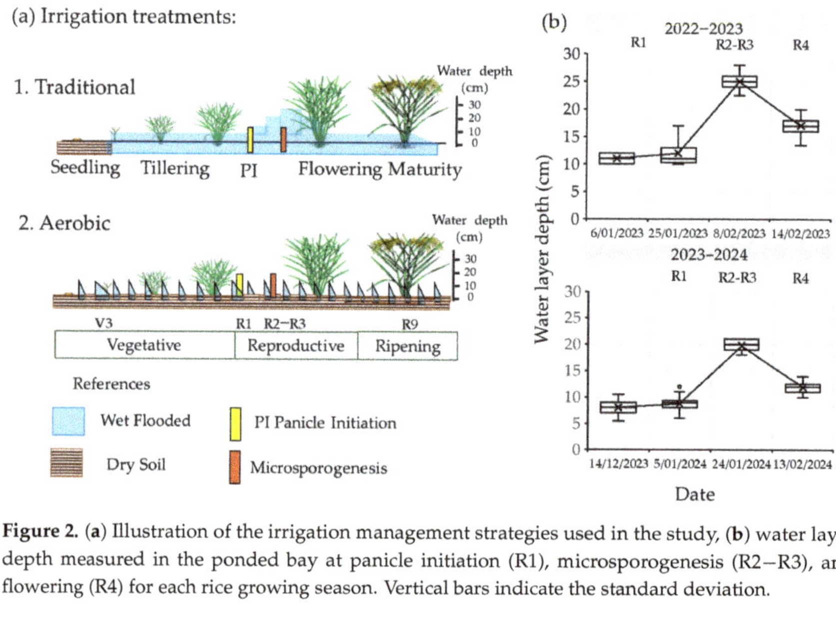

Figure 2. (**a**) Illustration of the irrigation management strategies used in the study, (**b**) water layer depth measured in the ponded bay at panicle initiation (R1), microsporogenesis (R2–R3), and flowering (R4) for each rice growing season. Vertical bars indicate the standard deviation.

In the aerobic rice management, water was applied to the bay and drained to let the soil dry and return to an aerobic condition before the next irrigation event took place. Irrigation frequency varied during the season depending on water needs in the contiguous bay and other operations like fertilizer applications. However, irrigation was scheduled to try to avoid soil water potential values below −15 kpa at 0.10 m depth (Figure 3). This was monitored with Watermarks (Model 200SS, Irrometer Company Inc., Riverside, CA, USA) that were installed in pairs at the beginning of each growing season. Watermark readings showed that soil matric potential decreased in the aerobic bay between irrigation events (Figure 3). On average, water was applied to the aerobically managed bay every 3–5 days. At the end of the growing season when grain moisture was below 26%, all the bays at the farm were drained off for harvesting the crop on dry soil.

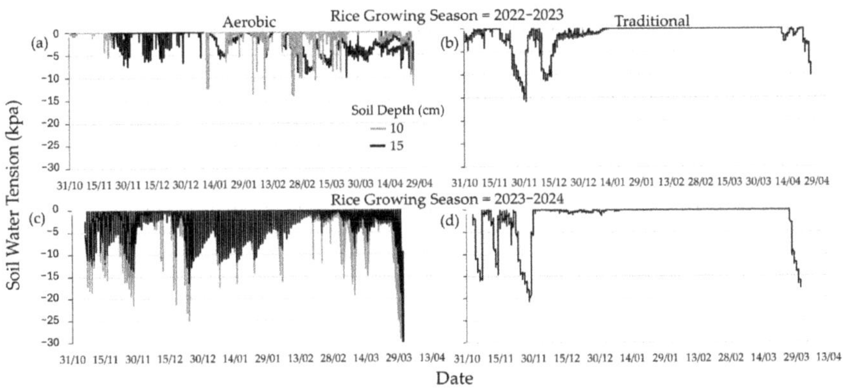

Figure 3. Soil water tension measured with watermark sensors at 10 cm (gray line) and 15 cm (black line) during both rice growing seasons in the aerobically (**a**,**c**) and traditionally (**b**,**d**) managed bays.

2.4. Multispectral Measurements

Multispectral images were collected four times across the growing season in the first and second year of study, corresponding approximately with the phenological stages of tillering, PI, flowering and maturity. Images were collected from an unmanned aerial vehicle (UAV; Matrice 100, DJI, Shenzhen, China) with the Altum camera (AgEagle Aerial Systems Inc., Wichita, KS, USA) that captures information from the blue, green, red, red edge, and near-infrared (NIR) spectral bands centered at 475, 560, 668, 717 and 840 nm, respectively. Flights were automated using the Pix4Dcapture app that was set up to collect images at 120 m altitude with an 80% overlap. These settings provided a pixel resolution of ~5.6 cm. The spectral bands used to calculate the vegetation indices in this study were the green, red, red edge and NIR bands. An image of a calibration panel was collected before and after each flight to remove the effects of sunlight variation. Information from the radiometric target, camera gain and exposure settings and a Downwelling Light Sensor (DLS2) that provides information about the illumination at the moment each image was collected, was used for the radiometric corrections when the images were processed with the Agisoft software (version 1.8.0, Agisoft Metashape, Saint Petersburg, Russia). A schematic illustration of the methodology followed in this study is provided in Figure 4. The Geo-TIFF files generated for each set of images with information of the multispectral bands were post-processed in QGIS (version 3.26). Four areas of interest of ~397 m^2 each (radius = 11.25 m), where soil moisture and plant measurements were conducted at different times during the growing seasons, were delimited in both bays. Since there was no treatment to cause variability in the present study, the location of the four areas of interest within each bay were selected based on the (low) variability in NDVI observed in the previous growing season from satellite imagery. The zonal statistical tool was used in QGIS to obtain the average reflectance of the five spectral bands captured with the camera on each measurement date for each area of interest. This information was exported from the attribute table and used to compute the vegetation indices in Excel. The formulation of the multispectral indices used in this study are shown in Table 3.

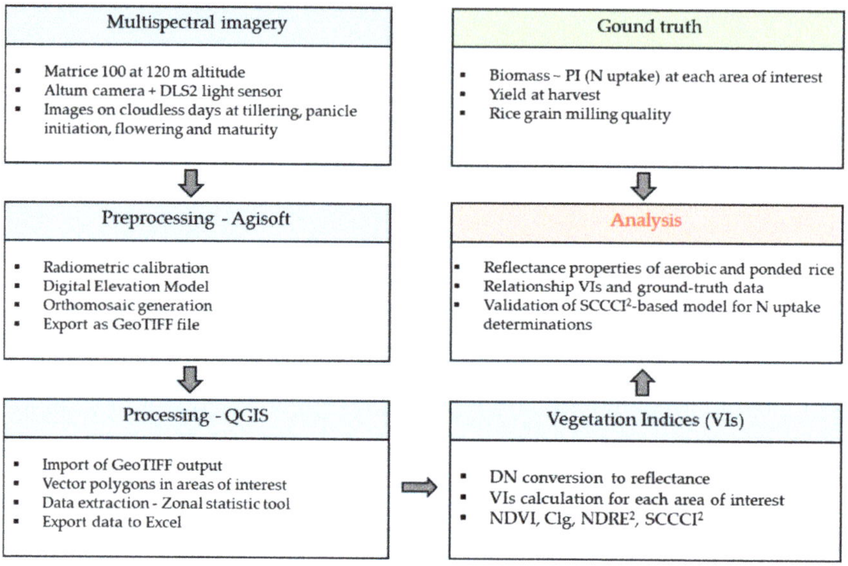

Figure 4. Schematic illustration of the methodology followed in this study. DN: digital number.

Table 3. Vegetation Indices computed from the multispectral images with their respective formulations. G, R, RE, and NIR correspond to reflectance in the green, red, red edge and near-infrared regions of the spectrum.

Vegetation Index	Formulation	References
Squared of simplified canopy chlorophyll content index	$SCCCI^2 = \left(\frac{NDRE}{NDVI}\right)^2 = \left(\frac{\left(\frac{NIR-RE}{NIR+RE}\right)}{\left(\frac{NIR-R}{NIR+R}\right)}\right)^2$	[24]
Squared of normalized difference red edge	$NDRE^2 = \left(\frac{NIR-RE}{NIR+RE}\right)^2$	[35]
Chlorophyll green	$Clg = \left(\frac{NIR}{G}\right) - 1$	[36]
Normalized difference vegetation index	$NDVI = \left(\frac{NIR-R}{NIR+R}\right)$	[37]

2.5. Plant Measurements

In both rice growing seasons, four above-ground biomass samples were collected per bay within each area of interest delimited for the remote sensing measurements around PI. Panicle initiation was defined as the moment when 30% of the main tillers had a developing panicle 1–3 mm long [38]. Plant samples were collected within 1 m^2 in each area of interest by manually cutting plants at ground level. Samples were dried in an oven at 60 °C until they reached a constant weight. Five whole plants including leaves and tillers weighing ~100 g were randomly picked per sample for the NIR tissue test at the Australian grain storage facility (AGS) to determine the nitrogen plant concentration (N%) [39]. Nitrogen uptake was estimated using the following equation:

$$N\ uptake = N\% \times DM \quad (1)$$

where DM is the plant biomass above ground expressed in kg of dry matter per hectare at PI.

Rice grain yield was assessed in late and mid-April in 2023 and 2024, respectively. In the first year of study, eight samples were manually harvested at the areas of interest of each bay using a sampling ring with an area of 0.2 m^2 [39]. In the second growing season when the crop was direct drilled, however, a quadrant of 1 m^2 (5 rows $\times$ 1 m) was used in both bays. A total of 16 samples were manually harvested in the two rice growing seasons when grain moisture was lower than 22% according to rice industry recommendations. Grain yields were normalized to 14% moisture. The whole bays were harvested with a commercial John Deere header with yield harvest monitor.

Grain moisture content was measured with FOSS (Infratec) equipment [40]. Each sample was threshed using a Pfeuffer grain sample cleaner and a visual check was performed for contaminants.

Grain milling quality was measured using the PaddyVision® [41] instrument to quickly assess dry or wet paddy rice at the Australian Grain Storage lab, part of Sunrice Group in Leeton, Australia [42]. A total of ~6000 grains were scanned across the sixteen samples (373 grains per sample) collected at harvest in the two evaluated rice growing seasons. The key quality parameters assessed were the percentage of cracked and immature grains, and the quality score, which is estimated from the previous two parameters.

2.6. MultispeQ Measurements

To monitor the possible occurrence of abiotic stress on the aerobic rice, chlorophyll fluorescence and photosynthesis-related parameters were periodically measured in both monitored bays with a hand-held MultispeQ device (V 2.0, Photosynq Inc., East Lansing, MI 48823, USA). Measurements were taken using the protocol Photosynthesis RIDES 2.0, between 11 a.m. and 3 p.m. in the flag leaf of 10 to 20 plants per area of interest within each bay on the same dates when the multispectral images were collected. Measurements were taken five meters from the soil moisture sensors starting from the North side and following counterclockwise to avoid duplicate measurements. A total of 206 individual leaf measurements were taken over the two rice growing seasons. Among the parameters measured with the MultispeQ, only seven were reported here: ambient temperature, relative humidity, leaf temperature differential, chlorophyll SPAD, maximum quantum yield of PSII under light conditions (Fv/Fm), ratio of the incoming light energy used for the photochemistry (Phi2), ratio of the incoming light energy that is lost to unregulated processes (PhiN0), and ratio of the incoming light that is dissipated as heat (PhiNPQ).

2.7. Data Analysis

The statistical analysis was conducted using R software (R version 4.2.2) [43]. Analyses of variance were performed to evaluate the influence of irrigation techniques and rice growing season on multispeQ measurements, above-ground biomass, N%, N uptake, rice grain yield and grain milling quality parameters (total cracked %, immature%, and quality score). Means were compared using Tukey's tests. Linear regression models were used to analyze the relationships between vegetation indices and rice grain yield and grain milling quality parameters. The best fitting model was selected based on the coefficient of determination (R^2) and Akaike information criterion (AIC). Pearson correlation analysis was performed using the cor.test function (R Core Team, 2022). The quadratic and linear regression models were fitted using the lm function in R [44].

Previously published models for N uptake estimates at PI using vegetation indices were assessed by comparing observed and predicted N uptake. A linear–plateau model was fitted for the relationship between $NDRE^2$ and N uptake recently published in Carracelas et al. [24], using the nls function in R [45,46]. A linear model was obtained between $NDRE^2$ and N uptake up to a plateau threshold of 166 ± 6.3 kg N uptake ha^{-1} (Figure S1). The obtained model equation with an R^2 = 0.80 and significance of $p < 0.001$ was:

$$N\ uptake\left(\text{kg N ha}^{-1}\right) = 311.NDRE^2 + 29.7 \quad (2)$$

The R^2, Bias, the root mean square error (RMSE), mean absolute percentage error (MAPE), and the concordance correlation coefficient (CCC) were used to evaluate the performance of different models in estimating N uptake at PI. The CCC was calculated using desctools package in R [43]. The RMSE, Bias and MAPE were calculated using metrics package in R [47] with the following equations:

$$RMSE = \sqrt{\frac{1}{n}\sum_{i=1}^{n}(xi - yi)^2} \quad (3)$$

$$Bias = \frac{1}{n}\sum_{i=1}^{n}(xi - yi) \quad (4)$$

$$MAPE = \frac{1}{n}\sum_{i=1}^{n}\frac{(xi - yi)}{xi} \quad (5)$$

where xi and yi are the predicted and measured values, respectively, and n is the number of observations.

3. Results

3.1. Canopy Reflectance and Seasonal Evolution of the Vegetation Indices

Early in the growing season, canopy reflectance in the blue, green, red, and red-edge regions of the spectrum was higher in the aerobic than in the ponded rice in both growing seasons while the opposite was observed for the NIR region in the second rice season (Figure 5). The differences in canopy reflectance in the visible region were more evident in the first rice growing season when crop development in the aerobic rice was lacking behind the ponded rice than in the second growing season. No differences were observed in the visible region between water management strategies later in the growing season except for flowering in 2023/24 and maturity in the 2022/23 growing season, when a slightly lower canopy reflectance was observed in the green and red edge regions in the aerobic rice than in the ponded rice (Figure 5). Canopy reflectance in the NIR was generally lower in the aerobic rice than in the ponded rice at maturity in the first rice season and at tillering and PI in the second rice season. However, these differences were minimized around flowering in both growing seasons (2 March 2023 and 15 February 2024), and maturity in the second growing season.

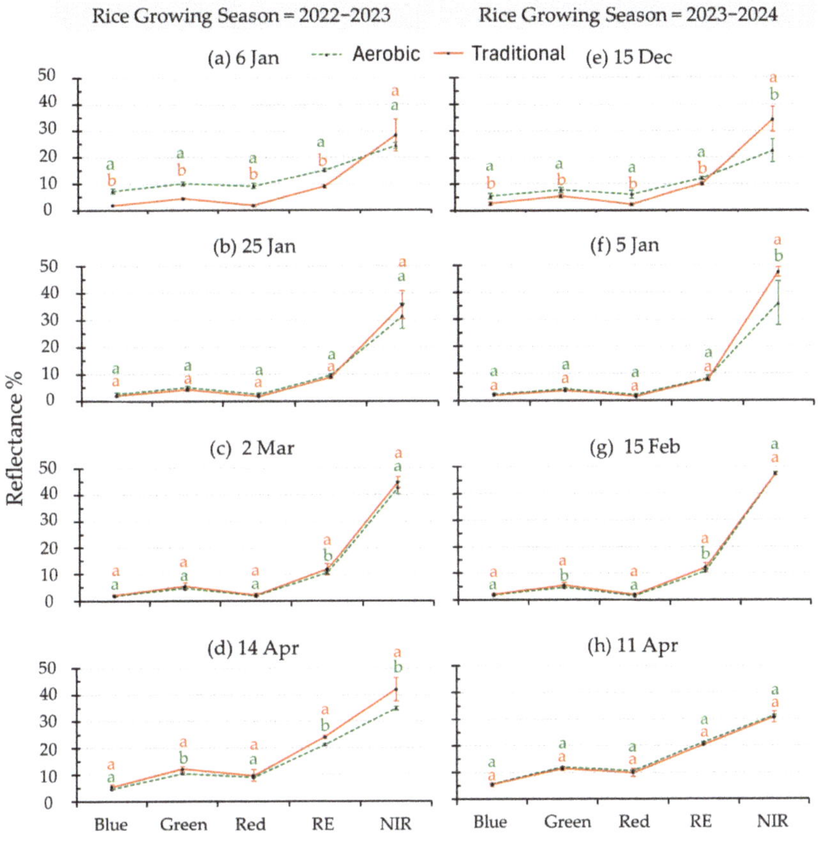

Figure 5. Canopy reflectance in the blue, green, red, red edge (RE) and near-infrared (NIR) regions measured on (**a**) 6 January 2023 (tillering stage in the aerobic rice and PI in ponded rice), (**b**) 25 January 2023 (PI stage in the aerobic rice and microsporogenesis in ponded rice), (**c**) 2 March 2023 (flowering in both water management strategies), (**d**) 14 April 2023 (maturity), (**e**) 15 December 2023 (tillering stage in both water management strategies), (**f**) 5 January 2024 (PI), (**g**) 15 February 2024 (flowering), and, (**h**) 11 April 2024 (maturity). Vertical bars indicate the standard deviation. Different letters for each reflectance region indicate statistically significant differences within irrigation treatments ($p < 0.01$).

The evolution of the vegetation indices SCCCI^2, NDRE^2, Clg and NDVI reported of interest for rice crop monitoring are illustrated for the aerobically and traditionally managed bays in Figure 6. Mean values of all the vegetation indices were lower in the aerobic rice than in the ponded rice on measurement dates occurring before flowering. This was observed in both rice growing seasons despite differences in the moment when PI occurred in each water management strategy were only detected in the first rice growing season. However, the differences were not statistically significant for the NDVI at PI and for the SCCCI^2 at PI in the 2022/23 growing season. The observed differences between water management strategies decreased for all the vegetation indices from flowering onwards. The aerobic rice showed slightly higher mean values than the ponded rice at flowering and showed practically no differences with the ponded rice at maturity (Figure 6).

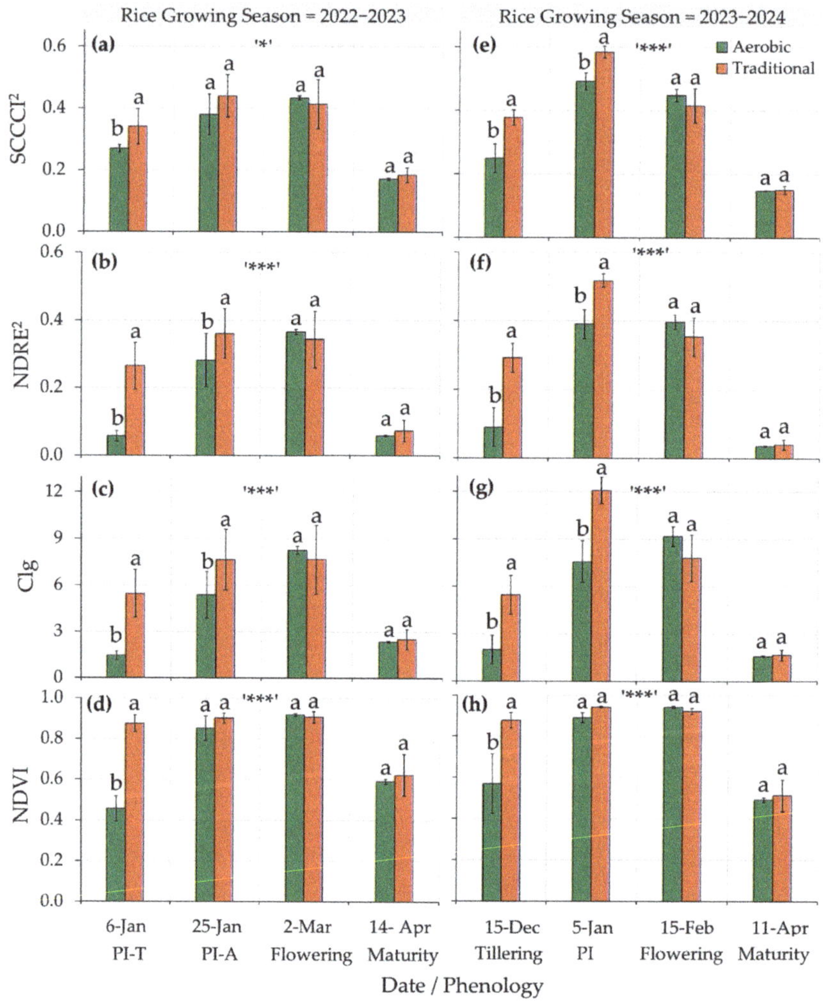

Figure 6. Mean values of the (**a**,**e**) squared simplified canopy chlorophyll content index ($SCCCI^2$), (**b**,**f**) squared normalized difference red edge index ($NDRE^2$), (**c**,**g**) chlorophyll green index (Clg), and (**d**,**h**) normalized difference vegetation index (NDVI) for each measurement date. Vertical bars indicate the standard deviation. The phenological stages at which the aerobic and ponded rice were at each measurement date is also indicated (PI-T = ponded rice at panicle initiation; PI-A = aerobic rice at panicle initiation). Different letters within irrigation treatments for each date indicate statistically significant differences for the interaction between irrigation and date effect by rice season ($p < 0.001$ '***', $p < 0.05$ '*').

3.2. *Biomass, N% and N Uptake at PI*

Average above-ground dry biomass and N uptake in 2023 were 3311 kg ha^{-1} and 100 Kg N ha^{-1}, respectively, while in 2024 above-ground dry biomass was 5733 kg ha^{-1} and N uptake was 155 Kg N ha^{-1} (Table 4).

The irrigation effect was not statistically significant for any of the parameters evaluated ($p > 0.1$). However, there was a statistically significant interaction between the irrigation strategy and growing season ($p < 0.01$) (Table 4 and Figure 7). In the first growing season, the ponded rice reached PI before (6 January 2023) the aerobic rice (~25 January 2023). The traditionally and aerobically managed bays had similar above-ground biomass, N%, and N uptake at PI with no significant differences observed between them (Figure 7). In the second growing season, no noticeable differences were observed in the date PI occurred between water management strategies (~5 January 2024). In this growing season, plants from the traditionally irrigated bay had significantly higher biomass, N% and N uptake than plants from the aerobically managed bay (Figure 7).

Table 4. Plant biomass (DM kg ha^{-1}), Nitrogen percent (N%) and Nitrogen uptake (Kg N ha^{-1}) parameters for the water strategy and rice growing season. The interaction irrigation x season and the mean value for each parameter is also indicated.

Treatments	Biomass (DM kg ha^{-1})	Nitrogen Percent (N%)	N Uptake (Kg N ha^{-1})
Irrigation			
Aerobic	3686	3.03	112
Traditional	5358	2.79	144
Irrigation effect	ns	ns	ns
Season			
2022–2023	3311 ^{b}	3.01	100 ^{b}
2023–2024	5733 ^{a}	2.81	155 ^{a}
Season effect	.	ns	*
Irrigation × Season effect	***	**	**
Mean	4522	2.91	128

Means followed by different letters are significantly different. Asterisks indicate statistical significance at $p < 0.001$ '***', $p < 0.01$ '**', $p < 0.05$ '*', $p < 0.1$ '.', 'ns' indicates non-significant differences.

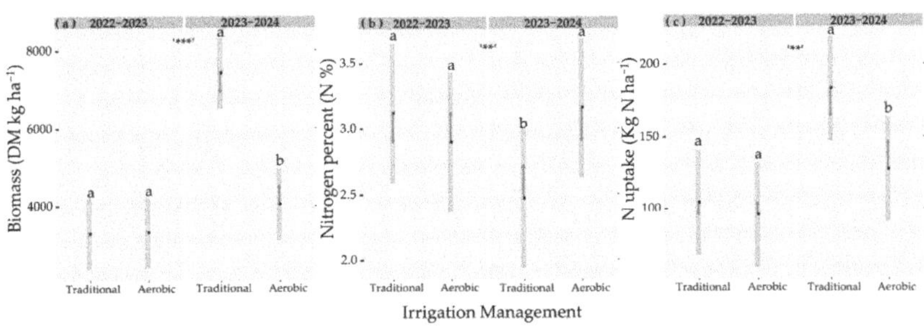

Figure 7. Above-ground plant biomass (**a**), N% (**b**), and N uptake (**c**), at panicle initiation grouped by irrigation management strategy and rice growing season. Black dots represent the mean values, blue bars indicate the standard error, and red arrows indicate the confidence interval. Different letters within irrigation treatments for each rice growing season indicate statistically significant differences for the interaction between irrigation and rice season effect, with the following probability: $p < 0.001$ '***', $p < 0.01$ '**'.

3.3. Nitrogen Uptake Estimates

A strong linear relationship was observed between N uptake and SCCCI^2 when data for both sites and rice growing seasons were plotted together (Figure 8). The equation and the statistically significant slope parameter obtained was very similar (1.9 kg N uptake ha^{-1} higher per 0.1 increase in SCCCI^2) to the one reported in Carracelas et al. [24] that was developed in a different site, variety and growing season and that we aimed to validate. The observed versus predicted plot using the model developed in Carracelas et al. [24] (N uptake (kg ha^{-1}) = 273.SCCCI^2, $R^2 = 0.73$, $p < 0.001$) is illustrated in Figure 9g, along with published single-vegetation-index linear models for N uptake estimates using the chlorophyll-sensitive indices Clg, Clre and NDRE^2.

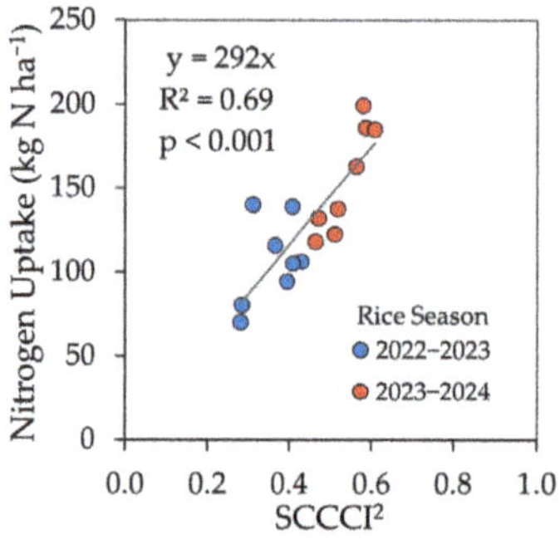

Figure 8. Relationship obtained in the study between N uptake and $SCCCI^2$. Model parameters are shown only when statistically significantly different.

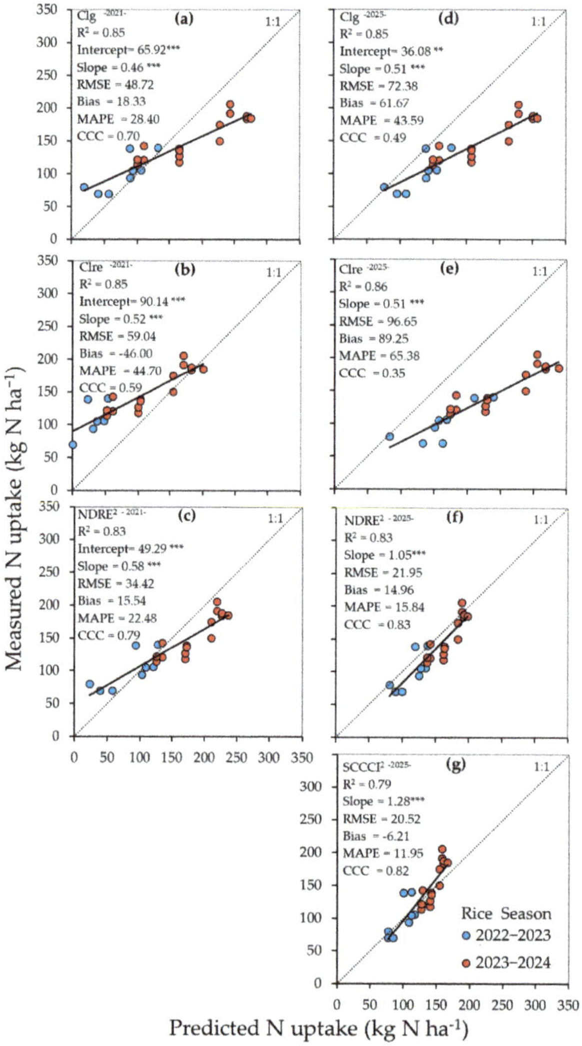

Figure 9. Measured versus predicted nitrogen (N) uptake at PI using the published linear models (**a**) y = −91.66.Clg + 33.95, R^2 = 0.88, (**b**) y = −104.64.Clre + 54.86, R^2 = 0.85 and (**c**) y = 567.47.$NDRE^2$ + 70.56,

R^2 = 0.83 reported in Brinkhoff et al. [48], and (**d**) y = 23.79.Clg, R^2 = 0.70, (**e**) y = 60.81.Clre, R^2 = 0.73 (**f**), y = 311.NDRE^2 + 29.7, R^2 = 0.80, and (**g**), y = 273.SCCCI^2, R^2 = 0.73 reported in Carracelas et al. [24]. Dot line = 1:1. Parameters were statistically significant at $p < 0.01$ '**' and $p < 0.001$ '***'. RMSE = root mean square error. R^2 = coefficient of determination. MAPE = mean absolute percentage error. CCC = concordance correlation coefficient. Model parameters are shown only when they are statistically significantly different.

Among the models assessed, the model using the SCCCI^2 had the lowest RMSE (20.5 kg N ha^{-1}), Bias (−6.21 kg N ha^{-1}), and MAPE (11.95%) (Figure 9g). Both the 2021 and 2025 NDRE^2 models had an R^2 of 0.83. However, compared with the SCCCI^2 model, the RMSE observed for the 2021 NDRE^2 model was 68% higher while only 7% higher for the 2025 NDRE^2 model (Figure 9c,f). In both cases, the Bias and MAPE were higher than those observed for the SCCCI^2 model. The SCCCI^2 and the 2025 NDRE^2 were the models with the highest CCC values, 0.82 and 0.83, respectively. The models using the indices Clg and Clre, had the highest RMSE, Bias and MAPE (Figure 9a,b,d,e).

3.4. Environmental Conditions at Canopy Level and Photosynthesis-Related Parameters

The MultispeQ measurements showed that temperatures at canopy level were similar for the aerobically and ponded managed bays, but relative humidity was significantly higher in the latter (Table 5). No statistically significant difference was observed between water management strategies in leaf temperature differential, which on average for all measurements was around −5.0 °C. No statistically significant difference was observed either between water management strategies in Phi2, PhiN0, PhiNPQ and the SPAD measurements (Table 5).

Table 5. Temperature, relative humidity, leaf to air temperature differential, ratio of the incoming light energy used for the photochemistry (Phi2), ratio of the incoming light energy that is lost to unregulated processes (PhiN0), ratio of the incoming light that is dissipated as heat (PhiNPQ) and SPAD obtained from the MultispeQ measurements.

Treatments	Ambient Parameters		Leaf Temperature Differential (°C)	Fv/Fm	Fraction of Energy Light Captured by PhotosystemII			SPAD
	Temperature	Humidity			Phi2	PhiN0	PhiNPQ	
Irrigation								
Aerobic	31.4	40.2 [b]	−4.98	0.68	0.32	0.3	0.38	51.3
Traditional	31.6	42.8 [a]	−5.30	0.67	0.32	0.29	0.4	52.4
Irrigation effect	ns	**	ns	ns	ns	ns	ns	ns
Rice Season								
2022–2023	32.2 [a]	38.0 [b]	−5.29	0.71 [a]	0.41 [a]	0.3	0.29 [b]	52.7
2023–2024	30.7 [b]	45.0 [a]	−4.98	0.64 [b]	0.22 [b]	0.29	0.49 [a]	51
Season effect	***	***	ns	***	***	ns	***	ns
Irrigation × Season effect	ns	ns	ns	ns	ns	ns	ns	ns
Mean	31.5	41.5	−5.14	0.67	0.32	0.30	0.39	51.9

Means followed by different letters are significantly different. Asterisks indicate statistical significance at $p < 0.001$ '***', $p < 0.01$ '**' and 'ns' indicates non-significant differences.

Significantly higher temperatures and lower relative humidity values were recorded in the first growing season than in the second on measurement dates (Table 5). The Phi2 and PhiNPQ measurements were also significantly higher and lower, respectively, in the first growing season than in the second. No statistically significant differences were observed between rice growing seasons for the other parameters measured.

3.5. Grain Yield and Milling Quality

Rice grain yield was significantly higher (10.5%) in the aerobic bay than in the ponded bay Grain milling quality parameters and the overall quality score were not significantly influenced by the water strategy (Table 6).

Table 6. Rice grain yield (Mg ha^{-1}) and grain milling quality parameters for the water strategy and rice growing season. The interaction irrigation x season and the mean value for each parameter is also indicated.

Treatments	Rice Grain Yield (14%) Mg ha^{-1}	Grain Milling Quality		
		Total Cracked %	Immature %	Quality Score
Irrigation				
Aerobic	15.03 ^{a}	12.9	2.76	8.43
Traditional	13.60 ^{b}	22.8	3.43	7.37
Irrigation effect	**	ns	ns	ns
Season				
S1 2022–2023	14.35	10.6 ^{b}	0.34 ^{b}	8.90
S2 2023–2024	14.29	25.1 ^{a}	5.85 ^{a}	6.91
Season effect	ns	*	***	ns
Irrigation × Season effect	ns	*	ns	**
Mean	14.32	17.85	3.09	7.90

Means followed by different letters are significantly different. Asterisks indicate statistical significance at $p < 0.001$ '***', $p < 0.01$ '**', $p < 0.05$ '*' and 'ns' indicates non-significant differences.

A significant interaction was observed between the irrigation water strategy and rice growing season for the percentage of cracked grain and quality score (Table 6), indicating these parameters were higher and lower, respectively, in the ponded rice than in the aerobic rice in the second rice growing season (Figure 10). The percentage of cracked grains and immature grains were significantly higher in the second rice growing season than in the first one, and therefore, the quality score was lower.

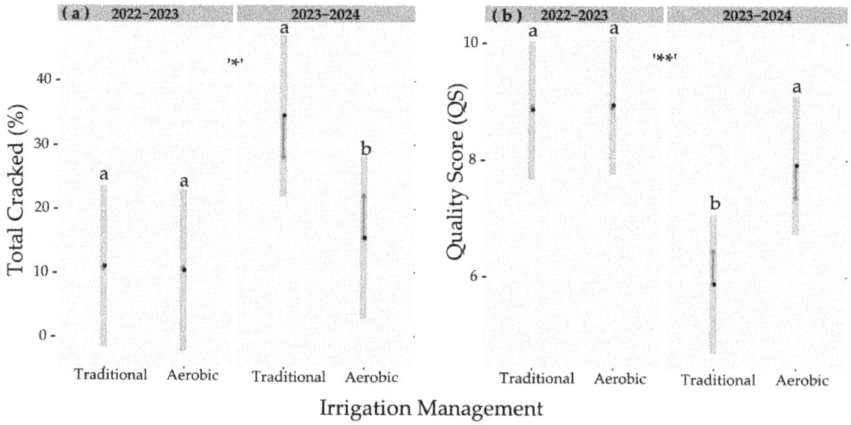

Figure 10. Total cracked grain percentage (**a**), and quality score (**b**), for the water management strategies within each rice growing season. Black dots represent the mean values, blue bars indicate the standard error, and red arrows indicate the confidence interval. Different letters between water strategies within each rice growing season indicate statistically significant differences with the following probability: $p < 0.05$ '*', $p < 0.01$ '**'.

3.6. Relationship Between Chlorophyll Sensitive Vegetation Indices and Grain Quality Parameters

All the assessed vegetation indices had a strong correlation with the percentage of cracked and immature grain, and the quality score (Figure 11).

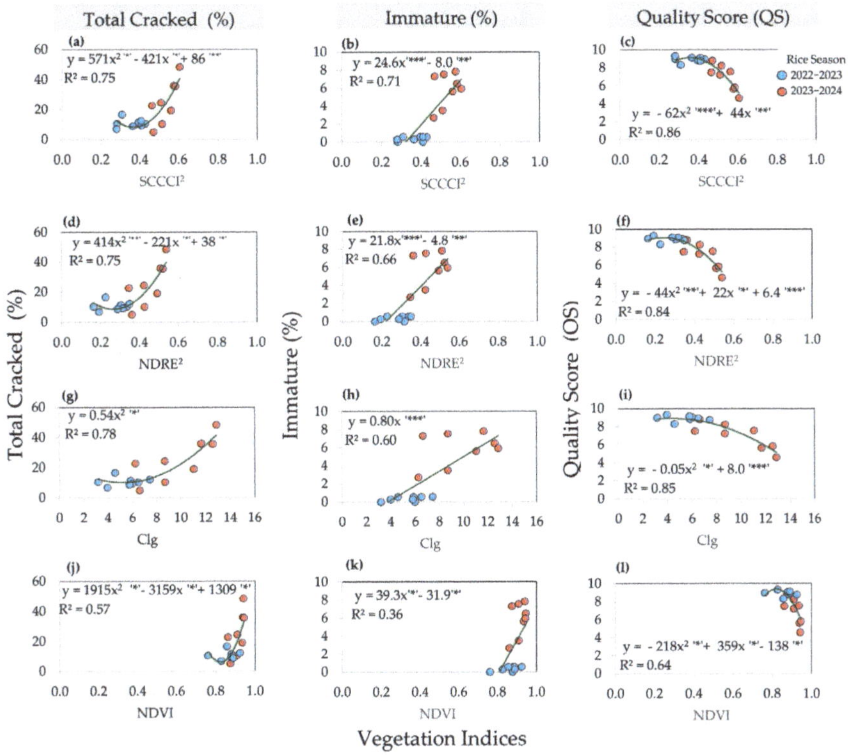

Figure 11. Relationships between percentage of total cracked grain, percentage of immature grain and quality score (QS) and vegetation indices (**a–c**) SCCCI^2 (squared simplified canopy chlorophyll content index), (**d–f**) NDRE^2 (squared normalized red edge vegetation index), (**g–i**) Clg (chlorophyll green index), (**j–l**) NDVI (normalized difference vegetation index), obtained at panicle initiation. Asterisks indicate statistical significance at $p < 0.001$ '***', $p < 0.01$ '**', $p < 0.05$ '*'.

All indices were linearly correlated with the immature percentage. For this grain quality parameter, the SCCCI^2 had the highest R^2 (0.71), followed by the NDRE^2 ($R^2 = 0.66$) and Clg ($R^2 = 0.61$) indices. The lowest R^2 value was observed for the NDVI. The relationship observed for the percentage of cracked grain and overall quality score fitted a polynomial quadratic model. The NDRE^2, SCCCI^2, and Clg indices had a high (>0.75) and similar R^2 among them for both quality parameters. The NDVI was the index with the lowest R^2 for both percentage of cracked grain and quality score.

No significant relationship was observed between rice grain yield and any of the vegetation indices ($p > 0.1$).

4. Discussion

Differences in canopy reflectance in the visible and NIR regions between high-yielding aerobic and ponded rice were more evident early in the growing season when the crop had not reached full canopy cover than later in the season. At tillering, aerobic rice had a higher reflectance in the visible and red edge regions than ponded rice, but the opposite was observed in the NIR region, particularly in the second rice growing season. The differences observed in canopy reflectance early in the growing season were more likely to be related to the advanced development of the above ground part of the plant generally observed in ponded rice with respect to aerobic rice or similar water management practices [5,6,21], thus less background exposed in the image. This could be associated with a promotion of root development in non-ponded strategies during the vegetative phase [49]. Carrijo et al. [50] reported that root length density was significantly higher (100%) under alternate wetting and drying safe in relation to continuously ponded, with no yield loss or plant water stress, as deeper roots accessed deeper water at 25–35 cm. The differences observed in canopy reflectance between aerobic and ponded rice in the visible and NIR regions early in the growing season, and mainly in the NIR region at tillering, PI and maturity, could also be attributed to the aerobic rice being water stressed on days when images were captured [51,52]. However, the MultispeQ measurements showed that leaf to air temperature differential on days when images were collected was similar and negative (leaves cooler than air temperature) in both water management strategies. Further, no statistically significant differences were observed between aerobic and ponded rice in chlorophyll fluorescence (Fv/Fm) and the photosynthesis-related parameters Phi2, PhiN0 and PhiNPQ in any date of measurement, suggesting the aerobic rice was not water stressed on those dates. The fact that grain yield at the monitored locations was higher in the aerobic rice than in the ponded rice at harvest (Table 6), corroborates the inference that aerobic rice did not experience water stress levels detrimental for crop production in any rice growing season. This was achieved by limiting the lowest soil water potential to −14 kPa and −25 kPa in the first and second growing seasons, respectively. These values were close to the soil water potential threshold (−19 kPa) reported by Bo et al. [14] and ≥−20 kpa reported by Carrijo et al. [8] at which rice yield was not reduced in the alternate wetting and drying strategy.

Differences in canopy reflectance between ponded and aerobic rice led to obtaining lower values of the vegetation indices SCCCI^2, NDRE^2, Clg and NDVI in the aerobic rice than in the ponded rice up to flowering (March 2 in the first growing season and February 15 in the second). At this phenological stage, differences in canopy reflectance were minimal between the aerobic and ponded rice, and the vegetation indices were either similar for both water management practices or slightly higher in the aerobic rice. As expected, the differences in NDVI observed between ponded and aerobic rice were smaller than those observed for the chlorophyll-sensitive indices because of the tendency of the NDVI to saturate earlier in the growing season in rice crops not deficient in N [35,53–55].

In agreement with that observed by Carracelas et al. [24] who first reported the use of the SCCCI^2 for N uptake monitoring in experimental plots, the SCCCI^2 was significantly correlated with N uptake at PI in a commercial rice farm. The R^2 of the linear relationship was slightly lower in this study (0.69; $p < 0.001$) than in Carracelas et al. [24] (0.73; $p < 0.001$), but still high despite the fact that differences in N uptake in the present study were caused by the crop management and the inherent variability within the bays as opposed to having N rate treatments. Furthermore, the equation and the statistically significant slope parameter obtained for SCCCI^2 were similar in both studies, even though it was conducted in other rice season with a different variety. Differences in predicted N uptake using either of these models are extremely low and would not affect the N recommended topdressing rates based on the guidelines tables developed for N topdressing for main rice varieties in Australia [56].

4.1. Validation of the SCCCI^2 Model for N Uptake Estimates

When assessing the performance of the model recently reported in a study by Carracelas et al. [24] that relies on SCCCI^2 measurements and that was developed in the same rice growing region of this study, but in a different growing season and using a different variety (Sherpa), there was a good agreement between the measured and estimated N uptake (Figure 9). The RMSE, Bias and MAPE obtained with the SCCCI^2 model (20.52, −6.21 Kg N ha^{-1} and 11.95%) were notably lower than those obtained when using the Clg and Clre models reported by Brinkhoff et al. [48] (Figure 9a,b) and Carracelas et al. [24] (Figure 9d,e). Other single-vegetation-index models reported in the literature using the NDRE and the ratio index NIR/RE [57] and ratio NIR/R [58], provided worse results with RMSE and Bias values ≥113 and ≥112 Kg N ha^{-1}, respectively. Models using the NDRE^2 index [24,48] (Figure 9f,c) showed more similar results to those obtained with the SCCCI^2 model, but especially the model obtained from Carracelas et al. [24] (Figure S1), that had a RMSE of 21.95, Bias of 14.96 Kg N ha^{-1}, MAPE of 15.84% and the highest CCC of all the models, 0.83 The better result observed with the NDRE^2 model reported in Carracelas et al. [24] was more likely because it was developed using images collected at PI at five different times of the day (decreasing the variability observed in NDRE measurements through the day) on similar weather conditions (clear-sky days) and with the same multispectral camera used in this study.

To better confirm the validity of the SCCCI^2 model for N uptake estimates, the results obtained from using this single-vegetation-index model were compared with the results obtained using a stepwise multiple linear regression model (SMLR) developed by Zha et al. [59] from UAV-based multispectral images based on the Green Optimized Soil Adjusted Vegetation Index (GOSAVI), Modified Normalized Difference Red Edge (MNDRE2), and Modified Chlorophyll Absorption in Reflectance Index1 (MCARI1). The N uptake estimates using the SMLR model (Plant N uptake = −198.601 + 353.387 GOSAVI + 132.397 MNDRE2 − 91.552 MCARI1) had an RMSE of 19.68 and a Bias of −4.96 Kg N ha^{-1} (R^2 = 0.77, $p < 0.001$), which were only 0.84 kg N ha^{-1} and 1.25 kg N ha^{-1} lower than those obtained using the SCCCI^2 model. These results validate the single-vegetation-index model reported by Carracelas et al. [24] and confirm the SCCCI^2 as a useful vegetation index for rice N uptake estimations at PI. The best performance of the SCCCI^2 model in estimating N uptake in comparison with the other single-vegetation-index models tested here could be due to the low influence that water depth and time of image collection have on this index with respect to the other chlorophyll-sensitive indices used in this study [24].

Additional validation studies across different locations, rice growing seasons, and varieties are required to further improve the reported SCCCI^2 model for broadly monitoring cultivated areas on commercial farms within the rice industry.

4.2. Remote Sensing Monitoring of Grain Milling Quality

All the vegetation indices explored at PI in this study were significantly correlated with the percentage of cracked grain, percentage of immature grain and quality score when data for both growing seasons were pooled together. Tsukaguchi et al. [30], reported a correlation between the Clg index, immature grains and chalky grain rate, which has been shown to be affected by basal N fertilization [60,61] and negatively associated with N uptake at heading [62]. Zhao et al. [63] also reported that optimal nitrogen application (~270 kg N ha^{-1}) improved rice grain quality, while excessive nitrogen application (345 kg N ha^{-1}) reduced it by lowering the average grain filling rate and prolonging the grain filling period. In the present study, no differences in chalky grain rate were observed between seasons and water management strategies.

The NDVI was the index that had the lowest R^2 for the relationship with all grain quality parameters, which agrees with that observed by Tsukaguchi et al. [30]. This result was most likely due to the early saturation of the NDVI in no N deficient crops in comparison with the other indices (Figures 6 and 11) as also reported in previous studies [35,53,55,64], the role that plant N availability plays in grain quality [25–28] and the fact that the $SCCCI^2$, $NDRE^2$ and Clg indices are all more sensitive than NDVI to leaf chlorophyll content and indirectly to N content. All the chlorophyll-sensitive indices had a similar R^2 when plotted against the percentage of cracked grain (0.75–0.78) and the quality score (0.84–0.86). However, the strongest relationship with the percentage of immature grain was observed for the $SCCCI^2$. The findings of this study suggest that grain milling quality parameters could potentially be associated with N uptake at PI since all $SCCCI^2$, $NDRE^2$ and Clg were correlated with N uptake at this phenological stage. The percentage of grain cracking and quality score increased and decreased, respectively, when the $SCCCI^2$ reached a value between 0.4 and 0.5 (Figure 11). This value of the $SCCCI^2$ corresponded to a N uptake between 150 and 160 Kg N ha^{-1}, which could possibly indicate that N uptake values at PI above this threshold could lead to a higher percentage of grain cracked and grain milling quality. Previous research has indeed shown that high doses of N topdressing may increase the cracking percentage [65] and induce changes in the grain protein composition [25,27]. However, grain quality parameters are influenced not only by plant N status, but also genetic factors, agronomic practices [28], weather conditions before and after heading [28,30,66], harvest time [67], and post-harvest handling practices [28].

The study showed that there is a relationship between spectral indices sensitive to chlorophyll and grain quality parameters, although further research on this topic is needed to better understand how weather conditions and crop management at PI relates to remotely sensed indices.

5. Conclusions

The findings of the present study show that advanced canopy development in ponded rice with respect to high-yielding aerobic rice leads to differences in canopy reflectance (mainly in the NIR region) up to flowering when these differences are minimized. These canopy reflectance variations led to having lower values of the $SCCCI^2$, $NDRE^2$, and Clg indices in the aerobic than in the ponded rice even though yield was not reduced in the non-ponded strategy.

This study validated the $SCCCI^2$ model for N uptake estimates at panicle initiation in a different variety, site, and growing seasons from the ones the model was developed. The better performance of the $SCCCI^2$ model in estimating N uptake in comparison with single-vegetation-index models using other chlorophyll-sensitive vegetation indices and the similar performance to that of a stepwise multiple linear regression model confirm the suitability of the $SCCCI^2$ for N uptake monitoring.

Moreover, this work suggests that grain milling quality parameters could potentially be associated with N uptake at PI since all the chlorophyll-sensitive vegetation indices tested here ($SCCCI^2$, $NDRE^2$, and Clg) were highly and similarly correlated with grain milling quality parameters. Among the quality parameters assessed, the best correlation for all the chlorophyll-sensitive indices was observed with the quality score, with the $SCCCI^2$ providing the highest R^2 (0.85). Further research is required to better understand how the management of the crop at panicle initiation could affect the overall grain milling quality score and how this relates to remotely sensed indices and weather conditions.

Supplementary Materials: The following supporting information can be downloaded at: https://www.mdpi.com/article/10.3390/rs17152598/s1, Figure S1: (a) Lineal-plateau significant relationship between $NDRE^2$ and N uptake at PI using a data set from a previous study conducted in Australia by Carracelas et al. [24], to determine the lineal response and threshold N uptake plateau. (b) Lineal model parameters from the relationship between N uptake and $NDRE^2$ up to the reported N uptake plateau threshold value of 166 ± 6.3 kg N uptake ha^{-1}.

Author Contributions: Conceptualization, G.C., J.H. and C.B.; methodology, G.C., J.H. and C.B.; formal analysis, G.C., J.H. and C.B.; investigation G.C., J.H. and C.B.; writing—original draft preparation, G.C., J.H. and C.B.; writing—review and editing, G.C., J.H. and C.B. All authors have read and agreed to the published version of the manuscript.

Funding: This research received no external funding.

Data Availability Statement: The data presented in this study are available on request from the corresponding author.

Acknowledgments: We acknowledge Deakin University, Australia, and the National Agricultural Research Institute, INIA, Uruguay. The farmers, C. and H. Morsehead, are gratefully acknowledged for their support and for running the experiments on their farm. We acknowledge J. Mann from Yenda Producers and M. Groat from Rice Extension for their support during the rice season. We thank A. Roel from INIA, Uruguay, for reading an earlier version of the manuscript and providing valuable feedback. G. Magalhaes, B. Tondato, and R. Maia from Deakin University are acknowledged for their help with field sampling. We also acknowledged J. Deeves, M. Talbot, and B. O Leary for laboratory analysis performed in AGS, Leeton, NSW, Australia.

Conflicts of Interest: The authors declare no conflicts of interest.

References

1. Reinke, R.; Beecher, G.; Dunn, B.; Sneil, P. Temperate rice in Australia. In *Advances in Temperate Rice Research*; Jena, K.K., Hardy, B., Eds.; International Rice Research Institute (IRRI): Los Baños, Philippines, 2012; pp. 1–14, ISBN 978-971-22-0289-6. Available online: http://books.irri.org/9789712202896_content.pdf (accessed on 19 December 2024).
2. Tuong, T.P.; Bouman, B.A.M. Rice production in water scarce environments. In *Water Productivity in Agriculture: Limits and Opportunities for Improvement*; Kijne, J.W., Barker, R., Molden, D., Eds.; CABI Publishing: Wallingford, UK, 2003; pp. 53–67. [CrossRef]
3. Mekonnen, M.M.; Hoekstra, A.Y. Sustainability: Four billion people facing severe water scarcity. *Sci. Adv.* **2016**, *2*, e1500323. [CrossRef] [PubMed]
4. Bajwa, A.A.; Chauhan, B.S. Rice Production in Australia. In *Rice Production Worldwide*; Chauhan, B., Jabran, K., Mahajan, G., Eds.; Springer: Cham, Switzerland, 2017; pp. 169–184. [CrossRef]
5. Bouman, B.A.M.; Lampayan, R.M.; Tuong, T.P. *Water Management in Irrigated Rice: Coping with Water Scarcity*; International Rice Research Institute (IRRI): Los Baños, Philippines, 2007; 54p, Available online: http://books.irri.org/9789712202193_content.pdf (accessed on 13 July 2024).
6. Bouman, B.A.M.; Humphreys, E.; Tuong, T.P.; Barker, R. Rice and Water. *Adv. Agron.* **2007**, *92*, 187–237. [CrossRef]
7. Dunn, B.W.; Gaydon, D.S. Rice growth yield and water productivity responses to irrigation scheduling prior to the delayed application of continuous flooding in southeast Australia. *Agric. Water Manag.* **2011**, *98*, 1799–1807. [CrossRef]
8. Carrijo, D.R.; Lundy, M.E.; Linquist, B.A. Rice yields and water use under alternate wetting and drying irrigation: A meta-analysis. *Field Crops Res.* **2017**, *203*, 173–180. [CrossRef]
9. Carracelas, G.; Hornbuckle, J.; Rosas, J.; Roel, A. Irrigation management strategies to increase water productivity in *Oryza sativa* (rice) in Uruguay. *Agric. Water Manag.* **2019**, *222*, 161–172. [CrossRef]
10. Bouman, B.A.M.; Yang, X.; Wang, H.; Wang, Z.; Zhao, J.; Chen, B. Performance of aerobic rice varieties under irrigated conditions in North China. *Field Crops Res.* **2006**, *97*, 53–65. [CrossRef]
11. Kato, Y.; Okami, M.; Katsura, K. Yield potential and water use efficiency of aerobic rice (*Oryza sativa* L.) in Japan. *Field Crops Res.* **2009**, *113*, 328–334. [CrossRef]
12. Froes de Borja Reis, A.; Estevam Munhoz de Almeida, R.; Cocco Lago, B.; Trivelin, P.C.; Linquist, B.; Favarin, J.L. Aerobic rice system improves water productivity, nitrogen recovery and crop performance in Brazilian weathered lowland soil. *Field Crops Res.* **2018**, *218*, 59–68. [CrossRef]
13. Champness, M.; Ballester, C.; Hornbuckle, J. Effect of Soil Moisture Deficit on Aerobic Rice in Temperate Australia. *Agronomy* **2023**, *13*, 168. [CrossRef]

14. Bo, Y.; Wang, X.; van Groenigen, K.J.; Linquist, B.A.; Müller, C.; Li, T.; Yang, J.; Jägermeyr, J.; Qin, Y.; Zhou, F. Improved alternate wetting and drying irrigation increases global water productivity. *Nat. Food* **2024**, *5*, 1005–1013. [CrossRef] [PubMed]
15. Groat, M. *The Future of Australian Rice Production. A Focus on Water Use Efficiency in the Australian Temperate Rice System. A Report for Nuffield Australia Farming Scholars*; Nuffield Australia: Sydney, NSW, Australia, 2020; pp. 1–44. Available online: https://www.mdba.gov.au/basin-plan (accessed on 5 May 2025).
16. Williams, R.L.; Angus, J.F. Deep floodwater protects high-nitrogen rice crops from low-temperature damage. *Aust. J. Exp. Agric.* **1994**, *34*, 927–932. [CrossRef]
17. Humphreys, E.; Lewin, L.G.; Khan, S.; Beecher, H.G.; Lacy, J.M.; Thompson, J.A.; Batten, G.D.; Brown, A.; Russell, C.A.; Christen, E.W.; et al. Integration of approaches to increasing water use efficiency in rice-based systems in southeast Australia. *Field Crops Res.* **2006**, *97*, 19–33. [CrossRef]
18. Masseroni, D.; Moller, P.; Tyrell, R.; Romani, M.; Lasagna, A.; Sali, G.; Facchi, A.; Gandolfi, C. Evaluating performances of the first automatic system for paddy irrigation in Europe. *Agric. Water Manag.* **2018**, *201*, 58–69. [CrossRef]
19. Champness, M.; Ballester-Lurbe, C.; Filev-Maia, R.; Hornbuckle, J. Smart sensing and automated irrigation for sustainable rice systems: A state of the art review. *Adv. Agron.* **2023**, *177*, 259–285. [CrossRef]
20. Zoffoli, G.; Gangi, F.; Ferretti, G.; Masseroni, D. The potential of a coordinated system of gates for flood irrigation management in paddy rice farm. *Agric. Water Manag.* **2023**, *289*, 108536. [CrossRef]
21. Ouyang, W.; Yin, X.; Yang, J.; Struik, P.C. Roles of canopy architecture and nitrogen distribution in the better performance of an aerobic than a lowland rice cultivar under water deficit. *Field Crops Res.* **2021**, *271*, 108257. [CrossRef]
22. Sun, T.; Fang, H.; Liu, W.; Ye, Y. Impact of water background on canopy reflectance anisotropy of a paddy rice field from multi-angle measurements. *Agric. For. Meteorol.* **2017**, *233*, 143–152. [CrossRef]
23. Carracelas, G.; Ballester, C.; Marchesi, C.; Roel, A.; Hornbuckle, J. Assessing Drone-Based Remote Sensing Indices for Monitoring Rice Nitrogen Plant Status Under Different Irrigation Techniques. *Agronomy* **2024**, *14*, 2976. [CrossRef]
24. Carracelas, G.; Hornbuckle, J.; Ballester, C. Impact of Standing Water Level and Observation Time on Remote-Sensed Canopy Indices for Rice Nitrogen Status Monitoring. *Remote Sens.* **2025**, *17*, 1045. [CrossRef]
25. Wood, R.M. The Impact of Crop Management Practices on Rice Grain Quality in South Eastern Australia. Doctoral Thesis, Charles Sturt University, Bathurst, NSW, Australia, 2019; pp. 1–186.
26. Wood, R.M.; Balindong, J.; Waters, D.L.E.; Mawson, A.J.; Blanchard, C.L.; Dunn, B.W.; Oli, P. Effect of agronomic management on rice grain quality Part I: A review of Australian practices. *Cereal Chem.* **2021**, *98*, 222–233. [CrossRef]
27. Wood, R.M.; Dunn, B.W.; Balindong, J.; Waters, D.L.E.; Blanchard, C.L.; Mawson, A.J.; Oli, P. Effect of agronomic management on rice grain quality Part II: Nitrogen rate and timing. *Cereal Chem.* **2021**, *98*, 234–248. [CrossRef]
28. Oli, P.; Talbot, M.; Snell, P. Understanding Pre- and Post-Milling Crack Formation in Rice Grain. *Trans. ASABE* **2021**, *64*, 1795–1804. [CrossRef]
29. Tsukaguchi, T.; Taniguchi, Y.; Ito, R. The effects of nitrogen uptake before and after heading on grain protein content and the occurrence of basal- and back-white grains in rice (*Oryza sativa* L.). *Plant Prod. Sci.* **2016**, *19*, 508–517. [CrossRef]
30. Tsukaguchi, T.; Miyamae, T.; Morikawa, A.; Yonezawa, R.; Sekine, D.; Fujihara, Y. Estimation of grain quality of rice (*Oryza sativa* L.) by UAV-acquired vegetation index and climate factors. *Plant Prod. Sci.* **2023**, *26*, 297–308. [CrossRef]
31. Hou, Y.; Bao, H.; Rimi, T.I.; Zhang, S.; Han, B.; Wang, Y.; Yu, Z.; Chen, J.; Gao, H.; Zhao, Z.; et al. Rice Quality and Yield Prediction Based on Multi-Source Indicators at Different Periods. *Plants* **2025**, *14*, 424. [CrossRef] [PubMed]
32. Hornbuckle, J.; Christen, E. Physical Properties of Soils in the Murrumbidgee and Coleambally Irrigation Areas. CSIRO Land and Water Technical Report 17/99. 1999. Available online: https://www.researchgate.net/publication/267717578_Physical_Properties_of_Soils_in_Murrumbidgee_and_Coleambally_Irrigation_Areas (accessed on 20 April 2025).
33. Albrecht, W.A. The Foundation of Soil Fertility. In *The Albrecht Papers, Volume 1: Foundation Concepts; Soil Science Society of America*; Acres U.S.A.: Viroqua, WI, USA, 1975.
34. Reams, C.A. *Biological Ionization and Soil Fertility: The Reams Biological Theory of Ionization*; Reams Biological Institute: Winter Haven, FL, USA, 1980.
35. Brinkhoff, J.; Dunn, B.W.; Robson, A.J.; Dunn, T.S.; Dehaan, R.L. Modeling mid-season rice nitrogen uptake using multispectral satellite data. *Remote Sens.* **2019**, *11*, 1837. [CrossRef]
36. Gitelson, A.A.; Gritz, Y.; Merzlyak, M.N. Relationships between leaf chlorophyll content and spectral reflectance and al-gorithms for non-destructive chlorophyll assessment in higher plant leaves. *J. Plant Physiol.* **2003**, *160*, 271–282. [CrossRef] [PubMed]
37. Rouse, J.W.; Haas, R.H.; Schell, J.A.; Deering, D.W. Monitoring Vegetation Systems in the Great Plains with ERTS. In Proceedings of the Third Earth Resources Technology Satellite-1 Symposium, Washington, DC, USA, 10–14 December 1973; pp. 309–317.
38. Dunn, T.; Dunn, B. Identifying Panicle Initiation in Rice; Primefact 1278, Fourth Edition. June 2018. Available online: https://www.dpi.nsw.gov.au/__data/assets/pdf_file/0003/449823/Identifying-panicle-initiation-in-rice.pdf (accessed on 6 December 2024).

39. Dunn, B. Nitrogen Tissue Test for Rice Panicle Initiation. 2012. Available online: https://www.agrifutures.com.au/wp-content/uploads/publications/12-047.pdf (accessed on 25 January 2025).
40. FOSS Analytics. Infratec 1241 Grain Analyzer. Hilleroed, Denmark. Available online: https://www.fossanalytics.com/en/search?query=Infratec%201241%20Grain%20Analyzer (accessed on 5 April 2025).
41. PaddyVision. Rapid Analysis of Paddy Rice Quality. Available online: https://paddyvision.com/ (accessed on 1 March 2025).
42. Talbot, M. Paddyvision®: Rapid Analysis of Paddy Rice Quality at the Point of Receival. IREC Farmers' Newsletter No. 210—Spring 2023. Available online: https://irec.org.au/wp-content/uploads/IREC-quicktake-Spring-210-1.pdf (accessed on 1 March 2025).
43. R Core Team. *R: A Language and Environment for Statistical Computing*; R Foundation for Statistical Computing: Vienna, Austria, 2022. Available online: http://www.R-project.org/ (accessed on 1 January 2025).
44. Chambers, J.M. Linear Models. In *Statistical Models*; Chambers, J.M., Hastie, T.J., Eds.; Routledge: New York, NY, USA, 1992.
45. Bates, D.M.; Watts, D.G. *Nonlinear Regression Analysis and Its Applications*; John Wiley & Sons, Inc.: Hoboken, NJ, USA, 1988. [CrossRef]
46. Bates, D.M.; Chambers, J.M. Nonlinear Models. In *Statistical Models*; Chambers, J.M., Hastie, T.J., Eds.; Wadsworth & Brooks/Cole: Pacific Grove, CA, USA, 1992; ISBN 9780203738535.
47. Hamner, B.; Frasco, M.; Ledell, E. Title Evaluation Metrics for Machine Learning. Available online: https://github.com/mfrasco/Metrics (accessed on 10 May 2025).
48. Brinkhoff, J.; Dunn, B.W.; Robson, A.J. Rice nitrogen status detection using commercial-scale imagery. *Int. J. Appl. Earth Obs. Geoinf.* **2021**, *105*, 102627. [CrossRef]
49. Yang, C.; Yang, L.; Yang, Y.; Ouyang, Z. Rice root growth and nutrient uptake as influenced by organic manure in continuously and alternately flooded paddy soils. *Agric. Water Manag.* **2004**, *70*, 67–81. [CrossRef]
50. Carrijo, D.R.; Akbar, N.; Reis, A.F.B.; Li, C.; Gaudin, A.C.M.; Parikh, S.J.; Linquist, B.A. Impacts of variable soil drying in alternate wetting and drying rice systems on yields, grain arsenic concentration and soil moisture dynamics. *Field Crops Res.* **2018**, *222*, 101–110. [CrossRef]
51. Panigrahi, N.; Das, B.S. Canopy Spectral Reflectance as a Predictor of Soil Water Potential in Rice. *Water Resour. Res.* **2018**, *54*, 2544–2560. [CrossRef]
52. Visitacion, G.J.; Saludes, R.B.; Luyun, R.A.; Mochica Pinca, Y.M.; Ferdie Eusebio Mv Jose, S.; Sur, C.; Journal, P. Statistical Analysis of Crop Water Stress in Rainfed Rice (Oryza sativa L.) Using Spectral and Non-spectral Indices. *Philipp. J. Sci.* **2022**, *151*, 587–603. [CrossRef]
53. Dunn, B.W.; Dunn, T.S.; Orchard, B.A. Nitrogen rate and timing effects on growth and yield of drill-sown rice. *Crop Pasture Sci.* **2016**, *67*, 1149–1157. [CrossRef]
54. Rehman, T.H.; Borja Reis, A.F.; Akbar, N.; Linquist, B.A. Use of normalized difference vegetation index to assess N status and predict grain yield in rice. *Agron. J.* **2019**, *111*, 2889–2898. [CrossRef]
55. Rehman, T.H.; Lundy, M.E.; Linquist, B.A. Comparative Sensitivity of Vegetation Indices Measured via Proximal and Aerial Sensors for Assessing N Status and Predicting Grain Yield in Rice Cropping Systems. *Remote Sens.* **2022**, *14*, 2770. [CrossRef]
56. Dunn, B.; Dunn, T. *Agronomy and Remote Sensing to Maximise Rice Water Productivity*; Agrifutures: Wagga Wagga, NSW, Australia, 2024; pp. 24–234. Available online: https://agrifutures.com.au/wp-content/uploads/2024/12/24-234-agronomy-and-remote-sensing.pdf (accessed on 10 May 2025).
57. Inoue, Y.; Sakaiya, E.; Zhu, Y.; Takahashi, W. Diagnostic mapping of canopy nitrogen content in rice based on hyperspectral measurements. *Remote Sens. Environ.* **2012**, *126*, 210–221. [CrossRef]
58. Huang, S.; Miao, Y.; Zhao, G.; Yuan, F.; Ma, X.; Tan, C.; Yu, W.; Gnyp, M.L.; Lenz-Wiedemann, V.I.S.; Rascher, U.; et al. Satellite Remote Sensing-Based In-Season Diagnosis of Rice Nitrogen Status in Northeast China. *Remote Sens.* **2015**, *7*, 10646–10667. [CrossRef]
59. Zha, H.; Miao, Y.; Wang, T.; Li, Y.; Zhang, J.; Sun, W.; Feng, Z.; Kusnierek, K. Improving unmanned aerial vehicle remote sensing-based rice nitrogen nutrition index prediction with machine learning. *Remote Sens.* **2020**, *12*, 215. [CrossRef]
60. Wakamatsu, K.; Sasaki, O.; Uezono, I.; Tanaka, A. Effect of the amount of nitrogen application on occurrence of whiteback kernels during ripening of rice [*Oryza sativa*] under high-temperature conditions. *Jpn. J. Crop Sci.* **2008**, *77*, 424–433. [CrossRef]
61. Takata, S.; Sakata, M.; Kameshima, M.; Yamamoto, Y.; Miyazaki, A. Varietal difference in the relation between the occurrence of white immature kernels caused by a high temperature during the ripening period and the amount of basal nitrogen application in rice. *Jpn. J. Crop Sci.* **2010**, *79*, 150–157. [CrossRef]
62. Tanaka, K.; Miyazaki, M.; Uchikawa, O.; Araki, M. Effects of the nitrogen nutrient condition and nitrogen application on kernel quality of rice. *Jpn. J. Crop Sci.* **2010**, *79*, 450–459. [CrossRef]
63. Zhao, C.; Liu, G.; Chen, Y.; Jiang, Y.; Dai, Q.; Huo, Z.; Shi, Y.; Zhao, L.; Liao, P.; Wang, W.; et al. Excessive Nitrogen Application Leads to Lower Rice Yield and Grain Quality by Inhibiting the Grain Filling of Inferior Grains. *Agriculture* **2022**, *12*, 962. [CrossRef]

64. Tsukaguchi, T.; Kobayashi, H.; Fujihara, Y.; Chono, S. Estimation of spikelet number per area by UAV-acquired vegetation index in rice (*Oryza sativa* L.). *Plant Prod. Sci.* **2022**, *25*, 20–29. [CrossRef]
65. Lang, Y.-Z.; Yang, X.-D.; Wang, M.-E.; Zhu, Q.-S. Effects of lodging at different filling stages on rice yield and grain quality. *Rice Sci.* **2012**, *19*, 315–319. [CrossRef]
66. Thompson, J.F.; Mutters, R.G. Effect of Weather and Rice Moisture at Harvest on Milling Quality of California Medium-Grain Rice. *Trans. ASABE* **2006**, *49*, 435–440. [CrossRef]
67. Takita, T. Evaluation method for varietal differences of crack formation in rice grains. *Tohoku Agric. Res.* **1999**, *52*, 15–16.

remote sensing

MDPI

Article

Assessing Model Trade-Offs in Agricultural Remote Sensing: A Review of Machine Learning and Deep Learning Approaches Using Almond Crop Mapping

Mashoukur Rahaman *, Jane Southworth, Yixin Wen and David Keellings

Department of Geography, University of Florida, 3141 Turlington Hall, P.O. Box 117315, Gainesville, FL 32611-7315, USA; jsouthwo@ufl.edu (J.S.); yixin.wen@ufl.edu (Y.W.); djkeellings@ufl.edu (D.K.)

* Correspondence: m.rahaman@ufl.edu

Abstract

This study presents a comprehensive review and comparative analysis of traditional machine learning (ML) and deep learning (DL) models for land cover classification in agricultural remote sensing. We evaluate the reported successes, trade-offs, and performance metrics of ML and DL models across diverse agricultural contexts. Building on this foundation, we apply both model types to the specific case of almond crop field identification in California's Central Valley using Landsat data. DL models, including U-Net, MANet, and DeepLabv3+, achieve high accuracy rates of 97.3% to 97.5%, yet our findings demonstrate that conventional ML models—such as Decision Tree, K-Nearest Neighbor, and Random Forest—can reach comparable accuracies of 96.6% to 96.8%. Importantly, the ML models were developed using data from a single year, while DL models required extensive training data spanning 2008 to 2022. Our results highlight that traditional ML models offer robust classification performance with substantially lower computational demands, making them especially valuable in resource-constrained settings. This paper underscores the need for a balanced approach in model selection—one that weighs accuracy alongside efficiency. The findings contribute actionable insights for agricultural land cover mapping and inform ongoing model development in the geospatial sciences.

Keywords: agriculture; land cover classification; segmentation; California central valley; computational efficiency; temporal analysis

1. Introduction

The agricultural sector has been a cornerstone of human civilization, providing essential resources such as food, raw materials, and employment [1]. It is a vital economic engine that sustains both local communities and broader economies, especially in regions where agriculture is the predominant activity [2]. However, the role of agriculture is increasingly under pressure due to global challenges such as climate change, water scarcity, and the need for sustainable resource management [3,4]). These pressures highlight the importance of optimizing agricultural practices and ensuring efficient resource allocation to secure food production in the face of growing uncertainties [5].

In this context, almond production in California's Central Valley stands out as a critical component of the state's agricultural landscape. California produces over 80% of the world's almonds, with the Central Valley playing a pivotal role due to its favorable climate and extensive irrigation infrastructure [6]). This region is responsible for more than half of the United States' fruits, vegetables, and nuts, making it one of the most productive agricultural areas globally [7,8]. Almonds, in particular, represent a significant share of the state's agricultural output, contributing over 10% to the overall agricultural income [9]. Given this, accurately monitoring and managing almond crop areas is essential for sustaining production levels, ensuring resource efficiency, and addressing the growing challenges posed by climate variability and extreme weather events.

Recent expansion of almond and other perennial crop acreage in California's Central Valley has attracted significant attention for its intensive reliance on groundwater, especially during extended droughts [10]. Remote sensing studies have indicated a shift towards nut crops such as almonds between 2007 and 2016, coinciding with increased groundwater pumping [11]. Independent research has also linked this agricultural intensification to substantial declines in groundwater levels and related environmental concerns [12]. Persistent groundwater overdraft has led to land subsidence, loss of aquifer storage, and deteriorating water quality, which carry significant implications for the sustainability of almond cultivation in this region [13,14]. Against this backdrop, improved spatial monitoring of almond acreage via remote sensing provides essential tools for informing sustainable water management and policy responses in the Central Valley.

Remote sensing technology has emerged as a powerful tool in this endeavor, offering the ability to monitor crops over large areas with high temporal and spatial resolution [15]. The use of satellite-based remote sensing, in particular, has revolutionized agricultural monitoring by providing continuous data that can be used to assess crop health, estimate yields, and detect changes in land cover [16–18]. Among these technologies, Landsat data has proven to be especially valuable due to its extensive temporal coverage and ability to capture detailed spectral information [19,20]. With this rich dataset, the classification of crops such as almonds has become increasingly feasible, enabling more accurate predictions of crop extent and health [21,22]. However, the challenge remains in selecting the most effective models for processing these vast datasets.

Machine learning (ML) and deep learning (DL) techniques have shown great promise in remote sensing applications, including land cover classification [23]. ML algorithms, such as Random Forest (RF) and Support Vector Machines (SVMs), have been widely used for crop classification due to their ability to handle large datasets and produce accurate results with relatively low computational costs [24–26]. These models rely on labeled training data and use statistical methods to identify patterns within the data, making them well-suited for applications where data availability is constrained [27,28]). In contrast, DL models, which utilize complex neural networks with multiple layers, have demonstrated exceptional performance in handling high-dimensional datasets, such as multi-spectral imagery from satellite sensors [29]. DL techniques like Convolutional Neural Networks (CNNs) and more advanced architectures such as U-Net and DeepLabv3+ can automatically extract features from raw data, often yielding higher classification accuracies than traditional ML models [30].

While DL models offer superior accuracy, they come with significantly higher computational demands and require large amounts of labeled training data, often spanning multiple years, to perform optimally [31,32]. This difference in resource requirements highlights a crucial trade-off between accuracy and efficiency. In resource-limited scenarios, where computational power, time, and data availability may be constrained, conventional ML models present a compelling alternative to more complex DL methods [33,34]. This study aims to explore these trade-offs in the context of almond crop classification in California's Central Valley. Given the critical importance of almond production in California's agricultural landscape, this study aims to leverage the power of remote sensing technology alongside both traditional ML and DL models to enhance the identification and monitoring of almond crop locations. The primary objective is not only to achieve high classification accuracy but also to evaluate the trade-offs between computational efficiency and precision. By comparing the performance of predictive models developed from Landsat data, this research seeks to offer practical insights into optimizing resource allocation, improving sustainability practices, and informing policy decisions related to agricultural management.

Building upon previous research, this study addresses two central research questions: (1) How accurately can remote sensing technologies, when paired with conventional ML and DL models, classify almond crop locations in California's Central Valley? (2) What are the comparative benefits of using ML models versus DL models, particularly in resource-constrained environments where computational efficiency is critical? In alignment with these questions, the research posits the following hypotheses: (1) Remote sensing technologies, combined with ML and DL methods, will significantly improve the accuracy of almond crop classification. (2) While DL models may provide slightly higher predictive accuracy due to their complex architectures, conventional ML models will prove to be more efficient in terms of computational costs and time, offering a competitive alternative in scenarios where resources are limited. The novelty of this study lies in its comprehensive evaluation of both ML and DL models for almond crop classification, with a specific focus on balancing accuracy with efficiency. By incorporating historical data and employing advanced DL architectures, alongside more resource-friendly ML models, this research provides a rigorous performance comparison that has direct implications for agricultural sustainability.

2. Methods

2.1. Study Area

The Central Valley of California is situated between the Sierra Nevada to the east and the littoral mountain range to the west (Figure 1). It contains approximately 6.5 million inhabitants [35]. In the broad, alluvial-filled structural trough, more than 250 distinct crops are grown with an estimated annual value of more than USD 20 billion [36]. The Central Valley experiences a Mediterranean climate, characterized by most of its rainfall occurring from November to March [37]. The climate experiences very little precipitation during late winter, summer, and early fall, which results in an irrigated area of 52,000 km^2, being among the largest irrigated regions globally [38,39]. The Central Valley consists of three distinct regions: The San Joaquin Valley in the central area, the Sacramento Valley in the northern part, and the semi-arid Tulare Basin in the southernmost section. The region features a coastal range with significant coastal urban areas to the west, Shasta National Forest in the north, Sierra Nevada Mountains to the east, and Mojave Desert to the southeast, all of which define the Central Valley [37] (Figure 1).

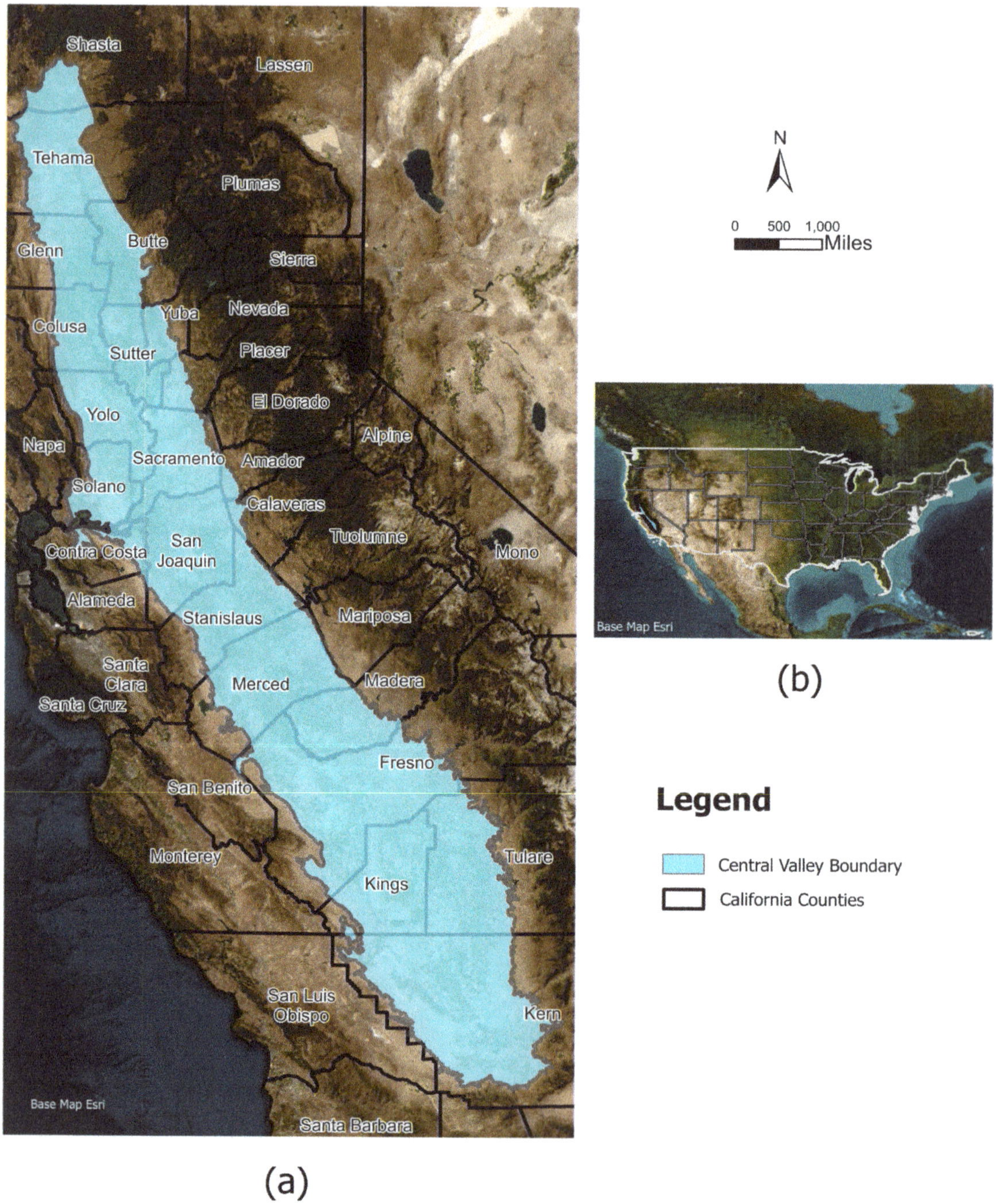

Figure 1. The location of the study area with (**a**) a map of California's Central Valley, highlighted in light blue, showing all counties within the study boundary used for almond crop classification, and (**b**) a locator map of the continental United States, indicating the geographic location of the Central Valley within the national context.

2.2. Data Input

2.2.1. Satellite Image Data

Landsat imagery from 2008 to 2022 was used in this study, and three spectral bands were selected based on their high discriminatory power for vegetation analysis (https://gisgeography.com/landsat-8-bands-combinations/ accessed on 26 July 2025). The chosen band combination—SWIR1, NIR, and a visible band (either red or blue depending on the Landsat generation)—was employed due to its proven effectiveness in capturing vegetation health dynamics (Table 1). The Near-Infrared (NIR) region, spanning approximately 0.7 to 1.3 μm, has consistently been identified as optimal for crop monitoring because of the strong absorption of visible light by chlorophyll and the substantial reflectance of NIR radiation by healthy plant foliage. Additionally, Shortwave Infrared (SWIR1) is sensitive to vegetation water content, while visible bands contribute to detecting plant pigments and overall canopy condition (https://eos.com/make-an-analysis/agriculture-band/ accessed on 26 July 2025).

Table 1. Landsat data obtained for each year of the analysis, indicating the Landsat sensor, image date, and bands extracted. All images have a pixel size of 30 m by 30 m.

Satellite	Date	Bands Extracted—Bandnumber, Name, Wavelength & Resolution
Landsat 5	2008–2009	Band 3 Visible Red (0.63–0.69 μm) 30 m Band 4 Near-Infrared (0.76–0.90 μm) 30 m Band 5 Near-Infrared (1.55–1.75 μm) 30 m
Landsat 7	2012	Band 3 Red (0.63–0.69 μm) 30 m Band 4 Near-Infrared (0.77–0.90 μm) 30 m Band 5 Short-Wave Infrared (1.55–1.75 μm) 30 m
Landsat 8–9	2013–2022	Band 2—Blue (0.45–0.51 μm) 30 m; Band 5—Near-Infrared (0.85–0.88 μm) 30 m; Band 6—SWIR1 (1.57–1.65 μm) 30 m

For each year between 2008 and 2022, a single Landsat scene was selected within the July–August window to coincide with the almond canopy's mature and spectrally stable phase. This mid-season acquisition strategy minimized phenological variability and ensured consistent image quality across years. To maintain spectral coherence across Landsat generations, a common set of three bands was utilized based on their proven relevance in vegetation analysis. For Landsat 8, the selected bands included Band 6 (SWIR1: 1.57–1.65 μm), Band 5 (NIR: 0.85–0.88 μm), and Band 2 (Blue: 0.45–0.51 μm). For Landsat 7, Band 5 (SWIR1: 1.55–1.75 μm), Band 4 (NIR: 0.77–0.90 μm), and Band 3 (Red: 0.63–0.69 μm) were employed, while the same configuration was adopted for Landsat 5, with Band 5 (SWIR1: 1.55–1.75 μm), Band 4 (NIR: 0.76–0.90 μm), and Band 3 (Red: 0.63–0.69 μm). This consistent band selection across sensors was instrumental in enhancing the detection of vegetation vigor and crop structural attributes throughout the temporal scope of the study.

To ensure compatibility with PyTorch 2.0.0-based semantic segmentation architectures, all satellite imagery was preprocessed into three-band RGB composites and converted into an 8-bit unsigned integer format. This transformation was necessary as the majority of pre-trained deep learning models in PyTorch are optimized for three-channel RGB inputs, reflecting the structure of natural color images. Although Landsat data provide a wide range of spectral bands and are commonly distributed in 16-bit format, direct use of all bands would have required significant architectural modifications to the model, increased computational demands, and potentially reduced training efficiency. Furthermore, the use of all available spectral bands was intentionally avoided to mitigate risks of model overfitting and to maintain computational tractability, particularly given the limited size and spatial coverage of the training dataset. Many Landsat bands are spectrally correlated, and their inclusion can introduce redundancy without improving model performance. Instead, this study employed a targeted band selection strategy focused on three bands: Shortwave Infrared 1 (SWIR1), Near-Infrared (NIR), and one visible band (either red or blue). This combination was selected based on its demonstrated effectiveness in prior remote sensing studies for capturing vegetation characteristics such as chlorophyll concentration, canopy structure, and moisture content—key parameters for land cover and crop classification. To optimize the number of training samples and preserve spatial context, a chip size of 64 × 64 pixels was adopted. This approach allowed efficient training and reduced memory requirements. Consequently, Landsat imagery acquired between 2008 and 2021 was standardized into 8-bit, three-band RGB chips for model training and validation. Representative examples of these image chips are provided in Figure 2.

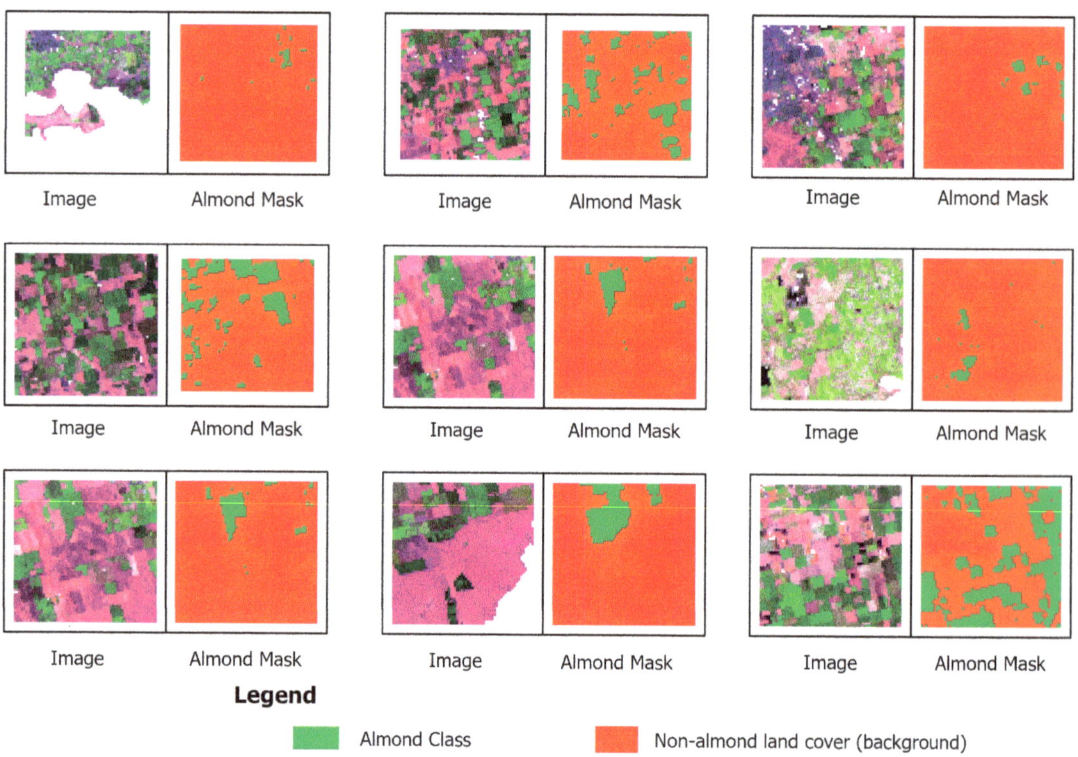

Figure 2. Sample Landsat image chips and corresponding almond masks used for model training. Each pair shows a satellite image with a 5,4,3 as R,G,B color composite which highlights vegetation as shades of green, bare soil as magenta, and urban as purple (**left**) and its binary classification mask (**right**), where green denotes almond land cover and red indicates non-almond land cover (background). The masks were derived from USDA Cropland Data Layer labels.

2.2.2. Crop Data for Almond Locations and Training

The USDA creates the Cropland Data Layer (CDL) annually for the continental US using moderate resolution satellite imagery and extensive agricultural ground truth [40]. The crop-specific data layer is freely available online (Available at https://nassgeodata.gmu.edu/CropScape/ accessed on 26 July 2025). The USDA Cropland Data Layer (CDL) was selected for this study due to its standardized nationwide coverage, temporal consistency, and seamless integration with remote sensing platforms such as Google Earth

Engine. Despite known regional limitations, its widespread use in academic research and compatibility with national agricultural statistics make it a practical and reproducible reference for multi-year crop classification. CDL data spanning from 2008 to 2021 were obtained for training the DL models (Figure 3). These images served as the ground truth for almond class identification. Additionally, a 2022 image (Figure 4) was used for validation data collection for the DL-based approaches.

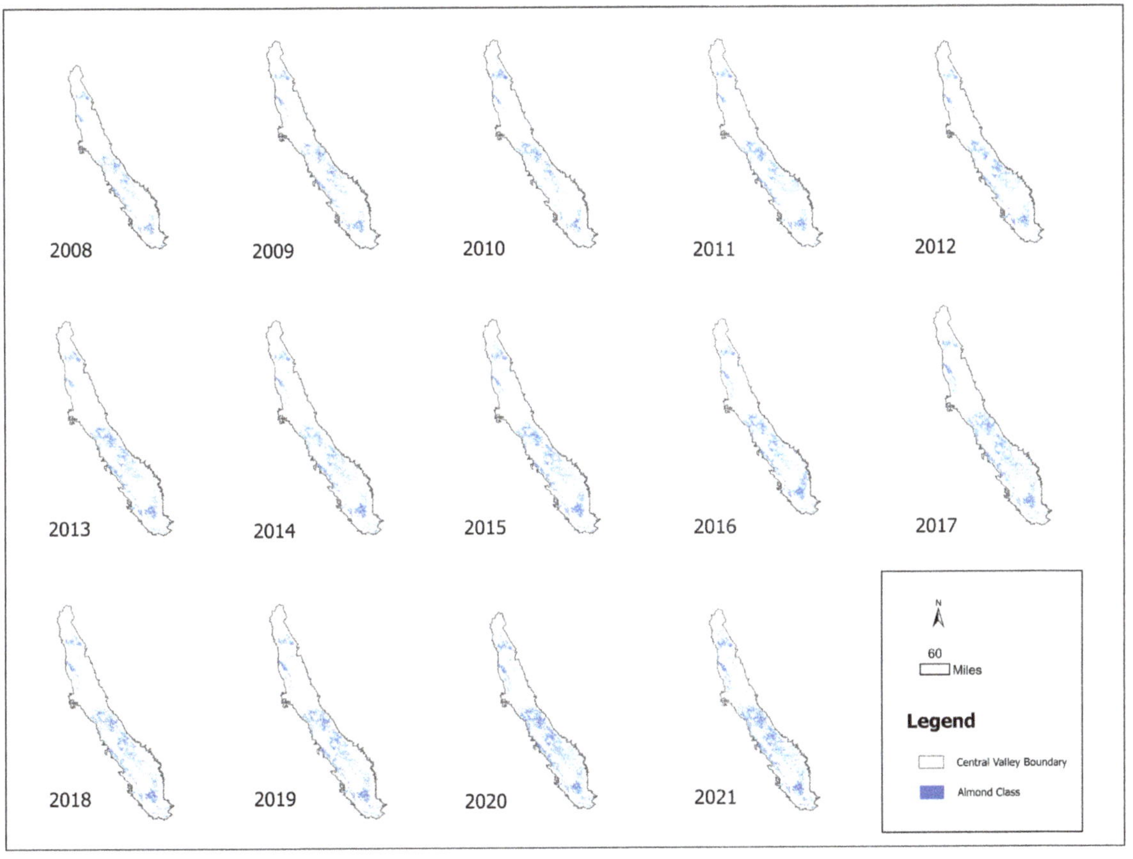

Figure 3. The annual Cropland Data Layer for almond for the study region for the 2008–2021 period for use as training data for the DL classifiers in the study.

The training data utilized in the ML process were derived from the CDL data layers for 2022. Two separate classes were designed, one for the almond class and the other for all other classes. Polygons were created for both the almond and non-almond classes based on the 2022 CDL data layer (Figure 4). In the context of ML, a single mask was employed for this purpose. ML relies on fewer computational resources and less intricate models in comparison to DL in this study. Shapefiles derived from past CDL layers were utilized to create training data for ML models.

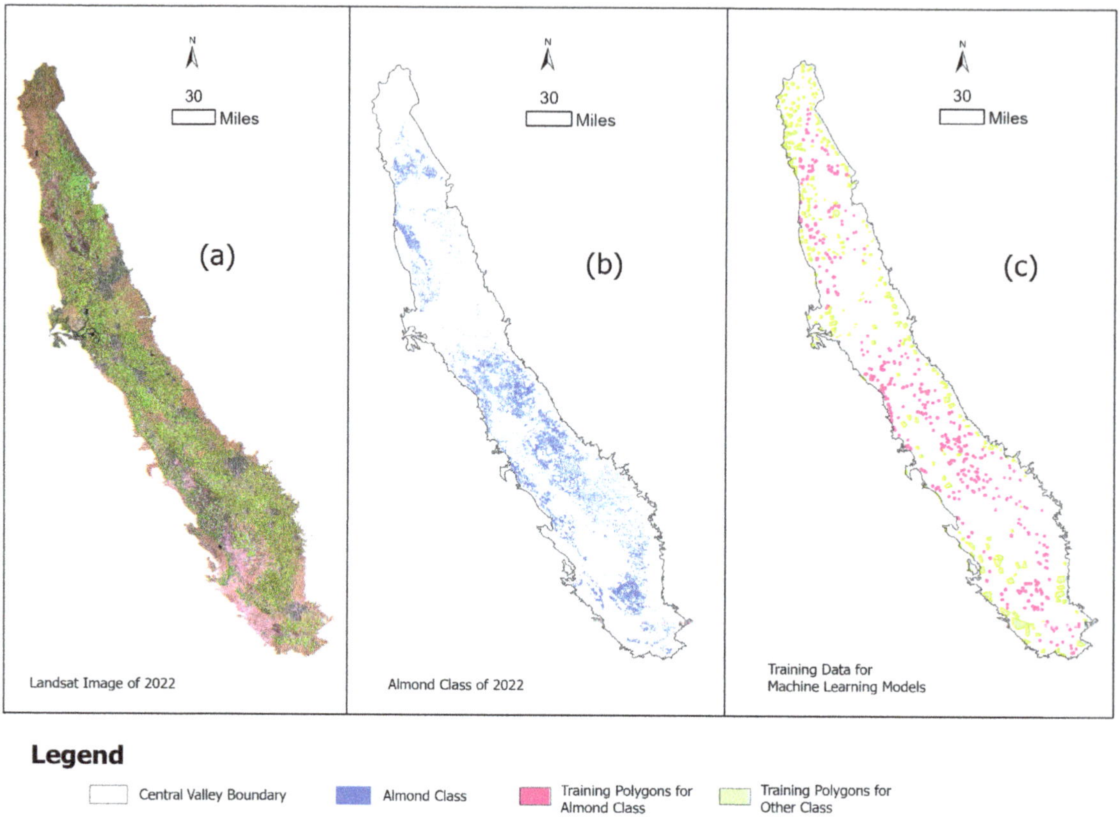

Figure 4. Overview of 2022 Landsat imagery, derived almond field classification, and training data polygons for machine learning model development in California's Central Valley. (**a**) False-color Landsat mosaic of the Central Valley study area. (**b**) Spatial distribution of almond fields in 2022 as extracted from the classification. (**c**) Locations of training polygons used for machine learning models, with pink indicating almond class and green indicating all other land cover.

In the context of ML, the training data utilized consisted exclusively of the produced CDL data layers from the year 2022, while the input for the ML models involved the Landsat image of the same year. Conversely, for DL, the training phase incorporated CDL data layers spanning the years 2008 to 2021, with the year 2022 being reserved for validation purposes. In summary, our approach involved utilizing a single image as input for the ML models, while for the DL models, we employed a dataset consisting of images spanning the years 2008 to 2021 for training purposes. We aim to evaluate the performance of ML models in scenarios when the available data is limited, in comparison to DL methodologies, and when there is a desire to conserve computational resources and minimize time expenditure for the research. In the discussion we will also consider these data requirements in the model comparison as the DL model requires a significant increase in data inputs compared to the ML models.

2.3. Model Selection and Setup

A frequent occurrence in ML and DL models is overfitting, which occurs when the model performs admirably on the training data but inadequately on validation or unobserved data. To combat overfitting, a variety of techniques and strategies have been implemented. We augmented the dimensions of the training dataset to acquire a more comprehensive understanding of the issue at hand. The DL processors were designed with dimensions of 64 × 64 in order to optimize the chip count, which is approximately 6610 in total. Additionally, a DL model utilizing the Resnet-50 architecture was implemented to increase the intricacy of the model (Table 2). Data augmentation methods that generate variants of the training data using random operations (such as rotation, inversion, and cropping) were also implemented. This effectively expands the training dataset, thereby enhancing the model's ability to generalize. Finally, experiments were conducted involving various hyperparameters, including the learning rate, sample size, and epoch count to determine the optimal configurations that strike a balance between training accuracy and model complexity.

Table 2. Hyperparameters of machine learning and deep learning models.

Models	Hyperparameters
Deep Learning (General Structure)	ENCODER = "resnet50" ENCODER_WEIGHTS = 'imagenet' CLASSES = ["Almond"] ACTIVATION = 'sigmoid' DEVICE = 'cuda' Epoch = 150 chip_size = 64, stride_x = 8, stride_y = 8, crop = 12, n_channels = 3
Linear Regression (LR)	LinearRegression ()
Logistic Regression (LGR)	LogisticRegression ()
Decision Tree (DT)	DecisionTreeClassifier ()
Gaussian Mixture Model (GMM)	GaussianMixture (n_components = 3)
Gradient Boosting (GB)	GradientBoostingClassifier (n_estimators = 100, learning_rate = 0.1, max_depth = 3)
K-Means Clustering (K-Means)	KMeans (n_clusters = 100)
K-Nearest Neighbors (KNN)	KNeighborsClassifier (n_neighbors = 3)
Multi-Layer Perceptron (MLP)	MLPClassifier (hidden_layer_sizes = (150, 100, 50), max_iter = 100, activation = 'relu', solver = 'adam')
Naive Bayes (NB)	MultinomialNB ()
Support Vector Machine (SVM)	SVC (C = 1.0, kernel = 'rbf', gamma = 'scale')

Extreme Gradient Boosting (XGB)	params = {
	'max_depth': 3,
	'learning_rate': 0.1,
	'n_estimators': 50
	}
	XGBClassifier(** params, tree_method = 'gpu_hist', predictor = 'gpu_predictor', gpu_id = 1)

Random Forest (RF)	RandomForestClassifier (n_estimators = 500, oob_score = True, verbose = 1)

Note: SVC = Support Vector Classifier; rbf = Radial Basis Function; stride_x / stride_y = patch stride in pixels; n_channels = number of input channels; n_components = number of Gaussian components; n_clusters = number of clusters; n_neighbors = number of neighbors; n_estimators = number of trees; gpu_hist = GPU-accelerated histogram algorithm; gpu_predictor = GPU-based prediction; gpu_id = GPU identifier; oob_score = out-of-bag validation; ** *params = Python syntax for keyword arguments from a dictionary.*

2.3.1. ML Models

In the comparative analysis of ML and DL models for large-scale agricultural land cover classification, we evaluated a diverse set of ML algorithms, focusing specifically on almond crop classification using 2022 imagery and training data from the CDL layers. The ML models included in this study are Linear Regression, Logistic Regression, Naive Bayes, Gaussian Mixture Model, K-Nearest Neighbors, Decision Trees, Random Forest, Gradient Boosting, XGBoost, and Multi-Layer Perceptron.

Linear Regression is a simple predictive model, but its limitations in capturing non-linear relationships restrict its performance in complex applications [41,42]. Logistic Regression extends this by modeling binary outcomes [43,44], while Naive Bayes, despite its assumption of feature independence, performs well in image classification tasks [45,46]. Support Vector Machines (SVMs) are particularly effective with small training datasets and excel in classification accuracy [47,48], while K-Nearest Neighbors (KNN) relies on proximity-based classification and is straightforward to implement [49]. K-Means, an unsupervised method, is frequently employed for clustering [50]. The Gaussian Mixture Model (GMM) offers probabilistic clustering by combining Gaussian distributions [51]. Decision Trees (DTs) are highly interpretable models used in classification tasks [52], and Random Forest (RF) enhances DT by aggregating multiple trees to reduce overfitting and improve predictive accuracy [53]. Gradient Boosting (GB) and XGBoost (XGB), both ensemble learning methods, iteratively improve prediction by minimizing loss functions [54,55]. Finally, Multi-Layer Perceptron (MLP), a type of neural network, captures non-linear relationships through its multiple layers [56].

2.3.2. DL Models

In this study, we employed a variety of DL models for almond crop classification, trained using CDL data from 2008 to 2021. The models include UNet, UNet++, Multi-Scale Attention Network, LinkNet, Feature Pyramid Network, Pyramid Scene Parsing Network, DeepLabv3, and DeepLabv3+. These models were trained using ResNet-50 and ResNet-18 backbones, leveraging pre-trained ImageNet weights for transfer learning to enhance performance in agricultural image classification.

U-Net, a popular segmentation model, utilizes an encoder–decoder structure with skip connections to maintain high segmentation accuracy, particularly for tasks with limited training data [57,58]. UNet++ extends this by incorporating densely connected sub-networks to improve feature extraction across multiple scales [59,60]. The Multi-Scale Attention Network (MANet) focuses on efficiently segmenting high-resolution images by utilizing attention mechanisms to highlight critical features while suppressing irrelevant data [61,62]. Similarly, LinkNet modifies UNet's structure with residual connections, enhancing computational efficiency without sacrificing accuracy [63]. Feature Pyramid Network (FPN) addresses multiscale object detection by leveraging a pyramid structure that combines high-level semantic features with low-level spatial features, improving scale invariance [64,65]. Pyramid Scene Parsing Network (PSPNet) further enhances segmentation by pooling contextual information from multiple scales, allowing for better pixel classification across diverse regions [66,67]. DeepLabv3 and DeepLabv3+ use Atrous Spatial Pyramid Pooling (ASPP) to capture multi-scale context while maintaining high-resolution feature maps, which are particularly useful for segmenting objects at varying scales [68]. DeepLabv3+ improves upon its predecessor by incorporating an explicit decoder module to better capture fine-grained details, especially at object boundaries [69].

2.3.3. Computing Requirements for Analysis and Available Resources

The computational demands of land use and land cover (LULC) classification, such as almond crop classification using Landsat data, vary significantly between ML and DL methods. ML models like DT, RF, and SVM typically rely on structured, well-labeled datasets with features extracted from spectral bands (e.g., Landsat Bands 1–7) or vegetation indices (e.g., NDVI and SAVI). While these models are relatively computationally efficient compared to DL, they still require considerable processing power during feature selection, hyperparameter tuning, and training on large datasets, like Landsat imagery, which spans vast geographic areas with 30-m spatial resolution [70,71]. CPU-based systems can handle these tasks, though cloud computing platforms can further optimize both processing and storage needs [72]. In contrast, DL models, such as CNNs, demand substantially more computational resources due to their ability to integrate both spectral and spatial features for more detailed analysis. For LULC tasks like almond crop classification, DL models benefit from large, multi-temporal datasets to capture crop growth cycles and patterns, requiring significant processing power [70,73]. The heavy computational burden of DL arises from its use of neural network layers, with numerous parameters requiring iterative optimization during training. This typically necessitates high-performance GPUs or custom hardware like TPUs to expedite training and inference [74,75]. To ensure accuracy, DL models often require extensive preprocessing steps, including normalization and data augmentation [73]. When incorporating temporal Landsat data to monitor seasonal changes, cloud platforms such as Google Earth Engine (GEE) are crucial, offering access to large satellite archives and the necessary processing capabilities for both ML and DL tasks. However, the high computational demands of DL can lead to increased cloud costs, particularly with large-scale training that requires hyperparameter tuning and cross-validation [76].

2.3.4. Accuracy Assessment of ML and DL Models

In order to precisely evaluate the performance of the models, many metrics and methodologies are utilized which are related to the type of problem at hand, i.e., image classification. These commonly used accuracy assessments are precision, recall (sensitivity), F1-score, and overall accuracy.

Precision: Simply put, precision is the quotient obtained by dividing the number of accurately predicted positive cases by the total number of occurrences predicted as positive [77]

$$\text{Precision} = \text{True Positives}/(\text{True Positives} + \text{False Positives})$$

The term "precision" is commonly described as the extent to which an individual's score achieved on one occasion is replicated on a subsequent occasion, a concept also known as test–retest reliability in classical language [78]. The metric evaluates the classifier's capacity to accurately detect positive examples within its set of predicted positive occurrences. The measure is frequently employed in image classification and other classification applications. The metric under consideration measures the precision of a classifier's positive predictions, specifically in the context of object identification in images, by comparing them to the overall number of positive predictions made. Precision is a term that is used to describe the level of accuracy or exactness in a measurement or calculation.

Recall: This refers to the proportion of accurately anticipated positive instances compared to the overall number of actual positive instances [79].

$$\text{Recall} = \text{True Positives}/(\text{True Positives} + \text{False Negatives})$$

Recall, alternatively referred to as sensitivity or True Positive Rate, holds significant importance as a performance statistic within the realm of image classification and various other classification endeavors. The metric assesses the classifier's capacity to accurately identify all pertinent instances, namely the ratio of correctly identified positive examples (e.g., items of interest) to the total number of actual positive instances, as determined by the model. Here, the metric denotes the proportion of almond pixels that have been correctly classified among the entire set of almond pixels.

F1-score: This represents the harmonic mean of precision and recall.

$$\text{F1} = 2 \times ((\text{Precision} \times \text{Recall})/(\text{Precision} + \text{Recall}))$$

The F1-score is usually used for the optimization of a model towards either precision or recall [80]. The F1-score assesses feature discrimination against target groups statistically. It generates a feature's score by comparing sample mean variation to sample variation [81].

Overall Accuracy: Accuracy is a metric used to assess the correctness of a model. It is calculated as the ratio of the number of pixels that are accurately classified to the total number of testing pixels.

$$\text{Overall Accuracy} = (\text{TP} + \text{TN})/(\text{TP} + \text{TN} + \text{FP} + \text{FN})$$

In image classification, true positives (TPs) are the number of correctly predicted positive instances. In the context of image classification, this means the number of pixels correctly classified as the target class. True Negatives (TN) refer to the count of accurately predicted negative instances. In the specific domain of image classification, this pertains to the quantity of pixels accurately identified as belonging to the non-target category. False positives (FP) are the number of instances that were incorrectly predicted as positive. In image classification, these are the pixels that were predicted to belong to the target class but do not. False negatives (FN) refer to situations that have been inaccurately classified as negative when they should have been classified as positive. Within the context of image classification, the term "misclassified pixels" refers to those pixels that are erroneously categorized as not belonging to the intended target class. The macro technique is employed to calculate Macro-Precision, Macro-Recall, and Macro-F1-score. The macro technique computes the arithmetic mean for many types of indicators [82].

3. Results

3.1. ML Model Performance

A predicted almond classification was created for 2022 for each model used (Figure 5). In addition, in the table provided (Table 3), we can observe the performance metrics of different ML models on almond classification using our 2022 Landsat image. The performance metrics reported are precision, recall, F1-score, and overall accuracy. Looking individually at each model in terms of these four metrics, we can compare models.

Linear Regression (LR) has moderate precision and recall values compared to other models. Its overall accuracy is better than the GMM but lower than the ensemble and tree-based models. LGR's performance is similar to LR but with slightly lower overall accuracy. KNN has similar precision, recall, and F1-score to the DT and GB models but slightly lower recall. Its overall accuracy is slightly higher than GB but lower than DT. The KM model exhibits inferior overall accuracy, precision, recall, and F1-score in comparison to the other models. The GMM exhibits the lowest overall accuracy. GMM has the lowest precision but the highest recall, suggesting it identifies most of the positive cases but with a high false positive rate. NB has a decent recall but lower precision, indicating a higher number of false positives. It has the second lowest overall accuracy. MLP's performance metrics are close to those of KNN but with a slightly lower recall. The SVM algorithm has superior performance in terms of overall accuracy. However, it exhibits the lowest scores in terms of F1-score and recall. Furthermore, in terms of precision, the score of the mentioned model is higher than that of K-Means but lower than all other models (Table 3).

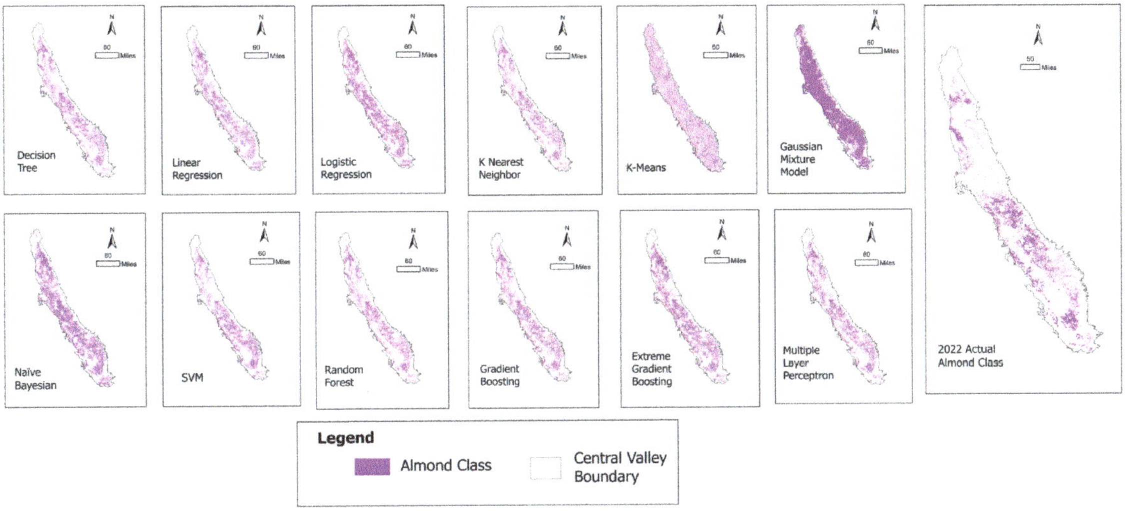

Figure 5. A comparison of the predicted almond class location for 2022 for each of the twelve machine learning models used relative to the actual almond class location from the Cropland Data Layer product.

Table 3. Accuracy assessment results of the twelve different ML models for almond classification for the 2022 Landsat image. The highest model scores in each category are bolded.

ML Model	Precision	Recall	F1-Score	Overall Accuracy
Linear Regression—LR	0.63	0.66	0.65	95.647
Logistic Regression—LGR	0.65	0.72	0.68	95.546
K-Nearest Neighbor—KNN	**0.70**	0.72	0.71	96.662
K-Means Clustering—K-Means	0.59	0.67	0.62	94.145
Gaussian Mixture Model—GMM	0.62	**0.90**	0.67	92.209
Naive Bayesian—NB	0.65	0.77	0.69	95.264
Support Vector Machine—SVM	0.61	0.58	0.59	96.100
Decision Tree—DT	**0.70**	0.74	**0.72**	96.650
Random Forest—RF	0.71	0.73	**0.72**	**96.798**
Gradient Boosting—GB	**0.70**	0.74	**0.72**	96.615
Extreme Gradient Boosting—XGB	0.67	0.74	0.69	95.93
Multiple Layer Perceptron—MLP	**0.70**	0.73	0.71	96.632

DT's overall accuracy is the highest compared to all the models, except for RF and KNN. It has moderate precision, recall, and F1-score compared to others. The RF model achieves the highest overall accuracy among all the models. It has the highest precision and a comparable recall and F1-score to the top-performing models. GB shows nearly identical performance to DT in terms of precision, recall, and F1-score but has slightly less overall accuracy than the DT model. EGB has moderate precision and recall, with an F1-score and overall accuracy that are decent but not the best among the models. In summary, if your primary concern is high overall accuracy, the RF model seems to be the best choice, followed closely by the DT and KNN models. The MLP model exhibits the highest precision among the three models considered. However, when considering recall, F1-score, and total accuracy, the MLP model demonstrates outstanding performance but does not achieve the top score. However, if you are looking for a model with a high recall (identifying most of the true positive cases), the GMM would be the best despite its lower overall accuracy (Table 3).

3.2. DL Model Performance

A predicted almond classification was created for 2022 for each DL model used (Figure 6). The spatial results shown for the DL models (Figure 6) are much more consistent than those shown for the ML models (Figure 5), which varied more in terms of area of almonds predicted, as well as in locations. In addition, in the table provided (Table 4), we can observe the performance metrics of all the different DL models on almond classification. The performance metrics reported are precision, recall, F1-score, and overall accuracy. Looking individually at each model in terms of these four metrics, we can compare models (Table 4).

Table 4. Accuracy assessment results of the different DL models for almond classification for the 2022 Landsat image. The highest model scores in each category are bolded.

Year	Precision	Recall	F1-Score	Overall Accuracy
U-Net	0.79	**0.66**	**0.70**	97.465
UNet++	0.79	0.62	0.66	97.394
MANet	0.78	0.62	0.67	97.338
LinkNet	**0.80**	0.64	0.69	97.455
FPN	**0.80**	0.61	0.66	97.422
PSPNet	**0.80**	0.60	0.65	97.404
DeepLabv3	0.79	0.62	0.66	97.380
DeepLabv3+	**0.80**	**0.66**	**0.70**	**97.502**

U-Net's overall accuracy was the highest among all models except for DeepLav3+. Its precision score was on par with many other models but lower than Linknet, FPN, PSPNet, and DeepLabv3+. It had relatively low recall, and it was higher only when compared to UNet++, MANet, and DeepLav3. Its F1-score was in mid-range, with three models scoring lower and four models scoring higher. UNet++ had moderately high overall accuracy, but it was smaller than UNet. Its precision was equal to U-Net and DeepLav3 but lower than several other models, and it had one of the lowest recall scores and was only higher than PSPNet. Its F1-score was in the lower mid-range and only higher than that of PSPNet. MANet's overall accuracy was the lowest among all the models, albeit marginally. Its precision was slightly lower than most other models, and its recall was equal to UNet++ and DeepLabv3 but lower than others. It had a mid-range F1-score, with three models scoring lower and four models scoring higher. LinkNet had a high overall accuracy, very close to the top-performing models. Its precisions score it was the highest, tied with FPN,

PSPnet, and DeepLabv3+. Recall was in the mid-range, higher than UNet++, MANet, FPN, PSPNet, and DeepLabv3, and its F1-score was in the higher mid-range, with only UNet and DeepLabv3+ scoring higher. FPN had an overall accuracy that was high, only slightly lower than LinkNet. Its precision score was also one of the highest, tied with LinkNet, PSPNet, and DeepLabV3+. Its recall score was in the lower mid-range, higher only than PSPNet, and its F1-score was also in the lower mid-range, higher than only PSPNet. PSPNet had an overall accuracy in the mid-range, higher than MANet and DeepLabv3 but lower than others. Its precision was among the highest, tied with LinkNet, FPN, and DeepLabv3+. However, its recall and F1-score were the lowest among all models. DeepLav3 had a mid-range overall accuracy, higher than MANet and PSPNet but lower than others. Its precision was equal to U-Net and UNet++ but lower than others. Recall scores were in the lower mid-range, only higher than UNet++, MANet, and PSPNet, and the F1-score was similarly in the lower mid-range, higher than only UNet++ and PSPNet. Finally, DeepLabv3+ had the highest overall accuracy among all the models, suggesting it classified the highest percentage of instances correctly. In addition, this model tied for the highest precision, indicating that it had the highest percentage of correct positive predictions and had the highest recall score, tied with U-Net, suggesting that it identified the highest proportion of actual positives correctly. This model also tied for the highest F1-score with U-Net, showing that it had the best balance between precision and recall. From these results (Table 4), it appears that the DeepLabv3+ model is the top performer in almond classification based on the 2022 Landsat image, with the highest overall accuracy and the best or tied best scores in all other metrics. LinkNet also performed notably well, especially in terms of precision, which was the highest and tied with others. It is essential to note that all models have achieved high overall accuracy above 97%, indicating a generally successful classification outcome across all methodologies. Moreover, the selection of the most appropriate model may depend on whether precision or recall is prioritized, as this decision should align with the specific objectives and constraints of the study.

Looking at the results spatially (Figures 7 and 8), as well as through the accuracy assessment results (Tables 3 and 4), we can better evaluate the model performance for both the ML and DL models. In Figure 7, we present the outcomes of different ML modeling approaches applied to the classification of almond crops in our study area for the year 2022. The spatial analyst feature from ArcGIS Pro was utilized to contrast predicted almond locations against ground truth data for the Cropland Data Layer product. The maps are distinguished by a color scheme that indicates the accuracy of each model's predictions: green areas represent true positives where the model correctly identified the presence of almond crops, magenta areas denote false positives where almond presence was incorrectly predicted, and blue areas represent false negatives where the model failed to detect actual almond crops (Figures 7 and 8). The visual analysis of the classified images indicates that models RF, XGB, and MLP demonstrate a more accurate prediction with the highest proportion of correct matches (green areas). Conversely, models like KNN, K-Means, and GMM show a substantial number of false positives (magenta), indicating overestimation of almond presence. Linear Regression (LR) and Logistic Regression (LGR) models, despite their simplicity, demonstrated a higher incidence of false negatives (highlighted in blue), indicating a tendency to overlook actual almond crop areas. In Figure 8, the performance of the DL models in the classification of almond crops for the year 2022 are mapped. The visual representation of the classification outcomes reveals a nuanced view of each model's predictive capabilities. Models such as UNet++, PSPNet, and MANet exhibit a considerable overlap of green and magenta areas. This suggests that while these models are adept at identifying almond crops, there is also a tendency to overestimate their presence, resulting in a higher rate of false positives. Conversely, models like LinkNet and FPN are marked by extensive blue areas, indicating a higher rate of false negatives, where they have missed detecting actual almond crops. Notably, DeepLabv3 and DeepLabv3+ appear to strike a balance in prediction accuracy. The interspersion of green, magenta, and blue suggests that these models achieve an equilibrium between precision and recall, reducing the likelihood of both false positives and false negatives.

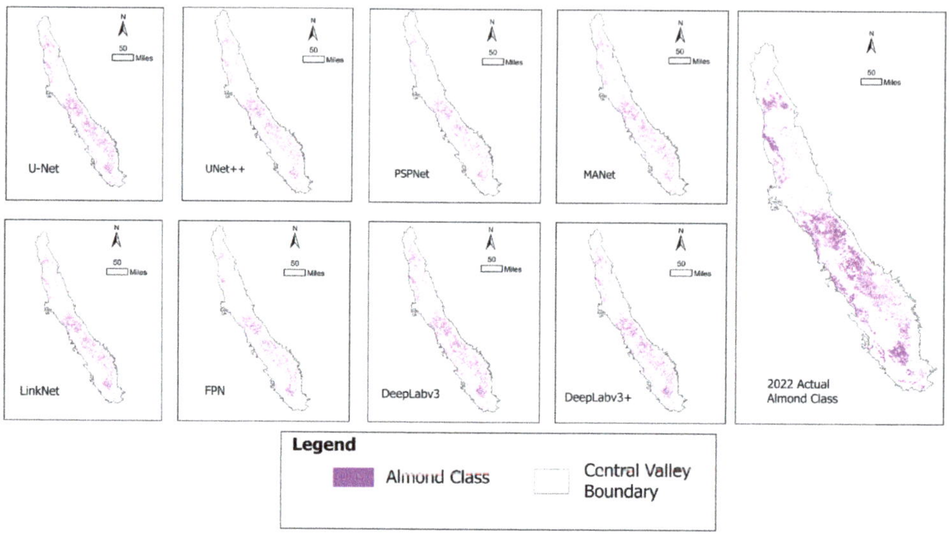

Figure 6. A comparison of the predicted almond class location for 2022 for each of the eight deep learning models used relative to the actual almond class location from the Cropland Data Layer product. Precision or recall should be prioritized depending on the specific requirements of your project.

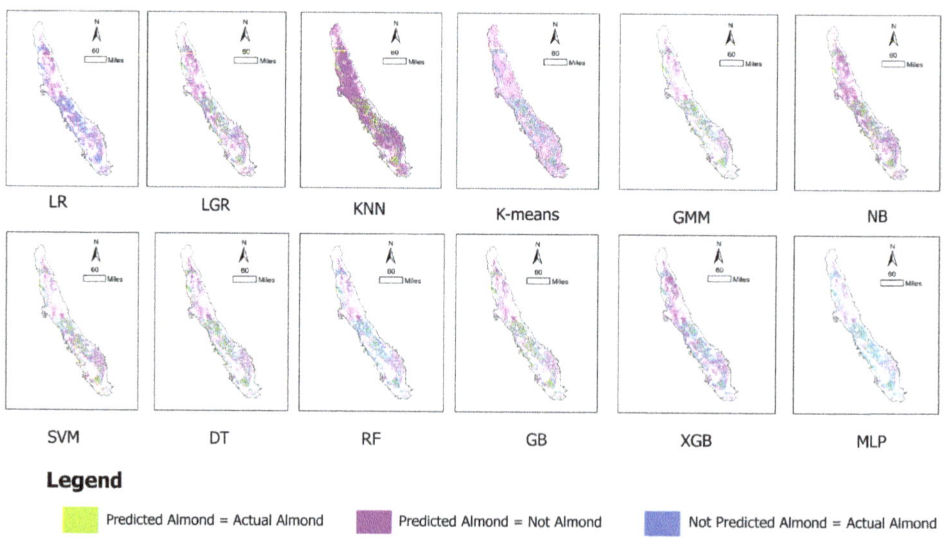

Figure 7. A comparison of the classified image for 2022 for almonds across the study area as a function of the different machine learning models used. Green: Correct predictions where the model identified almond crops that are indeed present (true positives). Magenta: Incorrect predictions where the model identified almond crops that are not actually present (false positives). Blue: Missed predictions where the model failed to identify actual almond crops (false negatives).

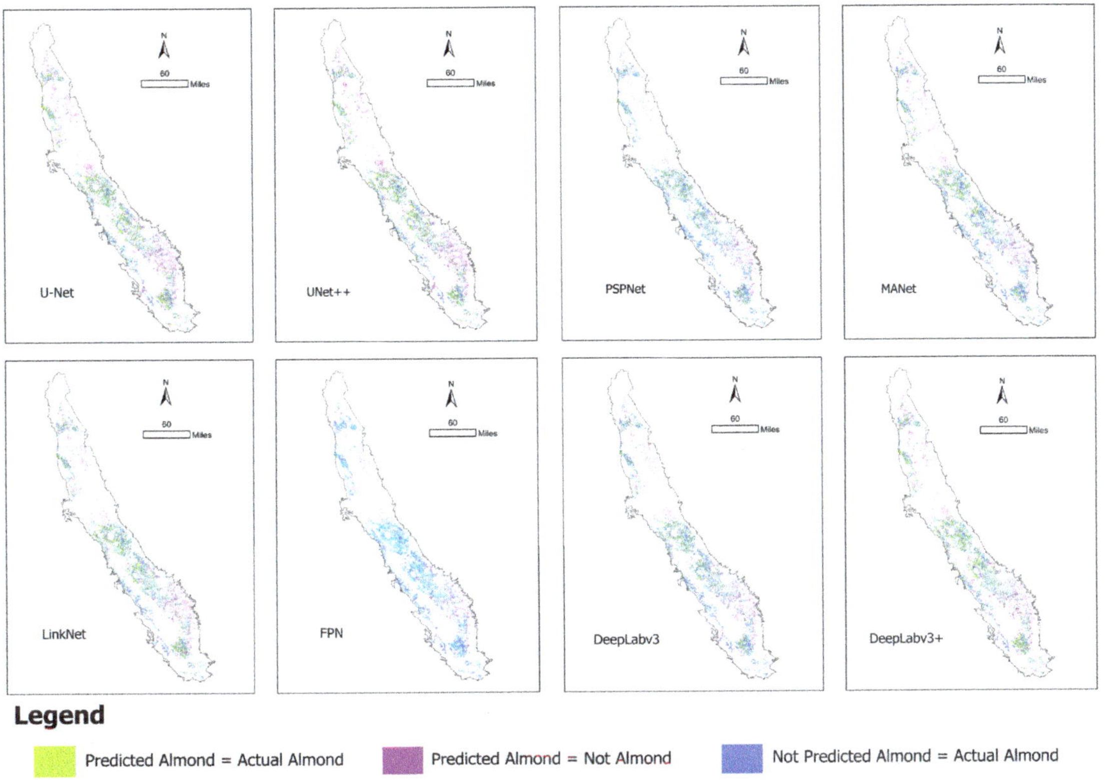

Figure 8. A comparison of the classified image for 2022 for almonds across the study area as a function of the different deep learning models used. Green: Correct predictions where the model identified almond crops that are indeed present (true positives). Magenta: Incorrect predictions where the model identified almond crops that are not actually present (false positives). Blue: Missed predictions where the model failed to identify actual almond crops (false negatives).

Using the RF model, due to the analysis highlighting its usefulness in this study, a change analysis can be conducted to look at almond crop coverage across the study area from 2008, to 2015, to 2022. The model results (Figure 9; Table 5) highlight the significant expansion of almonds as a crop between 2008 and 2015 and a more stable landscape afterward.

Table 5. Performance metrics of Random Forest model for different years.

Random Forest	Precision	Recall	F1-Score	Overall Accuracy
2008	0.72	0.60	0.63	98.634
2015	0.69	0.80	0.73	97.158
2022	0.71	0.73	0.72	96.798

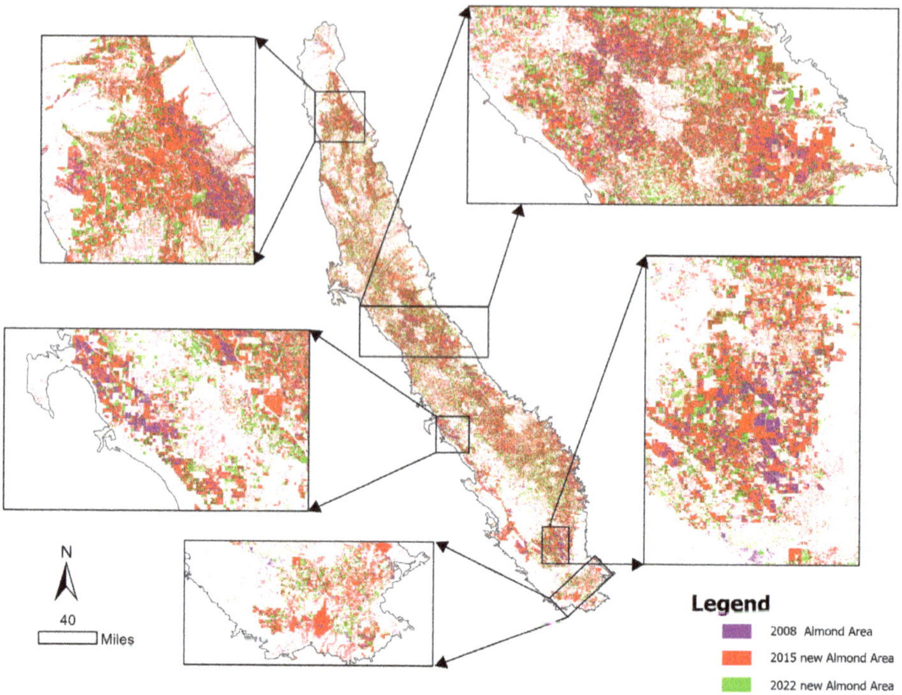

Figure 9. Random Forest model predictions of almond crop distribution in California's Central Valley for the Years 2008, 2015, and 2022, highlighting the expansion of almond coverage over time.

4. Discussion

Based on this evaluation and comparison of modeling approaches within agricultural land cover classification analyses, it was determined that the RF model was the most accurate and efficient of all the ML and DL models tested. In applying our top-performing RF ML model to the task of change in almond production across our study area over the past 25 years, we saw that much of the expansion in almond crops occurred between 2008 and 2015, with smaller changes (both into and out of almond) occurring since 2015. This peak in almond production is reflected across the study area from the CDL data (Figure 3), and this analysis with the RF model allows us to extend the known locations of almonds to 2022 as a prediction. Continued extensive almond production has implications for the environment, specifically the increase in water usage, as reflected by the increased area, and the continued and maintained almond area, meaning a continued use of irrigation in this landscape. Given the recent droughts and extensive fires related to these droughts, as well as a drop in the water table across California (due to both long-term drought and increased water use for irrigation), these implications and the use of such mapping studies are of real importance for resource management [36,83–87].

We analyzed and compared the performance of a variety of ML and DL models in the task of almond classification based on Landsat images from 2022 (Figures 5–8; Tables 3 and 4). The metrics of focus in our analysis encompassed precision, recall, F1-score, and overall accuracy, each of which paints a distinct picture of the models' capabilities and efficiency in classifying agricultural crops accurately. Upon reviewing the assessment results of the different ML models (Table 3), we found a closely contested field with RF slightly edging out others in terms of overall accuracy (96.798%), followed very closely by KNN (96.662%) and MLP (96.632%). These top performers exhibit a balance in precision, recall, and F1-score, signaling robust performance across different facets of the classification task. In contrast, the GMM lags noticeably with an overall accuracy of 92.209%, albeit boasting the highest recall of 0.90. While this high recall indicates a proficient identification of positive cases, the diminished precision and F1-score pinpoint a vulnerability in accurately distinguishing negative cases, hence yielding a larger number of false positives. The K-Means model demonstrated comparatively lower levels of overall accuracy, precision, recall, and F1-score when compared to the alternative models.

Transitioning to DL models, DeepLabv3+ emerges as the most balanced model, securing the highest overall accuracy (97.502%) alongside leading scores in precision and recall (Table 4). It exhibits a profound adeptness in both identifying true positives and avoiding false positives, thereby securing a high F1-score, which illustrates a well-rounded performance. Despite trailing slightly, models like Unet and Linknet also demonstrate commendable precision and overall accuracy, hinting at their proficient predictive capacity in almond classification. The findings emphasize the efficacy and nuanced understanding captured through the DL models as seen through high precision rates above 78% across all models. Spatial patterns are also important to evaluate when juxtaposed with the outputs of the ML models (Figure 7). The DL models' maps (Figure 8) also reflect a more refined ability to discern complex patterns within the data, which is characteristic of the deeper architectural layers present in DL models. This advanced pattern recognition translates into a higher rate of true positives and a reduction in both false positives and false negatives, showcasing the DL models' enhanced detection capabilities for almond crop classification (Tables 3 and 4). The discussion of these findings in the broader context of agricultural monitoring via remote sensing is significant. It underscores the necessity to weigh the precision and computational demands of DL models against their classification accuracy. It also raises the point that selecting the most appropriate model is crucial as it must align with the specific requirements of the remote sensing task at hand.

In comparing the results of our research to those of other researchers, we can evaluate if specific types of questions or analyses result in comparable findings. For example, several studies have examined the accuracy of various ML models for image classification [88–91]. These studies consistently indicate that the RF model exhibits superior accuracy compared to other ML models. Our findings align with these results as we observed that RF outperformed other ML models in terms of overall accuracy (Table 3). The findings of Singh et al. (2022) [92] and Lamba et al. (2021) [93], who detected plant diseases, and Feizizadeh et al. (2021) [94], who monitored land use/cover change, all of which used comparative ML and DL approaches, are consistent with the results of RF when compared to DL models. These studies demonstrated that DL models outperformed RF in terms of performance. Furthermore, several studies, including those by Kirola et al. (2022) [95] and Sujatha et al. (2021) [96], have successfully identified plant diseases through the utilization of comparative ML and DL techniques. Their studies have shown that the RF algorithm exhibits exceptional performance among ML algorithms. However, when compared to DL models, RF tends to exhibit subpar performance, which is consistent with our own research findings. In another comparative study conducted by Yao et al. (2022) [97] on on the classification methodology of crops, it was observed that the RF model exhibited superior performance compared to the DNN model when the RF model was integrated with the DNN model. In general, our results support that DL models outperform traditional ML models in this domain, although an individual researcher would need to evaluate if the additional data needs are worth the slight increase in accuracy.

Despite the model output numbers and statistics, the evaluation and comparison of ML versus DL model approaches must also consider other issues in its comparison. When considering the identification of almond classes, the decision between using ML and DL entails a compromise between computational expenditure and accuracy. ML techniques have the capability to yield findings that are reasonably precise while requiring significantly less computer resources compared to DL approaches. Hence, in scenarios where there are constraints on computational resources or a need for cost-effectiveness, ML may be the more favorable option. In situations where achieving high levels of precision and accuracy is of utmost importance, particularly when it is necessary to discern small differences between various categories of almonds, the utilization of DL techniques can be advantageous. In deciding between using ML and DL models, it is crucial to consider the project's distinct requirements and limitations while also striking a balance between processing resources and the necessary level of accuracy. Overall, then, the choice between using ML and DL for almond class identification depends on finding the right trade-off between computational resources and desired accuracy [98]. ML offers cost-effective solutions, while DL provides a powerful tool for achieving precision and handling complex data. The decision should be tailored to the specific goals and constraints of the project to strike an optimal balance between computational resources, budget, time constraints, and the desired level of accuracy. This selection of the RF ML model in this final change analysis was well validated by the findings of our research and the consideration of model efficiency, time and resources available. This research highlighted the comparison on ML and DL approaches and highlighted their usefulness within agricultural studies. Also of importance is the discussion between ML and DL models, which is one found throughout the literature on remote sensing applications [99].

As such, our experimental scope deliberately focused on established convolutional networks (U-Net, MA-Net, and DeepLabv3+) and classical ML classifiers (RF, KNN, and MLP) rather than emerging Transformer-based or hybrid architectures. While self-attention models like Vision Transformer [100], Swin Transformer [101], and data-efficient variants such as DeiT [102] have demonstrated state-of-the-art segmentation performance, they typically require extensive pretraining on large, labeled datasets and substantial GPU resources. Our Landsat-based dataset, though rich in spatial and temporal coverage, remains moderate in size, and our computational environment aligns with the resource profiles common in agricultural applications. By grounding our comparison in widely adopted convolutional and ML methods, we ensure direct comparability with the broader remote-sensing literature [88] and deliver reproducible insights that can immediately inform resource-constrained precision-agriculture workflows.

Over the past decade, agricultural land cover classification has evolved from isolated, single-model studies to dynamic, multi-source workflows that fuse optical, radar, and thermal imagery with in situ sensor networks via cloud-native platforms such as Google Earth Engine and Amazon SageMaker [103,104]. At the same time, high-resolution CubeSat constellations and UAV systems now supply sub-meter imagery, enabling the detection of fine-scale phenological changes and early stress indicators in cropping systems [105]. Looking ahead, hybrid frameworks that integrate deep-learning segmentation with process-based crops and hydrological models promise near-real-time yield forecasting and adaptive irrigation management [106]. Embedding these analytics into decision support systems—with automated anomaly detection, optimized resource scheduling, and risk-alert dashboards—will shift precision agriculture from retrospective mapping to proactive, adaptive management. Finally, as edge computing and federated learning enable privacy-preserving, distributed model updates, these platforms will become ever more responsive to climate variability, resource constraints, and sustainability targets [107].

Accurate classification in agriculture is vital. Precise classification can improve yield forecasts, refine harvest timing, and support sustainable farming by providing details on crop health. It also enables targeted farming interventions, which can minimize resource waste and enhance environmentally friendly practices. In this context, the high accuracy rates demonstrated by both ML and DL models in our study highlight their potential in precision agriculture and suggest a move towards more data-driven precision farming approaches. Our findings reveal considerable proficiency in both ML and DL models for crop classification, with DL models, particularly DeepLabv3+, showing a slight edge in accuracy for almond classification. This points to the significant capabilities of advanced algorithms in this field and paves the way for their inclusion in agricultural practices. Looking ahead, future research should examine the application of these models in real-world settings, assessing their adaptability to the ever-changing agricultural environment and various crop species. This will help advance agriculture into a future that benefits from technological advancements while maintaining ecological sustainability.

5. Conclusions

In this study, we conducted a comprehensive comparison of twelve ML and eight DL classifiers for almond orchard mapping using 2022 Landsat imagery and subsequently deployed the top-performing RF model to quantify land cover change from 2008 to 2022. Our results show that RF achieved an overall accuracy of 96.8%, closely rivaling the best DL model, DeepLabv3+, which recorded 97.5% accuracy. The DL architectures—particularly DeepLabv3+, U-Net, and LinkNet—demonstrated superior boundary delineation and reduced misclassification of mixed pixels, whereas ML methods such as KNN and MLP provided near-comparable performance with substantially lower computational and data preprocessing demands.

Temporal analysis with the RF model revealed significant expansion of almond coverage from 2008 to 2022, with approximately 82% of almond expansion occurring between 2008 and 2015, after which net annual gains declined below 1%. These patterns align closely with USDA CDL records and underscore the environmental implications of sustained irrigation in California's Central Valley amid prolonged drought and groundwater depletion. Our use of medium-resolution, freely available Landsat data illustrates that high-accuracy classification is attainable without the prohibitive costs associated with very-high-resolution sensors or extensive GPU infrastructure.

Nevertheless, the study is constrained by the 30 m spatial resolution of the Landsat imagery, which may underdetect small or recently established orchards, and by the exclusion of emerging transformer-based and hybrid segmentation frameworks, models that often require large pretraining datasets and advanced hardware. Future work will extend this framework by integrating multisensor datasets (e.g., Sentinel-2 and SAR), evaluating data-efficient transformer variants under moderate sample sizes, and coupling land cover outputs with socio-economic and hydrological variables to enable real-time yield forecasting and optimized water management strategies. Collectively, these efforts aim to advance precision agriculture tools that balance predictive accuracy with operational feasibility in resource-limited settings.

Author Contributions: Conceptualization, M.R. and J.S.; methodology, M.R.; software, M.R.; validation, M.R. and J.S.; formal analysis, M.R.; investigation, M.R. and J.S.; resources, M.R. and J.S.; data curation, M.R.; writing—original draft preparation M.R. and J.S.; writing—review and editing, M.R., J.S., Y.W. and D.K.; visualization, M.R.; supervision, J.S., Y.W. and D.K.; project administration, J.S., Y.W. and D.K. All authors have read and agreed to the published version of the manuscript.

Funding: This research received no external funding.

Data Availability Statement: Data will be available upon request.

Acknowledgments: During the preparation of this work, the authors used ChatGPT-4o in order to improve readability and language of the model comparisons and selection criteria, as well as for title creation. After using this tool, the authors reviewed and edited the content as needed and take full responsibility for the content of the publication.

Conflicts of Interest: The authors declare no conflicts of interest.

References

1. Singh, V. Agriculture and Food Resources. In *Textbook of Environment and Ecology*; Springer: Singapore, 2024; pp. 155–174. [CrossRef]
2. Dethier, J.J.; Effenberger, A. Agriculture and development: A brief review of the literature. *Econ. Syst.* **2012**, *36*, 175–205. [CrossRef]
3. Hultgren, A.; Carleton, T.; Delgado, M.; Gergel, D.R.; Greenstone, M.; Houser, T.; Hsiang, S.; Jina, A.; Kopp, R.E.; Malevich, S.B.; et al. Impacts of climate change on global agriculture accounting for adaptation. *Nature* **2025**, *642*, 644–652. [CrossRef] [PubMed]
4. Lyu, H.; Xing, H.; Duan, T. Optimizing Water Resource Allocation for Food Security: An Evaluation of China's Water Rights Trading Policy. *Sustainability* **2024**, *16*, 10443. [CrossRef]
5. Farah, A.A.; Mohamed, M.A.; Musse, O.S.H.; Nor, B.A. The multifaceted impact of climate change on agricultural productivity: A systematic literature review of SCOPUS-indexed studies (2015–2024). *Discov. Sustain.* **2025**, *6*, 397. [CrossRef]
6. Heesun, W. What Almond Growers Want This Year: Rain and Bees. 2016. Available online: https://www.cnbc.com/2016/01/26/prices-of-mighty-almonds-down-amid-el-nino-related-cold-rain.html (accessed on 21 July 2025).
7. Faunt, C.C.; Belitz, K.; Hanson, R.T. Développement d'un modèle tridimensionnel de la texture sédimentaire des dépôts de remplissage de la Vallée Centrale, Californie, Etats-Unis. *Hydrogeol. J.* **2010**, *18*, 625–649. [CrossRef]
8. Kocis, T.N.; Dahlke, H.E. Availability of high-magnitude streamflow for groundwater banking in the Central Valley, California. *Environ. Res. Lett.* **2017**, *12*, 084009. [CrossRef]
9. Parker, L.E.; Abatzoglou, J.T. Shifts in the thermal niche of almond under climate change. *Clim. Change* **2018**, *147*, 211–224. [CrossRef]
10. Liu, P.-W.; Famiglietti, J.S.; Purdy, A.J.; Adams, K.H.; McEvoy, A.L.; Reager, J.T.; Bindlish, R.; Wiese, D.N.; David, C.H.; Rodell, M. Groundwater depletion in California's Central Valley accelerates during megadrought. *Nat. Commun.* **2022**, *13*, 7825. [CrossRef]
11. Gebremichael, M.; Krishnamurthy, P.K.; Ghebremichael, L.T.; Alam, S. What drives crop land use change during multi-year droughts in California's central valley? Prices or concern for water? *Remote Sens.* **2021**, *13*, 650. [CrossRef]
12. Levy, Z.F.; Jurgens, B.C.; Burow, K.R.; Voss, S.A.; Faulkner, K.E.; Arroyo-Lopez, J.A.; Fram, M.S. Critical Aquifer Overdraft Accelerates Degradation of Groundwater Quality in California's Central Valley during Drought. *Geophys. Res. Lett.* **2021**, *48*, e2021GL094398. [CrossRef]
13. Faunt, C.C.; Traum, J.A.; Boyce, S.E.; Seymour, W.A.; Jachens, E.R.; Brandt, J.T.; Sneed, M.; Bond, S.; Marcelli, M.F. Groundwater Sustainability and Land Subsidence in California's Central Valley. *Water* **2024**, *16*, 1189. [CrossRef]
14. Lees, M.; Knight, R. Quantification of record-breaking subsidence in California's San Joaquin Valley. *Commun. Earth Environ.* **2024**, *5*, 677. [CrossRef]
15. Zhong, L.; Hu, L.; Zhou, H. Deep learning based multi-temporal crop classification. *Remote Sens. Environ.* **2019**, *221*, 430–443. [CrossRef]
16. Atzberger, C. Advances in remote sensing of agriculture: Context description, existing operational monitoring systems and major information needs. *Remote. Sens.* **2013**, *5*, 949–981. [CrossRef]
17. Veloso, A.; Mermoz, S.; Bouvet, A.; Le Toan, T.; Planells, M.; Dejoux, J.-F.; Ceschia, E. Understanding the temporal behavior of crops using Sentinel-1 and Sentinel-2-like data for agricultural applications. *Remote Sens. Environ.* **2017**, *199*, 415–426. [CrossRef]
18. Wang, X.; Zhang, J.; Xun, L.; Wang, J.; Wu, Z.; Henchiri, M.; Zhang, S.; Zhang, S.; Bai, Y.; Yang, S.; et al. Evaluating the Effectiveness of Machine Learning and Deep Learning Models Combined Time-Series Satellite Data for Multiple Crop Types Classification over a Large-Scale Region. *Remote Sens.* **2022**, *14*, 2341. [CrossRef]
19. Wulder, M.A.; Masek, J.G.; Cohen, W.B.; Loveland, T.R.; Woodcock, C.E. Opening the archive: How free data has enabled the science and monitoring promise of Landsat. *Remote Sens. Environ.* **2012**, *122*, 2–10. [CrossRef]
20. Roy, D.P.; Wulder, M.A.; Loveland, T.R.; Woodcock, C.E.; Allen, R.G.; Anderson, M.C.; Helder, D.; Irons, J.R.; Johnson, D.M.; Kennedy, R.; et al. Landsat-8: Science and product vision for terrestrial global change research. *Remote Sens. Environ.* **2014**, *145*, 154–172. [CrossRef]
21. Qu, C.; Li, P.; Zhang, C. A spectral index for winter wheat mapping using multi-temporal Landsat NDVI data of key growth stages. *ISPRS J. Photogramm. Remote Sens.* **2021**, *175*, 431–447. [CrossRef]
22. Wei, L.; Yu, M.; Liang, Y.; Yuan, Z.; Huang, C.; Li, R.; Yu, Y. Precise crop classification using spectral-spatial-location fusion based on conditional random fields for UAV-borne hyperspectral remote sensing imagery. *Remote Sens.* **2019**, *11*, 2011. [CrossRef]
23. Ahmed, M.; Mumtaz, R.; Anwar, Z.; Shaukat, A.; Arif, O.; Shafait, F. A Multi–Step Approach for Optically Active and Inactive Water Quality Parameter Estimation Using Deep Learning and Remote Sensing. *Water* **2022**, *14*, 2112. [CrossRef]

24. Kavzoglu, T.; Colkesen, I. A kernel functions analysis for support vector machines for land cover classification. *Int. J. Appl. Earth Obs. Geoinf.* **2009**, *11*, 352–359. [CrossRef]
25. Belgiu, M.; Drăgu, L. Random forest in remote sensing: A review of applications and future directions. *ISPRS J. Photogramm. Remote Sens.* **2016**, *114*, 24–31. [CrossRef]
26. Noi, P.T.; Kappas, M. Comparison of Random Forest, k-Nearest Neighbor, and Support Vector Machine Classifiers for Land Cover Classification Using Sentinel-2 Imagery. *Sensors* **2017**, *18*, 18. [CrossRef]
27. Wang, P.; Fan, E.; Wang, P. Comparative analysis of image classification algorithms based on traditional machine learning and deep learning. *Pattern Recognit. Lett.* **2021**, *141*, 61–67. [CrossRef]
28. Bahrami, H.; Homayouni, S.; Safari, A.; Mirzaei, S.; Mahdianpari, M.; Reisi-Gahrouei, O. Deep learning-based estimation of crop biophysical parameters using multi-source and multi-temporal remote sensing observations. *Agronomy* **2021**, *11*, 1363. [CrossRef]
29. Shirmard, H.; Farahbakhsh, E.; Müller, R.D.; Chandra, R. A review of machine learning in processing remote sensing data for mineral exploration. *Remote. Sens. Environ.* **2022**, *268*, 112750. [CrossRef]
30. Zhu, L.; Huang, L.; Fan, L.; Huang, J.; Huang, F.; Chen, J.; Zhang, Z.; Wang, Y. Landslide susceptibility prediction modeling based on remote sensing and a novel deep learning algorithm of a cascade-parallel recurrent neural network. *Sensors* **2020**, *20*, 1576. [CrossRef]
31. Ball, J.E.; Anderson, D.T.; Chan, C.S. Comprehensive survey of deep learning in remote sensing: Theories, tools, and challenges for the community. *J. Appl. Remote Sens.* **2017**, *11*, 042609. [CrossRef]
32. Kussul, N.; Lavreniuk, M.; Skakun, S.; Shelestov, A. Deep Learning Classification of Land Cover and Crop Types Using Remote Sensing Data. *IEEE Geosci. Remote Sens. Lett.* **2017**, *14*, 778–782. [CrossRef]
33. Maxwell, A.E.; Warner, T.A.; Fang, F. Implementation of machine-learning classification in remote sensing: An applied review. *Int. J. Remote Sens.* **2018**, *39*, 2784–2817. [CrossRef]
34. Lary, D.J.; Alavi, A.H.; Gandomi, A.H.; Walker, A.L. Machine learning in geosciences and remote sensing. *Geosci. Front.* **2016**, *7*, 3–10. [CrossRef]
35. Herckes, P.; Marcotte, A.R.; Wang, Y.; Collett, J.L. Fog composition in the Central Valley of California over three decades. *Atmos. Res.* **2015**, *151*, 20–30. [CrossRef]
36. Faunt, C.C.; Sneed, M.; Traum, J.; Brandt, J.T. Water availability and land subsidence in the Central Valley, California, USA. *Hydrogeol. J.* **2016**, *24*, 675–684. [CrossRef]
37. Schauer, M.; Senay, G.B. Characterizing crop water use dynamics in the Central Valley of California using Landsat-derived evapotranspiration. *Remote Sens.* **2019**, *11*, 1782. [CrossRef]
38. Lo, M.H.; Famiglietti, J.S. Irrigation in California's Central Valley strengthens the southwestern U.S. water cycle. *Geophys. Res. Lett.* **2013**, *40*, 301–306. [CrossRef]
39. Aktas, T.; Thy, P.; Williams, R.B.; McCaffrey, Z.; Khatami, R.; Jenkins, B.M. Characterization of almond processing residues from the Central Valley of California for thermal conversion. *Fuel Process. Technol.* **2015**, *140*, 132–147. [CrossRef]
40. Sun, J.; Di, L.; Sun, Z.; Shen, Y.; Lai, Z. County-level soybean yield prediction using deep CNN-LSTM model. *Sensors* **2019**, *19*, 4363. [CrossRef]
41. Maulud, D.; Abdulazeez, A.M. A Review on Linear Regression Comprehensive in Machine Learning. *J. Appl. Sci. Technol. Trends* **2020**, *1*, 140–147. [CrossRef]
42. Ray, S. A Quick Review of Machine Learning Algorithms. In Proceedings of the 2019 International Conference on Machine Learning, Big Data, Cloud and Parallel Computing (COMITCon), Faridabad, India, 14–16 February 2019; pp. 35–39.
43. Akgun, A. A comparison of landslide susceptibility maps produced by logistic regression, multi-criteria decision, and likelihood ratio methods: A case study at İzmir, Turkey. *Landslides* **2012**, *9*, 93–106. [CrossRef]
44. Boateng, E.Y.; Abaye, D.A. A Review of the Logistic Regression Model with Emphasis on Medical Research. *J. Data Anal. Inf. Process.* **2019**, *07*, 190–207. [CrossRef]
45. Wickramasinghe, I.; Kalutarage, H. Naive Bayes: Applications, variations and vulnerabilities: A review of literature with code snippets for implementation. *Soft Comput.* **2021**, *25*, 2277–2293. [CrossRef]
46. Jiang, L.; Zhang, L.; Li, C.; Wu, J. A Correlation-Based Feature Weighting Filter for Naive Bayes. *IEEE Trans. Knowl. Data Eng.* **2019**, *31*, 201–213. [CrossRef]
47. Patle, A.; Chouhan, D.S. SVM kernel functions for classification. In Proceedings of the 2013 International Conference on Advances in Technology and Engineering (ICATE), Mumbai, India, 23–25 January 2013; pp. 1–9. [CrossRef]
48. Sheykhmousa, M.; Mahdianpari, M.; Ghanbari, H.; Mohammadimanesh, F.; Ghamisi, P.; Homayouni, S. Support Vector Machine Versus Random Forest for Remote Sensing Image Classification: A Meta-Analysis and Systematic Review. *IEEE J. Sel. Top. Appl. Earth Obs. Remote. Sens.* **2020**, *13*, 6308–6325. [CrossRef]
49. Abu Alfeilat, H.A.; Hassanat, A.B.; Lasassmeh, O.; Tarawneh, A.S.; Alhasanat, M.B.; Salman, H.S.E.; Prasath, V.S. Effects of Distance Measure Choice on K-Nearest Neighbor Classifier Performance: A Review. *Big Data* **2019**, *7*, 221–248. [CrossRef] [PubMed]

50. Ali, I.; Rehman, A.U.; Khan, D.M.; Khan, Z.; Shafiq, M.; Choi, J.G. Model Selection Using K-Means Clustering Algorithm for the Symmetrical Segmentation of Remote Sensing Datasets. *Symmetry* **2022**, *14*, 1149. [CrossRef]
51. Kawabata, T. Gaussian-input Gaussian mixture model for representing density maps and atomic models. *J. Struct. Biol.* **2018**, *203*, 1–16. [CrossRef]
52. Myles, A.J.; Feudale, R.N.; Liu, Y.; Woody, N.A.; Brown, S.D. An introduction to decision tree modeling. *J. Chemom.* **2004**, *18*, 275–285. [CrossRef]
53. Cutler, D.R.; Edwards, T.C., Jr.; Beard, K.H.; Cutler, A.; Hess, K.T.; Gibson, J.; Lawler, J.J. Random forests for classification in ecology. *Ecology* **2007**, *88*, 2783–2792. [CrossRef]
54. Bentéjac, C.; Csörgő, A.; Martínez-Muñoz, G. A comparative analysis of gradient boosting algorithms. *Artif. Intell. Rev.* **2021**, *54*, 1937–1967. [CrossRef]
55. Gupta, A.; Rajput, I.S.; Gunjan; Jain, V.; Chaurasia, S. NSGA-II-XGB: Meta-heuristic feature selection with XGBoost framework for diabetes prediction. *Concurr. Comput.* **2022**, *34*, e7123. [CrossRef]
56. Cigizoglu, H.K. Estimation and forecasting of daily suspended sediment data by multi-layer perceptrons. *Adv. Water Resour.* **2004**, *27*, 185–195. [CrossRef]
57. Su, Z.; Li, W.; Ma, Z.; Gao, R. An improved U-Net method for the semantic segmentation of remote sensing images. *Appl. Intell.* **2022**, *52*, 3276–3288. [CrossRef]
58. Li, C.; Fu, L.; Zhu, Q.; Zhu, J.; Fang, Z.; Xie, Y.; Guo, Y.; Gong, Y. Attention enhanced u-net for building extraction from farmland based on google and worldview-2 remote sensing images. *Remote Sens.* **2021**, *13*, 4411. [CrossRef]
59. Hoorali, F.; Khosravi, H.; Moradi, B. Automatic Bacillus anthracis bacteria detection and segmentation in microscopic images using UNet++. *J. Microbiol. Methods* **2020**, *177*, 106056. [CrossRef] [PubMed]
60. Zhou, Z.; Siddiquee, M.M.R.; Tajbakhsh, N.; Liang, J. UNet++: Redesigning Skip Connections to Exploit Multiscale Features in Image Segmentation. *IEEE Trans. Med. Imaging* **2020**, *39*, 1856–1867. [CrossRef] [PubMed]
61. Xu, Y.; Lam, H.K.; Jia, G. MANet: A two-stage deep learning method for classification of COVID-19 from Chest X-ray images. *Neurocomputing* **2021**, *443*, 96–105. [CrossRef]
62. Chen, B.; Xia, M.; Qian, M.; Huang, J. MANet: A multi-level aggregation network for semantic segmentation of high-resolution remote sensing images. *Int. J. Remote Sens.* **2022**, *43*, 5874–5894. [CrossRef]
63. Pravitasari, A.; Asnawi, M.; Nugraha, F.; Darmawan, G.; Hendrawati, T. Enhancing 3D Lung Infection Segmentation with 2D U-Shaped Deep Learning Variants. *Appl. Sci.* **2023**, *13*, 11640. [CrossRef]
64. Zhang, Y.; Ding, F.; Kwong, S.; Zhu, G. Feature pyramid network for diffusion-based image inpainting detection. *Inf. Sci.* **2021**, *572*, 29–42. [CrossRef]
65. Dhalla, S.; Maqbool, J.; Mann, T.S.; Gupta, A.; Mittal, A.; Aggarwal, P.; Saluja, K.; Kumar, M.; Saini, S.S. Semantic segmentation of palpebral conjunctiva using predefined deep neural architectures for anemia detection. *Procedia Comput. Sci.* **2023**, *218*, 328–337. [CrossRef]
66. Shi, T.; Guo, Z.; Li, C.; Lan, X.; Gao, X.; Yan, X. Improvement of deep learning Method for water body segmentation of remote sensing images based on attention modules. *Earth Sci. Inf.* **2023**, *16*, 2865–2876. [CrossRef]
67. Yin, Y.; Guo, Y.; Deng, L.; Chai, B. Improved PSPNet-based water shoreline detection in complex inland river scenarios. *Complex. Intell. Syst.* **2023**, *9*, 233–245. [CrossRef]
68. Zhang, X.Y.; Rahman, A.H.A.; Qamar, F. Semantic visual simultaneous localization and mapping (SLAM) using deep learning for dynamic scenes. *PeerJ Comput. Sci.* **2023**, *9*, e1628. [CrossRef]
69. Cai, C.; Tan, J.; Zhang, P.; Ye, Y.; Zhang, J. Determining Strawberries' Varying Maturity Levels by Utilizing Image Segmentation Methods of Improved DeepLabV3+. *Agronomy* **2022**, *12*, 1875. [CrossRef]
70. Abdulkareem, K.H.; Mohammed, M.A.; Gunasekaran, S.S.; Al-Mhiqani, M.N.; Mutlag, A.A.; Mostafa, S.A.; Ali, N.S.; Ibrahim, D.A. A review of fog computing and machine learning: Concepts, applications, challenges, and open issues. *IEEE Access* **2019**, *7*, 153123–153140. [CrossRef]
71. Wang, X.; Huang, J.; Feng, Q.; Yin, D. Winter wheat yield prediction at county level and uncertainty analysis in main wheat-producing regions of China with deep learning approaches. *Remote Sens.* **2020**, *12*, 1744. [CrossRef]
72. Nguyen, G.; Dlugolinsky, S.; Bobák, M.; Tran, V.; García, Á.L.; Heredia, I.; Malík, P.; Hluchý, L. Machine Learning and Deep Learning frameworks and libraries for large-scale data mining: A survey. *Artif. Intell. Rev.* **2019**, *52*, 77–124. [CrossRef]
73. Mukhamediev, R.I.; Popova, Y.; Kuchin, Y.; Zaitseva, E.; Kalimoldayev, A.; Symagulov, A.; Levashenko, V.; Abdoldina, F.; Gopejenko, V.; Yakunin, K.; et al. Review of Artificial Intelligence and Machine Learning Technologies: Classification, Restrictions, Opportunities and Challenges. *Mathematics* **2022**, *10*, 2552. [CrossRef]
74. Butt, U.A.; Mehmood, M.; Shah, S.B.H.; Amin, R.; Shaukat, M.W.; Raza, S.M.; Suh, D.Y.; Piran, J. A review of machine learning algorithms for cloud computing security. *Electronics* **2020**, *9*, 1379. [CrossRef]
75. Gill, S.S.; Xu, M.; Ottaviani, C.; Patros, P.; Bahsoon, R.; Shaghaghi, A.; Golec, M.; Stankovski, V.; Wu, H.; Abraham, A.; et al. AI for next generation computing: Emerging trends and future directions. *Internet Things* **2022**, *19*, 100514. [CrossRef]

76. Alsadie, D. A Comprehensive Review of AI Techniques for Resource Management in Fog Computing: Trends, Challenges, and Future Directions. *IEEE Access* **2024**, *12*, 118007–118059. [CrossRef]
77. Hasan, A.S.M.M.; Diepeveen, D.; Laga, H.; Jones, M.G.; Sohel, F. Image patch-based deep learning approach for crop and weed recognition. *Ecol. Inf.* **2023**, *78*, 102361. [CrossRef]
78. Streiner, D.L.; Norman, G.R. 'Precision' and 'accuracy': Two terms that are neither. *J. Clin. Epidemiol.* **2006**, *59*, 327–330. [CrossRef] [PubMed]
79. Suji, R.J.; Godfrey, W.W.; Dhar, J. Exploring pretrained encoders for lung nodule segmentation task using LIDC-IDRI dataset. *Multimed. Tools Appl.* **2024**, *83*, 9685–9708. [CrossRef]
80. Wardhani, N.W.S.; Rochayani, M.Y.; Iriany, A.; Sulistyono, A.D.; Lestantyo, P. Cross-validation Metrics for Evaluating Classification Performance on Imbalanced Data. In Proceedings of the 2019 International Conference on Computer, Control, Informatics and its Applications (IC3INA), Tangerang, Indonesia, 23–24 October 2019; pp. 14–18. [CrossRef]
81. Ali, U.A.M.E.; Hossain, M.A. Feature Subspace Detection for Hyperspectral Images Classification using Segmented Principal Component Analysis and F-score. In Proceedings of the 2020 IEEE Region 10 Symposium (TENSYMP), Dhaka, Bangladesh, 5–7 June 2020; pp. 134–137. [CrossRef]
82. Zhao, J.; Wang, X.; Dou, X.; Zhao, Y.; Fu, Z.; Guo, M.; Zhang, R. A high-precision image classification network model based on a voting mechanism. *Int. J. Digit. Earth* **2022**, *15*, 2168–2183. [CrossRef]
83. Flint, L.E.; Flint, A.L.; Mendoza, J.; Kalansky, J.; Ralph, F.M. Characterizing drought in California: New drought indices and scenario-testing in support of resource management. *Ecol. Process* **2018**, *7*, 1. [CrossRef]
84. Hanak, E.; Lund, J.R. Adapting California's water management to climate change. *Clim. Change* **2012**, *111*, 17–44. [CrossRef]
85. He, X.; Wada, Y.; Wanders, N.; Sheffield, J. Intensification of hydrological drought in California by human water management. *Geophys. Res. Lett.* **2017**, *44*, 1777–1785. [CrossRef]
86. Melton, F.S.; Johnson, L.F.; Lund, C.P.; Pierce, L.L.; Michaelis, A.R.; Hiatt, S.H.; Guzman, A.; Adhikari, D.D.; Purdy, A.J.; Rosevelt, C.; et al. Satellite irrigation management support with the terrestrial observation and prediction system: A framework for integration of satellite and surface observations to support improvements in agricultural water resource management. *IEEE J. Sel. Top. Appl. Earth Obs. Remote Sens.* **2012**, *5*, 1709–1721. [CrossRef]
87. Wilson, T.S.; Sleeter, B.M.; Cameron, D.R. Future land-use related water demand in California. *Environ. Res. Lett.* **2016**, *11*, 054018. [CrossRef]
88. Cracknell, M.J.; Reading, A.M. Geological mapping using remote sensing data: A comparison of five machine learning algorithms, their response to variations in the spatial distribution of training data and the use of explicit spatial information. *Comput. Geosci.* **2014**, *63*, 22–33. [CrossRef]
89. Hudait, M.; Patel, P.P. Crop-type mapping and acreage estimation in smallholding plots using Sentinel-2 images and machine learning algorithms: Some comparisons. *Egypt. J. Remote Sens. Space Sci.* **2022**, *25*, 147–156. [CrossRef]
90. Wu, L.; Zhu, X.; Lawes, R.; Dunkerley, D.; Zhang, H. Comparison of machine learning algorithms for classification of LiDAR points for characterization of canola canopy structure. *Int. J. Remote Sens.* **2019**, *40*, 5973–5991. [CrossRef]
91. Islam, N.; Rashid, M.; Wibowo, S.; Xu, C.-Y.; Morshed, A.; Wasimi, S.A.; Moore, S.; Rahman, S.M. Early weed detection using image processing and machine learning techniques in an australian chilli farm. *Agriculture* **2021**, *11*, 387. [CrossRef]
92. Jwo, D.-J.; Chiu, S.-F. Deep learning based automated detection of diseases from apple leaf images. *Comput. Mater. Contin.* **2022**, *71*, 1849–1866. [CrossRef]
93. Lamba, M.; Gigras, Y.; Dhull, A. Classification of plant diseases using machine and deep learning. *Open Comput. Sci.* **2021**, *11*, 491–508. [CrossRef]
94. Feizizadeh, B.; Alajujeh, K.M.; Lakes, T.; Blaschke, T.; Omarzadeh, D. A comparison of the integrated fuzzy object-based deep learning approach and three machine learning techniques for land use/cover change monitoring and environmental impacts assessment. *GIsci Remote Sens.* **2021**, *58*, 1543–1570. [CrossRef]
95. Kirola, M.; Singh, N.; Joshi, K.; Chaudhary, S.; Gupta, A. Plants Diseases Prediction Framework: A Image-Based System Using Deep Learning. In Proceedings of the 2022 IEEE World Conference on Applied Intelligence and Computing (AIC), Sonbhadra, India, 17–19 June 2022. [CrossRef]
96. Sujatha, R.; Chatterjee, J.M.; Jhanjhi, N.Z.; Brohi, S.N. Performance of deep learning vs machine learning in plant leaf disease detection. *Microprocess. Microsyst.* **2021**, *80*, 103615. [CrossRef]
97. Yao, J.; Wu, J.; Xiao, C.; Zhang, Z.; Li, J. The Classification Method Study of Crops Remote Sensing with Deep Learning, Machine Learning, and Google Earth Engine. *Remote Sens.* **2022**, *14*, 2758. [CrossRef]
98. Gómez-Carmona, O.; Casado-Mansilla, D.; Kraemer, F.A.; López-de-Ipiña, D.; García-Zubia, J. Exploring the computational cost of machine learning at the edge for human-centric Internet of Things. *Future Gener. Comput. Syst.* **2020**, *112*, 670–683. [CrossRef]
99. Southworth, J.; Smith, A.C.; Safaei, M.; Rahaman, M.; Alruzuq, A.; Tefera, B.B.; Muir, C.S.; Herrero, H.V. Machine learning versus deep learning in land system science: A decision-making framework for effective land classification. *Front. Remote. Sens.* **2024**, *5*, 1374862. [CrossRef]

100. Dosovitskiy, A.; Beyer, L.; Kolesnikov, A.; Weissenborn, D.; Zhai, X.; Unterthiner, T.; Dehghani, M.; Minderer, M.; Heigold, G.; Gelly, S.; et al. An Image is Worth 16x16 Words: Transformers for Image Recognition at Scale. *arXiv* **2021**, arXiv:2010.11929. [CrossRef]
101. Liu, Z.; Lin, Y.; Cao, Y.; Hu, H.; Wei, Y.; Zhang, Z.; Lin, S.; Guo, B. Swin Transformer: Hierarchical Vision Transformer using Shifted Windows. In Proceedings of the 2021 IEEE/CVF International Conference on Computer Vision (ICCV), Montreal, QC, Canada, 10–17 October 2021; pp. 9992–10002. [CrossRef]
102. Touvron, H.; Cord, M.; Douze, M.; Massa, F.; Sablayrolles, A.; Jégou, H. Training data-efficient image transformers & distillation through attention. *arXiv* **2021**, arXiv:2012.12877. [CrossRef]
103. Ganjirad, M.; Bagheri, H. Google Earth Engine-based mapping of land use and land cover for weather forecast models using Landsat 8 imagery. *Ecol. Inf.* **2024**, *80*, 102498. [CrossRef]
104. Pande, C.B.; Srivastava, A.; Moharir, K.N.; Radwan, N.; Sidek, L.M.; Alshehri, F.; Pal, S.C.; Tolche, A.D.; Zhran, M. Characterizing land use/land cover change dynamics by an enhanced random forest machine learning model: A Google Earth Engine implementation. *Environ. Sci. Eur.* **2024**, *36*, 84. [CrossRef]
105. Brown, C.F.; Brumby, S.P.; Guzder-Williams, B.; Birch, T.; Hyde, S.B.; Mazzariello, J.; Czerwinski, W.; Pasquarella, V.J.; Haertel, R.; Ilyushchenko, S.; et al. Dynamic World, Near real-time global 10 m land use land cover mapping. *Sci. Data* **2022**, *9*, 251. [CrossRef]
106. Miller, L.; Pelletier, C.; Webb, G.I. Deep Learning for Satellite Image Time-Series Analysis: A review. *IEEE Geosci. Remote Sens. Mag.* **2024**, *12*, 81–124. [CrossRef]
107. Qichi, Y.; Lihui, W.; Jinliang, H.; Linzhi, L.; Xiaodong, L.; Fei, X.; Yun, D.; Xue, Y.; Feng, L. A novel alpine land cover classification strategy based on a deep convolutional neural network and multi-source remote sensing data in Google Earth Engine. *GIsci Remote Sens.* **2023**, *60*, 2233756. [CrossRef]

remote sensing

MDPI

Article

Panoptic Plant Recognition in 3D Point Clouds: A Dual-Representation Learning Approach with the PP3D Dataset

Lin Zhao [1,2,3], Sheng Wu [2,3], Jiahao Fu [2,3], Shilin Fang [2,3], Shan Liu [1,2,3,4,*] and Tengping Jiang [1,2,3,4]

1 Key Laboratory of Degraded and Unused Land Consolidation Engineering, Ministry of Natural Resources, Xi'an 710075, China
2 Jiangsu Center for Collaborative Innovation in Geographical Information Resource Development and Application, Nanjing 210023, China
3 State Key Laboratory of Climate System Prediction and Risk Management, Nanjing Normal University, Nanjing 210023, China
4 Technology Innovation Center for Integrated Applications in Remote Sensing and Navigation, Ministry of Natural Resources, Nanjing 210044, China
* Correspondence: liushan@njnu.edu.cn

Abstract

The advancement of Artificial Intelligence (AI) has significantly accelerated progress across various research domains, with growing interest in plant science due to its substantial economic potential. However, the integration of AI with digital vegetation analysis remains underexplored, largely due to the absence of large-scale, real-world plant datasets, which are crucial for advancing this field. To address this gap, we introduce the PP3D dataset—a meticulously labeled collection of about 500 potted plants represented as 3D point clouds, featuring fine-grained annotations for approximately 20 species. The PP3D dataset provides 3D phenotypic data for about 20 plant species spanning model organisms (e.g., *Arabidopsis thaliana*), potted plants (e.g., Foliage plants, Flowering plants), and horticultural plants (e.g., *Solanum lycopersicum*), covering most of the common important plant species. Leveraging this dataset, we propose the panoptic plant recognition task, which combines semantic segmentation (stems and leaves) with leaf instance segmentation. To tackle this challenge, we present SCNet, a novel dual-representation learning network designed specifically for plant point cloud segmentation. SCNet integrates two key branches: a cylindrical feature extraction branch for robust spatial encoding and a sequential slice feature extraction branch for detailed structural analysis. By efficiently propagating features between these representations, SCNet achieves superior flexibility and computational efficiency, establishing a new baseline for panoptic plant recognition and paving the way for future AI-driven research in plant science.

Keywords: point cloud dataset; plant phenotyping; point cloud segmentation; deep learning

1. Introduction

Three-dimensional (3D) plant phenotyping, a cornerstone of precision agriculture and smart farming [1], faces a critical challenge: the scarcity of high-quality, specialized datasets. Current phenotypic traits encode substantial geometric and semantic information, enabling high-fidelity plant reconstruction. However, conventional analysis methods remain confined to small-scale, destructive manual measurements [2]. Given the rapidly growing demand for high-throughput phenotyping [3], the development of standardized datasets and associated algorithms has emerged as an urgent priority for the field.

Despite recent advances in plant-phenotyping point cloud datasets (e.g., Soybean-MVS, PlanesT-3D, and Pheno4D), three fundamental challenges remain unresolved: (1) the intricate morphological architecture of plant organs [4], (2) the technical hurdles of data collection in dynamic growth conditions, and (3) the inadequate compatibility of conventional 3D processing approaches with plant biological properties [5–7]. These limitations severely constrain both the scope and precision of current datasets for deep learning applications in phenotyping [8]. In response, we introduce PP3D, a groundbreaking 3D point cloud benchmark specifically developed for plant phenotyping research, featuring three key advancements—(1) hybrid data acquisition: combining adaptive multi-view imaging with advanced stereo reconstruction techniques to produce high-resolution point clouds with preserved morphological fidelity; (2) comprehensive organ annotation: implementing rigorous annotation protocols to deliver precise semantic and instance segmentation at the stem and leaf level; (3) ecological diversity: encompassing approximately 20 agriculturally significant species with nearly 500 temporally tracked specimens.

To fully leverage the dataset's potential, we developed the SCNet benchmark framework with two novel architectural innovations: (1) a cylindrical feature extraction module that effectively encodes global plant architecture; (2) a sequential slice analysis module that maintains fine-scale morphological details. Our experimental results show that SCNet outperforms existing state-of-the-art methods by more than 10% in accuracy metrics. The key contributions of this work are threefold: First, we introduce the PP3D dataset, a novel resource that significantly advances plant phenotypic parameter extraction and supports the development of smart agriculture technologies. Second, we formulate the panoptic plant recognition task, unifying both semantic (stem/leaf classification) and instance-level (individual leaf identification) segmentation into a comprehensive framework. Third, we propose an innovative baseline method employing a flexible dual-representation approach that establishes new state-of-the-art performance on the PP3D dataset.

The remainder of this paper is organized as follows: Following this introduction, Section 2 provides a comprehensive review of related work in plant dataset and 3D point cloud analysis. Section 3 provides a comprehensive description of the dataset construction (see Figure 1) and its accompanying benchmark algorithms. Section 4 presents extensive experimental results, including benchmark comparisons, ablation studies, and real-world application scenarios. Finally, Section 5 concludes the paper with a summary of key contributions and discussions of future research directions.

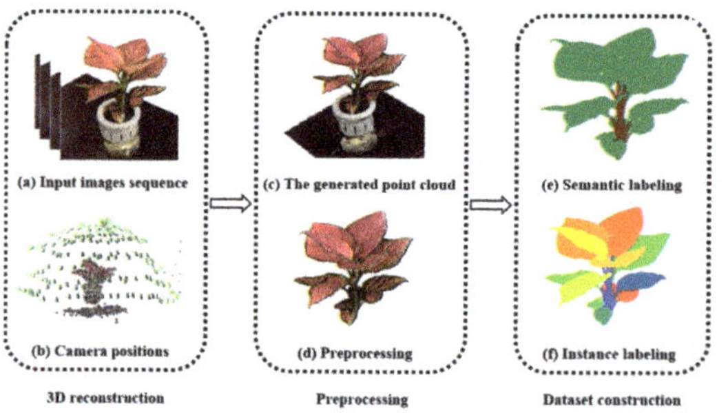

Figure 1. The PP3D dataset was developed through a systematic three-stage pipeline: (1) 3D reconstruction, (2) preprocessing, and (3) dataset compilation. For each plant specimen, this workflow was implemented through six sequential processing steps: initial image capture, feature point extraction, sparse reconstruction, dense reconstruction, point cloud refinement, and final data packaging. This structured approach ensured both the consistency and quality of all specimens within the PP3D collection.

2. Related Work

Recent advances in high-resolution plant point cloud datasets have greatly advanced the development of plant phenotyping algorithms. However, panoramic recognition—a crucial component of phenotyping analysis—still faces notable limitations. This section provides a focused review of existing plant point cloud datasets and state-of-the-art panoramic recognition techniques.

2.1. Three-Dimensional Plant Datasets

Current 3D segmentation datasets predominantly concentrate on common objects and scenes across various domains. In autonomous driving scenarios, benchmark datasets like SemanticKITTI [9], Panoptic nuScenes [10], and SemanticSTF [11] provide comprehensive annotations for typical transportation environment elements, including road surfaces, sidewalks, vehicles, and terrain features. The SensatUrban [12] dataset significantly advances urban scene understanding by offering detailed semantic labels for complex street environments. For indoor scene analysis, datasets such as S3DIS [13], ScanNet [14], and PartNet [15] facilitate fine-grained object recognition through their meticulous annotations of household items like tables, chairs, and cups. These diverse datasets have catalyzed significant progress in panorama segmentation, particularly in fine-grained scene understanding, establishing it as an increasingly prominent research direction.

The automatic acquisition of plant phenotypic information through high-precision 3D plant modeling has emerged as a significant research focus in recent years [16]. The accuracy of plant organ segmentation plays a pivotal role in determining both the quantification of plant structural characteristics and the reliability of phenotypic analysis. Deep learning-based approaches for plant point cloud segmentation have become the predominant methodology in this field [17,18]. However, the lack of comprehensive and high-quality datasets currently hinders further advancements in this research direction. To systematically evaluate the available resources, we conducted a comparative analysis of five existing plant point cloud datasets, as summarized in Table 1.

Soybean-MVS [19] is a comprehensive 3D point cloud dataset capturing soybean plants throughout their complete growth cycle for organ-level segmentation tasks. The dataset was constructed using multi-view stereo (MVS) reconstruction techniques, providing high-fidelity point clouds of 102 soybean samples spanning five distinct varieties across 13 developmental stages. Each sample is meticulously annotated with three key organ categories: leaf, main stem, and branch stem, enabling fine-grained structural analysis and phenotypic studies.

PlanesT-3D [20] is a high-quality 3D plant point cloud dataset comprising 34 samples across three species: 10 pepper plants, 10 rose-bush plants, and 14 ribes plants. Each sample consists of a complete, colorized 3D point cloud reconstructed using MVS techniques. The dataset provides detailed semantic annotations for two key organ categories: stem and leaf, with the latter further segmented at the instance level to enable fine-grained plant structure analysis.

Pheno4D [21] is a high-resolution, multi-temporal plant point cloud dataset acquired through high-precision 3D laser scanning, achieving sub-millimeter accuracy. The dataset features maize and tomato plants captured throughout their growth cycles. Each point cloud is manually annotated with three semantic categories: soil, stem, and leaf, where leaves are further segmented at the instance level (i.e., each individual leaf is delineated as a distinct object). This fine-grained annotation enables detailed morphological and phenotypic analysis of plant development over time.

Table 1. Comparative analysis of existing plant point cloud datasets.

Dataset	Soybean-MVS	PLANesT-3D	Pheno-4D	ROSE-X	Plant3D	PP3D (Ours)
Year	2023	2023	2021	2020	2017	2025
Plant Species	soybean	pepper, rosebush, and ribes	maize and tomato	rosebush	tomato, tobacco and sorghum	20 species
Acquisition Method	reconstructed using MVS technology	reconstructed from 2D color images of real plants through MVS	measured with a highly accurate 3D laser scanning system with a spatial precision of less than a tenth of a millimeter	acquired through X-ray scanning	mapped using high-precision 3D laser scanning	partly collected by MVS and partly by photogrammetry
Sensor	SLR digital camera	MVS system	Laser scanning	X-ray tomography	Laser scanning	MVS and photogrammetry system
Color	Yes	Yes	No	No	No	Yes
Number of Point Clouds	102	34	126	11	505	~500
Labeled Classes	leaf, main stem and stem	leaf and stem	soil, stem and leaf	leaf, stem, flower and pot	-	leaf and stem
Organ-level Label	No	Yes	Yes	No	-	Yes

ROSE-X [22] is a high-resolution 3D point cloud dataset comprising 11 rosebush specimens acquired through X-ray computed tomography (CT) scanning, enabling non-destructive structural analysis. The dataset achieves exceptional spatial resolution, with 0.5 mm inter-leaf spacing and 1 mm pixel spacing, capturing intricate plant architectures with high fidelity. Each sample is annotated with four semantic classes: leaf, stem, flower, and pot, providing comprehensive ground truth for plant organ segmentation tasks. The X-ray acquisition methodology particularly facilitates the study of internal plant structures while preserving specimen integrity.

Plant3D [23] represents a comprehensive 3D plant architecture dataset encompassing 505 high-resolution scans acquired through precision 3D laser scanning. The dataset systematically captures three agriculturally significant species—tomato, tobacco, and sorghum—under 35 controlled environmental conditions (including light variations, thermal stress, and drought) across 20 developmental time points. This multidimensional design enables the robust investigation of genotype-by-environment interactions and temporal growth patterns. Each scan provides detailed structural information, making the dataset particularly valuable for studies in plant phenomics, stress response analysis, and developmental biology.

While existing plant point cloud datasets (e.g., Pheno4D, Plant3D) capture multi-temporal growth phases, they remain limited in scale, diversity, and practical applicability due to constrained environmental conditions and temporal coverage. To address these limitations, this paper introduces PP3D, a novel large-scale dataset featuring the following: (1) high-precision 3D reconstructions of diverse crop species; (2) comprehensive growth-stage coverage under real agricultural conditions; (3) annotated structural phenotypes tailored for precision farming applications. This work aims to provide robust data infrastructure for advancing smart agriculture technologies, including automated growth monitoring, yield prediction, and resource optimization systems.

2.2. Three-Dimensional Panoptic Segmentation

Panoptic segmentation has emerged as a unified approach to scene understanding that combines semantic segmentation (class-level classification) and instance segmentation (object distinction) into a single framework. For LiDAR point cloud analysis, this task involves simultaneously classifying each point into semantic categories while assigning unique instance identifiers, thereby providing comprehensive 3D scene interpretation. The development of benchmark datasets such as Panoptic nuScenes [10] and the extended SemanticKITTI [24] has significantly advanced research in this domain, leading to two predominant methodological paradigms: detection-based approaches that first identify object instances before classification, and grouping-based methods that cluster points before assigning semantic and instance labels.

Detection-based panoptic segmentation methods typically follow a two-stage approach: first generating foreground object proposals, then refining and merging them with background semantic segmentation results. For instance, EfficientLPS [25] employs point similarity metrics to model geometric transformations between points and images, while augmenting Mask R-CNN with a semantic head to produce panoptic segmentation outputs. Building on this framework, Panoptic-TrackNet [26] extends EfficientPS [27] by incorporating a tracking head, thereby unifying panoptic segmentation with object tracking in a multi-task learning paradigm. While these approaches demonstrate promising results, their overall effectiveness remains fundamentally constrained by the performance of the underlying object detection component.

Grouping-based panoptic segmentation methods typically follow a two-stage approach, first predicting point-wise semantic labels, then applying clustering algorithms to generate instance predictions [28–30]. For instance, Panoptic-PolarNet [31] transforms the problem by projecting point cloud features into a polar Bird's Eye View (BEV) representation, where instance segmentation is achieved through centroid regression. Building on this BEV paradigm, Panoptic-PHNet [32] further improves clustering efficiency by introducing pseudo-heatmaps and a dedicated centroid grouping module. Another notable approach, DSNet [33], enhances the clustering process through its innovative Dynamic Shifting module, which adaptively adjusts point features for more accurate instance segmentation. These methods demonstrate how geometric transformations and adaptive feature learning can effectively address the instance grouping challenge in 3D point clouds.

Recent advances in agricultural point cloud analysis have developed innovative approaches for extracting phenotypic information from plant structures. Several notable methods have emerged to address the challenges of plant organ segmentation: PlantNet [18] introduced a dual-function network architecture capable of simultaneous semantic and instance segmentation for both dicotyledonous and monocotyledonous plants in the Plant3D dataset. PST [7] leveraged transformer architectures to achieve high-resolution semantic segmentation of rapeseed plants from handheld laser scanning (HLS) data. PSegNet [34] advanced the field through its novel double-neighborhood feature extraction and double-granularity feature fusion mechanisms, enabling joint semantic and leaf instance segmentation for multiple plant species.

Alternative approaches have focused on reducing annotation requirements while maintaining accuracy. Eff-3DPSeg [35] demonstrated that weakly supervised learning can achieve comparable performance to fully supervised methods for plant shoot segmentation across both soybean and Pheno4D datasets. Complementing these data-driven methods [36], proposed a graph-based framework that incorporates botanical structural constraints to enable rapid and accurate organ-level segmentation while preserving plant architecture integrity. These diverse methodologies collectively address key challenges in plant phenotyping, ranging from handling complex plant architectures to reducing dependency on labeled data, while consistently validating their approaches on established benchmark datasets.

While numerous methods have been developed for plant phenotyping using point cloud data, current approaches face two fundamental limitations that constrain their practical application: (1) the restricted scale and diversity of existing plant datasets, and (2) the inability of current models to achieve sufficiently fine-grained phenotypic analysis. These constraints highlight the critical need for developing more robust plant phenotyping frameworks specifically designed to meet the requirements of precision agriculture and smart farming systems. Future research should prioritize creating comprehensive datasets and advanced analytical models capable of supporting detailed, large-scale phenotypic characterization that can directly inform agricultural decision-making and automated management practices.

3. Materials and Methods

To address the growing demand for 3D plant segmentation amid limited available datasets, we introduce PP3D—a densely annotated point cloud dataset—along with an efficient segmentation method.

3.1. The PP3D Dataset

PP3D is a large-scale, publicly available dataset designed for panoramic plant identification using 3D point clouds. The dataset comprises photogrammetric point clouds reconstructed from approximately 2 million high-resolution images spanning 20 plant species and about 500 individual potted plants. Using advanced 3D reconstruction techniques, we generated dense, structurally accurate point clouds from these images, enabling detailed plant organ analysis. All point clouds in PP3D are annotated with dual-level labels: semantic segmentation (distinguishing stems and leaves) and instance segmentation (identifying individual organs). The data were collected throughout 2023 across multiple locations in Eastern China, capturing seasonal growth variations and environmental diversity. With its unprecedented scale, multimodal 3D representations, and fine-grained annotations, PP3D serves as a foundational resource for applications in plant phenotyping, agricultural robotics, and 3D computer vision. The PP3D dataset comprises thousands of high-precision 3D models of potted plants across multiple species. This section provides a comprehensive overview of the dataset construction pipeline, detailing each critical stage from initial data acquisition to final dataset preparation.

3.1.1. Images Acquisition

The development of the PP3D dataset commenced with RGB image acquisition of representative potted plant species from Eastern China, including but not limited to jade plants (*Crassula ovata*), money trees (*Pachira aquatica*), mint (*Mentha* spp.), and kalanchoe. During preliminary data collection, we faced several technical challenges. Initial image sets showed inadequate overlap because of inconsistent camera paths and poorly planned shooting angles. Compounding these issues, variable lighting conditions produced problematic glare and shadow artifacts that degraded image quality and consequently affected 3D reconstruction accuracy. To overcome these limitations, we implemented a standardized imaging protocol and refined our environmental controls.

We captured all images using smartphone cameras with high-resolution sensors. Our standardized imaging protocol used 4K UHD resolution (3840 $\times$ 2160 pixels) at 30 frames per second to precisely document plant morphological features. All images were saved in standard RGB format (red, green, blue channels) to ensure reliable color-based analysis. To improve dataset generalization, we systematically varied both environmental conditions and lighting parameters during image collection. For optimal 3D point cloud reconstruction quality, we created a specialized multi-ring scanning protocol. Each potted plant was systematically imaged using our tiered capture system, which employs carefully calibrated camera paths and viewing angles to fully document all plant structures—from stems and leaves to the entire canopy. Specifically, our imaging protocol followed three concentric circular paths: lower foliage, mid-canopy, and upper crown regions. Between consecutive shots in each zone, we consistently maintained 70–80% image overlap—ensuring robust feature matching for downstream 3D reconstruction. Our multi-level imaging protocol typically captures about 250 photos per plant, with extra coverage dedicated to the upper canopy. The three-ring scanning pattern minimizes leaf overlap issues, producing comprehensive 3D models of every leaf surface in the crown area.

3.1.2. Three-Dimensional Point Cloud Reconstruction

To ensure high-quality 3D point cloud reconstruction, we implemented a comprehensive preprocessing pipeline for the original plant images. The preprocessing consisted of four key steps: First, we conducted rigorous quality control by eliminating images affected by human-induced blurring, insufficient overlap, or other quality issues. This initial screening ensured only optimal images were retained for subsequent processing. Second, we applied fundamental image enhancement techniques, including noise reduction and lens distortion correction, to improve the overall image quality. Since most plant images were captured in indoor environments without GPS metadata, we established a unified coordinate system to reconstruct the relative positional relationships between camera stations. This approach enabled the accurate restoration of shooting positions and the effective integration of images from multiple viewing angles. Finally, to minimize interference during point cloud reconstruction, we performed precise foreground segmentation. This process selectively preserved the relevant elements (the plant, pot, and placement surface) while removing extraneous background clutter, thereby focusing the reconstruction on the essential botanical features. These preprocessing steps collectively enhanced the accuracy and reliability of our subsequent 3D reconstruction workflow.

Leveraging high-resolution, multi-angle plant images, we performed 3D point cloud reconstruction through feature matching and multi-view geometry analysis. Our methodology employed two complementary approaches: a direct sparse reconstruction method that rapidly generated raw point clouds but with limited detail in foliar structures, and a more computationally intensive high-fidelity pipeline that first created detailed 3D mesh models before converting them to dense, accurate point clouds. While the former provided efficient preliminary reconstructions, the latter delivered superior results with complete botanical features. This dual-method approach enabled us to successfully generate approximately 500 plant point clouds with varying levels of detail suitable for different analytical purposes.

Our study on 3D plant point cloud reconstruction revealed a significant dependence of reconstruction quality on the number of input photographs. To systematically investigate this relationship, we conducted a controlled experiment using a reference dataset of 250 uniformly captured plant images under consistent lighting conditions. Through randomized sampling from this dataset, we evaluated three representative sample sizes: 200 (Figure 2a), 100 (Figure 2b), and 75 (Figure 2c) images. The results demonstrate a clear quality trend: with 200 images, we achieved optimal reconstruction with complete plant morphology; reducing samples to 100 introduced visible artifacts including leaf surface holes; further reduction to 75 images exacerbated these issues, causing both significant scene distortion (particularly in the supporting table structure) and substantial increase in surface discontinuities. These observations suggest that reconstruction quality follows a logarithmic decay curve with respect to sample size reduction, providing important practical guidance for balancing data acquisition effort with reconstruction fidelity in plant phenotyping applications.

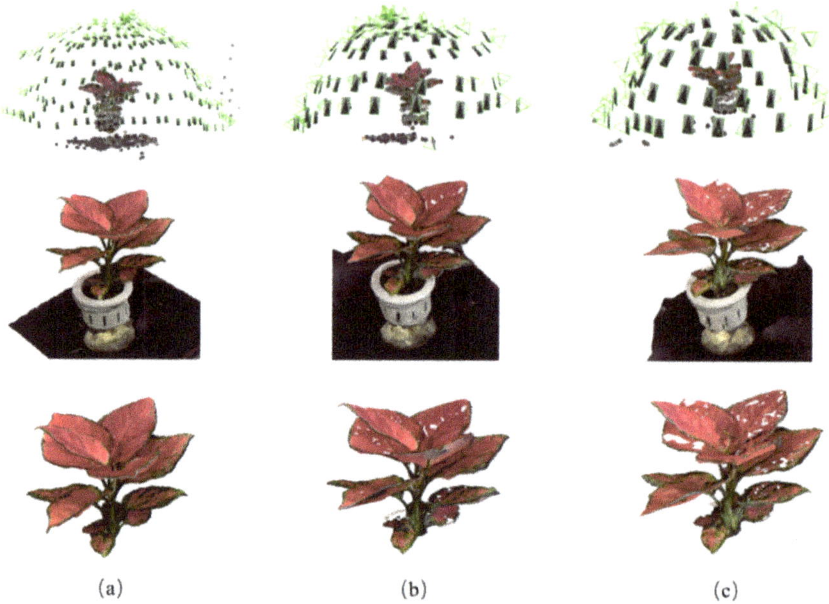

Figure 2. Comparative modeling results showing the impact of photographic sample size on reconstruction quality: (**a**) 200 images, (**b**) 100 images, and (**c**) 75 images. Note the progressive degradation in model completeness and surface quality with the decreasing image count.

For standard plant specimens similar to the reference sample shown, we collected 300 photographs per subject—exceeding the empirically determined optimal count of 250 images—to ensure robust reconstruction quality. This 20% buffer accounts for potential image quality variations during processing. To maintain dataset complexity, we additionally included a subset of structurally complex specimens requiring customized sampling strategies. For these challenging cases, we implemented adaptive image acquisition protocols where the sample size was dynamically adjusted based on the real-time assessment of morphological complexity, typically ranging between 300 and 400 photographs depending on foliage density and structural intricacy.

3.1.3. Point Cloud Annotation

To enable advanced organ-level analysis within our dataset, we implemented a comprehensive point-wise annotation system with hierarchical labeling. Each data point was meticulously annotated with: (1) a primary classification as either "stem" or "leaf", and (2) a unique organ-specific identifier distinguishing individual leaves within the same plant. As illustrated in Figure 3, our annotation pipeline leveraged the open-source software CloudCompare version 2.12.4 for precise manual segmentation of plant morphological structures. The annotated point clouds incorporate seven key data dimensions stored in TXT format: XYZ coordinates (spatial positioning), RGB values (true color information), C (binary stem/leaf classification), and I (unique leaf identification code). This multi-layered labeling scheme provides researchers with unprecedented granularity for structural and functional analyses of plant architecture.

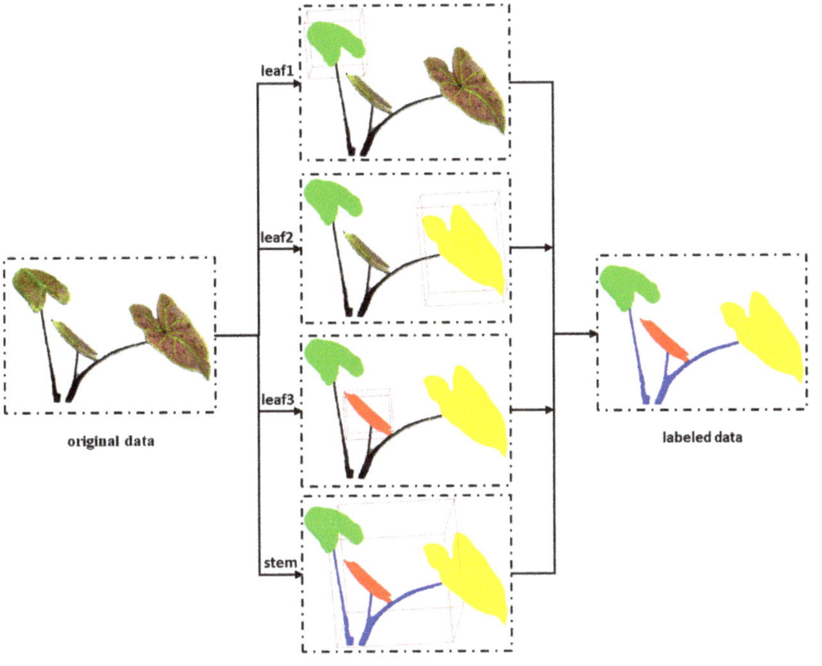

Figure 3. Data annotation pipeline for plant organ segmentation. Different colors represent different instances.

Prior to annotation, all point cloud data underwent a standardized preprocessing pipeline to ensure data quality and consistency. For each potted plant reconstruction, we first performed semi-automatic background and noise removal to isolate the complete botanical structure, preserving only the pot, branches, and leaves. Additionally, to address scale variations inherent in multi-view stereo reconstruction, we implemented a normalization procedure involving coordinate translation to the origin and proportional scaling based on pot dimensions. This spatial standardization enabled meaningful comparisons across specimens while maintaining biological accuracy in the reconstructed plant architectures.

The annotation process involved ten trained annotators holding degrees in geomatics with a minimum of 3 years of labeling experience. To ensure consistency, each sample was independently labeled by two annotators, with discrepancies resolved through the following: (1) third-annotator arbitration (for technical conflicts); (2) professional software verification (for spatial data). The PP3D dataset annotation was performed using CloudCompare, an open-source point cloud processing software. Following the import of preprocessed data, we conducted comprehensive manual annotation across hundreds of plant specimens, including: (1) the semantic segmentation of stems and leaves, (2) instance-level leaf labeling for panoramic segmentation tasks (Figure 4), and (3) the random down-sampling of each labeled organ to optimize data size while preserving structural fidelity. Each potted plant sample was independently annotated by two labelers at the instance level. Annotations with IoU < 0.85 were adjudicated by a third expert annotator. These annotations provide multi-granularity phenotypic information, from organ classification to individual leaf identification. Future work will expand the annotation schema to include finer morphological features (e.g., petioles, nodes) to support advanced plant architecture analysis [4]. The current annotation pipeline achieves an optimal balance between data richness and computational efficiency through strategic downsampling that maintains critical topological features.

Figure 4. Overview of the proposed PP3D dataset. The first and third columns are the original point clouds, colored by texture; the second and fourth columns are the ground truth; different colors represent different instances.

The current PP3D dataset partitions all data into training, validation, and test sets following a 7:1:2 ratio. The dataset is organized in a hierarchical structure where each plant specimen occupies a dedicated directory containing the following: (1) the complete plant point cloud file, and (2) an annotation subdirectory with separated stem and leaf point clouds. Each point is characterized by six attributes (XYZ coordinates and RGB color values) stored in a standardized tabular format. To facilitate initial evaluation, we have released representative samples from the dataset. Upon completion, we will publicly share the full-scale dataset to support advanced research in panoramic plant phenotyping and 3D morphological analysis.

3.2. The Proposed SCNet

In this section, we present the theoretical foundations and architectural principles underlying the proposed SCNet framework. The system primarily comprises three core components: (1) sequential slice feature extraction, (2) cylindrical feature extraction, and (3) a feature fusion module. Each component is carefully designed to address specific challenges in the feature representation pipeline.

3.2.1. Sequential Slice Feature Extraction

As depicted in Figure 5, the sequential slice feature extraction branch comprises three main modules: a sequential slice module for input processing, a multi-scale point encoder for hierarchical feature learning, and a recurrent neural network (RNN) with GRU to capture temporal dependencies.

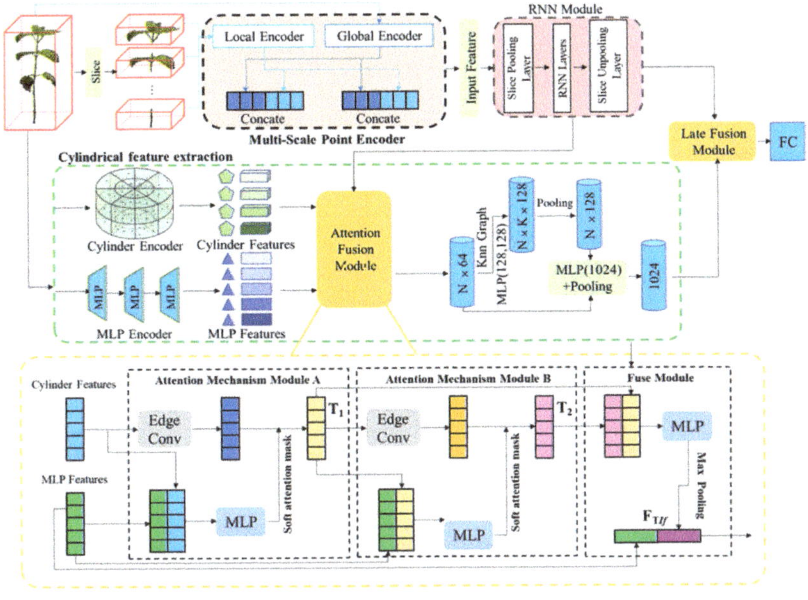

Figure 5. The pipeline of the proposed SCNet.

Sequential Slicer. Prior to inputting the potted plant point cloud into the network, we transform the raw point cloud into a sequential-slice representation. This operation is motivated by the natural growth pattern of potted plants, which typically exhibit rotational symmetry around the vertical axis. Given an input point cloud $P = (x, y, z)$, we first normalize the geometric coordinates to the range [0, 1] to ensure scale invariance. Subsequently, we partition P into M slices along the z-axis, where M is a user-defined hyperparameter (typically, $M = 1, 2, \ldots, m$). Formally, for a point cloud P containing N points, the slice operation is defined as follows:

$$S = S_M | S_M = p_i | p_i \in P, \lceil z_i \times m \rceil = M \quad (1)$$

where S_M denotes the M-th slice, where z_i represents the z-coordinate of point p_i. The slice assignment is determined through a ceiling operation ($\lceil \cdot \rceil$), which clusters all points into their respective slices along the z-axis. This process generates an ordered slice set $S = \{S_1, S_2, \ldots, S_M\}$, where each S_M contains points falling within a specified height interval. The primary objective of this operation is dimensionality reduction: by decomposing the 3D point cloud into a sequence of 2D slices, we enable more efficient feature extraction while preserving the structural hierarchy of the plant.

Point Encoder. To extract multi-scale features from each slice S_i, we first project it into 2D space along the z-axis and rasterize it into a grid-mapped image I_i with resolution r. The feature extraction process begins with PointNet [37] capturing global slice features, followed by PointConv [38] efficiently extracting local geometric patterns. We then employ MLPs to integrate these features across multiple scales, combining point-wise attributes with neighborhood context to enhance the representation. Finally, the concatenated global and multi-scale local features are prepared for subsequent processing.

RNN Layer. To model the sequential dependencies between slices, we employ an RNN architecture where each slice is treated as a timestamp, enabling information flow across the sequence. This transforms the unordered point cloud features into structured temporal representations. For enhanced feature discrimination, we introduce a sequence voting layer that performs weighted averaging to aggregate high-level features across all timestamps.

3.2.2. Cylindrical Feature Extraction

To enhance the network's descriptive capability, we designed a cylindrical feature extraction module that captures comprehensive contextual information from plant point clouds. Addressing the challenges of uneven point distribution and variable point density, our approach first employs a key point extraction layer to generate a fixed number of representative points while preserving critical structural information. These sampled points are then processed through a Multi-Layer Perceptron (MLP) comprising four fully connected layers, each equipped with ReLU activation and Batch Normalization.

The extracted features are subsequently mapped to cylindrical voxels based on the spatial occupancy of the original points. For feature refinement, we apply a spatial sparse 3D convolution layer [39] to aggregate local geometric patterns efficiently. In the final fusion stage, cylindrical voxel features are spatially aligned with slice point features and concatenated to form enriched fused point representations, effectively combining both local geometric details and global contextual information.

3.2.3. Feature Fusion Module

Attention-Guided Feature Fusion. SCNet employs an attention mechanism to dynamically integrate multi-modal features for enhanced plant representation. The network first embeds the MLP-derived features M_f and uses them to adaptively guide the learning of cylindrical features C_f through attention coefficients, effectively capturing the relative importance of different local structures in potted plants. Our framework incorporates two complementary attention modules: the Explicit Geometric Attention (Module A) focusing on the salient structural features of plant morphology, and the Implicit Geometric Attention (Module B) targeting latent spatial relationships and subtle geometric patterns. As illustrated in Figure 5, this dual-attention architecture enables the comprehensive fusion of 2D semantic information from images with 3D spatial features from point clouds, providing a more complete representation of plant characteristics.

Attention Mechanism Module A. To effectively fuse the cylindrical features C_f and MLP features M_f across different dimensional spaces, we first project M_f into an N-dimensional space matching C_f's dimensionality. The aligned features are then concatenated to form a relational feature representation $F_1\left(C_f, M_f\right)$. This combined feature undergoes adaptive transformation through an MLP, followed by normalization to the $[0, 1]$ range using a sigmoid function, ultimately generating a soft attention mask M that quantifies the relative importance of cylindrical features with respect to MLP features.

Attention Mechanism Module B. While explicit geometric features provide fundamental structural information, they alone cannot fully capture the rich semantic and geometric complexity of potted plants. To address this limitation, we conduct an in-depth exploration of latent implicit features through Module B. This enhanced attention mechanism processes refined point features in conjunction with both cylindrical features (C_f) and MLP-derived features (M_f), enabling the comprehensive learning of subtle spatial relationships and contextual patterns that are critical for accurate plant characterization.

4. Results and Discussion

4.1. Representative Baselines and Implementation Details

This section elaborates on the experimental setup, including four representative baseline methods and implementation details. For comprehensive evaluation on the PP3D dataset, we conducted panoptic segmentation experiments using the following:

(1) ASIS [40], a pioneering point cloud segmentation approach; (2) HAIS [41], which incorporates hierarchical aggregation; (3) ISBNet [42], known for its instance-aware segmentation; and (4) JSNet [43], featuring joint segmentation capabilities. These baselines were compared against our proposed SCNet architecture to validate its effectiveness. Detailed configurations include the following: hardware specifications (e.g., GPU models), software environment (e.g., PyTorch version 1.12.1), and hyperparameter settings (learning rate, batch size, etc.), ensuring reproducible experimental conditions.

ASIS [40] serves as a foundational framework for concurrent instance and semantic segmentation in 3D point clouds. This approach introduces a synergistic dual-task learning mechanism that creates mutual enhancement between both segmentation tasks through: (1) semantically informed instance embedding learning, where instance segmentation benefits from the semantic context by incorporating category-aware features into point-level embeddings, and (2) instance-aware semantic refinement, where semantic predictions are improved by aggregating features from points belonging to identical instances. This bidirectional information flow establishes a mutually reinforcing improvement cycle, significantly boosting performance on both tasks compared to processing them independently.

HAIS [41] proposes an efficient hierarchical clustering framework that leverages both point-level and cluster-level spatial relationships for instance segmentation. The method employs a two-stage aggregation process: (1) Point Aggregation: geometrically proximate points are clustered into primitive segments using learned feature affinities, forming initial instance proposals; (2) Set Aggregation: these primitive segments are progressively merged into complete instances through hierarchical feature propagation and spatial reasoning. This bottom-up approach eliminates the need for computationally expensive non-maximum suppression (NMS) post-processing while maintaining competitive performance, achieving an optimal balance between accuracy and computational efficiency in large-scale 3D scenes.

ISBNet [42] introduces an innovative cluster-free paradigm for instance segmentation that conceptualizes instances as learnable kernel representations. The framework comprises three key components: (1) Instance-aware Farthest Point Sampling strategically selects candidate seed points to ensure comprehensive spatial coverage; (2) a modified PointNet++ architecture employs local feature aggregation to encode robust candidate representations; and (3) dynamic convolution kernels generate instance masks by adaptively weighting point features. This approach eliminates traditional clustering bottlenecks while achieving superior recall rates through its discriminative kernel learning mechanism, particularly effective in handling densely distributed instances with complex geometries.

JSNet [43] proposes an integrated framework for joint instance and semantic segmentation through bidirectional feature interaction. The core innovation is the Joint Instance Semantic Segmentation (JISS) module that establishes the following: (1) a semantic-to-instance pathway that transforms categorical features into discriminative instance embeddings through learned projection, and (2) an instance-to-semantic pathway that enriches semantic features with structural information from instance groupings. This dual cross-pollination of features creates a synergistic effect where instance boundaries benefit from semantic consistency while semantic predictions gain structural precision from instance awareness. The framework demonstrates particular effectiveness in complex botanical scenes where morphological continuity and categorical distinction are equally critical.

All computational experiments were performed on a high-performance workstation featuring an NVIDIA GeForce RTX 3090 graphics (NVIDIA Corporation, Santa Clara, CA, USA) processor with 24GB GDDR6X memory. To ensure optimal compatibility with existing point cloud processing architectures, we preprocessed the PP3D dataset into dual standardized formats: (1) the STPL3D [44] vegetation-specific format and (2) the general-purpose S3DIS [13] point cloud format, both of which will be made publicly available to maximize research utility. For implementations involving HAIS and ISBNet frameworks, we adopted specific parameter configurations: each potted plant specimen was processed as an independent 3D scene with a scaling factor of 3 to preserve fine-scale morphological features, while the 'max_npoint' parameter was optimized to its maximum viable value (typically ranging between 50,000 and 100,000 points per specimen) to comprehensively capture intricate plant structures. Feature dimensions were uniformly maintained at [128, 512] throughout upscaling operations, with all remaining hyperparameters preserved at their default values to ensure the authentic representation of each network's baseline performance capabilities. This rigorous experimental protocol was designed to facilitate direct cross-method comparisons while addressing the unique challenges of 3D plant phenotyping analysis.

4.2. Qualitative and Quantitative Results

Figure 6 presents a comparative visualization of instance segmentation performance on the PP3D dataset, evaluating five state-of-the-art methods: ASIS [40], HAIS [41], ISBNet [42], JSNet [43], and our proposed SCNet. Experimental results demonstrate SCNet's superior robustness in leaf-level instance segmentation across diverse plant morphologies compared to the four baseline methods. Notably, while photogrammetric point clouds inherently suffer from precision limitations due to reconstruction artifacts, our SCNet maintains remarkable segmentation consistency. The method particularly excels in handling complex canopy structures, successfully distinguishing individual leaves even in challenging scenarios involving dense foliar overlap near stem junctions.

For comprehensive performance evaluation, Table 2 provides quantitative metrics comparing all five methods on the PP3D dataset. The results clearly indicate SCNet's dominance across all evaluation criteria, with HAIS [41] ranking second and ISBNet [42] showing the weakest performance. This exceptional performance originates from our novel dual-branch architecture design: (1) the sequential slice module employs adaptive slicing planes to achieve precise leaf boundary detection, while (2) the cylindrical feature extractor specializes in accurate leaf-stem junction identification, effectively eliminating non-foliar point clusters. The synergistic operation of these complementary feature extractors ensures reliable individual leaf identification regardless of point cloud quality variations.

Table 2. Segmentation performance evaluation on PP3D.

Methods	AP(%)		AP50(%)		AP25(%)	
	Stem	Leaf	Stem	Leaf	Stem	Leaf
ASIS [40]	16.7	14.1	34.8	28.2	42.5	39.9
HAIS [41]	54.4	53.0	70.2	62.6	74.0	71.5
ISBNet [42]	0.4	7.1	1.2	9.5	2.2	1.3
JSNet [43]	16.3	13.2	33.7	27.2	40.9	35.7
SCNet (ours)	60.0	56.1	71.3	64.5	75.4	72.2

GT

Figure 6. *Cont.*

Figure 6. Panoramic segmentation results of the photogrammetric plant point cloud. Different colors represent different instances.

Nevertheless, as shown in Figure 7, the proposed SCNet still presents three characteristic limitations in segmentation performance: (1) Morphological ambiguity between woody branches and slender petioles frequently causes erroneous region splitting due to their nearly identical local geometric features. (2) Partial occlusion artifacts and inconsistent neighborhood contexts often induce irreducible classification errors, particularly in dense foliar regions. (3) Morphologically anomalous leaves (e.g., highly lobed or curled specimens) are occasionally mislabeled as distinct botanical components. These findings underscore the persistent challenges inherent in our PP3D dataset, highlighting critical areas that demand immediate research attention and methodological refinement.

Figure 7. Illustration of several examples of typical mis-segmentation. First line: ground truths. Second line: predictions. Different colors represent different instances.

4.3. Experimental Analysis and Discussion

The Role of True RGB Color. Our photogrammetric point clouds typically contain richer information, offering additional features that can enhance network training. To evaluate the impact of color information on model performance, we conducted comparative experiments using four existing networks (ASIS [40], HAIS [41], ISBNet [42] and JSNet [43]) along with our proposed network. Following the approach of SensatUrban [12], we trained these networks using either point coordinates alone or both point coordinates and color information. The qualitative segmentation results without real RGB information are presented in Figure 8. As shown in Table 3, most existing networks demonstrate improved segmentation performance when trained with both geometric and color information. However, since our proposed network primarily relies on geometric partitioning, incorporating color information does not yield significant performance gains.

Figure 8. Plant point cloud segmentation results without real RGB information. Different colors represent different instances.

Table 3. Instance segmentation performance of the proposed PP3D dataset. # represents without real RGB information. Δ shows the change in performance without real RGB information.

Methods	AP(%)		AP50(%)		AP25(%)	
	Stem	Leaf	Stem	Leaf	Stem	Leaf
ASIS [40]	16.7	14.1	34.8	28.2	42.5	39.9
ASIS #	16.8	13.8	33.9	27.4	40.1	36.5
ASIS Δ	0.1	−0.3	−0.9	−0.8	−2.4	−3.4
HAIS [41]	54.4	53.0	70.2	62.6	74.0	71.5
HAIS #	50.2	49.6	63.8	55.2	69.1	63.2
HAIS Δ	−4.2	−3.4	−6.4	−7.4	−4.9	−8.3
ISBNet [42]	0.4	7.1	1.2	9.5	2.2	1.3
ISBNet #	0.4	7.0	1.1	8.8	2.0	1.3
ISBNet Δ	0.0	−0.1	−0.1	−0.7	−0.2	0.0
JSNet [43]	16.3	13.2	33.7	27.2	40.9	35.7
JSNet #	18.9	14.2	33.8	27.0	39.2	35.6
JSNet Δ	2.6	1.0	0.1	−0.2	−1.7	−0.1
SCNet (ours)	60.0	56.1	71.3	64.5	75.4	72.2
SCNet #	56.3	56.0	69.5	63.4	74.2	69.9
SCNet Δ	−3.7	−0.1	−1.8	−1.1	−1.2	−2.3

Study on the Weak Supervision. Weak supervision has emerged as a crucial research direction in point cloud segmentation, seeking to maintain satisfactory performance while significantly reducing annotation costs. Following the approach of SegGroup [45], we adopt an extremely sparse annotation scheme where only a single randomly selected point per instance serves as the ground truth label. Using WSIS [46] as our baseline, we evaluate its segmentation performance under an exceptionally low annotation rate of 0.1% on the PP3D dataset. As shown in Table 4, while the mIoU achieved by WSIS is slightly lower than fully supervised models, the results demonstrate that competitive segmentation accuracy can still be attained with minimal supervisory signals. As for the weak supervision segmentation, the qualitative results of the PP3D dataset are presented in Figure 9. This finding substantiates the practical viability of weak supervision approaches for point cloud segmentation tasks.

Table 4. A weakly supervised study on PP3D. We chose WSIS for the weakly supervised study and compared it with the fully supervised methods.

Methods	Annotation	AP(%)		AP50(%)		AP25(%)	
		Stem	Leaf	Stem	Leaf	Stem	Leaf
HAIS [41]	100%	54.4	53.0	70.2	62.6	74.0	71.5
HAIS [41]	0.10%	22.0	21.5	33.8	31.2	45.6	44.7
ISBNet [42]	100%	0.4	7.1	1.2	9.5	2.2	1.3
ISBNet [42]	0.10%	1.3	6.8	0.9	7.6	2.5	1.0
SCNet (ours)	100%	60.0	56.1	71.3	64.5	75.4	72.2
SCNet (ours)	0.10%	20.5	11.9	38.3	31.8	46.2	46.8
WSIS [46]	0.10%	30.5	26.5	46.7	41.2	50.1	45.8

Figure 9. The plant point cloud segmentation results of a weakly supervised study. Different colors represent different instances.

5. Conclusions

While deep learning has achieved remarkable progress in visual recognition tasks in recent years, the 3D organ-level segmentation of plants remains significantly underdeveloped. To bridge this critical gap, we present PP3D—a comprehensive point cloud dataset featuring panoptic segmentation of real-world potted plants with over 500 meticulously annotated plant organs. To the best of our knowledge, PP3D represents the largest and most diverse dataset of its kind, surpassing existing plant point cloud datasets in both scale (by 3×) and species coverage (by 5×). The dataset introduces novel challenges for plant panoptic segmentation from 3D point clouds, including complex occlusions and fine structural details. To establish a robust benchmark, we conduct an extensive evaluation of state-of-the-art panoptic segmentation algorithms alongside our proposed method. In future developments, we will first expand the PP3D dataset by incorporating additional point cloud samples to increase its scale. Next, we will enhance its diversity by including more plant species while implementing temporal modeling for perennial plants to improve both species coverage and temporal representation. Finally, we will refine and strengthen the annotation protocols to establish PP3D as a standard reference dataset in plant phenotyping research. We firmly believe this initiative will drive advancements in two critical domains: high-throughput plant phenotyping pipelines and smart agricultural systems, ultimately contributing to more sustainable and efficient agricultural practices.

Author Contributions: Conceptualization, S.L. and T.J.; Data curation, L.Z., S.W., S.L. and T.J.; Formal analysis, S.W., S.L. and T.J.; Funding acquisition, T.J.; Investigation, J.F., S.L. and T.J.; Methodology, L.Z., S.F. and T.J.; Project administration, T.J.; Resources, S.L. and T.J.; Software, L.Z., S.W., J.F., S.F. and T.J.; Supervision, T.J.; Validation, S.W., J.F., S.F. and S.L.; Visualization, S.L. and T.J.; Writing—original draft, L.Z., S.W., S.L. and T.J.; Writing—review and editing, S.L. and T.J. All authors have read and agreed to the published version of the manuscript.

Funding: This research was supported by the Open Fund of the Key Laboratory of Degraded and Unused Land Consolidation Engineering, Ministry of Natural Resources of China (grant number SXDJ2024-22), the Open Fund of Key Laboratory of Land Satellite Remote Sensing Application, Ministry of Natural Resources of China (grant number KLSMNR-K202305), the Science and Technology Research and Development Program of China State Railway Group Co., Ltd. (grant number Q2024G032), the National Natural Science Foundation of China (grant number 42401552), the Natural Science Foundation of Jiangsu Province, China (grant number BK20240598), the Natural Science Foundation of the Higher Education Institutions of Jiangsu Province, China (grant number 24KJB420005), and the Open Fund of Technology Innovation Center for Integrated Applications in Remote Sensing and Navigation, Ministry of Natural Resources of China (grant number TICIARSN-2023-06).

Data Availability Statement: Data underlying the results presented in this paper are not publicly available at this time but may be obtained from the authors upon reasonable request.

Acknowledgments: The authors acknowledge all the reviewers for their valuable comments.

Conflicts of Interest: The authors declare no conflicts of interest.

References

1. Song, H.; Wen, W.; Wu, S.; Guo, X. Comprehensive Review on 3D Point Cloud Segmentation in Plants. *Artif. Intell. Agric.* **2025**, *15*, 296–315. [CrossRef]
2. Wang, Z.; Chen, M.; Liu, Q. A review on multimodal communications for human-robot collaboration in 5G: From visual to tactile. *Intell. Robot.* **2025**, *5*, 579–606. [CrossRef]
3. Jin, S.C.; Su, Y.J.; Wu, F.F.; Pang, S.X.; Gao, S.; Hu, T.Y.; Guo, Q.H. Stem–leaf segmentation and phenotypic trait extraction of individual maize using terrestrial LiDAR data. *IEEE Trans. Geosci. Remote Sens.* **2018**, *57*, 1336–1346. [CrossRef]
4. Hu, H.; Wang, J.; Nie, S.; Zhao, J.; Batley, J.; Edwards, D. Plant pangenomics, current practice and future direction. *Agric. Commun.* **2024**, *2*, 100039. [CrossRef]

5. Li, D.; Shi, G.L.; Kong, W.J.; Wang, S.F.; Chen, Y. A leaf segmentation and phenotypic feature extraction framework for multiview stereo plant point clouds. *IEEE J. Sel. Top. Appl. Earth Obs. Remote Sens.* **2020**, *13*, 2321–2336. [CrossRef]
6. Jiang, T.; Liu, S.; Zhang, Q.; Xu, X.; Sun, J.; Wang, Y. Segmentation of individual trees in urban MLS point clouds using a deep learning framework based on cylindrical convolution network. *Int. J. Appl. Earth Obs. Geoinf.* **2023**, *123*, 103473. [CrossRef]
7. Du, R.; Ma, Z.; Xie, P.; He, Y.; Cen, H. PST: Plant segmentation transformer for 3D point clouds of rapeseed plants at the podding stage. *ISPRS J. Photogramm. Remote Sens.* **2023**, *195*, 380–392. [CrossRef]
8. Salve, D.A.; Ferreyra, M.J.; Defacio, R.A.; Maydup, M.L.; Lauff, D.B.; Tambussi, E.A.; Antonietta, M. Andean maize in Argentina: Physiological effects related with altitude, genetic variation, management practices and possible avenues to improve yield. *Technol. Agron.* **2023**, *3*, 14. [CrossRef]
9. Behley, J.; Garbade, M.; Milioto, A.; Quenzel, J.; Behnke, S.; Stachniss, C.; Gall, J. Semantickitti: A dataset for semantic scene understanding of lidar sequences. In Proceedings of the IEEE/CVF Conference on Computer Vision and Pattern Recognition (CVPR), Long Beach, CA, USA, 16–20 June 2019; pp. 9297–9307.
10. Fong, W.K.; Mohan, R.; Hurtado, J.V.; Zhou, L.; Caesar, H.; Beijbom, O.; Valada, A. Panoptic nuscenes: A large-scale benchmark for lidar panoptic segmentation and tracking. *IEEE Robot. Autom. Lett.* **2022**, *7*, 3795–3802. [CrossRef]
11. Xiao, A.; Huang, J.; Xuan, W.; Ren, R.; Liu, K.; Guan, D.; Saddik, A.E.; Lu, S.; Xing, E.P. 3d semantic segmentation in the wild: Learning generalized models for adverse-condition point clouds. In Proceedings of the IEEE/CVF Conference on Computer Vision and Pattern Recognition (CVPR), Vancouver, BC, Canada, 17–24 June 2023; pp. 9382–9392.
12. Hu, Q.Y.; Yang, B.; Khalid, S.; Xiao, W.; Trigoni, N.; Markham, A. Towards semantic segmentation of urban-scale 3D point clouds: A dataset, benchmarks and challenges. In Proceedings of the IEEE/CVF Conference on Computer Vision and Pattern Recognition (CVPR), Nashville, TN, USA, 20–25 June 2021; pp. 4977–4987.
13. Armeni, I.; Sax, S.; Zamir, A.R.; Savarese, S. Joint 2d-3d-semantic data for indoor scene understanding. *arXiv* **2017**, arXiv:1702.01105.
14. Dai, A.; Chang, A.X.; Savva, M.; Halber, M.; Funkhouser, T.; Nießner, M. Scannet: Richly-annotated 3d reconstructions of indoor scenes. In Proceedings of the IEEE/CVF Conference on Computer Vision and Pattern Recognition (CVPR), Honolulu, HI, USA, 21–26 July 2017; pp. 5828–5839.
15. Mo, K.; Zhu, S.; Chang, A.X.; Yi, L.; Tripathi, S.; Guibas, L.J.; Su, H. Partnet: A large-scale benchmark for fine-grained and hierarchical part-level 3d object understanding. In Proceedings of the IEEE/CVF Conference on Computer Vision and Pattern Recognition (CVPR), Long Beach, CA, USA, 2–8 July 2019; pp. 909–918.
16. Zhu, R.; Sun, K.; Yan, Z.; Yan, X.; Yu, J.; Shi, J.; Hu, Z.; Jiang, H.; Xin, D.; Zhang, Z.; et al. Analysing the phenotype development of soybean plants using low-cost 3D reconstruction. *Sci. Rep.* **2020**, *10*, 7055. [CrossRef] [PubMed]
17. Jiang, T.; Wang, Y.; Liu, S.; Zhang, Q.; Zhao, L.; Sun, J. Instance recognition of street trees from urban point clouds using a three-stage neural network. *ISPRS J. Photogramm. Remote Sens.* **2023**, *199*, 305–334. [CrossRef]
18. Li, D.; Shi, G.L.; Li, J.S.; Chen, Y.L.; Zhang, S.Y.; Xiang, S.Y.; Jin, S.C. PlantNet: A dual-function point cloud segmentation network for multiple plant species. *ISPRS J. Photogramm. Remote Sens.* **2022**, *184*, 243–263. [CrossRef]
19. Sun, Y.; Zhang, Z.; Sun, K.; Li, S.; Yu, J.; Miao, L.; Zhang, Z.; Li, Y.; Zhao, H.; Hu, Z.; et al. Soybean-MVS: Annotated three-dimensional model dataset of whole growth period soybeans for 3D plant organ segmentation. *Agriculture* **2023**, *13*, 1321. [CrossRef]
20. Mertoğlu, K.; Şalk, Y.; Sarıkaya, S.K.; Turgut, K.; Evrenosoğlu, Y.; Çevikalp, H.; Gerek, Ö.N.; Dutagaci, H.; Rousseau, D. PLANesT-3D: A New Annotated Data Set of 3D Color Point Clouds of Plants. In Proceedings of the Signal Processing and Communications Applications Conference (SIU), Istanbul, Turkey, 2–8 July 2023; pp. 1–4.
21. Schunck, D.; Magistri, F.; Rosu, R.A.; Cornelißen, A.; Chebrolu, N.; Paulus, S.; Léon, J.; Klingbeil, L. Pheno4D: A spatio-temporal dataset of maize and tomato plant point clouds for phenotyping and advanced plant analysis. *PLoS ONE* **2021**, *16*, e0256340. [CrossRef]
22. Dutagaci, H.; Rasti, P.; Galopin, G.; Rousseau, D. ROSE-X: An annotated data set for evaluation of 3D plant organ segmentation methods. *Plant Methods* **2020**, *16*, 28. [CrossRef]
23. Conn, A.; Pedmale, U.V.; Chory, J.; Navlakha, S. High-resolution laser scanning reveals plant architectures that reflect universal network design principles. *Cell Syst.* **2017**, *5*, 53–62. [CrossRef]
24. Behley, J.; Milioto, A.; Stachniss, C. A benchmark for LiDAR-based panoptic segmentation based on KITTI. In Proceedings of the 2021 IEEE International Conference on Robotics and Automation (ICRA), Xi'an, China, 30 May–5 June 2021; pp. 13596–13603.
25. Sirohi, K.; Mohan, R.; Büscher, D.; Burgard, W.; Valada, A. Efficientlps: Efficient lidar panoptic segmentation. *IEEE Trans. Robot.* **2021**, *38*, 1894–1914. [CrossRef]
26. Hurtado, J.V.; Mohan, R.; Burgard, W.; Valada, A. Mopt: Multi-object panoptic tracking. *arXiv* **2020**, arXiv:2004.08189.
27. Mohan, R.; Valada, A. Efficientps: Efficient panoptic segmentation. *Int. J. Comput. Vis.* **2021**, *129*, 1551–1579. [CrossRef]
28. Gasperini, S.; Mahani, M.A.N.; Marcos-Ramiro, A.; Navab, N.; Tombari, F. Panoster: End-to-end panoptic segmentation of lidar point clouds. *IEEE Robot. Autom. Lett.* **2021**, *6*, 3216–3223. [CrossRef]

29. Milioto, A.; Behley, J.; McCool, C.; Stachniss, C. Lidar panoptic segmentation for autonomous driving. In Proceedings of the IEEE/RSJ International Conference on Intelligent Robots and Systems (IROS), Las Vegas, NV, USA, 25–29 October 2020; pp. 8505–8512.
30. Razani, R.; Cheng, R.; Li, E.; Taghavi, E.; Ren, Y.; Bingbing, L. Gp-s3net: Graph-based panoptic sparse semantic segmentation network. In Proceedings of the IEEE/CVF International Conference on Computer Vision (ICCV), Montreal, QC, Canada, 10–17 October 2021; pp. 16076–16085.
31. Zhou, Z.; Zhang, Y.; Foroosh, H. Panoptic-polarnet: Proposal-free lidar point cloud panoptic segmentation. In Proceedings of the IEEE/CVF Conference on Computer Vision and Pattern Recognition (CVPR), Nashville, TN, USA, 20–25 June 2021; pp. 13194–13203.
32. Li, J.K.; He, X.; Wen, Y.; Gao, Y.; Cheng, X.Q.; Zhang, D. Panoptic-phnet: Towards real-time and high-precision lidar panoptic segmentation via clustering pseudo heatmap. In Proceedings of the IEEE/CVF Conference on Computer Vision and Pattern Recognition (CVPR), New Orleans, LA, USA, 18–24 June 2022; pp. 11799–11808.
33. Hong, F.Z.; Zhou, H.; Zhu, X.; Li, H.S.; Liu, Z.W. Lidar-based panoptic segmentation via dynamic shifting network. In Proceedings of the IEEE/CVF Conference on Computer Vision and Pattern Recognition (CVPR), Nashville, TN, USA, 20–25 June 2021; pp. 13090–13099.
34. Li, D.; Li, J.S.; Xiang, S.Y.; Pan, A.Q. PSegNet: Simultaneous semantic and instance segmentation for point clouds of plants. *Plant Phenomics* **2022**, *2022*, 9787643. [CrossRef] [PubMed]
35. Luo, L.Y.; Jiang, X.T.; Yang, Y.; Samy, E.R.A.; Lefsrud, M.G.; Hoyos-villegas, V.; Sun, S.P. Eff-3dpseg: 3d organ-level plant shoot segmentation using annotation-efficient deep learning. *Plant Phenomics* **2023**, *5*, 0080. [CrossRef] [PubMed]
36. Mirande, K.; Godin, C.; Tisserand, M.; Charlaix, J.; Besnard, F.; Hétroy-Wheeler, F. A graph-based approach for simultaneous semantic and instance segmentation of plant 3D point clouds. *Front. Plant Sci.* **2022**, *13*, 1012669. [CrossRef]
37. Qi, C.R.; Su, H.; Mo, K.; Guibas, L.J. PointNet: Deep Learning on Point Sets for 3D Classification and Segmentation. In Proceedings of the IEEE/CVF Conference on Computer Vision and Pattern Recognition (CVPR), Honolulu, HI, USA, 21–26 July 2017; pp. 652–660.
38. Wu, W.; Qi, Z.; Li, F. PointConv: Deep convolutional networks on 3d point clouds. In Proceedings of the IEEE/CVF conference on computer vision and pattern recognition (CVPR), Long Beach, CA, USA, 15–20 June 2019; pp. 9621–9630.
39. Ibrahim, M.; Akhtar, N.; Anwar, S.; Mian, A. SAT3D: Slot attention transformer for 3D point cloud semantic segmentation. *IEEE Trans. Intell. Transp. Syst.* **2023**, *24*, 5456–5466. [CrossRef]
40. Wang, X.; Liu, S.; Shen, X.; Shen, C.; Jia, J. Associatively segmenting instances and semantics in point clouds. In Proceedings of the IEEE/CVF Conference on Computer Vision and Pattern Recognition (CVPR), Long Beach, CA, USA, 15–20 June 2019; pp. 4096–4105.
41. Chen, S.; Fang, J.; Zhang, Q.; Liu, W.; Wang, X.G. Hierarchical aggregation for 3d instance segmentation. In Proceedings of the IEEE/CVF International Conference on Computer Vision (ICCV), Seoul, Republic of Korea, 27 October–2 November 2019; pp. 15467–15476.
42. Ngo, T.D.; Hua, B.S.; Nguyen, K. ISBNet: A 3D point cloud instance segmentation network with instance-aware sampling and box-aware dynamic convolution. In Proceedings of the IEEE/CVF Conference on Computer Vision and Pattern Recognition (CVPR), Vancouver, BC, Canada, 18–19 June 2023; pp. 13550–13559.
43. Zhao, L.; Tao, W. JSNet: Joint instance and semantic segmentation of 3d point clouds. In Proceedings of the AAAI Conference on Artificial Intelligence (AAAI), New York, NY, USA, 7–12 February 2020; pp. 12951–12958.
44. Chen, M.; Hu, Q.; Yu, Z.; Thomas, H.; Feng, A.; Hou, Y.; McCullough, K.; Ren, F.; Soibelman, L. STPLS3D: A large-scale synthetic and real aerial photogrammetry 3D point cloud dataset. *arXiv* **2022**, arXiv:2203.09065.
45. Tao, A.; Duan, Y.; Wei, Y.; Lu, J.; Zhou, J. Seg-group: Seg-level supervision for 3d instance and semantic segmentation. *IEEE Trans. Image Process* **2020**, *31*, 4952–4965. [CrossRef]
46. Tang, L.; Hui, L.; Xie, J. Learning inter-superpoint affinity for weakly supervised 3d instance segmentation. In Proceedings of the Asian Conference on Computer Vision (ACCV), Macao, China, 4–8 December 2022; pp. 1282–1297.

remote sensing

MDPI

Article

TWDTW-Based Maize Mapping Using Optimal Time Series Features of Sentinel-1 and Sentinel-2 Images

Haoran Yan [1], Ruozhen Wang [1], Jiaqian Lian [1], Xinyue Duan [1], Liping Wan [1], Jiao Guo [1,2] and Pengliang Wei [1,2,*]

[1] College of Mechanical and Electronic Engineering, Northwest Agriculture and Forestry University, Yangling 712100, China; 15666756171@nwafu.edu.cn (H.Y.); wng_rz1224.kin@nwafu.edu.cn (R.W.); lianjiaqian@nwafu.edu.cn (J.L.); duanxinyue@nwafu.edu.cn (X.D.); geek@nwafu.edu.cn (L.W.); gjiao@nwafu.edu.cn (J.G.)

[2] Shaanxi Key Laboratory of Agricultural Information Perception and Intelligent Service, Yangling 712100, China

* Correspondence: weipengliang@nwafu.edu.cn

Highlights

What are the main findings?

- The optimal multi-source time series features for distinguishing maize from non-maize were determined using a two-step Jeffries Matusita distance-based global search strategy (i.e., twelve spectral bands, the NDVI, the EVI, VV, and VH collected from the maize jointing to tasseling stages).
- Optimal multi-source time series features-driven TWDTW not only significantly reduced computational complexity, but also outperformed both TWDTW using full-season optimal multi-source features and traditional supervised machine learning models reliant on large multi-category sample sizes.

What is the implication of the main findings?

- A new and comprehensive strategy for feature optimization in maize mapping tasks could be provided.
- A guide for enhancing TWDTW performance in large-scale crop mapping could be provided.

Abstract

Time-Weighted Dynamic Time Warping (TWDTW), adapted from speech recognition, is used in agricultural remote sensing to model crop growth, particularly under limited ground sample conditions. However, most related studies rely on full-season or empirically selected features, overlooking the systematic optimization of features at each observation time to improve TWDTW's performance. This often introduces a large amount of redundant information that is irrelevant to crop discrimination and increases computational complexity. Therefore, this study focused on maize as the target crop and systematically conducted mapping experiments using Sentinel-1/2 images to evaluate the potential of integrating TWDTW with optimally selected multi-source time series features. The optimal multi-source time series features for distinguishing maize from non-maize were determined using a two-step Jeffries Matusita (JM) distance-based global search strategy (i.e., twelve spectral bands, Normalized Difference Vegetation Index, Enhanced Vegetation Index, and the two microwave backscatter coefficients collected during the maize jointing to tasseling stages). Then, based on the full-season and optimal multi-source time series features, we compared TWDTW with two widely used temporal machine learning models in agricultural remote sensing community. The results showed that TWDTW outperformed traditional supervised temporal machine learning models. In particular, compared with

TWDTW driven by the full-season optimal multi-source features, TWDTW using the optimal multi-source time series features improved user accuracy by 0.43% and 2.30%, and producer accuracy by 7.51% and 2.99% for the years 2020 and 2021, respectively. Additionally, it reduced computational costs to only 25% of those driven by the full-season scheme. Finally, maize maps of Yangling District from 2020 to 2023 were produced by optimal multi-source time series features-based TWDTW. Their overall accuracies remained consistently above 90% across the four years, and the average relative error between the maize area extracted from remote sensing images and that reported in the statistical yearbook was only 6.61%. This study provided guidance for improving the performance of TWDTW in large-scale crop mapping tasks, which is particularly important under conditions of limited sample availability.

Keywords: maize mapping; Sentinel-1/2 images; optimal time series; optimal features; TWDTW

1. Introduction

Maize is one of the three major food crops in the world, with a global production exceeding 1.22 billion tons in 2024, according to data from the United States Department of Agriculture (USDA). However, while it alleviates the hunger problem in modern society, its large-scale cultivation has already negatively influenced the natural environment [1–3]. Consequently, accurate and timely mapping of maize planting areas is essential for sustainable agriculture and national food security, while also serving as a key data source for maize growth monitoring and yield estimation [4–6].

Nowadays, for practical crop mapping tasks, three main types of remote sensing imagery are commonly utilized: Synthetic Aperture Radar (SAR), optical imagery, and their integration [7–11]. For staple food crops, regardless of the data source, the most frequently adopted methods in practical crop mapping tasks are the traditional supervised machine learning algorithms (e.g., random forests, RF) [12–14] or phenology-based methods (e.g., simple threshold-based method) [15–17]. Although these methods have contributed to practical crop mapping tasks, they still have some disadvantages, as described below.:

First, traditional supervised machine learning algorithms depend on localized ground samples, limiting model transferability and product timeliness. This limitation arises from the fact that models trained on localized ground samples often fail to capture the essential change patterns of land cover, and their performance tends to degrade significantly [18–22] when applied to target regions that are geographically distant from the reference area and lack ground truth data, as similar land cover types may exhibit distinct phenological behaviors due to regional environmental differences [23–25]. Additionally, the collection of non-target crop ground samples (i.e., negative samples) is also relatively challenging, as it requires a wider range of land cover types than the target crop to ensure sufficient diversity and representativeness for robust model training. The issue becomes more pronounced in deep learning models with strong feature extraction capabilities. These challenges indicate that traditional supervised machine learning methods alone are insufficient for timely updates and spatial scalability of crop mapping data products [26–29].

Second, simple phenology-based methods can only extract pixels that exhibit obvious features consistent with the prior phenological information of the target crop during specific growth stages. This is mainly because these methods are built on the prior threshold of vegetation index corresponding to the key growth stages of crops [30–32]. As a consequence, they tend to focus solely on surface-level spectral characteristics of land cover at specific observation period (i.e., in multi-temporal data, land cover features are typically treated as temporally independent), while ignoring the temporal variation patterns of land cover across different observation periods. Therefore, while simple phenology-based methods avoid the need for manually annotated samples, they are less applicable in regions characterized by diverse crop types and overlapping phenological stages.

In light of this, accurately capturing the temporal phenology patterns of target crops becomes essential for timely updates and spatial expansion of crop mapping data products, particularly under conditions of limited ground sample availability. Currently, to address this challenge and better incorporate the time-varying growth characteristics of crops, various time series similarity measurement methods—originally developed in the speech recognition community—have been adapted for use in agricultural remote sensing community [33–35]. Compared with simple phenology-based methods, these approaches have demonstrated improved crop mapping performance and achieved results comparable to those of traditional supervised machine learning models [36–38]. Among them, Dynamic Time Warping (DTW) is one of the most representative techniques [33,34,38,39].

However, when the DTW is directly applied to crop phenological time series, it can result in excessive stretching or compression along the temporal axis, thereby limiting its effectiveness in crop mapping. To overcome this limitation, the method has been further refined in the agricultural remote sensing community through the introduction of time weighting—resulting in the Time Weighted Dynamic Time Warping (TWDTW) approach—which retains the nonlinear alignment capability of DTW while effectively capturing the seasonal dynamics of crop growth [40]. For example, Shen et al. applied TWDTW combined with Landsat and Sentinel imagery to successfully produce a high-resolution (30 m) maize distribution map across 22 major provinces in China, achieving an overall accuracy of 79%, with user accuracy and producer accuracy of 81.59% and 76.15%, respectively, demonstrating the practical utility and effectiveness of TWDTW for large-scale crop monitoring [41]. Nonetheless, it is worth noting that most related studies rely on crop time series features covering the full season for analysis, often overlooking the potential of phenological features during key growth stages [34,42–44]. While full-season time series provide more comprehensive phenological information, they also increase algorithmic complexity and may reduce transferability due to interannual phenological variations. Moreover, for the input features of TWDTW, either only single-source features are used, or the selection of multi-temporal and multi-source features is driven by empirical rules [40,43–45]. The potential benefits of systematically optimizing features at each observation time to improve TWDTW performance have largely been overlooked [46–49]. For example, Chaves et al. applied TWDTW to full-season MODIS time series to map interannual cropping practice changes in the Brazilian Cerrado, without considering feature optimization at key growth stages [48]. Similarly, Wei et al. conducted early-season crop mapping in Northeast China using Sentinel-2 time series, where feature selection was based on commonly used spectral bands and vegetation indices rather than systematic optimization [49].

Consequently, in the case of limited ground sample availability, this study took maize, one of the three major staple crops, as the target crop, and systematically explored the potential of multi-source remote sensing images (i.e., Sentinel-1 and Sentinel-2 images) covering key phenological periods of maize to drive TWDTW for maize mapping at a regional scale. Meanwhile, in order to provide a benchmark, we selected the traditional supervised machine learning model (i.e., RF) and the temporal deep learning model (i.e., Long Short-Term Memory, LSTM) for comparison, since their robust performances for crop mapping have been verified [14,50–52]. Then, the advantages of using TWDTW driven by the optimal multi-source time series features were evaluated from multiple perspectives, including optimization of feature and temporal combinations from multi-source remote sensing images, computational complexity, and maize mapping results. The specific objectives of this study are as follows: (1) to propose a strategy for identifying the optimal multi-source time series features for maize; (2) to evaluate, based on optimal multi-source features with different temporal lengths, the comparative performance of TWDTW against commonly used temporal machine learning models in terms of classification accuracy and computational efficiency; (3) to validate the applicability of TWDTW driven by the optimal multi-source time series features across different years.

Meanwhile, the following three research questions (RQs) can be addressed based on the three objectives outlined above:

(RQ1) What are the key time series and feature combinations used for maize mapping based on Sentinel-1/Sentinel-2 images?

(RQ2) How did TWDTW performance differ when driven by Sentinel-1/Sentinel-2 images covering maize key phenological periods versus the full-season?

(RQ3) How accurate were the maize mapping results produced by TWDTW using Sentinel-1/Sentinel-2 images covering key maize phenological periods across different years?

2. Study Area and Datasets

2.1. Study Area

Yangling District is located in the west-central part of Shaanxi Province and is an important agricultural area within the Guanzhong region of the province. Its specific geographical location is shown in Figure 1. The region has a terrain that rises from south to north and a semi-humid continental monsoon climate with four distinct seasons. Moreover, its abundant water resources and fertile soils make it highly suitable for cultivating a wide range of crops, forming a typical double-cropping system dominated by winter wheat and summer maize. Maize and wheat are the main crops, while vegetables, fruit trees, and other cash crops are also cultivated. Agricultural land is relatively concentrated, irrigation conditions are favorable, and the region is highly representative of agricultural production in the Guanzhong Plain. In addition, the region contains non-crop land covers such as built-up areas, roads, rivers, grasslands, and woodlands. Consequently, this region was selected to explore the potential of combining TWDTW with optimally selected multi-source time series features for maize mapping. The specific phenology stages of maize in Yangling District are presented in Table 1. Maize is usually sown in early June, and seedlings emerge and begin growing from late June to early July. The jointing and tasseling stages occur from mid-late July to early September, during which the crop grows vigorously, especially at the tasseling stage, when plant height increases rapidly, and vegetation cover expands significantly. Maturity typically occurs in late September.

Table 1. Phenology period of maize in Yangling District.

Month	Ten-Day	Maize
6	Upper	Sowing
	Middle	Seedlings
	Lower	Long leaves
7	Upper	—
	Middle	—
	Lower	Jointing

Table 1. *Cont.*

Month	Ten-Day	Maize
8	Upper	—
	Middle	Tasseling
	Lower	—
9	Upper	—
	Middle	Grouting
	Lower	Maturity

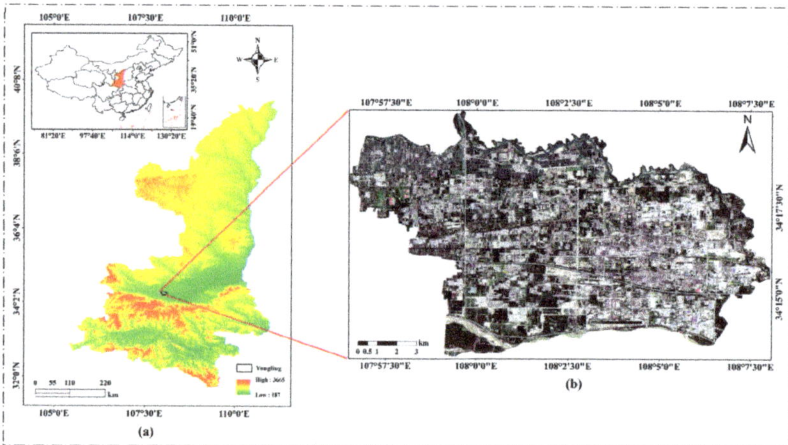

Figure 1. The specific location of the study area. (**a**) Geographical location of the study area. (**b**) Sentinel-2 true color optical imagery of Yangling District acquired on 10 September 2021.

2.2. Datasets

To ensure uniform coverage of the entire maize growth cycle from sowing to maturity, a unified set of 12 observation dates was selected at approximately 10-day intervals. For each observation date, Sentinel-1 and Sentinel-2 images were acquired as close in time as possible. In cases where missing acquisitions or cloud contamination occurred, temporal linear interpolation was applied to reconstruct the missing values, thereby maintaining the consistency and continuity of the multi-source time series. Further details are provided in Sections 2.2.1 and 2.2.2.

2.2.1. Sentinel-1 Images

Based on the Google Earth Engine (GEE) platform, Sentinel-1 Ground Range Detected (GRD) data products in Interferometric Wide Swath (IW) mode were acquired for the study area from 2020 to 2023, covering the full phenology period of maize. Both VH (vertical transmit, horizontal receive) and VV (vertical transmit, vertical receive) polarization modes were included. All images were radiometrically calibrated and geometrically corrected, resampled to a 10-meter spatial resolution and reprojected to the WGS84 geographic coordinate system. The specific retained Sentinel-1 image dates after preliminary selection via the GEE are listed in Figure 2.

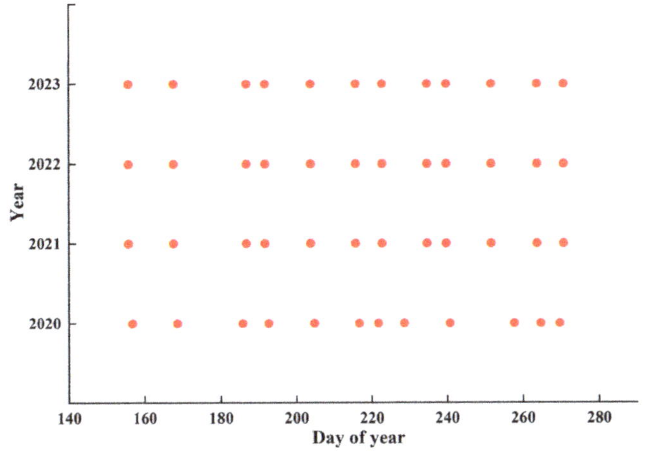

Figure 2. Captured Sentinel-1 image observation dates in different years.

Additionally, because of Sentinel-1's orbital configuration, swath width constraints, and occasional platform failures, data gaps may occur in the Yangling District, which lies at the edge of the satellite's coverage. To fill data gaps in parts of the study area, a temporal linear interpolation method based on the nearest valid values before and after the missing observation dates was used, since the effectiveness of this method in agricultural remote sensing community has been widely and convincingly demonstrated [50]. The specific missing regions of the study area before and after interpolation are shown in Figure 3.

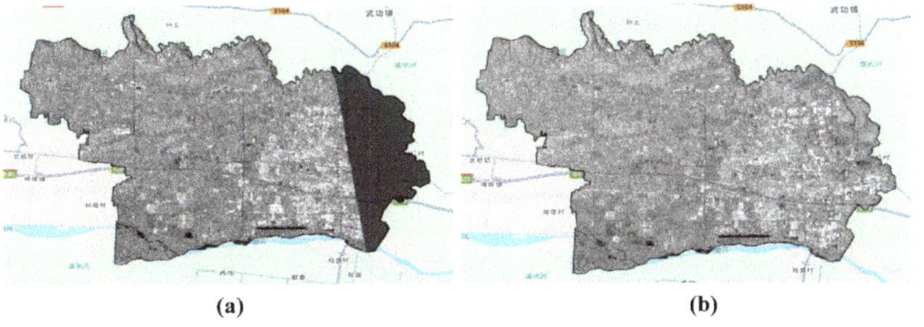

Figure 3. Comparison of Sentinel-1 image gaps in the study area before and after interpolation. (**a**) Original image. (**b**) Interpolated image.

2.2.2. Sentinel-2 Images

Sentinel-2 MSI Level-2A (i.e., Bottom of Atmosphere, BOA) data products were obtained via the GEE platform for the study area from 2020 to 2023, covering the full phenology period of maize. The 12 spectral bands, namely B1, B2, B3, B4, B5, B6, B7, B8, B8A, B9, B11, and B12, were used in this study. Similarly, all images were radiometrically calibrated and atmospherically corrected, resampled to a 10-meter spatial resolution and reprojected to the WGS84 geographic coordinate system. The specific retained Sentinel-2 image dates after preliminary selection via the GEE are listed in Figure 4.

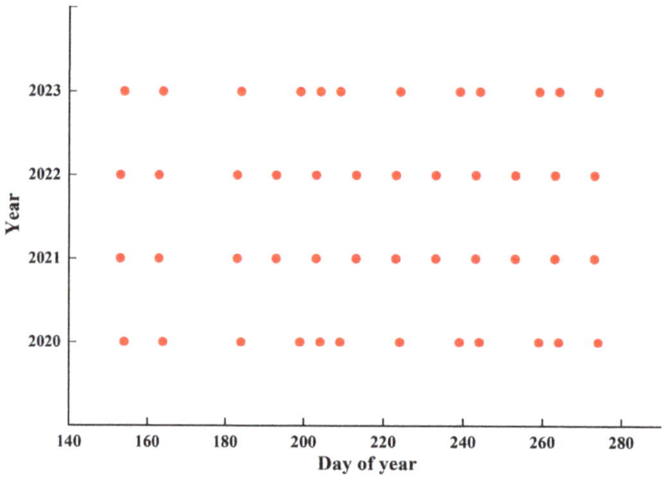

Figure 4. Captured Sentinel-2 image observation dates in different years.

Additionally, due to the susceptibility of optical imagery to weather conditions, several Sentinel-2 scenes were affected by varying degrees of cloud contamination. To address the resulting data gaps and enhance data quality, the same temporal linear interpolation method described in Section 2.2.1 was applied to reconstruct the missing values. For the scenes with localized cloud contamination, only the invalid pixels were interpolated. For the scenes with extensive cloud contamination, the whole image was interpolated using the nearest cloud-free images before and after the affected observation dates. The processed images can exhibit improved continuity and clarity, as shown in Figure 5.

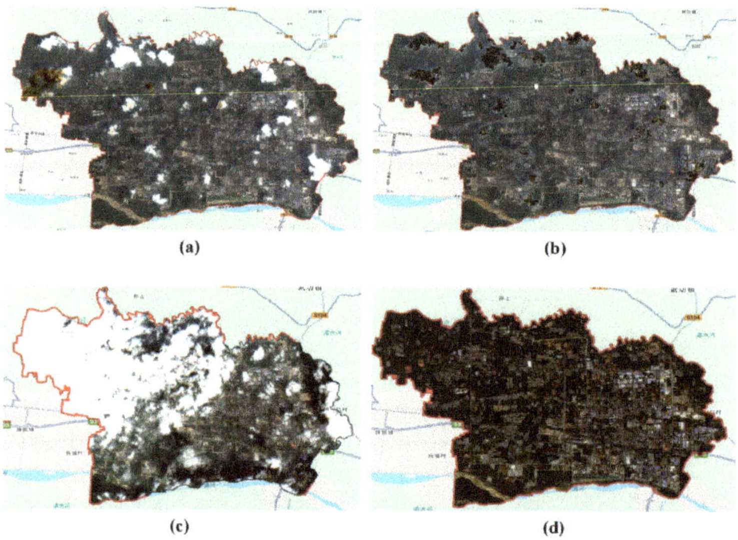

Figure 5. Comparison of Sentinel-2 imagery before and after interpolation. (**a**) Original imagery in a localized cloud-covered area. (**b**) Interpolated imagery in a localized cloud-covered area. (**c**) Original imagery in a large-scale cloud-covered area. (**d**) Interpolated imagery in a large-scale cloud-covered area.

2.2.3. Vegetation Index

For the vegetation index, this study employed two widely used indices in the agricultural remote sensing community: the Normalized Difference Vegetation Index (NDVI) and the Enhanced Vegetation Index (EVI). The *NDVI* can reflect vegetation coverage and growth conditions by comparing the reflectance difference between the near-infrared and red bands, while the *EVI* is designed to optimize the vegetation signal by correcting for atmospheric and canopy background effects [53], making it particularly well suited for monitoring vegetation conditions. Their specific calculation formulas are explained in Equations (1) and (2):

$$NDVI = \frac{NIR - RED}{NIR + RED} \tag{1}$$

$$EVI = G \times \frac{NIR - RED}{NIR + C_1 \times RED - C_2 \times BLUE + L} \tag{2}$$

where *NIR* and *RED* represent the reflectance values of the near-infrared and red bands contained in the Sentinel-2 image, respectively. *G* represents the gain factor, which is typically set to 2. *BLUE* represents the reflectance values of the blue band contained in the Sentinel-2 image. C_1 and C_2 represent constants (C_1 = 6, C_2 = 7.5), and *L* represents the background correction factor, which is typically set to 1.

2.2.4. Ground Reference Data

To assess the accuracy of maize mapping, ground reference data from multiple sources were utilized. For the years 2020 and 2021, existing maize map products provided by [54] were used as open-source ground reference datasets. These products achieved overall accuracies ranging from 0.87 to 0.95, and the estimated maize planting areas were highly consistent with statistical yearbook records (R^2 ranging from 0.83 to 0.95), thereby confirming their reliability. These datasets, available for download at (https://doi.org/10.6084/m9.figshare.22689751.v17, accessed on 15 May 2025), were utilized to validate the results of this study, as illustrated in Figure 6.

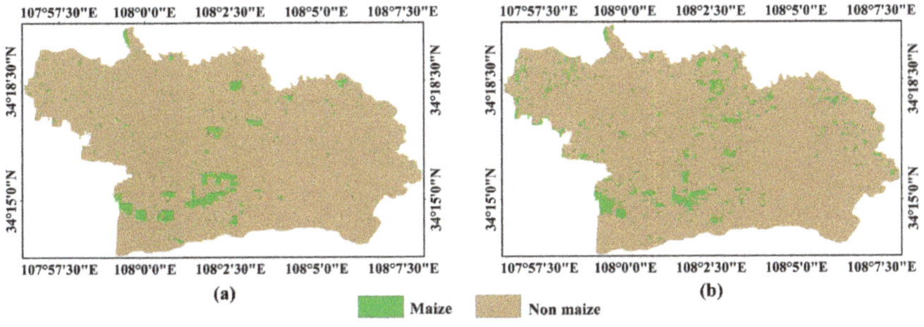

Figure 6. Open-source ground reference data for 2020 and 2021. (**a**) Open-source ground reference data for 2020. (**b**) Open-source ground reference data for 2021.

In addition, high-quality open-source maize maps were unavailable for the years 2022 and 2023. Therefore, based on field surveys and visual interpretation of Sentinel-2 true color imagery, 20 representative maize fields were manually annotated for each of the two years. Specifically, the annotated fields included 3519 pixels in 2022 and 4710 pixels in 2023. The detailed distribution of maize samples for both years is shown in Figure 7.

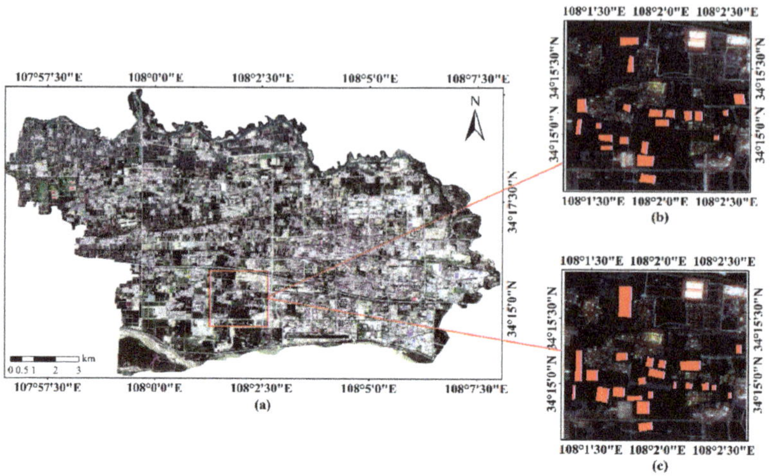

Figure 7. Visual interpretation plot map of maize in Yangling District for 2022 and 2023. (**a**) Sentinel-2 RGB image acquired on 10 September 2021. (**b**) Visual interpretation results for 2022 (Sentinel-2, 10 September). (**c**) Visual interpretation results for 2023 (Sentinel-2, 15 September).

3. Methodology

An overview of the framework investigated in this study, which consists of three principal modules, is shown in Figure 8: (1) optimal multi-source time series feature determination, (2) performance comparison between RF, LSTM, and TWDTW driven by different feature sets, and (3) TWDTW-based maize mapping using optimal time series features. The details of each module are present in the following sections.

All analyses and machine learning models in this study were conducted in the PyCharm environment using Python 3.8. The TWDTW method was implemented with customized code. Throughout the analysis, we primarily employed commonly used libraries including NumPy 2.2.6, Rasterio 1.4.3, GeoPandas 1.1.1, Scikit-learn 1.7.1, PyTorch 2.8.0+cpu, Joblib 1.5.2, and TQDM 4.67.1.

3.1. Determination of Optimal Multi-Source Time Series Features

3.1.1. Determination of the Optimal Multi-Source Features in Different Observation Dates Using Global Search Strategy

A quantitative separability indicator-based global search strategy was used to determine the optimal multi-source features for each observation date. This approach could ensure that the selected feature combinations consistently provide high separability between maize and non-maize across different observation dates. The specific determination of the optimal multi-source features for each observation date is illustrated in Figure 9. For each observation date, all possible feature combinations were generated. The separability indicators of these combinations were then calculated to identify the optimal multi-source features for that specific date. Finally, a comprehensive analysis was conducted on the optimal multi-source feature combinations identified for each observation date. The feature combinations that most frequently yielded the highest separability indicators across all observation dates were identified as the most representative for distinguishing maize from non-maize, and considered the optimal multi-source features.

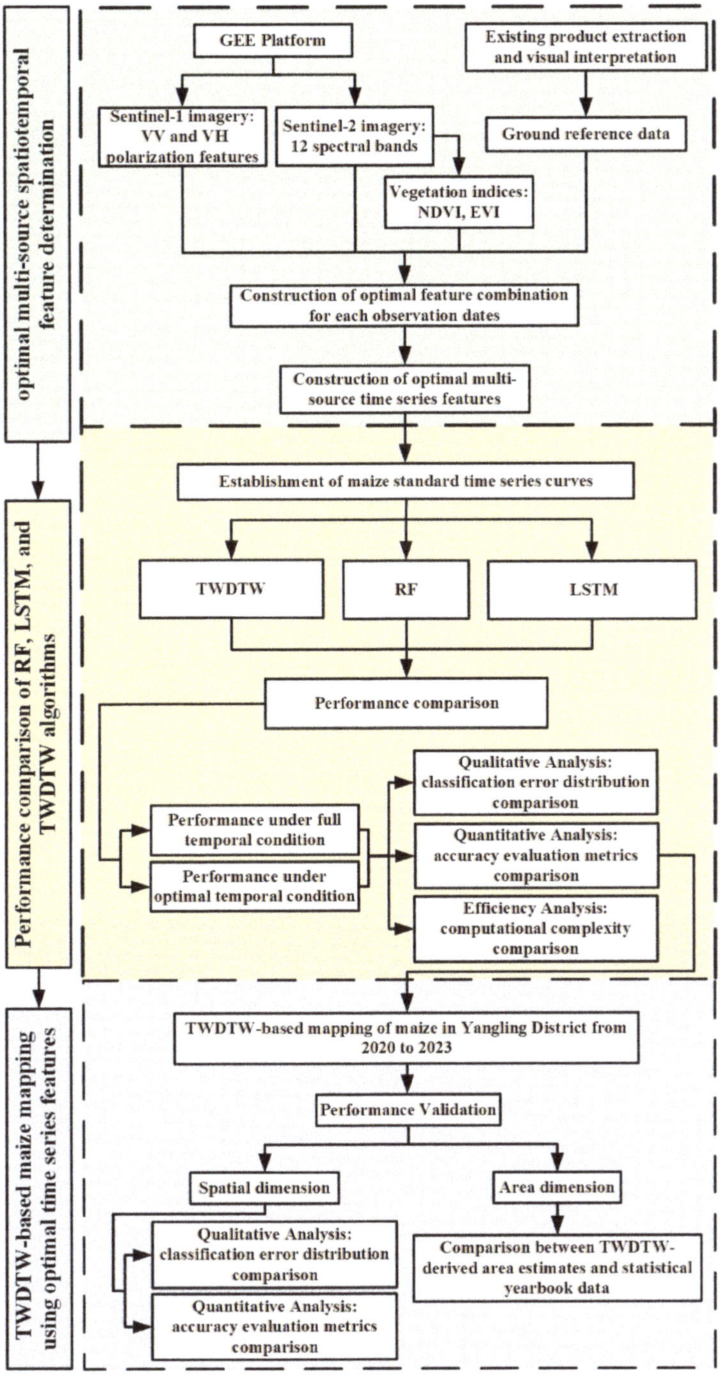

Figure 8. The overview of the proposed framework.

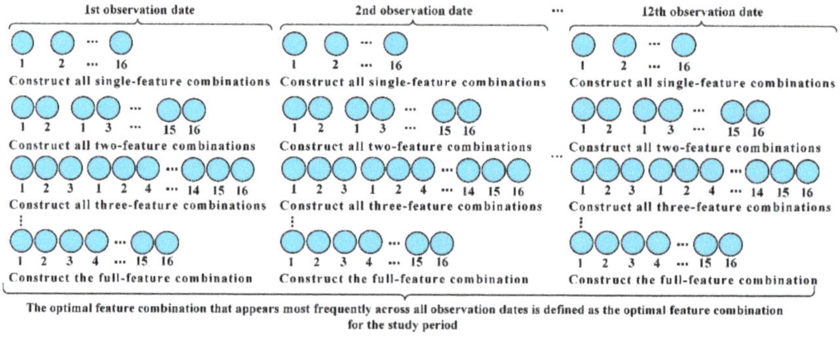

Figure 9. Flow chart for constructing the optimal multi-source features for each observation date.

Moreover, the Jeffries–Matusita (JM) distance was used in this study as the separability indicator, since it can overcome the limitations of traditional metrics in high-dimensional feature spaces, can effectively measure the differences between the joint probability distributions of multiple classes, and has demonstrated robust performance in the agricultural remote sensing community [55,56]. The specific calculation formulas are explained in Equations (3) and (4):

$$J_{ij} = 2\left(1 - e^{-d_{ij}}\right) \tag{3}$$

$$d_{ij} = \frac{1}{8}(m_j - m_i)^{T}\left[\frac{\Sigma_i + \Sigma_j}{2}\right]^{-1}(m_j - m_i) + \frac{1}{2}\ln\frac{\left|\frac{\Sigma_i + \Sigma_j}{2}\right|}{\sqrt{|\Sigma_i|\,|\Sigma_j|}} \tag{4}$$

where m_i represent the mean of feature vector of class i, Σ_i and denotes the covariance matrix of class i. The superscript T represents the transpose of the matrix. The value range of JM distance is [0, 2]—the two classes can be completely distinguished when it equals 2, and they are completely confused when it equals 0.

3.1.2. Determination of the Optimal Multi-Source Time Series Features Using Global Search Strategy

Based on the optimal multi-source feature combination identified in Section 3.1.1, the optimal multi-source time series feature was subsequently derived. Specifically, each observation date was used as a start point, with the sequence length progressively extended until including the final observation date (i.e., the 12th date). Figure 10 illustrates this process using the first observation date as the start point. On this basis, all Sentinel-1 images within the observation range of each Sentinel-2 time series combination were used in combination with the corresponding Sentinel-2 images, forming multi-source time series datasets. This process can also be used when other observation dates are used as start point. Similarly, for each time series combination, the JM distance between maize and non-maize was assessed using the optimal multi-source feature combination identified in Section 3.1.1. Consequently, by combining the optimal multi-source features for common observation dates, the optimal recognition period for maize mapping could be found through this type of method.

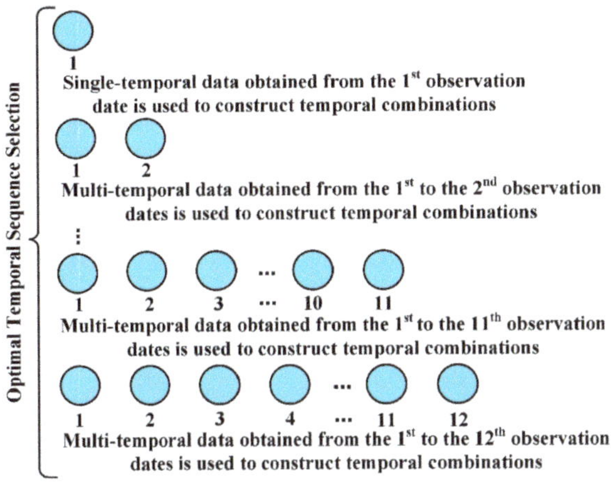

Figure 10. Flow chart of the optimal multi-source time series feature construction. Taking, as an example, the first observation date that was used as the start date, this process could also be used when other observation dates were used as the start dates.

3.2. *Performance Comparison Between RF, LSTM, and TWDTW Driven by Different Feature Sets*

Based on the full-season optimal multi-source features and the constructed optimal multi-source time series features, this study compared the differences in maize mapping performance between TWDTW, RF, and LSTM.

For the supervised RF and LSTM, 10% of the maize and non-maize pixel samples from 2021, corresponding to a total of 160,833 pixels, were randomly selected as the training set to build the models. Data from both 2020 and 2021 were used to evaluate model performance, as comprehensive ground reference datasets were available for these two years. Additionally, the model structure and parameter settings were the same as those reported in the references [57,58].

For the TWDTW, the algorithm achieves alignment of temporal sequences by nonlinearly warping the time axis and incorporating time-based weights, as shown in Figure 11. This helps to more accurately capture the seasonal growth patterns shared between the standard temporal curve of the target crop and the temporal curve of the pixels to be classified. For the time-weighted component, a Gaussian function was employed to penalize the time offset (i.e., excessive distortion on the time axis). The specific calculation formula is explained in Equation (5):

$$w(t) = e\left(-\frac{(t-\mu)^2}{2\sigma^2}\right) \quad (5)$$

where t is the time offset, representing the time difference between two time points in the sequence. μ represents the central time point, which is usually set to 0 and indicates an ideal match in time. σ represents the standard deviation, which is related to the data density of the time series. σ is typically set as 1.2 to 2 times the median time interval between adjacent time points, which could be used to control the decay rate of the time weight and improve the adaptability to different time series distributions.

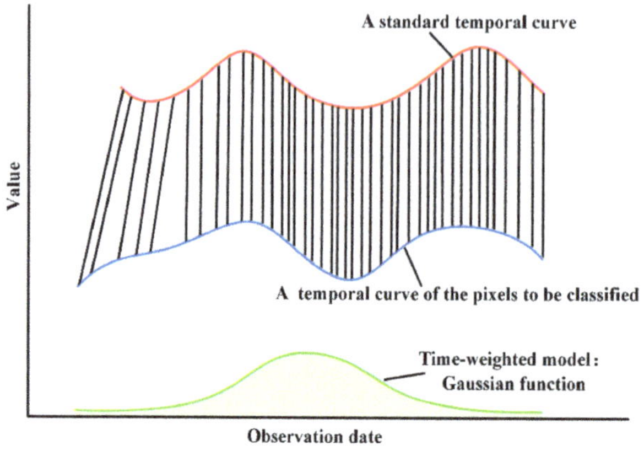

Figure 11. The calculation process of the TWDTW.

Moreover, different from that is driven by the single feature, the TWDTW employed in this study is driven by the optimal multi-source time series features, allowing for a more comprehensive characterization of temporal dynamics, illustrated in Figure 12. Specifically, multiple standard time series curves for maize were generated by averaging the values of maize samples for each time series feature included in the optimal multi-source time series features. The individual TWDTW distance between the pixel to be classified and the corresponding standard maize time series curve was then calculated for each time series feature. The final TWDTW distance is computed by summing the individual TWDTW distances corresponding to each time series feature within the optimal multi-source time series features. A smaller TWDTW distance indicates a higher similarity between the pixel and the standard time series curve, thereby implying a greater likelihood that the pixel belongs to the maize category.

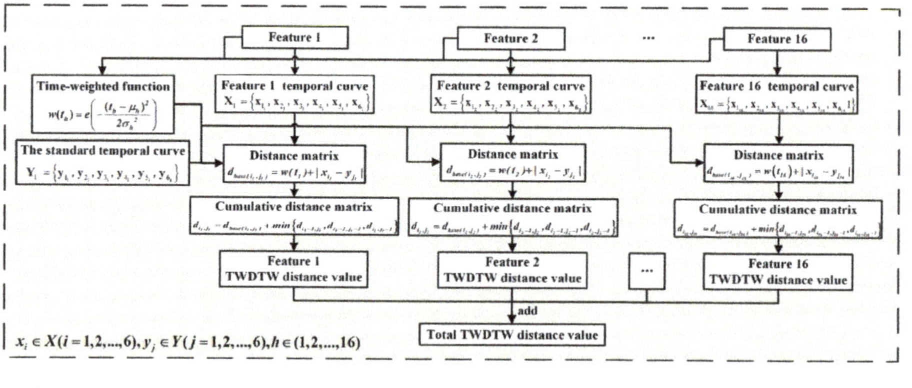

Figure 12. TWDTW-based maize mapping using optimal multi-source time series features.

In this study, the standard maize time series curve constructed from 2021 was used as a reference for cross-year mapping. Time series data from other years were directly matched against this curve to perform maize mapping, which could be used to evaluate the temporal generalization ability of the TWDTW method.

3.3. Maize Mapping via TWDTW Driven by Optimal Multi-Source Time Series Features

The optimal standard time series curves for maize, constructed in 2021, were used as the benchmark. Subsequently, based on TWDTW and optimal multi-source time series features, maize maps of Yangling District were generated for the years 2020 to 2023. Meanwhile, the extraction results were assessed from both spatial and statistical perspectives, including spatial consistency and extracted area-based comparisons.

For maize spatial distribution verification, the evaluation indicators were calculated by comparing differences between mapping results and ground reference datasets. The indicators mainly included user accuracy, producer accuracy, F1-score, etc. Please refer to [59] for the specific mathematical expression. For verification of the maize area statistics, average relative error was calculated by comparing the differences between the maize area derived from Sentinel-1/2 images and the maize area recorded in the statistics data.

4. Experimental Results and Discussions

4.1. Optimal Multi-Source Time Series Features of Sentinel-1 and Sentinel-2 Images for Maize Mapping

(RQ1) In order to construct the optimal multi-source time series features of Sentinel-1 and Sentinel-2 images for maize mapping, the two-step JM distance-based global search strategy was used to select optimal multi-source features at each observation date and to build optimal multi-source time series features by incorporating temporal information. Subsequently, the standard time series curve for maize was established using the constructed dataset.

4.1.1. Determination of the Optimal Features of Sentinel-1/2 Images Across Multiple Observation Dates

Based on the framework presented in Section 3.1.1, the optimal features of Sentinel-1/2 across multiple observation dates were systematically selected. The JM distances from different feature combinations for each observation date were calculated. Due to the large number of feature combinations, only the top eight groups with the highest JM distance values for each observation date are displayed (i.e., 26 feature combinations, as listed in Table 2), and their JM distances calculated from different observation dates are shown in Figure 13. Different colors represent the values of JM distance calculated from different feature combinations.

Table 2. The top eight feature combinations with the highest JM distance values for each observation date. Across all observation dates, a total of 26 unique feature combinations were identified.

Feature Combination	Feature Name
1	B1\B2\B3\B4\B5\B6\B7\B8\B8A\B9\B11\B12\NDVI\EVI\VV\VH
2	B1\B2\B3\B4\B5\B6\B7\B8\B8A\B9\B11\B12\NDVI\EVI\VV
3	B1\B2\B3\B4\B5\B7\B8\B8A\B9\B11\B12\NDVI\EVI\VV\VH
4	B1\B2\B3\B4\B5\B6\B7\B9\B11\B12\NDVI\EVI\VV\VH
5	B1\B2\B3\B4\B5\B6\B7\B8\B11\B12\NDVI\EVI\VV\VH
6	B2\B3\B4\B5\B6\B7\B8\B8A\B9\B11\B12\NDVI\EVI\VV\VH
7	B1\B3\B4\B5\B6\B7\B8\B8A\B9\B11\B12\NDVI\EVI\VV\VH
8	B1\B2\B3\B4\B5\B6\B7\B8\B8A\B9\B11\B12\NDVI\EVI\VH
9	B1\B2\B4\B5\B6\B7\B8\B8A\B9\B11\B12\NDVI\EVI\VV\VH
10	B1\B2\B3\B4\B5\B7\B8\B8A\B9\B11\B12\NDVI\EVI\VV
11	B1\B2\B3\B4\B6\B7\B8\B8A\B9\B11\B12\NDVI\EVI\VV\VH
12	B1\B2\B3\B4\B5\B6\B7\B8\B11\B12\NDVI\EVI\VV\VH

Table 2. *Cont.*

Feature Combination	Feature Name
13	B1\B2\B3\B5\B6\B7\B8\B8A\B9\B11\B12\NDVI\EVI\VV\VH
14	B1\B3\B4\B5\B6\B7\B8\B8A\B9\B11\B12\NDVI\EVI\VV
15	B1\B2\B4\B5\B6\B7\B8\B8A\B9\B11\B12\NDVI\EVI\VV
16	B1\B2\B3\B4\B5\B7\B8\B8A\B9\B11\B12\NDVI\EVI\VV
17	B1\B3\B4\B5\B7\B8\B8A\B9\B11\B12\NDVI\EVI\VV\VH
18	B1\B2\B3\B4\B5\B6\B8\B8A\B9\B11\B12\NDVI\EVI\VV\VH
19	B1\B2\B3\B4\B5\B6\B7\B8\B8A\B9\B11\B12\NDVI\EVI
20	B1\B2\B3\B4\B5\B6\B7\B8\B8A\B9\B11\B12\NDVI\VV\VH
21	B1\B2\B3\B4\B5\B6\B7\B8\B8A\B9\B11\B12\NDVI\VV
22	B1\B2\B3\B4\B5\B7\B8\B8A\B9\B11\B12\NDVI\VV\VH
23	B1\B2\B3\B4\B5\B6\B7\B8\B11\B12\NDVI\VV\VH
24	B1\B2\B3\B4\B5\B6\B7\B9\B11\B12\NDVI\VV\VH
25	B1\B2\B3\B4\B5\B6\B7\B8\B8A\B9\B11\B12\NDVI\VH
26	B1\B2\B3\B4\B5\B6\B8\B8A\B9\B11\B12\NDVI\VV\VH

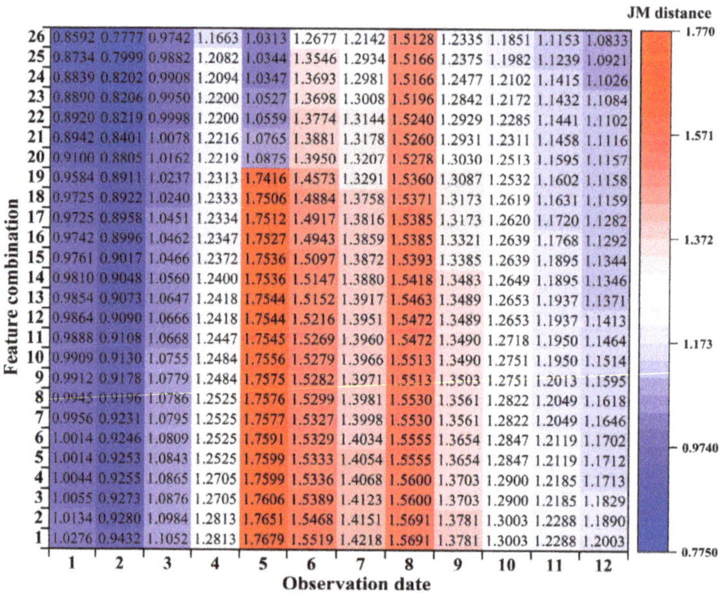

Figure 13. The JM distance distribution between maize and non-maize calculated from different feature combinations under different observation dates.

It can be seen that the feature combination including all features presented in this study (Feature combination 1) most frequently yielded the highest JM distance across multiple observation dates (Figure 13). These features include twelve multispectral bands and two vegetation indices (NDVI and EVI) from Sentinel-2 images, as well as two polarization features (VV and VH) from Sentinel-1 images. In addition, it was worth noting that JM distances calculated from different feature combinations on the fifth and sixth observation dates, which corresponds to the early jointing stage of maize, were significantly higher than those from other observation dates. This is mainly because maize undergoes rapid changes in its external morphological characteristics during the early jointing stage, particularly in plant height and leaf expansion. Therefore, at this phenology stage, spectral signals from other land cover types are primarily governed by surface diffuse reflection from the soil, whereas maize exhibits strong volumetric scattering due to its developing canopy structure.

This contrast results in increased spectral separability between maize and other land cover types.

In summary, the combination of twelve spectral bands, NDVI, EVI, and VV and VH polarization features represented the optimal multi-source features for distinguishing maize from non-maize throughout the maize growth period. Therefore, based on this feature combination, this study further explored the optimal temporal combination for maize discrimination.

4.1.2. Determination of the Optimal Multi-Source Time Series Features of Sentinel-1/2 Images

Based on the optimal feature combinations obtained from Section 4.1.1 and the framework presented in Section 3.1.2, the optimal multi-source time series features of Sentinel-1/2 were selected. The JM distances derived from the optimal multi-source features constructed over different temporal windows corresponding to various growth stages of maize are shown in Figure 14. The curves in different colors represent the JM distance distributions calculated from time series that gradually incorporate data from various initial observation dates to the maturity stage of maize.

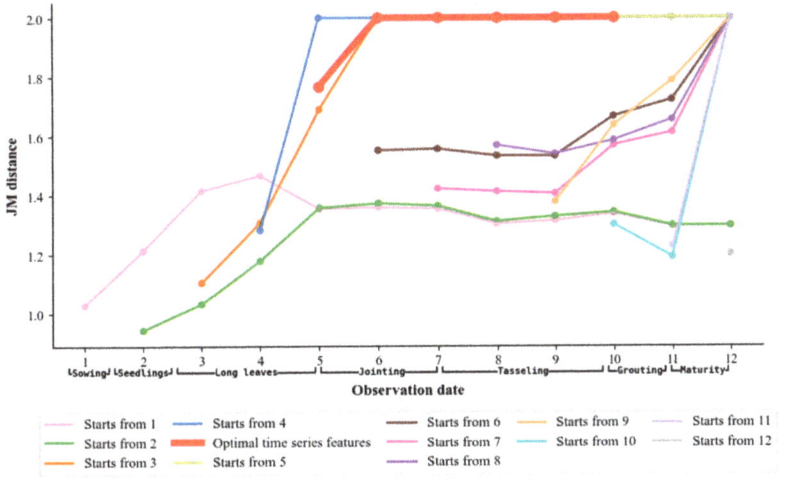

Figure 14. The JM distance distribution between maize and non-maize calculated from the optimal multi-source features constructed over different temporal windows.

It can be seen that the JM distance was not the highest when images covering the entire maize growth cycle were used (Figure 14). This may be attributed to the presence of non-maize crops in the region that share similar phenology stages with maize, especially during the early growth stages, which limits the separability between maize and non-maize even with the introduction of multi-temporal data. This also indicates that directly using data spanning the entire growth cycle of maize may not be optimal for accurate maize mapping. In contrast, when only the optimal multi-source features from Sentinel-1/2 images during the mid-to-late growth stages were used, the JM distance between maize and non-maize reached its maximum (i.e., 2) and tended to stabilize. Specifically, based on the optimal multi-source features, when the fifth observation date was used as the starting point, the JM distance between maize and non-maize was effectively improved and remained constant at 2 as optimal multi-source features from subsequent dates were progressively added to the time series. However, when the time series features exclude the fifth observation date, the separability decreases significantly. This indicates the importance of multi-source features from the fifth observation date (i.e., the observation dates aligned with the maize jointing stage) in accurately distinguishing maize from non-maize. Moreover, when the

10th observation date was used as the starting point, and subsequent data after the 10th observation date were incorporated into the time series feature set, the JM distance exhibited a declining trend. This means that the data from observation dates after the 10th observation date had little impact on distinguishing maize from non-maize.

Therefore, to utilize shorter yet effective time series features, this study selected the optimal multi-source features from the 5th to 10th observation dates, corresponding to the period from late July to mid-September. This period aligns with the jointing to tasseling stages of maize. The observation dates of the final optimal multi-source time series features are shown in Figure 15. Compared with using full-season time series data, this temporal combination not only maintained a high level of class separability but also reduced redundant observations. Further details and validations are presented in the subsequent sections.

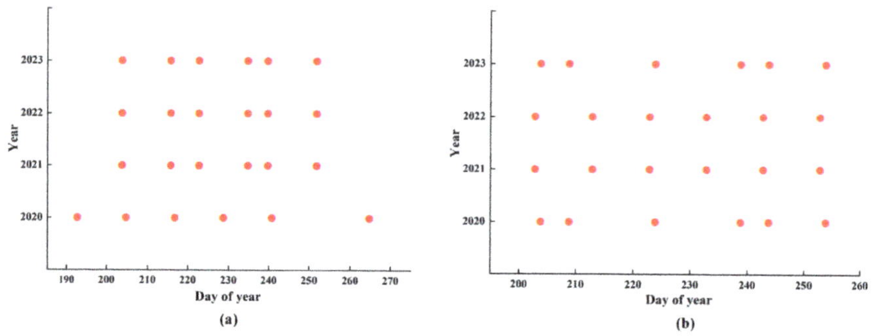

Figure 15. The observation dates distribution of the final optimal multi-source time series features for different years. (**a**) The optimal Sentinel-1 observation date distribution for maize mapping. (**b**) The optimal Sentinel-2 observation date distribution for maize mapping.

4.1.3. Construction and Analysis of Standard Time Series Curves Based on the Optimal Multi-Source Time Series Features from Sentinel-1/2 Images

Based on the optimal multi-source time series features identified in Sections 4.1.1 and 4.1.2, sixteen standard maize time series curves were constructed, each covering six observation dates spanning from the jointing to tasseling stages. To improve the robustness and representativeness of the standard time series curves, maize sample pixels were further screened, rather than relying solely on the mean values of each feature across all samples.

Specifically, the data from 2021 was taken as an example. For each observation date, the initial mean and a two-standard-deviation range for each feature were derived from the 98,982 maize pixels in the reference data. These statistics were used to represent the overall distribution of maize features in the study area. Subsequently, to remove potential outliers and reduce extreme variability, only maize pixels with feature values falling within ± 1 standard deviation of the mean were retained, resulting in 73,682 pixels. Finally, the mean values of different features were recalculated based on the filtered data to construct the optimized maize standard time series curve. The specific changes in the maize standard time series curves based on different features before and after optimization in Yangling District in 2021 are shown in Figure 16.

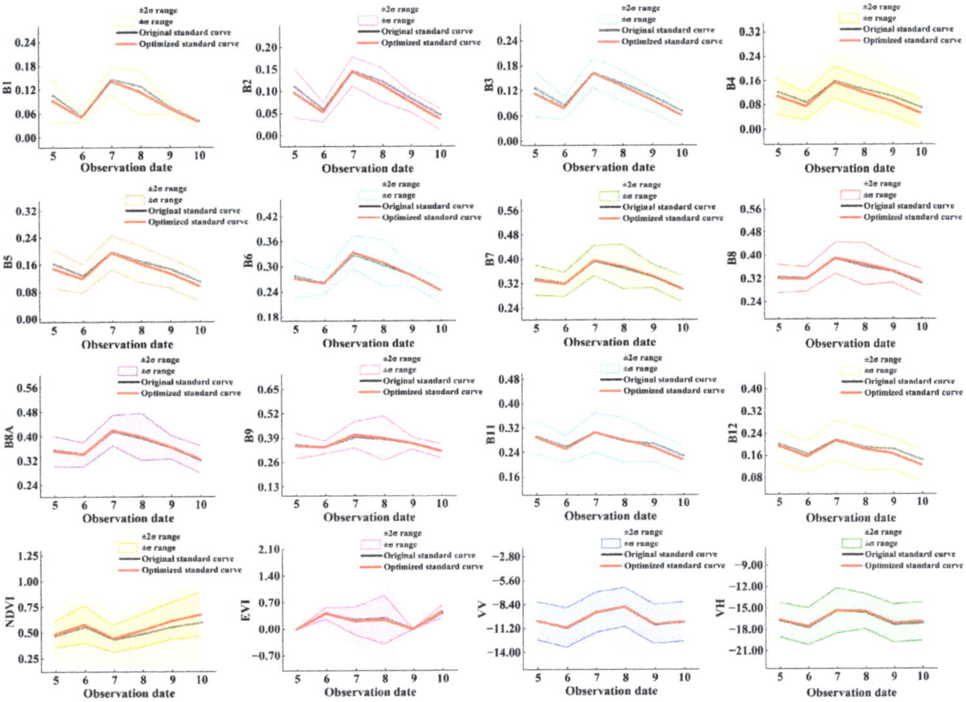

Figure 16. Comparison of maize standard time series curves before and after optimization in Yangling District, 2021.

It can be seen that this data filtering method could reduce the fluctuation of pixel features used to construct the maize standard time series curve, without substantially altering its mean values (Figure 16). This indicates that the method can effectively suppress the influence of outliers and noise, thereby improving the stability and representativeness of the resulting time series curves, which is crucial for improving the accuracy of TWDTW-based crop mapping. Furthermore, the minimal change in mean values before and after optimization provides a certain degree of validation for the reliability of the overall maize distribution in the open-source reference data. This also indicates that constructing a maize standard curve based on the open-source reference data used in this study is feasible.

4.2. Comparison Between TWDTW and Traditional Supervised Machine Learning Methods for Maize Mapping Across Different Time Series Lengths

(RQ2) Based on the full-season optimal multi-source features and the constructed optimal multi-source time series features, this study compared the performance of TWDTW, RF, and LSTM methods in maize mapping within Yangling District, focusing on both mapping accuracy and computational complexity.

4.2.1. Comparison of Maize Mapping Methods Using Full-Season Optimal Multi-Source Features

The TWDTW, traditional supervised RF, and LSTM methods, driven by the full-season optimal multi-source features, produced maize mapping results for Yangling District across different years, as shown in Figure 17. Their corresponding accuracy evaluation results for different years are presented in Table 3, and the detailed confusion matrices can be found in Tables S1 and S2 of the Supplementary Material.

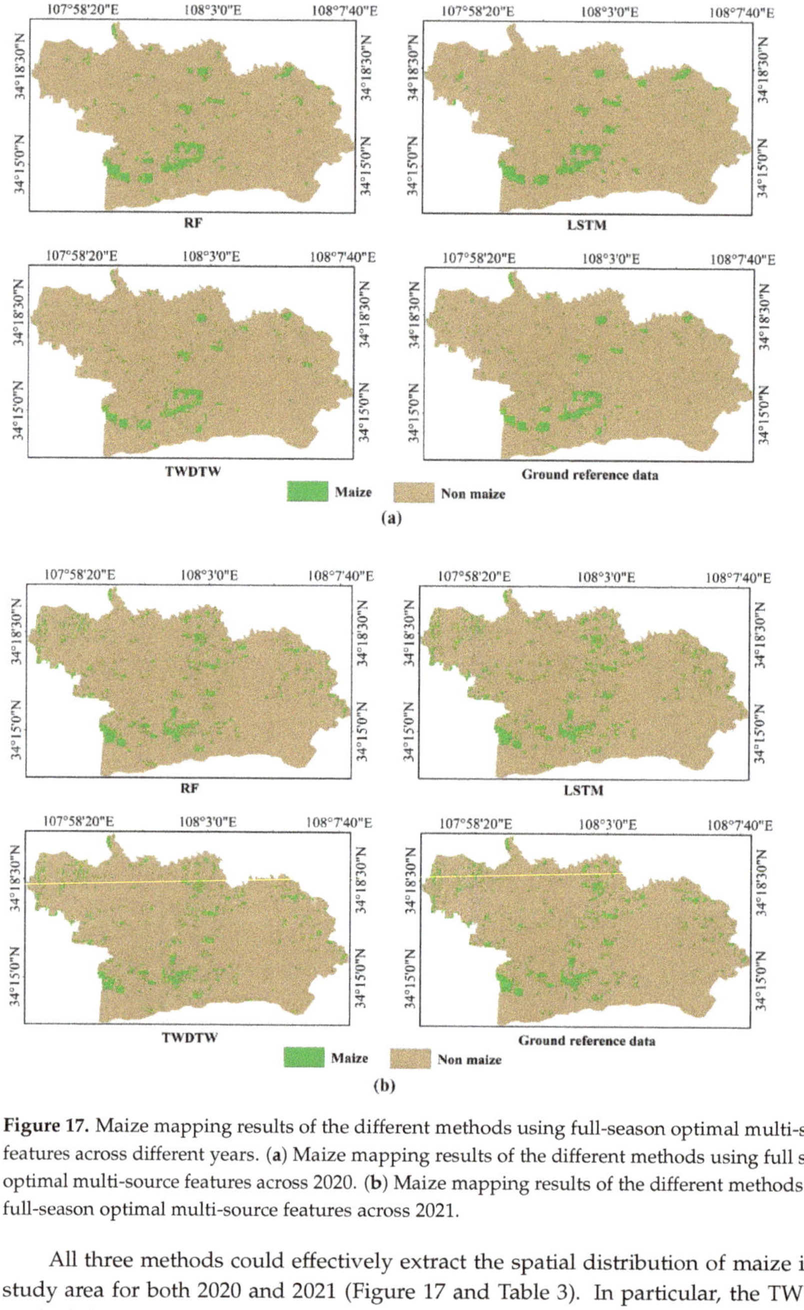

Figure 17. Maize mapping results of the different methods using full-season optimal multi-source features across different years. (**a**) Maize mapping results of the different methods using full season optimal multi-source features across 2020. (**b**) Maize mapping results of the different methods using full-season optimal multi-source features across 2021.

All three methods could effectively extract the spatial distribution of maize in the study area for both 2020 and 2021 (Figure 17 and Table 3). In particular, the TWDTW method, driven by the full-season optimal multi-source features, significantly outperformed the supervised RF and demonstrated certain advantages over the supervised LSTM in terms of mapping accuracy and stability. Specifically, for the maize mapping in 2021, the TWDTW achieved an overall accuracy of 98.01%, with a user accuracy of 97.65%, a producer accuracy of 89.92%, and an F1 score of 0.9369. Compared with the supervised RF and LSTM, these indicators showed improvements of 2.46% and 7.88% in user accuracy, 10.99% and 4.76% in producer accuracy, and 0.0739 and 0.0629 in F1 score, respectively.

The superior performance of TWDTW was also evident in the maize mapping results for 2020, further confirming its robustness across different temporal contexts. However, it was worth noting that the overall accuracies of TWDTW were slightly lower than those of the supervised RF and LSTM models. This was mainly due to the significant class imbalance in the study area, where non-maize pixels greatly outnumbered maize pixels. As a result, when the model tends to classify more pixels as non-maize, the overall accuracy becomes higher.

Moreover, representative maize-intensive and maize-sparse regions in Yangling District were selected for visual comparison to assess the effectiveness of TWDTW driven by full-season optimal multi-source features in mapping spatial details, as shown in Figures 18 and 19.

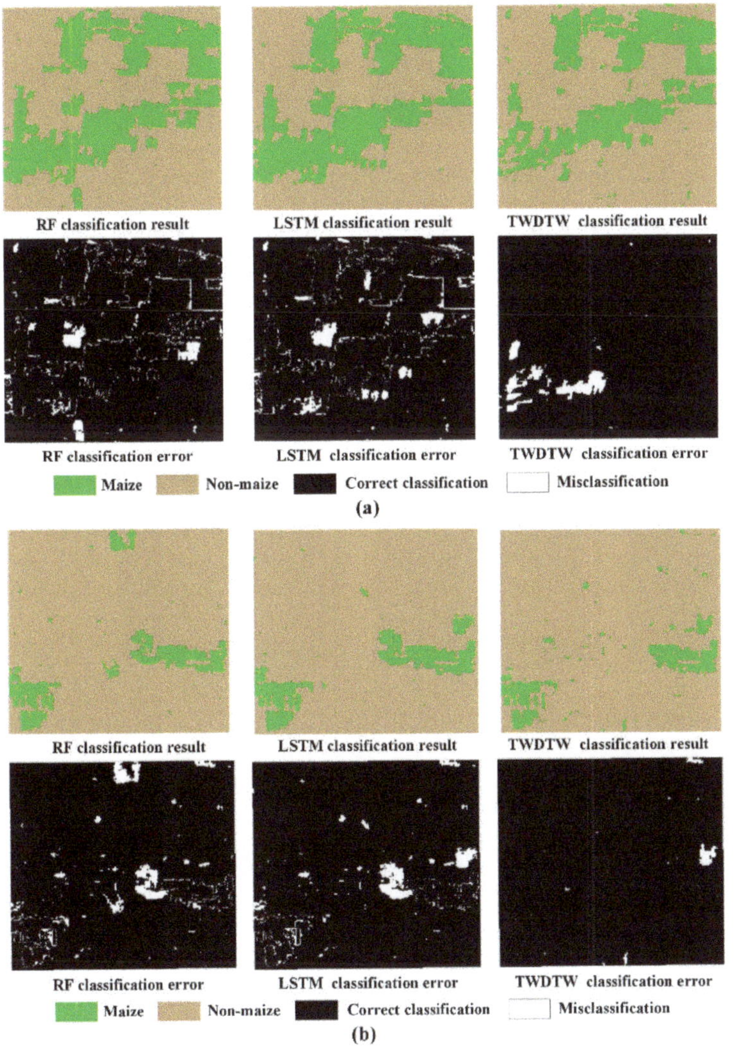

Figure 18. Maize mapping results in different maize planting regions for 2020, generated by different methods based on full-season optimal multi-source features. (**a**) Maize-intensive regions mapping results generated by different methods using full-season optimal multi-source features. (**b**) Maize-sparse regions mapping results generated by different methods using full-season optimal multi-source features.

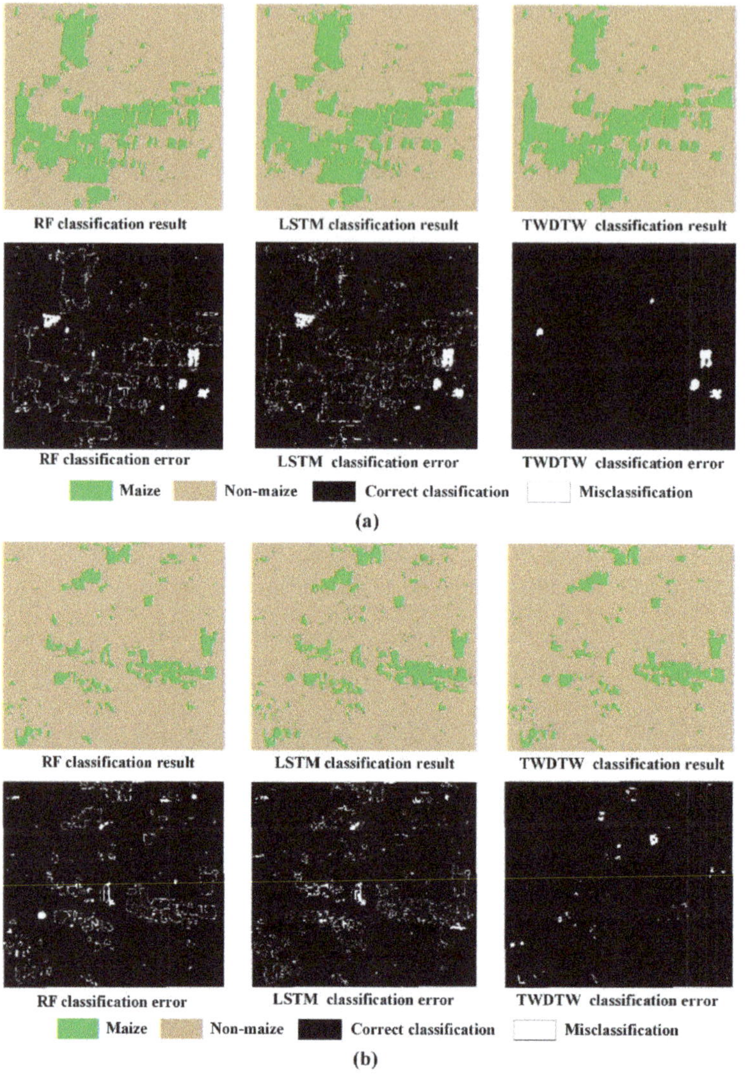

Figure 19. Maize mapping results in different maize planting regions for 2021, generated by different methods based on full-season optimal multi-source features. (**a**) Maize-intensive regions mapping results generated by different methods using full-season optimal multi-source features. (**b**) Maize-sparse regions mapping results generated by different methods using full-season optimal multi-source features.

Table 3. Comparison of maize mapping accuracies of the different methods using full-season optimal multi-source features across different years.

Year	Method	OA	Maize UA	Maize PA	Maize F1 Score
2020	RF	94.67%	96.05%	70.68%	0.8144
	LSTM	95.34%	89.17%	81.75%	0.8530
	TWDTW	93.66%	99.51%	84.28%	0.9134
2021	RF	98.22%	95.19%	78.93%	0.8630
	LSTM	98.25%	89.77%	85.16%	0.8740
	TWDTW	98.01%	97.65%	89.92%	0.9369

For maize mapping across different planting regions and years, it can be seen that in both maize-intensive and maize-sparse regions, the supervised RF and LSTM models performed well in extracting large-scale contiguous maize planting regions (Figures 18 and 19). However, under the influence of mixed pixels along field boundaries, noticeable misclassification along field boundary and salt-and-pepper noise were observed in the mapping results of RF and LSTM. In contrast, TWDTW could effectively preserve field integrity with fewer boundary errors and significantly reduce salt-and-pepper noise in the mapping results, thereby producing the most accurate maize maps across different planting regions in Yangling District. This is mainly because, unlike the supervised RF and LSTM models that rely primarily on spectral values, TWDTW exploits the temporal trajectory of crop growth and aligns it with standard crop time series curves. This trajectory-based comparison emphasizes the temporal consistency and shape of vegetation dynamics, thereby reducing the sensitivity to spectral distortions introduced by non-crop components within mixed pixels. As a result, TWDTW can partially overcome the effects of mixed pixels.

In summary, compared with the supervised RF and LSTM, commonly used in the agricultural remote sensing community, TWDTW driven by the full-season optimal multi-source features can also be applied to maize mapping and demonstrate certain advantages in maize mapping performance. However, the use of full-season optimal multi-source features may increase the computational complexity of TWDTW, which can also lead to a certain amount of redundant information. Therefore, in the following section, we would further explore the maize mapping performance of TWDTW driven by optimal multi-source time series features.

4.2.2. Comparison of Maize Mapping Methods Using Optimal Multi-Source Time Series Features

Moreover, the TWDTW, traditional supervised RF, and LSTM methods, driven by the optimal multi-source time series features, produced maize mapping results for Yangling District across different years, as shown in Figure 20. Their corresponding accuracy evaluation results for different years are presented in Table 4, and the detailed confusion matrices can be found in Tables S3 and S4 of the Supplementary Material.

Table 4. Comparison of maize mapping accuracies of the different methods using optimal multi-source time series features across different years.

Year	Method	OA	Maize UA	Maize PA	Maize F1 Score
2020	RF	98.02%	98.40%	89.64%	0.9372
	LSTM	94.89%	89.64%	83.64%	0.8373
	TWDTW	99.66%	99.94%	91.79%	0.9569
2021	RF	99.03%	98.95%	87.26%	0.9274
	LSTM	98.81%	94.21%	88.74%	0.9139
	TWDTW	99.61%	99.95%	92.91%	0.9630

It can be seen that all three methods, driven by optimal multi-source time series features, were still able to effectively extract the spatial distribution of maize in the study area for both 2020 and 2021 (Figure 20 and Table 4). Similarly, the TWDTW method, driven by optimal multi-source time series features, significantly outperformed the supervised RF and LSTM in terms of mapping accuracy and stability. Specifically, for maize mapping in 2021, the overall accuracy, maize user's accuracy, producer accuracy, and F1 score of TWDTW reached 99.61%, 99.95%, 92.91%, and 0.9630, respectively. Notably, the maize user's accuracy, producer accuracy, and F1 score of TWDTW showed greater improvements than those of the supervised RF and LSTM. Its superior performance could also be observed in the maize mapping results for 2020. In addition, compared with the methods driven by full-season optimal multi-source features, those using the optimal multi-source features further improved maize mapping performance. These results confirmed the effectiveness of the selected temporal features in this study. In particular, the improvement observed in TWDTW was more pronounced.

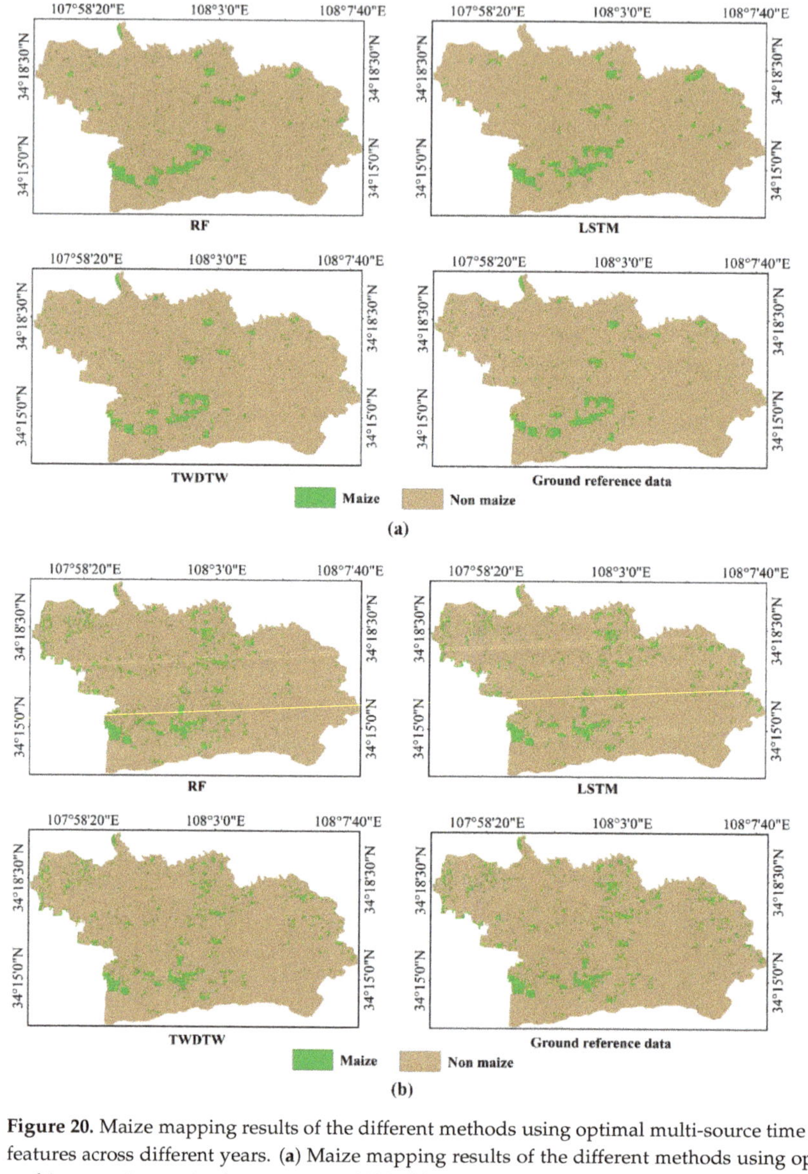

Figure 20. Maize mapping results of the different methods using optimal multi-source time series features across different years. (**a**) Maize mapping results of the different methods using optimal multi-source time series features across 2020. (**b**) Maize mapping results of the different methods using optimal multi-source time series features across 2021.

Moreover, to further evaluate the spatial details of maize mapping, different representative maize planting regions in Yangling District, selected in Section 4.2.1, were used for visual comparison to assess the effectiveness of TWDTW driven by optimal multi-source time series features, as shown in Figures 21 and 22.

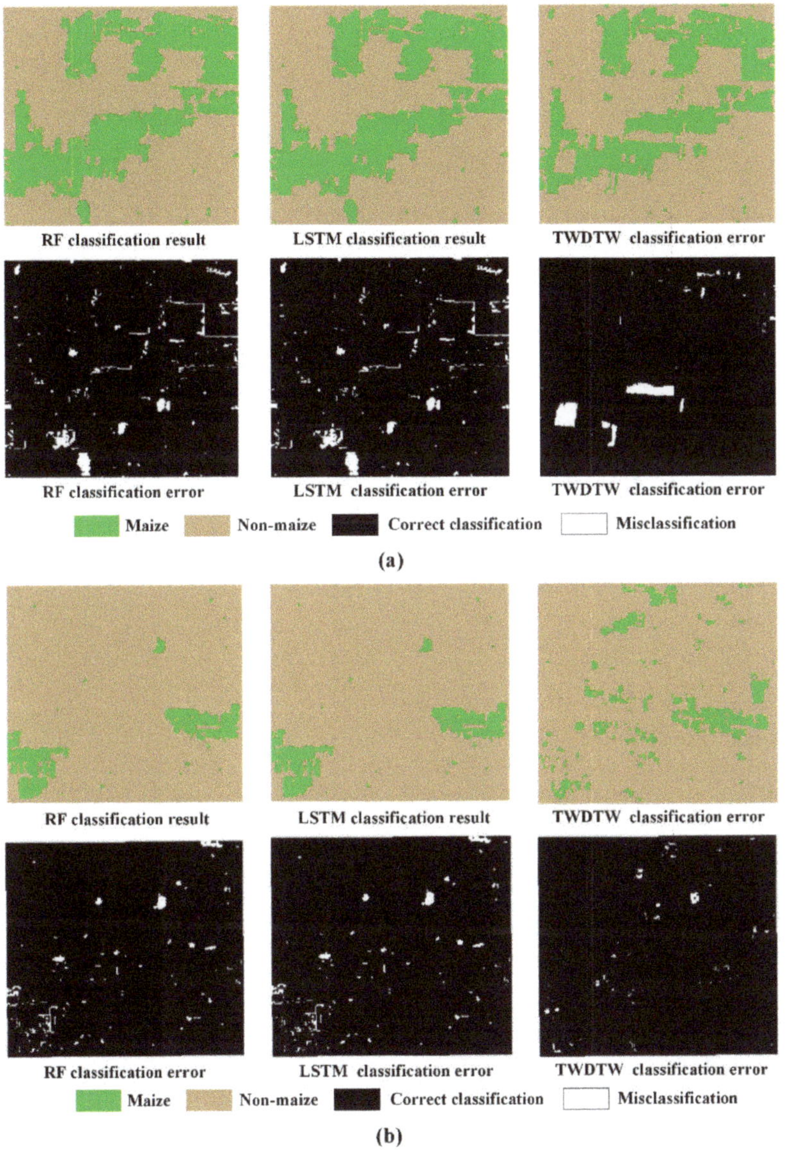

Figure 21. Maize mapping results in different maize planting regions for 2020, generated by different methods based on optimal multi-source time series features. (**a**) Maize-intensive regions mapping results generated by different methods using optimal multi-source time series features. (**b**) Maize-sparse regions mapping results generated by different methods using optimal multi-source time series features.

For maize mapping across different planting regions and years, it can be seen that the performance of the three methods driven by optimal multi-source time series features was generally consistent with that of the same methods driven by full-season optimal multi-source features in both maize-intensive and maize-sparse regions (Figures 21 and 22). RF and LSTM exhibited considerable misclassification in small and fragmented maize planting areas, whereas TWDTW produced the most accurate mapping results in these regions.

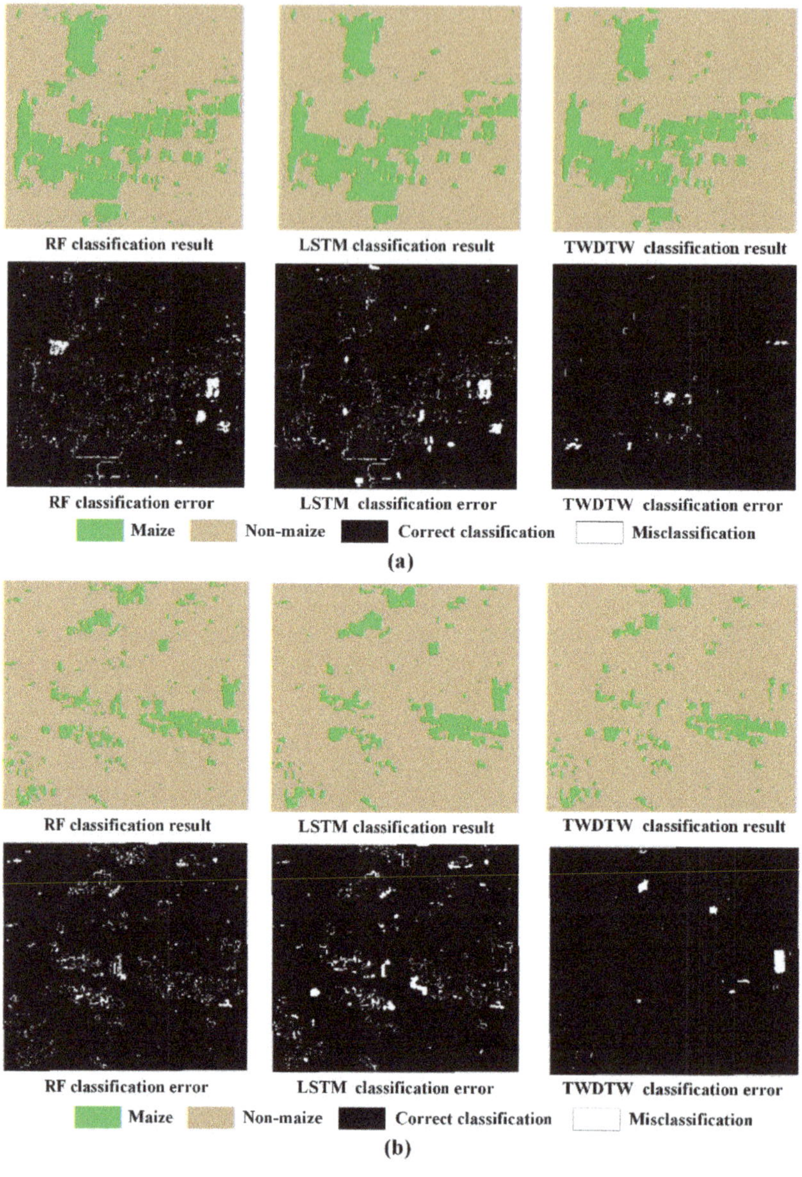

Figure 22. Maize mapping results in different maize planting regions for 2021, generated by different methods based on optimal multi-source time series features. (**a**) Maize-intensive regions mapping results generated by different methods using optimal multi-source time series features. (**b**) Maize-sparse regions mapping results generated by different methods using optimal multi-source time series features.

In summary, although RF and LSTM can also achieve good maize mapping results, they rely heavily on a large number of manually annotated ground samples across various land cover categories. This dependency limits their transferability and timeliness, especially in complex environments characterized by diverse non-target classes (e.g., bare soil, various non-target crops, and other types of vegetation). In contrast, TWDTW can classify pixels by matching them with the standard time series curve of the target crop, even with limited samples of the target crop. This process does not require the collection of non-target crop samples, which greatly facilitates practical crop mapping. In addition, it was worth noting that compared with the methods driven by full-season optimal multi-source features, those using optimal multi-source time series features achieved improved maize mapping performance while requiring less input data. This significantly reduces the computational complexity of the method, which would be proved in the subsequent content.

4.2.3. Comparison of Computational Complexity Between Full-Season and Optimal Time Series Schemes

For practical crop mapping tasks, while high mapping accuracy is important, computational complexity ultimately determines a model's feasibility for large-scale crop mapping. Consequently, we further compared computational complexity for models driven by optimal multi-source features under the full-season and optimal time series schemes, as listed in Table 5.

Table 5. Comparison of computational complexity between full-season and optimal time series schemes.

Temporal Length of the Optimal Multi-Source Features	Method	Processing Time (min)	Parameters
Optimal time series	RF	5.90	6,594,476
	LSTM	247.85	219,393
	TWDTW	38.73	—
Full season	RF	8.02	11,578,996
	LSTM	330.10	219,393
	TWDTW	154.22	—

Note: Processing time for TWDTW reflects the time consumed in generating the distance matrix; for RF and LSTM, it reflects the time required for producing classification results.

It can be seen that the inference time of all models was significantly reduced when using the optimal multi-source features from the optimal time series (Table 5). Specifically, for TWDTW, its inference time dropped from 154.22 min to 38.73 min, a reduction of approximately 74.9%. For the supervised RF and LSTM, their inference times dropped by about 24.9% and 26.4%, respectively. The reduced inference time of the RF model using optimal multi-source time series features could be attributed to a 43% reduction in parameters compared to the version based on full-season optimal multi-source features, resulting in significantly lower computational complexity. For the LSTM model, the number of parameters remained the same across the two feature modes, as it primarily depends on the input feature dimension (i.e., 16) at each time step, the hidden state size, and the number of LSTM layers, all of which were kept constant in this study. However, the longer time sequence in the full-season scheme increased computational complexity, resulting in a longer inference time. For the TWDTW, a non-parametric method with no trainable parameters, the computational cost is primarily determined by the time series length (n) and feature dimension (d), with a complexity of $O(n^2 \cdot d)$. Consequently, when optimal multi-source features from the reduced time series were used, the computational cost of TWDTW could theoretically be reduced to 25% of that under the full-season scheme, as the time series length was halved while the feature dimension per observation date remained constant at 16. Combined with the mapping performance results in Sections 4.2.1 and 4.2.2, this further confirms the practical value of TWDTW constructed with optimal multi-source features from the optimal time series for large-scale crop mapping tasks.

In general, the incorporation of optimal multi-source time series features effectively improved mapping performance while substantially lowering computational costs compared to the full-season scheme, particularly for the TWDTW, which relies less on ground samples (i.e., only representative samples of target crops are needed, without requiring samples of complex non-target crop categories).

4.3. Maize Mapping for Yangling District via TWDTW Driven by Optimal Multi-Source Time Series Features

(RQ3) The optimal multi-source time series features from different years were input into the TWDTW model calibrated using 2021 data, to generate 10 m maize maps of Yangling District from 2020 to 2023. They were validated from the perspectives of maize spatial distribution and maize area statistics.

4.3.1. Comparison Between the Mapping Results and Reference Datasets

The specific maize mapping results of Yangling District for different years are shown in Figure 23, and their accuracies, evaluated against ground reference data, are summarized in Table 6, and the detailed confusion matrices can be found in Table S5 of the Supplementary Material.

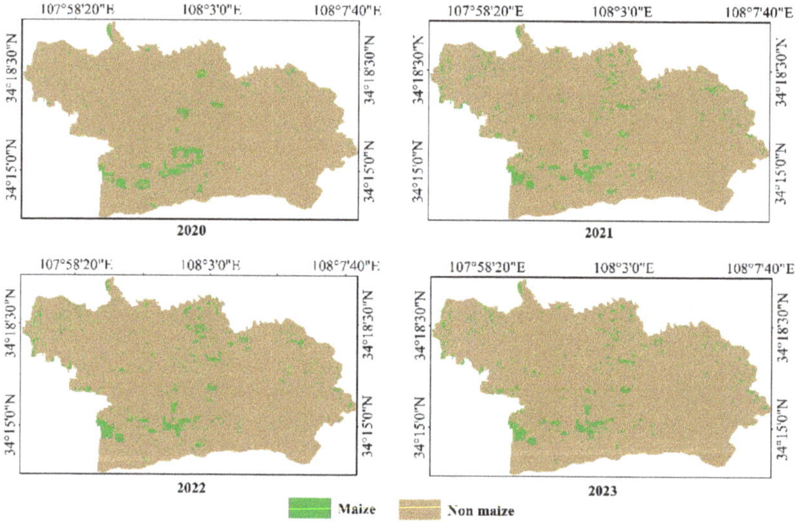

Figure 23. Maize maps of Yangling District from 2020 to 2023.

Table 6. Accuracies of maize maps in Yangling District from 2020 to 2023.

Year	OA	Maize UA	Maize PA	Maize F1 Score
2020	99.66%	99.94%	91.79%	0.9569
2021	99.61%	99.95%	92.91%	0.9630
2022	99.57%	99.93%	92.31%	0.9597
2023	99.35%	98.76%	88.09%	0.9312

It can be seen that the spatial distribution of maize cultivation in Yangling District remained relatively stable from 2020 to 2023, with cultivation primarily concentrated in the central and southern areas (Figure 23). The TWDTW-based maize maps, driven by optimal multi-source time series features, maintained high accuracy across the four years (Table 6). Specifically, the overall classification accuracies consistently remained above 99%, exhibiting minimal interannual variation. Additionally, user accuracy, producer accuracy, and F1 scores for maize classification were all stably above 85%, with the difference between user and producer accuracy remaining within a 10% range. This indicates that both omission and commission errors for maize were limited. However, the producer's accuracy for maize in 2023 dropped to 88.09%, possibly due to limitations in the accuracy of visual interpretation reference samples and interannual variations in crop phenology (i.e., There are certain differences in crop phenology across different years, especially when the years are widely spaced), which together led to misclassification in a small number of areas.

Moreover, visual analysis of maize extraction results from local areas in Yangling District across different years was conducted to assess the robustness of the TWDTW method, driven by optimal multi-source time series features, in capturing local details. Specifically, the visually interpreted regions, as described in Section 2.2.4, from 2022 and 2023 were used as reference areas to analyze the local maize mapping results over these years. For 2020 and 2021, the mapping results were directly validated using open-source ground reference data within the areas interpreted in 2022 and 2023. The specific maize mapping results across different years for this local area are shown in Figure 24.

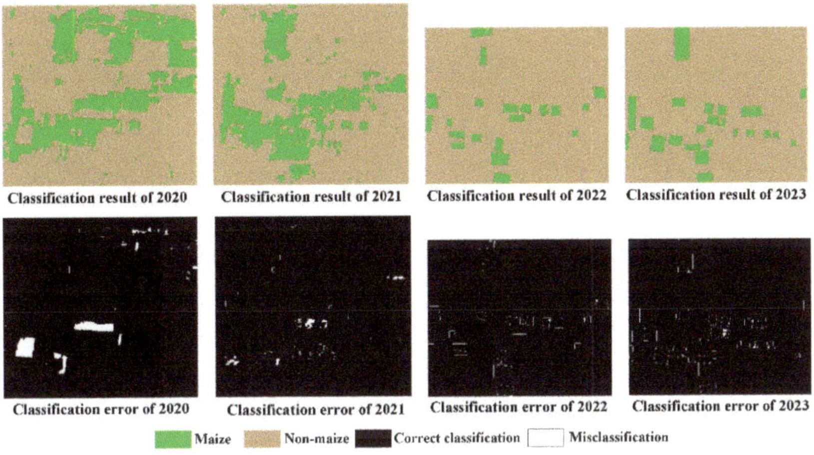

Figure 24. Maize mapping results in local regions of Yangling District from 2020 to 2023.

It can be seen that the maize mapping results generated by the TWDTW method, driven by optimal multi-source time series features, demonstrated relatively stable performance in the same local areas over multiple years, with minimal classification error (Figure 24). In particular, the maize mapping results for 2020 and 2021 better preserved field integrity, with fewer boundary errors, whereas slightly more misclassifications occurred along field edges in 2022 and 2023, especially in 2023, when this type of error was relatively more frequent. A possible reason is that interannual variations in crop phenology were smaller in 2020 and 2021, leading to better alignment between the standard curve and observed temporal trajectories (The maize standard curve was constructed using data from 2021). This allows TWDTW to more accurately discriminate field boundaries (i.e., mixed pixels), whereas larger phenological shifts in 2022 and 2023 resulted in slightly higher boundary misclassification.

In general, from the perspectives of maize spatial distribution, the TWDTW driven by optimal multi-source time series features can effectively extract the maize planting areas across different years in Yangling District.

4.3.2. Comparison Between the Extracted Maize Area and Subnational Statistics Data

Moreover, maize areas extracted from Sentinel-1/2 images across different years were compared with those recorded in statistical data, as reported in Table 7.

Table 7. Maize area comparison between extracted maize maps and statistical data across different years.

Year	Actual Area (kha)	Predicted Area (kha)	Relative Error (%)
2020	0.6500	0.6058	6.80%
2021	0.8800	0.8191	6.92%
2022	0.8700	0.8442	2.97%
2023	0.8000	0.7222	9.73%

It can be found that based on the TWDTW driven by optimal multi-source time series features, the four-year average relative error between the maize planting area of Yangling District derived from Sentinel-1/2 images and the statistical data was 6.61% (Table 7). This may be due to differences in the phenology periods of maize between 2021 and 2023. Therefore, when the maize standard curve constructed in 2021 was directly used for 2023, even with optimal multi-source time series features covering key growth periods, the mapping performance of TWDTW can still be limited to some extent.

In summary, in terms of maize spatial distribution and area statistics, the TWDTW driven by optimal multi-source time series features robustly extracted maize planting information from Sentinel-1/2 images over multiple years. Moreover, this method demonstrated satisfactory interannual transferability when the temporal gap between years was relatively short. It holds significant practical value for obtaining crop distribution information in regions with limited sample availability.

4.4. Advantages and Limitations

Compared with existing TWDTW-based crop mapping methods [43–45,60], this study employed the two-step JM distance-based global search strategy to select optimal multi-source features at each observation date and to build optimal multi-source time series features by incorporating temporal information. Then, the differences between TWDTW driven by full-season optimal multi-source features, TWDTW driven by optimal multi-source time series features, and commonly used supervised methods were systematically analyzed. The analysis focused on mapping performance across different years and computational complexity, avoiding the direct use of full-season time series data or empirically selected multi-temporal and multi-source features. Finally, the maize maps produced by the optimal multi-source time series features-driven TWDTW were consistent with existing research results, as the evaluation metrics calculated based on previous maize maps all reached desirable levels [54].

However, there were still some limitations in this study, described as follows:

(1) In this study, optimal multi-source time series features were fused by directly stacking feature dimensions. However, high feature dimensionality may still occur even after optimization. This issue can be addressed by introducing criteria for dimensionality reduction and selecting appropriate reduction methods.

(2) In this study, the standard corn growth curve established in 2021 was directly applied to TWDTW-based cross-year mapping for other years. However, when the phenology characteristics of maize vary significantly between years, the effectiveness of this transfer may be limited, especially when the temporal intervals are long (i.e., The situation of larger phenological shifts may occur). To address this limitation, the benchmark maize standard curve can be gradually refined by integrating multi-year sample datasets from diverse regions, which would improve the spatiotemporal generalization ability of TWDTW and enhance its robustness in large-scale regions.

(3) In this study, the advantages of the TWDTW for maize mapping, based on the constructed optimal multi-source time series features, were mainly validated in Yangling District, a plain area, while its applicability in other complex environments was ignored.

Such environments pose additional challenges, including terrain-induced spectral variability, crop mixtures within smallholder fields, and irregular planting patterns. These issues can be further investigated through transfer experiments using the TWDTW method with optimal multi-source time series features in different regions, thereby helping to validate and enhance its generalization capacity.

5. Conclusions

In the case of limited ground sample availability, this study systematically explored the potential of multi-source remote sensing images (i.e., Sentinel-1 and Sentinel-2), acquired during key phenological stages of maize, to support TWDTW-based mapping at the regional scale. First, the optimal multi-source time series features were identified through a two-step JM distance-based global search strategy. Then, based on optimal multi-source features with different temporal lengths, a systematic comparison of maize mapping performance and computational complexity between TWDTW and commonly used supervised machine learning models in agricultural remote sensing was conducted. Finally, maize maps of Yangling District from 2020 to 2023 were produced using optimal multi-source time series features-based TWDTW. The main conclusions are described as follows:

(1) For the maize mapping tasks in Yangling District, twelve spectral bands, the NDVI, EVI, as well as VV and VH corresponding to the maize jointing to tasseling stages, were the optimal multi-source time series features.

(2) For the maize mapping tasks in Yangling District, TWDTW driven by optimal multi-source time series features not only significantly reduced computational complexity, but also outperformed both TWDTW using full-season optimal multi-source features and traditional supervised machine learning models reliant on large multi-category sample sizes.

(3) Maize maps of Yangling District from 2020 to 2023, produced using TWDTW based on optimal multi-source time series features, consistently achieved overall accuracies above 90%, with an average relative error of only 6.61% compared to statistical yearbook data.

We hope this study may provide a new and comprehensive strategy for feature optimization in maize mapping tasks. In addition, it may offer guidance for enhancing TWDTW performance in large-scale crop mapping, which is particularly important when sample availability is limited.

Supplementary Materials: The following supporting information can be downloaded at: https://www.mdpi.com/article/10.3390/rs17173113/s1, Table S1: Confusion matrices of three methods using full-season optimal multi-source features in 2020. Table S2: Confusion matrices of three methods using full-season optimal multi-source features in 2021. Table S3: Confusion matrices of three methods using optimal multi-source time series features in 2020. Table S4: Confusion matrices of three methods using optimal multi-source time series features in 2021. Table S5: Confusion matrices of TWDTW based on optimal multi-source time series features from 2020 to 2023.

Author Contributions: Conceptualization, P.W.; methodology, H.Y.; validation, H.Y., R.W., J.L. and X.D.; formal analysis, P.W.; writing—original draft preparation, H.Y. and P.W.; writing—review and editing, H.Y., R.W., J.L., X.D., L.W., J.G. and P.W.; visualization, H.Y. and R.W.; supervision, P.W. and J.G.; project administration, P.W.; funding acquisition, P.W. All authors have read and agreed to the published version of the manuscript.

Funding: This study was supported by the National Key R&D Program Project of China (No. 2023YFC3305104).

Data Availability Statement: The raw data supporting the conclusions of this article will be made available by the authors on request.

Acknowledgments: The authors would like to thank the editor and anonymous reviewers for their helpful and constructive comments to improve this article.

Conflicts of Interest: The authors declare no conflicts of interest.

References

1. Shorachi, M.; Kumar, V.; Steele-Dunne, S. Sentinel-1 SAR backscatter response to agricultural drought in the Netherlands. *Remote Sens.* **2022**, *14*, 2435. [CrossRef]
2. Su, P.; Zhang, A.; Wang, J.; Xu, W. Plausible maize planting distribution under future global change scenarios. *Field Crops Res.* **2023**, *302*, 109079. [CrossRef]
3. Liu, X.; Peng, X.; Li, Y.; Gu, X.; Yu, L.; Wang, Y.; Cai, H. Environmental influences on evapotranspiration in wheat-maize rotation systems under diverse hydrological regimes in the Guanzhong Plain, China. *Agric. Water Manag.* **2024**, *306*, 109204. [CrossRef]
4. Guo, Y.; Fu, Y.; Chen, S.; Hao, F.; Zhang, X.; de Beurs, K.; He, Y. Predicting grain yield of maize using a new multispectral-based canopy volumetric vegetation index. *Ecol. Indic.* **2024**, *166*, 112295. [CrossRef]
5. Luo, Y.; Wang, H.; Cao, J.; Li, J.; Tian, Q.; Leng, G.; Niyogi, D. Evaluation of machine learning-dynamical hybrid method incorporating remote sensing data for in-season maize yield prediction under drought. *Precis. Agric.* **2024**, *25*, 1982–2006. [CrossRef]
6. Zhu, Q.; Wang, F.; Chen, S.; Peng, D.; Yi, Q.; He, L.; Liu, Z. Improved corn phenology monitoring using translation and weighting of characteristic points from time-series vegetation index. *Comput. Electron. Agric.* **2025**, *234*, 110297. [CrossRef]
7. Ni, R.; Tian, J.; Li, X.; Yin, D.; Li, J.; Gong, H.; Zhang, J.; Zhu, L.; Wu, D. An enhanced pixel-based phenological feature for accurate paddy rice mapping with Sentinel-2 imagery in Google Earth Engine. *ISPRS J. Photogramm. Remote Sens.* **2021**, *178*, 282–296. [CrossRef]
8. Wei, P.; Chai, D.; Huang, R.; Peng, D.; Lin, T.; Sha, J.; Sun, W.; Huang, J. Rice mapping based on Sentinel-1 images using the coupling of prior knowledge and deep semantic segmentation network: A case study in Northeast China from 2019 to 2021. *Int. J. Appl. Earth Obs. Geoinf.* **2022**, *112*, 102948. [CrossRef]
9. Birinyi, E.; Kristóf, D.; Hollós, R.; Barcza, Z.; Kern, A. Large-scale maize condition mapping to support agricultural risk management. *Remote Sens.* **2024**, *16*, 4672. [CrossRef]
10. Liu, N.; Zhao, Q.; Williams, R.; Duan, S.; Liu, X.; Barrett, B. Enhanced crop mapping using polarimetric SAR features and time series deep learning: A case study in Bei'an, China. *IEEE Trans. Geosci. Remote Sens.* **2025**, *63*, 5002917. [CrossRef]
11. Qiu, B.; Wu, F.; Hu, X.; Yang, P.; Wu, W.; Chen, J.; Chen, X.; He, L.; Joe, B.; Tubiello, F.; et al. A robust framework for mapping complex cropping patterns: The first national-scale 10 m map with 10 crops in China using Sentinel 1/2 images. *ISPRS J. Photogramm. Remote Sens.* **2025**, *224*, 361–381. [CrossRef]
12. Jiang, J.; Zhang, H.; Ge, J.; Zuo, L.; Xu, L.; Song, M.; Ding, Y.; Xie, Y.; Huang, W. The 20 m Africa rice distribution map of 2023. *Earth Syst. Sci. Data* **2025**, *17*, 1781–1805. [CrossRef]
13. Mai, J.; Feng, Q.; Fu, S.; Wang, R.; Zhang, S.; Zhang, R.; Liang, T. Enhancing crop type mapping in data-scarce regions through transfer learning: A case study of the Hexi Corridor. *Remote Sens.* **2025**, *17*, 1494. [CrossRef]
14. Shen, R.; Peng, Q.; Li, X.; Chen, X.; Yuan, W. CCD-Rice: A long-term paddy rice distribution dataset in China at 30 m resolution. *Earth Syst. Sci. Data* **2025**, *17*, 2193–2216. [CrossRef]
15. Nguyen, D.; Gruber, A.; Wagner, W. Mapping rice extent and cropping scheme in the Mekong Delta using Sentinel-1A data. *Remote Sens. Lett.* **2016**, *7*, 1209–1218. [CrossRef]
16. Dong, J.; Pang, Z.; Fu, Y.; Peng, Q.; Li, X.; Yuan, W. Annual winter wheat mapping dataset in China from 2001 to 2020. *Sci. Data* **2024**, *11*, 1218. [CrossRef]
17. Mei, Q.; Zhang, Z.; Han, J.; Song, J.; Dong, J.; Wu, H.; Xu, J.; Tao, F. ChinaSoyArea10 m: A dataset of soybean-planting areas with a spatial resolution of 10 m across China from 2017 to 2021. *Earth Syst. Sci. Data* **2024**, *16*, 3213–3231. [CrossRef]
18. Hao, P.; Di, L.; Zhang, C.; Guo, L. Transfer learning for crop classification with Cropland Data Layer data (CDL) as training samples. *Sci. Total Environ.* **2020**, *733*, 138869. [CrossRef]
19. Ge, S.; Zhang, J.; Pan, Y.; Yang, Z.; Zhu, S. Transferable deep learning model based on the phenological matching principle for mapping crop extent. *Int. J. Appl. Earth Obs. Geoinf.* **2021**, *102*, 102451. [CrossRef]
20. Yang, L.; Huang, R.; Zhang, J.; Huang, J.; Wang, L.; Dong, J.; Shao, J. Inter-continental transfer of pre-trained deep learning rice mapping model and its generalization ability. *Remote Sens.* **2023**, *15*, 2443. [CrossRef]
21. Hoppe, H.; Dietrich, P.; Marzahn, P.; Weiss, T.; Nitzsche, C.; von Lukas, U.; Wengerek, T.; Borg, E. Transferability of machine learning models for crop classification in remote sensing imagery using a new test methodology: A study on phenological, temporal, and spatial influences. *Remote Sens.* **2024**, *16*, 1493. [CrossRef]
22. Ma, Y.; Chen, S.; Ermon, S.; Lobell, D.B. Transfer learning in environmental remote sensing. *Remote Sens. Environ.* **2024**, *301*, 113924. [CrossRef]

23. Begue, A.; Vintrou, E.; Saad, A.; Hiernaux, P. Differences between cropland and rangeland MODIS phenology (start-of-season) in Mali. *Int. J. Appl. Earth Obs. Geoinf.* **2014**, *31*, 167–170. [CrossRef]
24. Liu, Y.; Shen, X.; Zhang, J.; Wang, Y.; Wu, L.; Ma, R.; Lu, X.; Jiang, M. Spatiotemporal variation in vegetation phenology and its response to climate change in marshes of Sanjiang Plain, China. *Ecol. Evol.* **2023**, *13*, e9755. [CrossRef]
25. Ontel, I.; Avram, S.; Gheorghe, C.; Niculae, M.; Pascu, I.; Rodino, S. Shifting vegetation phenology in protected areas: A response to climate change. *Ecol. Inform.* **2025**, *85*, 102962. [CrossRef]
26. Paris, C.; Bruzzone, L. A novel approach to the unsupervised extraction of reliable training samples from thematic products. *IEEE Trans. Geosci. Remote Sens.* **2021**, *59*, 1930–1948. [CrossRef]
27. Belgiu, M.; Bijker, W.; Csillik, O.; Stein, A. Phenology-based sample generation for supervised crop type classification. *Int. J. Appl. Earth Obs. Geoinf.* **2021**, *95*, 102264. [CrossRef]
28. Antonijević, O.; Jelić, S.; Bajat, B.; Kilibarda, M. Transfer learning approach based on satellite image time series for the crop classification problem. *J. Big Data* **2023**, *10*, 54. [CrossRef]
29. Shi, Q.; He, D.; Liu, Z.; Liu, X.; Xue, J. Globe230k: A benchmark dense-pixel annotation dataset for global land cover mapping. *J. Remote Sens.* **2023**, *3*, 0078. [CrossRef]
30. Huang, X.; Liu, J.; Zhu, W.; Atzberger, C.; Liu, Q. The optimal threshold and vegetation index time series for retrieving crop phenology based on a modified dynamic threshold method. *Remote Sens.* **2019**, *11*, 2725. [CrossRef]
31. Bhatti, M.; Gilani, H.; Ashraf, M.; Iqbal, M.; Munir, S. Field validation of NDVI to identify crop phenological signatures. *Precis. Agric.* **2024**, *25*, 2245–2270. [CrossRef]
32. Htitiou, A.; Möller, M.; Riedel, T.; Beyer, F.; Gerighausen, H. Towards optimising the derivation of phenological phases of different crop types over Germany using satellite image time series. *Remote Sens.* **2024**, *16*, 3183. [CrossRef]
33. Guan, X.; Huang, C.; Liu, G.; Meng, X.; Liu, Q. Mapping rice cropping systems in Vietnam using an NDVI-based time-series similarity measurement based on DTW distance. *Remote Sens.* **2016**, *8*, 19. [CrossRef]
34. Csillik, O.; Belgiu, M.; Asner, G.; Kelly, M. Object-based time-constrained dynamic time warping classification of crops using Sentinel-2. *Remote Sens.* **2019**, *11*, 1257. [CrossRef]
35. He, W.; Lei, M.; Ziyun, Y.; Lu, H. Evaluation of advanced time series similarity measures for object-based cropland mapping. *Int. J. Remote Sens.* **2023**, *44*, 3777–3800. [CrossRef]
36. Guo, Z.; Yang, K.; Liu, C.; Lu, X.; Cheng, L.; Li, M. Mapping national-scale croplands in Pakistan by combining dynamic time warping algorithm and density-based spatial clustering of applications with noise. *Remote Sens.* **2020**, *12*, 3644. [CrossRef]
37. Ye, J.; Bao, W.; Liao, C.; Chen, D.; Hu, H. Corn phenology detection using the derivative dynamic time warping method and Sentinel-2 time series. *Remote Sens.* **2023**, *15*, 3456. [CrossRef]
38. Huang, X.; Vrieling, A.; Dou, Y.; Belgiu, M.; Nelson, A. A robust method for mapping soybean by phenological aligning of Sentinel-2 time series. *ISPRS J. Photogramm. Remote Sens.* **2024**, *218*, 1–18. [CrossRef]
39. Wei, B.; Xie, Y.; Wang, X.; Jiao, J.; He, S.; Bie, Q.; Jia, X.; Xue, X.; Duan, H. Land cover mapping based on time-series MODIS-NDVI using a dynamic time warping approach: A case study of the agricultural pastoral ecotone of northern China. *Land Degrad. Dev.* **2020**, *31*, 1050–1068. [CrossRef]
40. Maus, V.; Câmara, G.; Cartaxo, R.; Sanchez, A.; Ramos, F.; de Queiroz, G. A time-weighted dynamic time warping method for land-use and land-cover mapping. *IEEE J. Sel. Top. Appl. Earth Obs. Remote Sens.* **2016**, *9*, 3729–3739. [CrossRef]
41. Shen, R.; Dong, J.; Yuan, W.; Han, W.; Ye, T.; Zhao, W. A 30 m resolution distribution map of maize for China based on Landsat and Sentinel images. *J. Remote Sens.* **2022**, *2022*, 9846712. [CrossRef]
42. Belgiu, M.; Csillik, O. Sentinel-2 cropland mapping using pixel-based and object-based time-weighted dynamic time warping analysis. *Remote Sens. Environ.* **2018**, *204*, 509–523. [CrossRef]
43. Peng, Q.; Shen, R.; Li, X.; Ye, T.; Dong, J.; Fu, Y.; Yuan, W. A twenty-year dataset of high-resolution maize distribution in China. *Sci. Data* **2023**, *10*, 658. [CrossRef] [PubMed]
44. Shen, R.; Pan, B.; Peng, Q.; Dong, J.; Chen, X.; Zhang, X.; Ye, T.; Huang, J.; Yuan, W. High-resolution distribution maps of single-season rice in China from 2017 to 2022. *Earth Syst. Sci. Data* **2023**, *15*, 3203–3222. [CrossRef]
45. Dong, J.; Fu, Y.; Wang, J.; Tian, H.; Fu, S.; Niu, Z.; Han, W.; Zheng, Y.; Huang, J.; Yuan, W. Early-season mapping of winter wheat in China based on Landsat and Sentinel images. *Earth Syst. Sci. Data* **2020**, *12*, 3081–3095. [CrossRef]
46. Suwanlee, S.R.; Keawsomsee, S.; Izquierdo-Verdiguier, E.; Moreno-Martínez, Á.; Ninsawat, S.; Somard, J. Synergistic use of multi-sensor satellite data for mapping crop types and land cover dynamics from 2021 to 2023 in Northeast Thailand. *Int. J. Appl. Earth Obs. Geoinf.* **2025**, *141*, 104673. [CrossRef]
47. Jensch, K.; Ghazaryan, G.; Ernst, S.; Hostert, P.; Nendel, C. Integrating Landsat, Sentinel-2 and Sentinel-1 time series for mapping intermediate crops. *Eur. J. Remote Sens.* **2025**, *58*, 2507738. [CrossRef]
48. Chaves, M.E.; Alves, M.D.C.; Sáfadi, T.; de Oliveira, M.S.; Picoli, M.C.; Simoes, R.E.; Mataveli, G.A. Time-weighted dynamic time warping analysis for mapping interannual cropping practice changes in large-scale agro-industrial farms in Brazilian Cerrado. *Sci. Remote Sens.* **2021**, *3*, 100021. [CrossRef]

49. Wei, P.; Ye, H.; Qiao, S.; Liu, R.; Nie, C.; Zhang, B.; Song, L.; Huang, S. Early crop mapping based on Sentinel-2 time-series data and the Random Forest algorithm. *Remote Sens.* **2023**, *15*, 3212. [CrossRef]
50. Xu, J.; Zhu, Y.; Zhong, R.; Lin, Z.; Xu, J.; Jiang, H.; Huang, J.; Li, H.; Lin, T. DeepCropMapping: A multi-temporaldeep learning approach with improved spatial generalizability for dynamic corn and soybean mapping. *Remote Sens. Environ.* **2020**, *247*, 111946. [CrossRef]
51. Hu, J.; Zhang, B.; Peng, D.; Huang, J.; Zhang, W.; Zhao, B.; Li, Y.; Cheng, E.; Lou, Z.; Liu, S.; et al. Mapping 10-m harvested area in the major winter wheat-producing regions of China from 2018 to 2022. *Sci. Data* **2024**, *11*, 1038. [CrossRef]
52. Li, M.; Feng, X.; Belgiu, M. Mapping tobacco planting areas in smallholder farmlands using Phenological-Spatial-Temporal LSTM from time-series Sentinel-1 SAR images. *Int. J. Appl. Earth Obs. Geoinf.* **2024**, *129*, 103826. [CrossRef]
53. Almalki, R.; Khaki, M.; Saco, P.; Rodriguez, J. Monitoring and mapping vegetation cover changes in arid and semi-arid areas using remote sensing technology: A review. *Remote Sens.* **2022**, *14*, 5143. [CrossRef]
54. Li, X.; Qu, Y.; Geng, H.; Xin, Q.; Huang, J.; Peng, S.; Zhang, L. Mapping annual 10-m maize cropland changes in China during 2017–2021. *Sci. Data* **2023**, *10*, 765. [CrossRef]
55. Wang, Y.; Qi, Q.; Liu, Y. Unsupervised segmentation evaluation using area-weighted variance and jeffries-matusita distance for remote sensing images. *Remote Sens.* **2018**, *10*, 1193. [CrossRef]
56. Li, H.; Qi, A.; Chen, H.; Chen, S.; Zhao, D. Hsiao framework in feature selection for hyperspectral remote sensing images based on jeffries-matusita distance. *IEEE Trans. Geosci. Remote Sens.* **2025**, *63*, 3527138. [CrossRef]
57. Zhu, A.; Zhao, F.; Pan, H.; Liu, J. Mapping rice paddy distribution using remote sensing by coupling deep learning with phenological characteristics. *Remote Sens.* **2021**, *13*, 1360. [CrossRef]
58. Lou, Z.; Peng, D.; Shi, Z.; Wang, H.; Zhang, Y.; Yan, X.; Chen, Z.; Ye, S.; Yu, L.; Hu, J.; et al. A 30 m resolution annual cropland extent dataset of Africa in recent decades of the 21st century. *Earth Syst. Sci. Data Discuss.* **2025**, *2025*, 1–29. [CrossRef]
59. Chai, D.; Newsam, S.; Huang, J. Aerial image semantic segmentation using DCNN predicted distance maps. *ISPRS J. Photogramm. Remote Sens.* **2020**, *161*, 309–322. [CrossRef]
60. Tu, Y.; Wu, S.; Chen, B.; Weng, Q.; Bai, Y.; Yang, J.; Yu, L.; Xu, B. A 30 m annual cropland dataset of China from 1986 to 2021. *Earth Syst. Sci. Data* **2024**, *16*, 2297–2316. [CrossRef]

remote sensing

MDPI

Article

Research on Rice Field Identification Methods in Mountainous Regions

Yuyao Wang [1], Jiehai Cheng [1,2,*], Zhanliang Yuan [1] and Wenqian Zang [3]

1 School of Surveying and Land Information Engineering, Henan Polytechnic University, Jiaozuo 454000, China; 212204010001@home.hpu.edu.cn (Y.W.); yuan6400@hpu.edu.cn (Z.Y.)
2 Henan Province Spatial Big Data Acquisition Equipment Development and Application Engineering Technology Research Center, Jiaozuo 454000, China
3 Aerospace Information Research Institute, Chinese Academy of Sciences, Beijing 100094, China; zangwq@aircas.ac.cn
* Correspondence: chengjiehai@hpu.edu.cn

Highlights

What are the main findings?

- A customized GCN was constructed using rice plots as graph nodes and integrating multidimensional rice features, achieving 98.3% accuracy for rice identification under complex mountainous terrain and improving computational efficiency.

What is the implication of the main finding?

- The integration of multidimensional features provided an ecologically consistent representation of rice, significantly enhancing classification accuracy under complex terrain.
- The method showed strong transferability and scalability, supporting large-area rice mapping and monitoring across diverse regions and cloudy conditions.

Abstract

Rice is one of the most important staple crops in China, and the rapid and accurate extraction of rice planting areas plays a crucial role in the agricultural management and food security assessment. However, the existing rice field identification methods faced the significant challenges in mountainous regions due to the severe cloud contamination, insufficient utilization of multi-dimensional features, and limited classification accuracy. This study presented a novel rice field identification method based on the Graph Convolutional Networks (GCN) that effectively integrated multi-source remote sensing data tailored for the complex mountainous terrain. A coarse-to-fine cloud removal strategy was developed by fusing the synthetic aperture radar (SAR) imagery with temporally adjacent optical remote sensing imagery, achieving high cloud removal accuracy, thereby providing reliable and clear optical data for the subsequent rice mapping. A comprehensive multi-feature library comprising spectral, texture, polarization, and terrain attributes was constructed and optimized via a stepwise selection process. Furthermore, the 19 key features were established to enhance the classification performance. The proposed method achieved an overall accuracy of 98.3% for the rice field identification in Huoshan County of the Dabie Mountains, and a 96.8% consistency compared to statistical yearbook data. The ablation experiments demonstrated that incorporating terrain features substantially improved the rice field identification accuracy under the complex topographic conditions. The comparative evaluations against support vector machine (SVM), random forest (RF), and U-Net models confirmed the superiority of the proposed method in terms of accuracy, local performance, terrain adaptability, training sample requirement, and computational cost, and demonstrated its effectiveness and applicability for the high-precision rice field distribution mapping in mountainous environments.

Keywords: rice field identification; optical remote sensing; SAR imagery; graph convolutional network; mountainous terrain

1. Introduction

Rice, as one of the world's three major staple crops, provides the essential food security for the vast majority of the global population. The accurate estimation of the rice yield is therefore critically important [1]. The rice planting area serves as a crucial indicator for estimating the rice yield and has attracted considerable attention from agricultural authorities. To achieve rapid and accurate extraction of rice planting areas, the precise rice field identification is essential [2,3]. Due to the geographical variability of rice cultivation regions, some regions have the characteristic of the mountainous and hilly terrain [4]. However, the current studies primarily focus on flat and open plain regions. Few studies have focused on identifying the rice fields under the mountainous and hilly terrainrice field identification. If the rice field distribution under the above conditions can be accurately identified and integrated with data from plain regions, it would be possible to generate a comprehensive nationwide rice field distribution map. Gaining a complete and accurate understanding of rice cultivation patterns plays a vital role in advancing agricultural modernization.

Due to the low elevation and relatively uniform terrain of plain regions, previous studies have generally achieved favorable results under such conditions. However, the rice field identification in mountainous regions, which have rugged topography and unique climatic conditions, often requires consideration of additional influencing factors.mountainous region. In recent years, some researchers have begun to explore methods for the rice field identification in mountainous regions [5–7]. Due to the higher elevation and frequent cloud covermountainous region, the remote sensing imagery for the rice field identification is often affected by the cloud contamination. Quan et al. [8] employed a combination of traditional unsupervised classification and visual interpretation to extract the rice planting areas in the Dabie Mountain region of Anqing City, Anhui Province, for the years 2016 and 2017. Although this study achieved the large-scale extraction of the rice planting areas, the use of the low-resolution Landsat imagery and the lack of consideration for cloud and fog effects on remote sensing imageries made it difficult to clearly identify rice fields under mountainous conditions. Sang et al. [9] successfully produced a high-resolution rice field distribution map by performing the cloud removal on optical remote sensing imageriesimageries using the Google Earth Engine (GEE) platform. Although the use of high-resolution cloud-free imageries improved the rice field classification accuracy, the method overly relied on the simplistic features. As a result, it results in the crops with similar spectral characteristics to be easily misclassified as rice. Therefore, mountainous regionit is essential to ensure that optical remote sensing imageries are clear and free of clouds to avoid omission caused by cloud and fog occlusion. Moreover, relying solely on the spectral features of remote sensing imageries without integrating additional features often fails to achieve satisfactory results.

Due to the highly undulating terrain in mountainous regions, the rice field distribution tends to be scattered and fragmented. Treating rice fields as independent individual units may result in misclassification or omission of small-sized rice fields [10]. For example, Liu et al. [11] extracted rice planting areas in certain mountainous regions of Guiyang City by combining the high-resolution optical imagery with time-series SAR imageries.

Chen et al. [12] used the classification features such as vegetation indices, red-edge features, and texture characteristics, and employed the random forest method to achieve the rice planting area extraction in the mountainous regions of southwest China. Zhang et al. [13] used multi-scale features based on multi-source remote sensing data using a convolutional neural network (CNN) to map the rice field distribution in the hilly regions of southern China. Although these studies achieved better identification results by integrating multi-source features, they overlooked the fragmented nature of the crop distribution in mountainous regions. Most research treated rice fields as independent individual units, failed to adequately consider the impact of spatial relationships between unitsunit on rice field identification in mountainous terrains, and neglected the spatial connections between rice fields and other crops/land use units. You et al. [14] incorporated spatial correlation features between units through spatial autocorrelation analysis, and the research demonstrated that this approach can effectively improve the rice field identification accuracy. Additionally, Zhang et al. [13] investigated the influence of different features on the rice field identification in mountainous regions by using a feature selection method, and the research revealed that terrain features had a significant impact on the rice field identification accuracy. Ma et al. [15] improved the rice field identification accuracy by combining the remote sensing imagery with geographic information data to enhance the weight of terrain features in the classification algorithm. These studies indicate that both the spatial correlation features between units and terrain characteristics play significant roles in rice extraction within mountainous regions. Therefore, it is essential to fully consider multi-dimensional features in the rice field identification tasks under the complex mountainous terrains. Building upon the strengths and addressing the limitations of previous research, this study proposes a method for the rice field identification in mountainous regions. The main contributions of this study are as follows:

(1) It demonstrates that integrating terrain features with spectral, texture, and polarization characteristics can effectively enhance rice identification accuracy in mountainous areas.
(2) It introduces a graph convolutional neural network that accounts for spatial correlation features between plots, and validates its effectiveness and applicability for rice identification tasks in mountainous regions.

2. Materials and Methods

The proposed framework, as illustrated in Figure 1, was composed of four stages: (1) To address the severe cloud and fog coverage in mountainous remote sensing imageries, a combined "coarse cloud removal + fine cloud removal" method for optical remote sensing imagery was designed. In the coarse removal stage, local and global features from SAR imageries and edge information from optical imageries were utilized to preliminarily remove cloud-covered areas. Based on this, the fine removal stage employed spectral information from neighboring temporal optical imageries to achieve more precise reconstruction of cloud regions, thereby producing high-quality cloud-free optical imageries. (2) Construction and optimization of the feature library. Multi-source remote sensing data, including multispectral imageries, SAR imageries, and DEM data, were comprehensively used to perform object-based multi-scale segmentation, by which the entire study area was divided into multiple unit objects. Spectral and texture features were extracted from multispectral imageries, polarization features from SAR imageries, and terrain features from DEM data to build an initial unit-level feature set. This feature set was subsequently filtered and optimized using a feature selection model, resulting in a unit-level feature library containing the optimal feature subset. (3) Model construction for the rice field identification in mountainous regions. A graph convolutional network (GCN) model was constructed as the rice field identification model. The combined features from the unitunit-level feature library were used as node attributes, and an adjacency matrix representing the relationships among nodes was established. Node features and the adjacency matrix were utilized to generate graph data. The graph data were input into the GCN for model training, through which spatial correlation features among units were learned, and the trained model was employed to identify rice planting areas in mountainous regions. (4) Rice field distribution mapping results and comparative analysis. The rice field distribution mapping results in the mountainous regions were obtained by using the proposed method. The multiple evaluation strategies including overall accuracy assessment, local visual inspection, and comparison with statistical yearbook data were used to assess the results. Additionally, the comparative experiments with the traditional methods were conducted to analyze the advantages and applicability of the proposed method under complex mountainous terrain conditions.

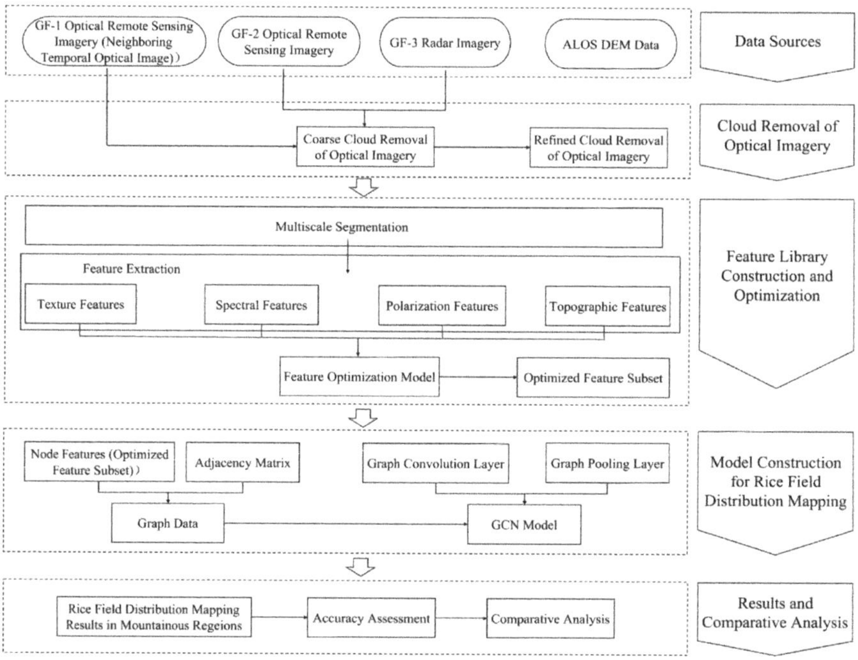

Figure 1. The Flowchart of the Proposed Framework.

2.1. Optical Remote Sensing Imagery Cloud Removal

Optical remote sensing imageries were easily affected by cloud coverage, especially under mountainous conditions where the weather changed frequently [16]. Achieving high-quality cloud removal of the imagery was a prerequisite for accurately identifying the rice fields in mountainous regions. To address this issue, a cloud removal method that integrated features from SAR imageries and neighboring temporal optical remote sensing imageries was proposed in this study. For brevity, the neighboring temporal optical imageries were hereafter referred to as "neighboring optical remote sensing imageries." The overall workflow of the optical remote sensing imagery cloud removal method was illustrated in Figure 2 and mainly consisted of the following two stages: (1) Coarse cloud removal of optical remote sensing imageries. A deep learning model was employed to automatically detect the cloud-covered areas in the cloud-covered optical remote sensing imageries. The SAR imageries and the cloud-covered optical remote sensing imageries were registered.A series of $n \times n$ imagery pairs (Imagery Patch Pair 1) were generated based on cloudy and cloud-free areas. Meanwhile, the SAR imageries were input into a Vision Transformer (ViT) model to obtain global feature maps. Both the SAR imageries and the cloud-covered optical remote sensing imageries were input into an edge generation model to produce the edge maps. Imagery Patch Pair 1, the global feature maps, and the edge maps were then fed into a conditional generative adversarial network (CGAN) to produce the coarse cloud-free imagery. (2) Refined cloud removal of optical remote sensing imageries. Neighboring optical remote sensing imageries corresponding to the cloud-covered imageries were collected and registered with the coarse cloud-free imageries. Based on the reconstructed and non-reconstructed areas, Imagery Patch Pair 2 was generated and was input into the CGAN to produce the refined cloud-free imagery.

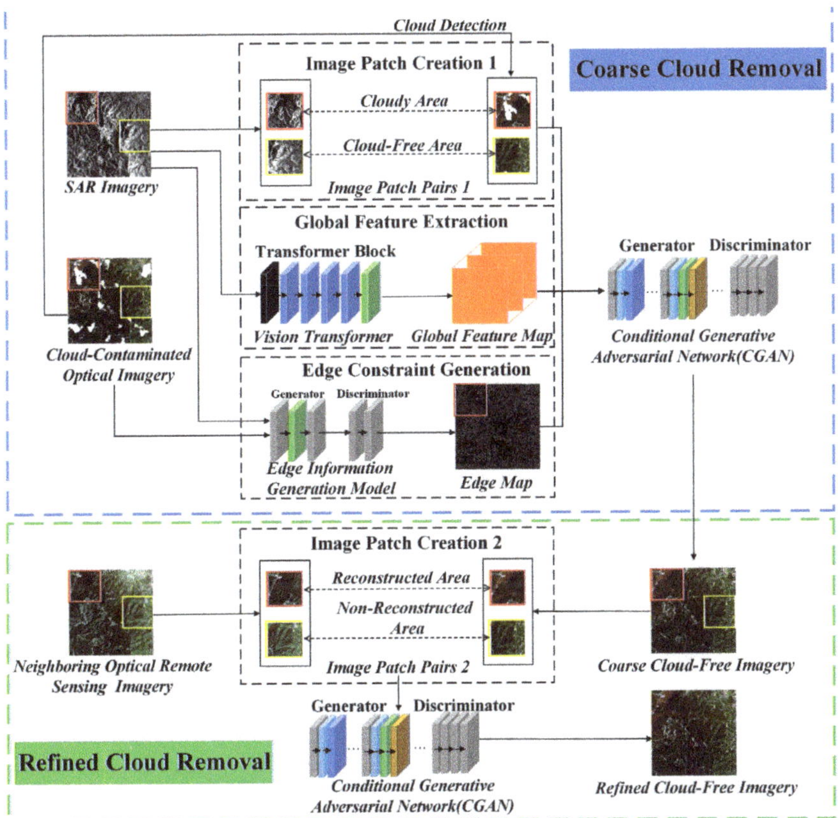

Figure 2. The Overall Workflow of the Optical Remote Sensing Imagery Cloud Removal Method.

2.2. Construction and Optimization of Feature Library

The construction and optimization of the feature library were conducted in three stages: (1) Imagery segmentation. Object-based imagery segmentation was performed using a multi-scale segmentation algorithm to divide the imagery into multiple unit objects. (2) Feature extraction. Various features were extracted based on the unit objects, primarily including spectral features, texture features, polarization features, and terrain features. (3) Feature selection. A feature selection model was applied to filter feature subsets that contributed significantly to the rice field identification task, and the optimal combination of features was ultimately retained.

2.2.1. Imagery Segmentation

To accommodate the subsequent graph convolutional network and reduce data volume, object-based multi-scale imagery segmentation was performed on the remote sensing imageries prior to feature extraction. The multi-scale segmentation algorithm had become a mainstream method due to its flexible adaptation to various land cover features in the high-resolution remote sensing imagery processing [17]. Its key parameters included scale, shape, and compactness, among which the scale parameter played a central role in determining the size and number of segmented objects [18]. During the imagery segmentation process, setting the scale parameter too small led to over-segmentation, producing a large number of redundant small regions and causing a single land cover type to be divided into multiple objects, thus compromising its integrity. Conversely, setting the parameter too large resulted in under-segmentation, where multiple land cover types were incorrectly merged into one object, affecting the accuracy of segmentation results and remote sensing interpretation. Therefore, a Segmentation Quality Index (SQI) was constructed in this study by weighting the spectral homogeneity within segmented objects and the spectral heterogeneity between objects to evaluate segmentation quality. Experimental results showed that the SQI reached its maximum value when the scale parameter was set to 170. The segmentation results generated with this parameter were considered optimal.

2.2.2. Feature Extraction

A wide variety of remote sensing imagery features were available for the rice field identification research, with existing studies typically involving spectral features, texture features, polarization features, and others. The selection and combination of features directly affected the performance of extraction models. Considering the rice field identification task under mountainous terrain conditions, terrain features were incorporated into the conventional feature set in this study to improve identification accuracy. Therefore, a feature set comprising spectral, texture, polarization, and terrain features was constructed as the preliminary feature set for subsequent experiments. The initial feature set included 55 features, as detailed in Table 1.

Table 1. The List of the Initial Feature Set.

Feature Type	Feature Description
Spectral Features	B1, B2, B3, B4, NDVI, CARI, RVI, DVI, EVI, SAVI, DNWI, VARI, PCA1, PCA2, HSV spectral-spatial features (12 types)
Texture Features	Entropy, Energy, Autocorrelation, Contrast, Dissimilarity, Variance, Small Gradient Dominance, Large Gradient Dominance, Uneven Distribution of Gray Levels, Uneven Distribution of Gradients, Gradient Energy, Gradient Variance, Correlation, Gray-level Entropy, Gradient Entropy, Mixed Entropy
Polarization Features	HH, VV, HV, VH
Terrain Features	DEM, Aspect, Slope, Curvature

2.2.3. Feature Selection

During the feature extraction process, coupling existed among spectral, texture, polarization, and terrain features, resulting in a high degree of redundant features that hindered subsequent classification models from effectively capturing key attributes. However, traditional feature selection methods often focused on a single type of feature and failed to consider the complementarity among multiple feature types. This limitation was particularly evident in complex mountainous environments, where relying on a single feature selection strategy could overlook critical features or introduce irrelevant ones, thereby negatively affecting rice field identification accuracy [19].

To address the above issues, a stepwise selection strategy was employed by combining the mutual information (MI) algorithm [20], Relief algorithm [21], gray wolf optimization (GWO) algorithm [22] and random forest (RF) algorithm to form the MI-Relief-GWO_RF model, referred to as the MRG_RF model. This model was used to obtain the optimal feature subset for the rice field identification task under mountainous conditions.

The MRG_RF model first applied the mutual information (MI) algorithm to obtain feature subsets with moderate to weak relevance to the target task. Then, the Relief algorithm ranked the initially preselected features based on their importance weights and filtered out features exceeding a predefined threshold. The retained features were passed on to the next selection stage. Subsequently, the gray wolf optimization (GWO) algorithm was combined with the random forest (RF) algorithm to form a wrapper-based feature selection method (GWO_RF). In this method, the GWO algorithm searched for the optimal subset within the feature subset space, while the RF algorithm served as the fitness function to evaluate the performance of feature subsets. The GWO_RF feature selection method was applied to the preselected feature set, ultimately obtaining the optimal feature subset.

(A) Mutual Information Algorithm

The mutual information (MI) algorithm is an information theory–based feature selection method that measures the correlation between two random variables by quantifying the divergence between their joint distribution and marginal distributions. For variables X and Y, the mutual information value can be calculated using Equation (1):

$$MI(X;Y) = \sum_{x\in X}\sum_{y\in Y} p(x,y)\log\frac{p(x,y)}{p(x)p(y)} \tag{1}$$

In the equation, $p(x,y)$ represents the joint probability distribution of variables X and Y; $p(x)$ and $p(y)$ denote the marginal probability distributions of X and Y, respectively; and $MI(X;Y)$ represents the mutual information between X and Y. A higher mutual information value indicates a stronger correlation between the two variables; when $MI(X;Y) = 0$, it signifies that X and Y are completely independent.

The initial feature set in this study included 55 features, which were subjected to preliminary screening using the mutual information algorithm. Mutual information values were divided into three correlation intervals: a value greater than 0.5 indicated a strong correlation (S) between the feature and the target variable; a value between 0.2 and 0.5 indicated a moderate correlation (W); and a value less than 0.2 indicated low or no correlation (N). Features were preliminarily filtered based on the strength of their correlation with the target variable, resulting in a preselected feature subset. Multiple experiments verified that the preliminary screening effectively removed features with weak correlations.

(B) Relief Algorithm

Although the preliminary screening reduced the feature dimensionality, the feature set still remained high-dimensional and computationally complex. Therefore, a secondary selection based on feature weights in the task was necessary. The Relief algorithm was applied for the secondary screening of the feature set for rice field identification in mountainous regions. Using the preliminarily screened feature set as input, feature weights were calculated, and features were further selected based on a weight threshold. The specific procedure was as follows: a sample S was randomly selected; *k* nearest neighbors NH were drawn from the same class as S, and kkk nearest neighbors NM were selected from other classes. Then, feature weights $\omega(X)$ were iteratively updated according to the formulas shown in Equations (2) and (3):

$$\omega(X) = \omega(X) - \sum_{j=1}^{k} \frac{diff(X,S,NH_j)}{m \times k} + \sum_{C \neq Class(S)} \frac{\frac{P(C)}{1-P(Class(S))} \times \sum_{j=1}^{k} diff(X,S,NM(C)_j)}{m \times k} \tag{2}$$

$$diff(X,S,S') = \begin{cases} \frac{|S[X]-S'[X]|}{\max(X)-\min(X)} & \text{if X is continuous} \\ 0 & \text{if X is discrete and } S[X] = S'[X] \\ 1 & \text{if X is discrete and } S[X] \neq S'[X] \end{cases} \tag{3}$$

In the equations, X represents a feature within the feature set; S and S' denote two randomly selected samples; k is the number of nearest neighbors; m is the number of iterations; C represents a class label; $P(C)$ is the probability that a feature belongs to class C; $S[X]$ denotes the feature value of sample S for feature X; NH_j is the j-th nearest neighbor sample within class C; $NM(C)_j$ is the j-th nearest neighbor sample outside class C; $\max(X)$ and $\min(X)$ denote the maximum and minimum values of feature X, respectively; $Class(S)$ is the class label of sample S; and $diff(X,S,S')$ represents the difference in feature X between samples S and S'.

Feature weights were divided into five ranges: 0.005, 0.010, 0.015, and 0.020. Using a random forest classifier, the identification accuracy and the number of features corresponding to each weight range were calculated to determine the optimal feature weight threshold. Experimental results showed that when the weight threshold was set to 0.010, the identification accuracy reached its highest value while maintaining relatively high computational efficiency. Therefore, balancing both identification accuracy and computational efficiency, the feature set with a weight threshold of 0.010 was ultimately selected as the secondary screened feature subset for subsequent analysis.

(C) GWO_RF Algorithm

After the Relief algorithm screening, some features still exhibited high redundancy and correlation. Therefore, the GWO_RF algorithm was employed for further selection and optimization of the feature subset.

The flowchart of the GWO_RF algorithm was shown in Figure 3, and the specific steps were as follows:

① Initialization of the population: A certain number of gray wolves were randomly generated, where each wolf's position represented a feature subset. Each element in the position vector corresponded to a feature and took a value of 0 or 1, indicating whether the feature was selected. The population size, maximum number of iterations, and related parameters (e.g., a, A, C) were also set.

② Fitness evaluation: The classification accuracy of each feature subset corresponding to a gray wolf was assessed using the random forest algorithm. The classification accuracy served as the fitness value of the gray wolf, with higher fitness indicating stronger classification capability of the feature subset.

③ Population ranking: The population was ranked based on fitness values, and the three wolves with the highest fitness were designated as the alpha (α) wolf (best solution), beta (β) wolf (second best), and delta (δ) wolf (third best), respectively.

④ Position update: The positions of ordinary gray wolves were updated based on the positions of the α, β, and δ wolves, as well as dynamically adjusted coefficients a and A, according to the following formulas:

$$D_\alpha = \left| C_1 \cdot \vec{X}_\alpha - \vec{X} \right| \tag{4}$$

$$D_\beta = \left| C_2 \cdot \vec{X}_\beta - \vec{X} \right| \quad (5)$$

$$D_\delta = \left| C_3 \cdot \vec{X}_\delta - \vec{X} \right| \quad (6)$$

$$\vec{X}(t+1) = \frac{\vec{X}_\alpha + \vec{X}_\beta + \vec{X}_\delta}{3} \quad (7)$$

$\vec{X}_\alpha$, $\vec{X}_\beta$, and $\vec{X}_\delta$ represent the positions of the α, β, and δ wolves, respectively; $\vec{X}$ denotes the current position of a gray wolf; C_1, C_2, and C_3 are random weighting coefficients; and D_α, D_β, and D_δ represent the distances between the gray wolf and the α, β, and δ wolves, respectively.

⑤ Iterative update: Steps ② through ④ were repeated. In each iteration, the positions and fitness values of the α, β, and δ wolves were updated, and the gray wolf population's positions were continuously optimized to progressively search for the global optimal feature subset.

⑥ Termination criteria: The algorithm terminated when the maximum number of iterations was reached or when there was no significant improvement in the wolves' fitness values.

⑦ Output: The optimal feature subset corresponding to the α wolf was ultimately obtained.

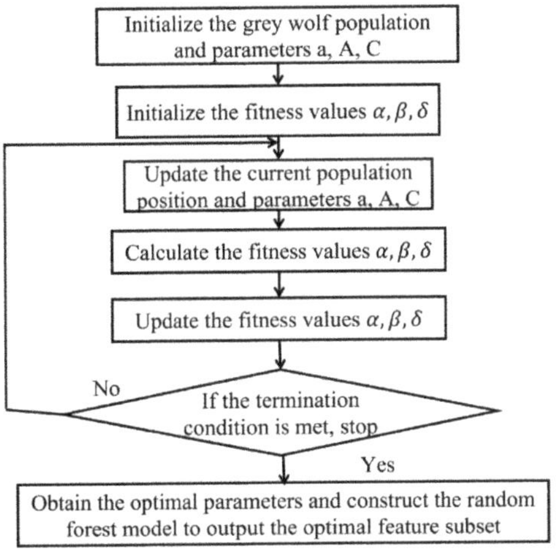

Figure 3. The Flowchart of the GWO_RF Algorithm.

2.3. Model Construction for Rice Field Distribution Mapping

Crops in mountainous regions typically exhibited scattered distribution and small unit sizes. Treating rice crops as independent individual units for identification easily led to misclassification and omission. To address this issue, a mountainous rice field identification model was constructed based on a graph convolutional network (GCN). This approach fully leveraged the powerful spatial correlation capabilities of the GCN model to extract spatial relational features among units from the overall spatial structure, thereby achieving accurate identification of rice crops [23]. The construction of the mountainous rice field identification model consisted of two parts: (1) graph data construction, and (2) model design.

2.3.1. Graph Data Construction

To utilize the graph convolutional network (GCN) for the identification task, the original data needed to be transformed into graph data format to meet the GCN's requirement for graph-structured input. The example of constructing spatially correlated graph data based on GCN was illustrated in Figure 4. Each segmented object (1, 2, 3, 4, 5, 6, 7, 8 shown in Figure 4) was represented as a node in the graph, and the adjacency relationships between segmented objects were used to define edges in the graph. This approach effectively embedded the spatial relationships among segmented objects into the graph data, providing fundamental support for subsequent identification tasks [24].

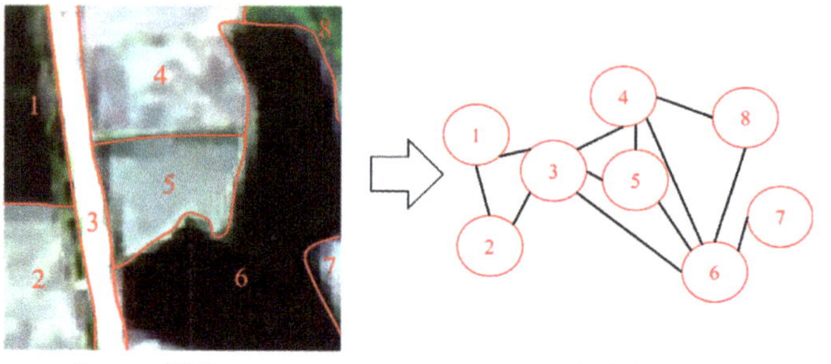

Figure 4. The Example of Constructing Spatially Correlated Graph Data.

In this study, the feature matrix was constructed using the optimal feature subset of the rice objects. As shown in Table 2, for a graph structure containing N nodes, the input feature matrix could be represented as $f \in R^{N \times M}$, $M = M_S + M_T + M_P + M_t$. Here, M_s denoted the dimensionality of spectral features for the segmented objects, M_T represented the dimensionality of texture features, M_P corresponded to the dimensionality of polarization features, and M_t indicated the dimensionality of terrain features.

Table 2. The Example of Feature Matrix X.

Features / Object ID	Feature 1, Feature 2, . . . , Feature M			
	Spectral Features	Texture Features	Polarization Feature	Topographic Features
1				
2				
N				

The adjacency matrix was constructed as follows: first, a zero matrix of size N×N was initialized, where N represents the total number of segmented objects. Then, neighborhood analysis was performed on the segmented objects; if two objects were adjacent, the corresponding entry in the zero matrix was set to 1, indicating the presence of an edge connecting them. Table 3 showed the adjacency matrix created based on the graph structure example in Figure 4.

Table 3. Example of Adjacency Matrix A.

Object \ Object	①	②	③	④	⑤	⑥	⑦	⑧
①	0	1	1	0	0	0	0	0
②	1	0	1	0	0	0	0	0
③	1	1	0	1	1	1	0	0
④	0	0	1	0	1	1	0	1
⑤	0	0	1	1	0	1	0	0
⑥	0	0	1	1	1	0	1	1
⑦	0	0	0	0	0	1	0	0
⑧	0	0	0	1	0	1	0	0

2.3.2. GCN Model Design

Section 2.3.1 briefly described the method for constructing graph convolutional network (GCN) data. This section focused on the design of the GCN architecture. The model primarily consisted of multiple graph convolutional layers, a fully connected layer, and a softmax classification layer. Specifically, each graph convolutional layer was further divided into two parts: the graph convolution operation and the pooling operation. Apart from receiving adjacency matrix inputs, this structure closely resembled the convolution and pooling layers found in conventional convolutional neural networks (CNNs). The overall architecture of the GCN is illustrated in Figure 5.

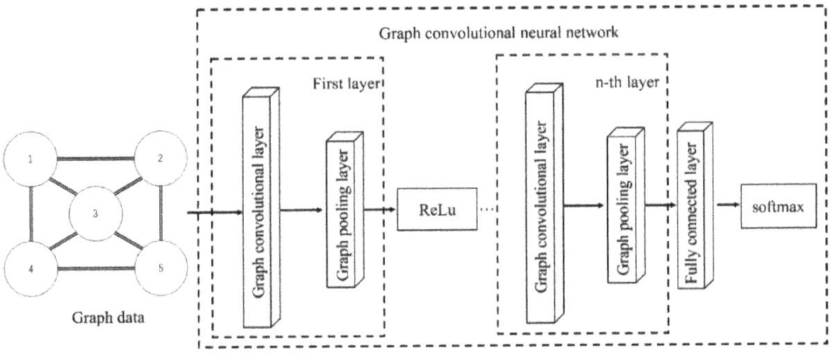

Figure 5. The Graph Convolutional Network Structure.

To fully leverage the advantages of the GCN model, this study explored and optimized the GCN structure from two aspects: first, the determination of the number of graph convolutional layers, and second, the selection of the neighborhood size k for nodes. The effects of these parameters on identification accuracy were analyzed separately. Experimental results demonstrated that the GCN model performed best with two graph convolutional layers, achieving the highest identification accuracy. Under this two-layer configuration, the accuracy peaked when the node neighborhood size k was set to 10. Therefore, this study adopted a GCN architecture with two graph convolutional layers and a neighborhood size k = 10, which was used consistently in subsequent experiments.

2.4. Accuracy Evaluation Methods

In remote sensing land cover classification experiments, accuracy assessment is a critical step for evaluating the reliability of the results. In this study, a combination of multiple evaluation methods was employed to comprehensively assess the classification accuracy. These methods included: generating a confusion matrix for accuracy evaluation [25], comparing statistical results of rice planting area, and conducting manual interpretation by comparing the classification results in local validation regions with high-resolution Google Earth imagery [26]. The evaluation focused on the following three aspects:

(1) Confusion Matrix-Based Accuracy Assessment

The confusion matrix is a core metric for evaluating classification accuracy in remote sensing land cover classification experiments and is widely recognized as an objective and reliable evaluation method [27]. By statistically comparing the one-to-one correspondence between object classes in the reference and predicted imageries, it assesses class consistency and enables the calculation of various evaluation metrics, including producer accuracy, user accuracy, overall accuracy, and F1 Score.

Producer Accuracy (PA) measures the proportion of correctly classified objects in a particular land cover class relative to the total number of reference objects in that class. It is calculated as:

$$PA = \left(\frac{t_i}{v_i}\right) \times 100\% \tag{8}$$

where t_i represents the number of correctly classified objects in class i, and v_i is the total number of reference objects in that class.

User Accuracy (UA) reflects the proportion of correctly classified objects within a class relative to the total number of objects that were classified into that class. It is given by:

$$UA = \left(\frac{t_i}{m_i}\right) \times 100\% \tag{9}$$

where m_i denotes the number of validation samples classified as class i.

F1 Score addresses the limitation of Overall Accuracy (OA), which only considers the diagonal elements of the confusion matrix. By integrating both precision and recall, the F1 Score offers a more comprehensive evaluation and serves as an important indicator of classification effectiveness. It is computed as:

$$F1 = \frac{2 \times PA \times UA}{PA + UA} \times 100\% \tag{10}$$

Overall Accuracy (OA) is defined as the ratio of correctly classified objects to the total number of validation objects. The total number is based on the reference land cover classification, which is determined from ground truth imagery or data from regions of interest. The diagonal elements of the confusion matrix represent correctly classified objects. The formula is:

$$OA = \left(\sum_{i=1}^{n} \frac{t_i}{V}\right) \times 100\% \tag{11}$$

where V is the total number of validation samples.

(2) Comparative Analysis with Statistical Yearbook Data

The total rice planting area in the study region was obtained from the 2022 edition of the Study Region Statistical Yearbook, specifically from the agriculture section. These official statistics were compared with the rice area derived from the classification results of this study. This comparison enabled an assessment of the experimental accuracy from a quantitative perspective.

(3) Visual Interpretation of Sample Regions

Manual visual interpretation was performed in selected regions of the classification results to assess the completeness of rice extraction and identify potential commission and omission errors. By subjectively analyzing the spatial consistency between the extracted rice distribution and high-resolution reference imagery, a comprehensive evaluation of classification accuracy and performance was conducted.

3. Experiments and Results

3.1. Study Area and Data

3.1.1. Study Area

The study area is located in Huoshan County, situated in the Dabie Mountains in the central-eastern region of China (as shown in Figure 6). The Dabie Mountains span the provinces of Hubei, Henan, and Anhui, serving as the watershed between the Yangtze and Huai Rivers. The terrain is dominated by low- to mid-altitude mountains and hilly landscapes, with significant variations in elevation. The region experiences abundant rainfall, hosts numerous rivers, and possesses ample water resources, all of which provide favorable conditions for agricultural production. As one of the major staple crops, rice is widely distributed across river valley plains, intermountain basins, and some gently sloping terraces.

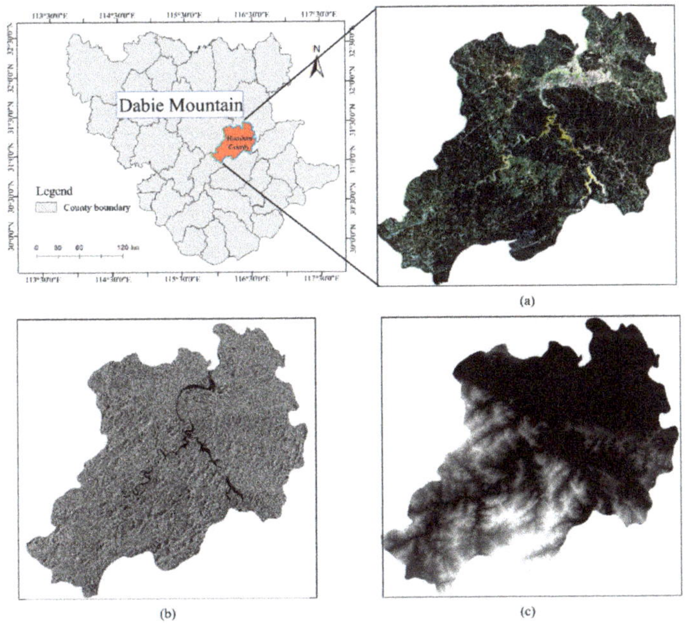

Figure 6. The Study Area and Data: (**a**) GF-2 Multispectral Imagery; (**b**) GF-3 Radar Imagery; (**c**) Global ALOS 12 m DEM Imagery.

Huoshan County is located in the eastern part of the Dabie Mountains and is characterized by terrain that generally slopes from west to east. Mountainous and hilly areas cover the majority of the county's land. Despite the significant topographic variation, rice cultivation in this region achieves high yields due to abundant water resources and favorable climatic conditions, making rice one of the most important staple crops in the area. Huoshan has also been designated as a key ecological agriculture demonstration zone in Anhui Province and has implemented a series of ecological agricultural policies to promote sustainable development and the green transformation of agricultural production [28]. Considering the widespread rice cultivation and the complex agricultural terrain in this region, Huoshan County was selected as the specific study area for this research.

3.1.2. Data

The data included the remote sensing data for the rice field identification and auxiliary data, as shown in Table 4.

Table 4. The Overview of the Data.

Data Type	Source	Acquisition Time	Spatial Resolution	Coverage	Remarks
Remote Sensing Data for Rice Identification	Gaofen-2 Multispectral Imagery	June 2022	1 m	Full Huoshan County	Some clouds; mountainous terrain, varying altitude
Auxiliary Data	ALOS Topographic Data (DEM)	July 2010	12.5 m	Full Huoshan County	Stable weather
	Gaofen-3 SAR Imagery	June 2022	5 m	Full Huoshan County	Microwave, weather-independent
	Gaofen-1 Multispectral Imagery	July 2022	2 m	Partial, low clouds	Clear-sky conditions

(1) Remote Sensing Data for the Rice field identification

Gaofen-2 multispectral imagery was selected as the remote sensing data for the rice field identification. With a spatial resolution of 1 m and a high revisit frequency, Gaofen-2 provides significant advantages for remote sensing monitoring in mountainous regions [29]. Due to Huoshan County's mountainous location, Gaofen-2 imagery covering the area exhibited varying degrees of cloud cover throughout the entire rice growing season. This study used imagery acquired in June 2022, during the vegetative growth stage of rice. At this stage, rice growth is stable and the plants are vigorously developing, making them more distinguishable and easier to accurately identify in remote sensing imagery [30].

(2) Auxiliary Data

① ALOS Topographic Data

ALOS topographic data were selected to provide terrain features for the rice field identification task. These data were derived from the PALSAR sensor onboard the ALOS satellite [31]. The ALOS data used in this study were acquired in July 2010, with the DEM data obtained under the Fine Beam Single (FBS) mode, achieving an actual spatial resolution of 12.5 m.

② Gaofen-3 SAR Imagery

Gaofen-3 SAR imagery was used for two primary purposes: (1) in the rice field identification task, it provided polarization features that helped distinguish rice from other land cover types; (2) in the cloud removal task, it supplied SAR features essential for the coarse cloud removal process. Gaofen-3 is China's first C-band, multi-polarization, high-resolution synthetic aperture radar (SAR) satellite, offering up to 12 imaging modes [32]. The Gaofen-3 SAR data used in this study were acquired in June 2022.

③ Gaofen-1 Multispectral Imagery

Gaofen-1 multispectral imagery was used as the neighboring optical remote sensing imagery. Although Gaofen-1 has a lower spatial resolution (2 m) compared to Gaofen-2, it provided cloud-free optical imagery from a nearby period during the rice growth season, making it suitable for the cloud removal task in this study. The Gaofen-1 multispectral imagery used was acquired in July 2022.

3.2. Experimental Setup

3.2.1. Sample Dataset

The rice field sample dataset used in this study consisted of two parts: a pre-trained dataset and a self-constructed dataset. Both datasets were derived from Gaofen-2 multispectral imagery and were visually interpreted with the assistance of high-resolution Google Earth imagery (0.3 m resolution). The pre-trained rice dataset contained 1581 imageries, each with a size of 256×256 pixels, primarily representing rice planting areas in plains. The self-constructed dataset comprised 866 imageries of the same size, selected from rice-growing regions within Huoshan County. Both datasets were split into training and testing sets in an 8:2 ratio. All samples were normalized to reduce the effects of illumination and other environmental factors. The rice field sample examples are shown in Figure 7.

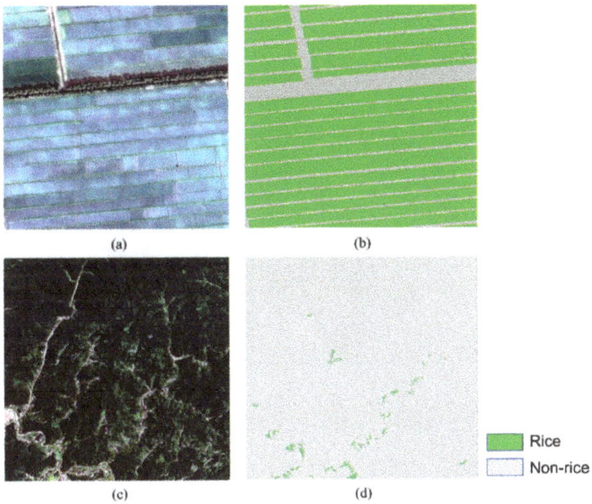

Figure 7. The Rice Field Sample Examples: (**a**) Pre-trained Imageries; (**b**) Pre-trained Labels; (**c**) Self-constructed Imageries; (**d**) Self-constructed Labels.

3.2.2. Experimental Environment and Parameter Settings

The experiments were conducted using the TensorFlow deep learning platform. The hardware configuration used in this study is listed in Table 5. The software environment included Microsoft Windows 10 as the operating system, Python 3.8 as the programming language, and TensorFlow_GPU 2.7.0 as the deep learning framework.

Table 5. The Hardware Platform.

Component	Specification	Frequency/Memory
CPU	i5-14600kf	3.5 GHz
Memory	32 G × 2	3600 MHz
GPU	NVIDIA GeForce GTX 4060ti	16 G

This study employed a GCN model for rice identification experiments, where the training parameter configuration played a decisive role in the final extraction accuracy. The loss function was set as cross-entropy loss, and the optimization algorithm used was the Adam optimizer. The number of epochs was set to 10,000 to ensure sufficient fitting of the network to the data. The learning rate was initialized at 0.001 with a decay schedule configured to gradually reduce the learning rate during training, thereby enhancing model optimization stability. To avoid memory overflow issues, the original dataset was divided into imagery blocks composed of several nodes for training. Among these, 75% were allocated as the training set and 25% as the validation set. From the training set, 6,000 samples were randomly selected for training, while 1000 samples were randomly selected from the validation set for validation.

3.3. Results

3.3.1. Cloud Removal Results of the Optical Remote Sensing Imagery

The cloud removal method proposed in this study was applied to the optical remote sensing imagery of Huoshan County, the study area. By effectively addressing the prevalent cloud cover issue in this region, the method successfully generated high-quality cloud-free imageries, thereby providing a reliable data foundation for subsequent rice field identification experiments. The cloud removal results were shown in Figure 8.

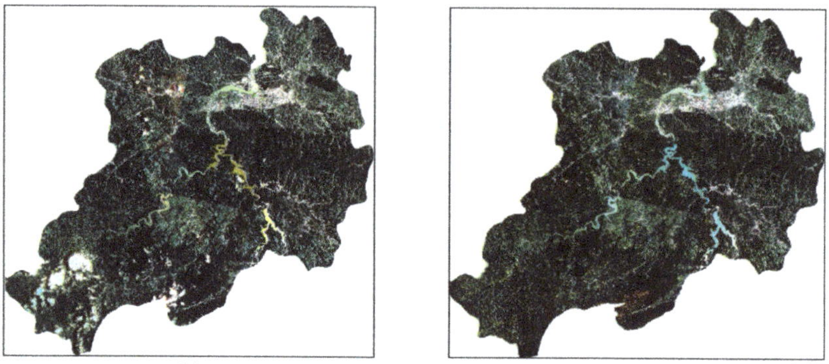

(a) (b)

Figure 8. The Cloud Removal Results of the Optical Remote Sensing Imagery: (**a**) Original Cloud-covered Imagery; (**b**) Cloud-free Imagery.

3.3.2. Feature Selection Results

Table 6 presented the results of the optimal feature subset. The spectral features dominated the subset, most of which were spectral indices calculated from near-infrared bands. This was because the rice crop exhibited the high reflectance and distinct differences from other land covers, which made the spectral features highly discriminative in identifying the rice field in mountainous regions. The texture features accounted for 26% of the total features. Due to the pronounced striping and regular arrangement characteristics of rice fields, these texture features demonstrated good capability in distinguishing rice fields from irregularly textured land covers, such as forests or shrubs.

Table 6. The Optimal Feature Subset.

Feature Type	Feature Content	Number of Features
Spectral Features	B3, B4, NDVI, EVI, SAVI, NDWI, VARI, PCA1, Maximum V, Standard Deviation of S	10
Texture Features	Entropy, Contrast, Correlation, Gray Level Variance, Gradient Entropy	5
Terrain Features	Slope, Aspect	2
Polarization Features	HV, HH	2
Total		19

3.3.3. Rice Field Identification Results

The graph-structured data of Huoshan County were input into the trained GCN model to identify the rice planting areas, and the results were visualized as shown in Figure 9. The identification accuracy is presented in Table 7. As shown in Table 7, the accuracy metrics of PA, UA, F1-Score, and OA were high.

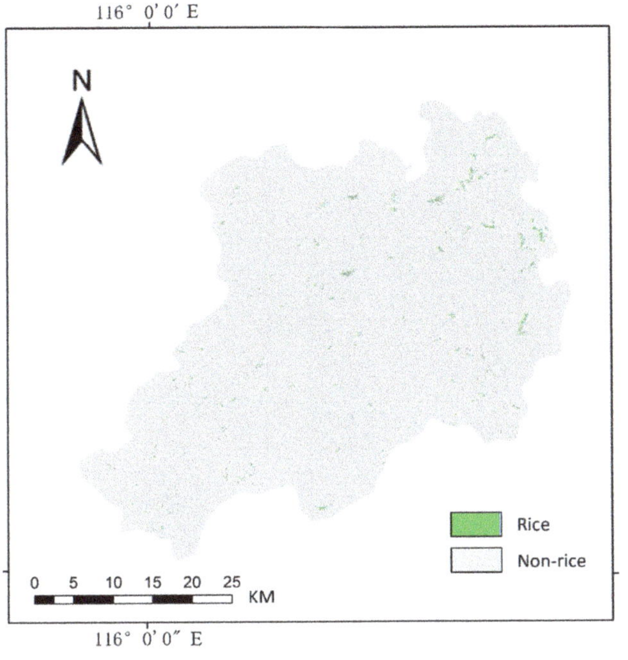

Figure 9. Rice field identification results.

Table 7. The Identification Accuracy of the Rice Fields.

Category		Truth (Objects) Rice	Truth (Objects) Non-Rice	PA	UA	F1-Score	OA
Classification (Objects)	Rice	5916	951	0.841	0.861	0.877	0.983
	No-Rice	1125	114,182	0.992	0.990	0.985	

To further verify the reliability of the identification results, the local region was visually interpreted by comparison with high-resolution Google Earth imagery, as shown in Figure 10. It can be observed that the rice fields can be correctly identified from easily confused land cover/use categories. This indicated that the proposed method had a low identification error, and the identified rice planting extent closely matched the actual distribution.

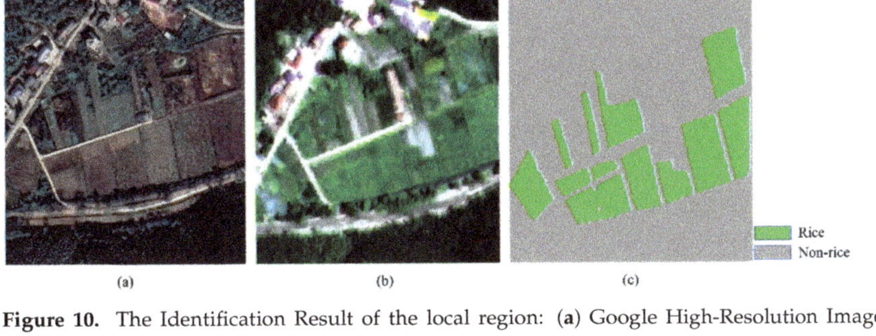

Figure 10. The Identification Result of the local region: (**a**) Google High-Resolution Imagery; (**b**) Original Imagery; (**c**) Identification Result.

The results of the rice field identification were compared and analyzed against the data from the statistical yearbook. Table 8 showed the rice planting area extracted from the identification results and the corresponding data from the 2022 Huoshan County Statistical Yearbook. According to the yearbook, the rice planting area in Huoshan County in 2022 was 179.1 km^2, while the proposed method estimated the rice planting area as 173.3 km^2. The data matching rate had reached 96.8%.

Table 8. The Data of the Identification and Statistical Yearbook.

Rice Recognition Area (km^2)	Huo Shan County Rice Area in 2022 (km^2)	Recognition Accuracy (%)
173.3	179.1	96.8

4. Discussion

4.1. Ablation Study on Terrain Features

To verify the impact of terrain features on the rice field identification under mountainous conditions, we designed a series of ablation experiments. We kept the experimental data and extraction models unchanged, and trained the rice field identification models for both plain(the example shown in Figure 11b) and mountainous(the example shown in Figure 11a) regions without including terrain features. Then, we incorporated terrain features into the models and compared the rice field identification accuracy across the two geographic settings.

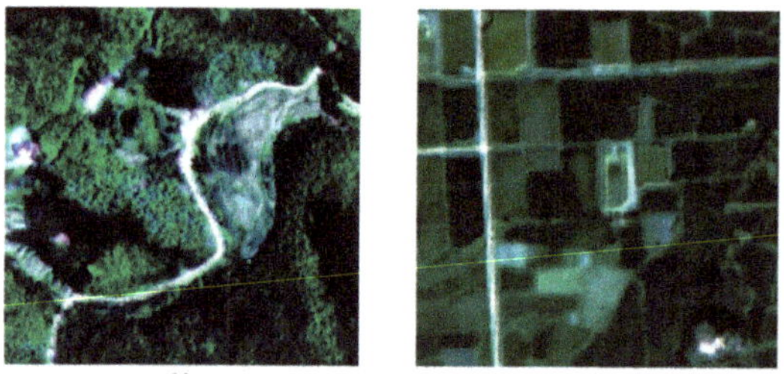

Figure 11. The Example of the Plain and Mountainous Region: (**a**) Mountainous region; (**b**) Plain Region.

Among the 19 optimized features, we removed the two terrain features—slope and aspect—and input the remaining features into the GCN model for training to obtain the identification results. Then, we reintroduced the terrain features and retrained the model to acquire updated the identification results. Table 9 compared the identification accuracies before and after incorporating the terrain features. The results showed that the identification accuracy improvement after incorporating the topographic features was 3.8% in the mountainous region, while in the plain region, the corresponding improvement was only 0.2%. This quantitatively demonstrated the importance of terrain features for the rice field identification in mountainous regions, which can effectively reduce the misclassification and omission errors.

Table 9. The Identification Accuracy Comparison before and after Incorporating Topographic Features.

Region	Overall Accuracy (OA)/%		Accuracy Improvement/%
	Before Incorporating Topographic Features	After Incorporating Topographic Features	
Mountainous Area	89.4	93.2	3.8
Plain region	94.6	94.8	0.2

Figure 12 showed the local identification results without and with the terrain features in the mountainous region.We observed that Object 1 and 2 were not identified in Figure 12b, but were successfully identified in Figure 12c. Object 1 and 2 was near the boundaries between the mountains and plains. This was partly due to the abundance of trees in the mountainous region, whose spectral characteristics closely resemble those of rice, making it difficult for other features to distinguish between these land cover types and resulting in the omissions of the rice fields along the edges. After incorporating terrain features such as slope and aspect, the model's identification performance under complex terrain conditions were improved significantly. These terrain features provided the crucial spatial context, and enabled the model to more accurately differentiate rice from other land covers.

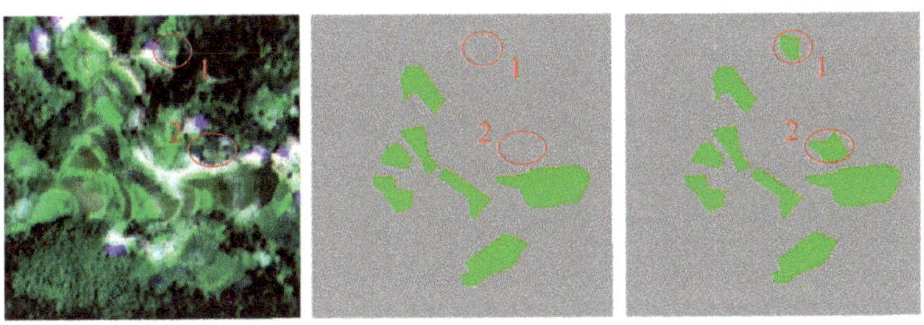

(a) (b) (c)

Figure 12. The Local Identification Results without and with the Terrain Features in the Mountainous Region: (**a**) Original Imagery; (**b**) The identification result Without the Terrain Features; (**c**) The identification result With the Terrain Features.

4.2. Comparison with Other Methods

Using the same sample dataset and feature set, we applied two machine learning methods—Support Vector Machine (SVM) [33] and Random Forest (RF) [34]—alongside the deep learning model U-Net [35] to identify the rice planting areas in Huoshan County. We then compared the results of these methods with our results.

4.2.1. Overall Accuracy Comparison

We compared the rice field identification accuracy of four methods across the entire study area, as summarized in Table 10. The results showed that the overall accuracy (OA) achieved by our proposed method significantly outperformed the other three classification algorithms. This demonstrated its superior performance for the rice field identification.

Table 10. The Overall Accuracy of different methods.

Model	Overall Accuracy (OA)/%
Proposed Method	98.3
SVM	92.1
RF	95.1
U-Net	96.7

Due to the limited number of rice field samples in mountainous regions, the experimental results may be affected by random data fluctuations. To verify whether the overall accuracy advantage of the proposed method was statistically significant rather than due to chance, we conducted the Wilcoxon Signed-Rank Test. This non-parametric test compared paired samples, where a p-value less than 0.05 indicated that the accuracy improvement of our method over comparative approaches was statistically significant, confirming its reliability. We selected 20 test regions and used the overall accuracy (OA) of each method within each region as paired samples. The p-values comparing our method with SVM, RF, and U-Net were calculated. Accuracy distributions for each method were visualized via box plots and interpreted alongside the significance test results to assess the statistical meaningfulness of the observed improvements.

The statistical significance of the differences between the proposed method and the comparison methods (SVM, RF, and U-Net) was assessed using the Wilcoxon signed-rank test at a significance level of 0.05. As summarized in Table 11, all comparisons yielded p-values below 0.05 ($p < 0.05$), which indicated that the improvements in overall accuracy (OA) achieved by the proposed method were statistically significant. Figure 13 presented the box plots of the OA distributions for the four methods across different test regions, where statistically significant differences were marked by asterisks (*). The proposed method consistently demonstrated higher accuracy and greater stability, which was reflected by its smaller interquartile range (IQR), indicating reduced variability in the results. These findings collectively confirmed the robustness and superiority of the proposed method for the rice field units identification in mountainous regions.

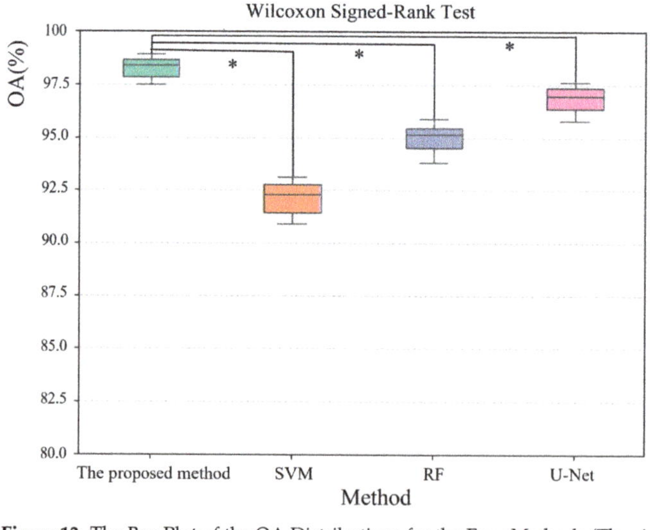

Figure 13. The Box Plot of the OA Distributions for the Four Methods (The significance level was set at $\alpha = 0.05$. An asterisk denotes a significant difference (* $p < 0.05$) between the proposed method and a comparator).

Table 11. The p-values of Comparing the Proposed Method with SVM, RF, and U-Net.

Method Comparison	Wilcoxon Signed-Rank Test (p-Value)
Proposed Method vs. SVM	0.015
Proposed Method vs. RF	0.027
Proposed Method vs. U-Net	0.041

4.2.2. Local Performance Comparison

The overall accuracy of the four methods across the entire study area had been presented above. To visually compare their classification performance in local regions, three representative regions were selected from the validation set. The rice field identification results for these regions, based on the four methods, were shown in Figure 14.

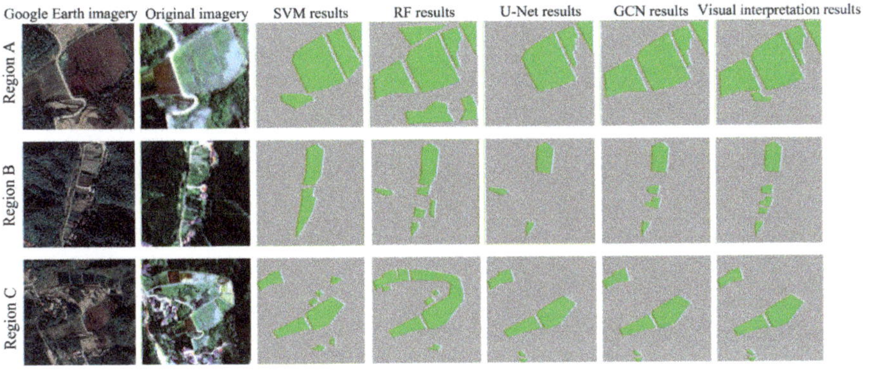

Figure 14. The Rice Field Identification Results of the Local Regions Using the Four Methods.

To ensure a fair comparison, all four methods were evaluated under identical conditions using object-based rice classification. This approach avoids issues such as mixed pixels and unclear boundaries, focusing instead on the accurate identification of rice objects. Region A features relatively flat terrain with forests, bare land, roads, and rice fields. Here, SVM and RF misclassified many bare land and forest areas as rice, likely due to their limited feature learning capacity and difficulty in balancing multiple feature classes, resulting in disordered identification outcomes. U-Net produced results similar to ours, as its deep feature extraction effectively captures rice characteristics, performing well in flat terrain but still missing some rice areas, possibly due to overfitting from model complexity. Region B represents typical mountainous terrain, primarily comprising forest, bare land, and rice. In this area, SVM and RF exhibited increased misclassification, tending to overestimate rice planting areas relative to the actual distribution. U-Net showed notable omission errors, as the spectral similarity between rice and adjacent forest areas caused misidentification. In contrast, our method accurately identified rice fields by emphasizing spatial relational features between land objects, which U-Net did not fully capture. Region C is a relatively flat area with pronounced spectral differences between land types. While SVM and U-Net both showed some omission and commission errors, U-Net's strong deep feature representation enabled successful identification of all rice units. Our method also identified all rice fields, which demonstrated its capability to effectively integrate multiple feature types. In summary, particularly in the complex mountainous terrain, our method leverages spatial relational features among land objects to precisely identify rice fields, and outperformed the other three methods.

4.2.3. Terrain Adaptability Comparison

To verify whether the proposed method can effectively utilize spatial relationships to improve rice field identification accuracy in mountainous terrain compared to the other three methods, the study area was divided into three slope categories based on slope values derived from the DEM: low slope (<10°), medium slope (10–30°), and high slope (>30°). The performance of each method was then evaluated separately within these slope classes. Although SVM and RF employed multiple feature types, they lacked the ability to leverage spatial features. The U-Net models were tested with convolution kernels of sizes 3 × 3, 5 × 5, and 7 × 7 to assess the effect of receptive field size, with the 5 × 5 kernel producing the best results. In contrast, our proposed method constructed graph structures based on adjacency matrices, which enabled the extraction of spatial topological information. Table 12 demonstrated that the proposed method achieved the most significant improvement in identification accuracy under the high-slope conditions, which outperformed SVM by 13.5%, RF by 10.9%, and U-Net by 7.4%. This finding indicates that incorporating spatial relational features between rice field units effectively enhances the rice field identification accuracy under complex mountainous terrain conditions. While U-Net performed better under the low-slope conditions due to its deep feature extraction being well-suited for simpler terrains, its accuracy declined sharply under the high-slope conditions, which reflected the limited adaptability of fixed receptive fields to complex topography. In contrast, both SVM and RF exhibited relatively poor performance under all slope conditions. This suggested that methods lacking the ability to leverage spatial information struggle to achieve satisfactory identification results under complex mountainous conditions.

Table 12. The Overall Accuracy of the Different Methods under Different Terrain Conditions.

Method	Low Slope (<10°)	Medium Slope (10–30°)	High Slope (>30°)	Average
Proposed Method	98.1%	95.6%	93.8%	95.8%
SVM	91.5%	87.6%	80.3%	87.1%
RF	91.8%	89.1%	82.9%	88.0%
U-Net	96.7%	93.3%	86.4%	92.1%

4.2.4. Comparison of Training Sample Requirements

In mountainous regions, rice field distribution tends to be sparse, and labeled data are often limited. To evaluate whether the proposed method demonstrates greater robustness compared to SVM, RF, and U-Net under small-sample training conditions, we kept the test set unchanged and conducted experiments using 100%, 75%, and 50% of the training samples. For SVM and RF, classifiers were directly trained with the reduced training sets. U-Net was trained using a standard fully supervised approach, with data augmentation applied to alleviate the issue of limited samples. Our proposed method leveraged adjacency information for semi-supervised training, which enabled feature learning from neighboring nodes even when some nodes lacked labels.

Table 13 showed that when the training samples were reduced to 50%, the identification accuracy of the proposed method decreased by only 10.3%, while SVM and RF experienced declines exceeding 24%, and U-Net dropped by 16.3%. This demonstrated that our method can effectively leverage limited labeled data for identification. In contrast, SVM and RF heavily depended on the quantity of training samples and suffer from severe overfitting when samples were insufficient. Similarly, U-Net, which strongly relied on supervised samples, also exhibited a significant decrease in identification accuracy under limited data conditions.

Table 13. The Comparison of Training Sample Requirements.

Method	Accuracy (%)			Accuracy Drop (%)
	100% Samples	75% Samples	50% Samples	
Proposed Method	98.3	95.1	88.0	−10.3
SVM	92.1	83.5	66.3	−25.8
RF	93.5	85.9	69.1	−24.4
U-Net	96.7	91.3	80.4	−16.3

4.2.5. Comparison of Computational Costs

To accurately identify scattered rice field units in mountainous regions, this study utilized high spatial resolution imagery. However, the use of high-resolution imageries inevitably leads to a significant increase in computational cost, demanding that the model possess efficient computing capabilities and optimization strategies. To assess the computational efficiency of the proposed method compared to other approaches, prediction times were measured on imagery blocks of sizes 256 × 256 and 512 × 512 pixels, and the number of model parameters was recorded. As traditional classifiers, SVM and RF require relatively low computational resources; U-Net, built upon deep convolutional architectures, incurs high computational overhead; while GCN, which models data through graph structures, exhibits computational complexity intermediate between the two.

Table 14 demonstrated that the proposed method's prediction speed was approximately 40% of that of U-Net, while utilizing fewer model parameters. This indicates that our method effectively reduces computational overhead without sacrificing identification accuracy, making it more suitable for processing large-scale high-resolution remote sensing imageries. In contrast, U-Net's deeper network architecture results in longer computation times, limiting its practicality for large-area high-resolution applications. Although SVM and RF offer faster prediction speeds, their identification accuracy is substantially lower than that of deep learning-based methods, rendering them less effective for fine-grained classification tasks.

Table 14. Computational Cost Comparison.

Method	Prediction Time /s (256 × 256)	Prediction Time (s) (512 × 512)	Model Parameters (M)
GCN	1.1	2.2	8
SVM	0.3	0.6	10
RF	0.4	0.8	15
U-Net	2.5	5.2	30

4.3. Key Advances and Highlights

The proposed method made several notable contributions to rice field identification in mountainous regions.

First, in terms of feature extraction, a systematic design was carried out according to the growth characteristics of rice and the complexity of mountainous terrain. Spectral features were derived from multiple vegetation indices, which effectively reflected the physiological and biochemical traits of rice (e.g., chlorophyll content, biomass, and vegetation coverage). These indices also reduced the influence of soil background and atmospheric conditions, thereby enhancing the discriminative power of rice signals [36,37]. Texture features captured the spatial patterns of paddy fields, which played an important role in distinguishing rice from other crops [38]. Polarization features, obtained from radar backscatter differences, further improved the separability between rice and other vegetation types [39]. Moreover, topographic features accounted for environmental constraints specific to mountainous conditions, such as elevation, slope, and aspect, which directly influenced water availability, illumination, and temperature gradients, thereby affecting the growth cycle and spatial distribution of rice [40]. The integration of these multidimensional features not only improved classification accuracy but also provided a representation more consistent with the ecological reality of rice in mountainous regions.

Second, in the model construction, the conventional graph convolutional network (GCN) architecture was customized for rice field identification. Superpixels (representing rice units) were used as graph nodes, and multi-source features were integrated into the graph structure, which effectively captured the complex spatial relationships of rice distribution in mountainous regions. These structural enhancements addressed the insufficient utilization of multi-dimensional features observed in previous studies, resulting in high-precision identification of rice fields under mountainous conditions and notable improvements in computational efficiency. Compared with traditional classifiers, the proposed method demonstrated a better ability to adapt to the spatial characteristics of rice distribution in complex terrain, providing an efficient and reliable approach for rice field identification in mountainous regions [41,42].

In addition, the proposed method demonstrated strong transferability. On the one hand, the data used in this study were not restricted to a specific region, making the approach applicable to both plain and mountainous regions. On the other hand, topographic factors were represented using a digital elevation model (DEM), which did not impose special requirements on slope gradients or landform types. Therefore, the method can be applied under various terrain conditions and across different regions, offering a valuable reference for large-scale rice field identification and monitoring.

4.4. Limitations and Future Research

This study proposed a rice field identification method under mountainous conditions, which could be effectively applied to high-resolution remote sensing imagery in mountainous rice-planting areas and achieved satisfactory identification results. However, several aspects remained to be improved.

(1) The cloud removal method based on the fusion of SAR and neighboring optical remote sensing imageries did not impose strict theoretical requirements on the selection of neighboring optical imageries. In principle, any cloud-free optical imagery acquired within the same rice growth stage could serve as neighboring imagery. In practice, however, neighboring optical imageries derived from different sensors, with varying spatial resolutions and spectral characteristics, might affect the cloud removal performance [43]. Future studies could investigate the influence of these factors on cloud removal results and attempt to establish criteria for the optimal selection of neighboring optical imageries.

(2) The rice field identification experiments in this study were conducted only in Huoshan County of the Dabie Mountains. Huoshan County is characterized by a typical mountainous terrain, with extensive rice cultivation, making it representative to a certain extent. Nevertheless, differences in topographic relief, elevation, and rice distribution across regions may influence the identification results [44]. Future research could expand the experimental scope to broader areas and incorporate transfer learning techniques to enhance the generalization capability of the proposed method, thereby further verifying its applicability across the Dabie Mountains and providing stronger technical support for precision agriculture monitoring in mountainous regions.

5. Conclusions

This study proposed a rice field identification method specifically designed for mountainous conditions, effectively leveraged multi-dimensional feature information to enhance identification performance. Compared with existing approaches, the proposed method demonstrated superior accuracy and broader applicability. The main conclusions are summarized as follows:

1. A coarse-to-fine cloud removal strategy was implemented, which comprehensively integrated features from SAR imagery and neighboring temporal optical remote sensing imageries to achieve effective cloud removal from optical data. The method achieved high performance and outperformed alternative methods in both accuracy and applicability.
2. By fully accounting for the growth environment and spatial distribution characteristics of rice in mountainous terrain, and by integrating multi-source features through graph structure modeling, the proposed approach successfully addressed the limitations of existing methods in handling complex terrain. This led to significant improvements in identification accuracy and model robustness.

Author Contributions: Conceptualization, Y.W.; Methodology, Y.W., Z.Y. and W.Z.; Validation, Z.Y.; Investigation, W.Z.; Resources, J.C. and Z.Y.; Data curation, Y.W.; Writing—original draft, J.C. and W.Z.; Writing—review & editing, W.Z.; Visualization, W.Z.; Project administration, Z.Y.; Funding acquisition, J.C. and Z.Y. All authors have read and agreed to the published version of the manuscript.

Funding: This research was funded by the National Natural Science Foundation of China (No. 41671514), the Surveying and mapping science and technology "double first-class" discipline creation project (No. GCCYJ202409, No. BZXCG202403), and the Natural Science Foundation of Henan Province (No. 162300410122).

Data Availability Statement: The original contributions presented in this study are included in the article. Further inquiries can be directed to the corresponding author.

Conflicts of Interest: The authors declare no conflicts of interest.

References

1. Rezaei, E.E.; Webber, H.; Asseng, S.; Boote, K.; Durand, J.L.; Ewert, F.; Martre, P.; MacCarthy, D.S. Climate change impacts on crop yields. *Nat. Rev. Earth Environ.* **2023**, *4*, 831–846. [CrossRef]
2. Yang, H.; Pan, B.; Wu, W.; Tai, J. Field-based rice classification in Wuhua county through integration of multi-temporal Sentinel-1A and Landsat-8 OLI data. *Int. J. Appl. Earth Obs. Geoinf.* **2018**, *69*, 226–236. [CrossRef]
3. Pang, J.; Zhang, R.; Yu, B.; Liao, M.; Lv, J.; Xie, L.; Li, S.; Zhan, J. Pixel-level rice planting information monitoring in Fujin City based on time-series SAR imagery. *Int. J. Appl. Earth Obs. Geoinf.* **2021**, *104*, 102551. [CrossRef]
4. Estacio, I.; Basu, M.; Sianipar, C.P.; Onitsuka, K.; Hoshino, S. Dynamics of land cover transitions and agricultural abandonment in a mountainous agricultural landscape: Case of Ifugao rice terraces, Philippines. *Landsc. Urban Plan.* **2022**, *222*, 104394. [CrossRef]
5. Zhao, S.; Liu, X.; Ding, C.; Liu, S.; Wu, C.; Wu, L. Mapping Rice Paddies in Complex Landscapes with Convolutional Neural Networks and Phenological Metrics. *GISci. Remote Sens.* **2019**, *57*, 37–48. [CrossRef]
6. Nduati, E.; Sofue, Y.; Matniyaz, A.; Park, J.G.; Yang, W.; Kondoh, A. Cropland Mapping Using Fusion of Multi-Sensor Data in a Complex Urban/Peri-Urban Area. *Remote Sens.* **2019**, *11*, 207. [CrossRef]
7. Jing, W.; Huang, J.; Zhang, K.; Li, X.; She, B.; Wei, C.; Gao, J.; Song, X. Rice Fields Mapping in Fragmented Area Using Multi-Temporal HJ-1A/B CCD Images. *Remote Sens.* **2015**, *7*, 3467–3488.
8. Quan, J.; Wang, Y.; Wang, X.; Tang, W.; Wang, Q. Extraction of rice planting areas in the Dabie Mountains using remote sensing images: A case study of Landsat 8. *Chin. Agric. Sci. Bull.* **2019**, *35*, 104–111.
9. Sang, G.; Tang, Z.; Mao, K.; Deng, G.; Wang, J.; Li, J. Extraction of high-resolution rice planting areas based on GEE cloud platform and Sentinel data: A case study of Hunan Province. *Acta Agron. Sin.* **2022**, *48*, 2409–2420.
10. Zhang, K.; Chen, Y.; Zhang, B.; Hu, J.; Wang, W. A Multitemporal Mountain Rice Identification and Extraction Method Based on the Optimal Feature Combination and Machine Learning. *Remote Sens.* **2022**, *14*, 5096. [CrossRef]
11. Liu, W.; Wang, J.; Luo, J.; Wu, Z.; Chen, J.; Zhou, Y.; Sun, Y.; Shen, Z.; Xu, N.; Yang, Y. Farmland Parcel Mapping in Mountain Areas Using Time-Series SAR Data and VHR Optical Images. *Remote Sens.* **2020**, *12*, 3733. [CrossRef]
12. Chen, A.; Li, Y. Rice identification at different growth stages in the southwest mountainous region based on Sentinel-2 imagery. *Trans. Chin. Soc. Agric. Eng.* **2020**, *36*, 8–15.
13. Zhang, W.; Liu, H.; Wu, W.; Zhan, L.; Wei, J. Mapping rice paddy based on machine learning with Sentinel-2 multi-temporal data: Model comparison and transferability. *Remote Sens.* **2020**, *12*, 1620. [CrossRef]

14. You, J.; You, S.; Tong, Y.; Li, J.; Pan, L.; Liu, H.; Zhang, L. A geostatistical approach for correcting bias in rice remote sensing identification. *Trans. Chin. Soc. Agric. Eng.* **2013**, *29*, 11–18.
15. Ma, H.; Wang, L.; Sun, W.; Yang, S.; Gao, Y.; Fan, L.; Yang, G.; Wang, Y. A new rice identification algorithm under complex terrain combining multi-characteristic parameters and homogeneous objects based on time series dual-polarization synthetic aperture radar. *Front. Ecol. Evol.* **2023**, *11*, 1093454. [CrossRef]
16. Chen, J.; Zhang, J.; Duan, Y.; Yang, Z.; Xia, X. A review of cloud detection and thick cloud removal in medium-resolution remote sensing imagery. *Remote Sens. Technol. Appl.* **2023**, *38*, 143–155.
17. Li, Y.; Hong, D.; Li, C.; Yao, J.; Chanussot, J. HD-Net: High-resolution decoupled network for building footprint extraction via deeply supervised body and boundary decomposition. *ISPRS J. Photogramm. Remote Sens.* **2024**, *209*, 51–65. [CrossRef]
18. Xiao, M.; Li, X.; Zhang, X.; Zhang, L. An image segmentation algorithm based on multi-scale region growing. *J. Jilin Univ. (Eng. Technol. Ed.)* **2017**, *47*, 1591–1597.
19. Alirezapour, H.; Mansouri, N.; Zade, B.M.H. A Comprehensive Survey on Feature Selection with Grasshopper Optimization Algorithm. *Neural Process. Lett.* **2024**, *56*, 28. [CrossRef]
20. Kong, Y.; Ni, D. Semi-Supervised Classification of Wafer Map Based on Ladder Network. In Proceedings of the 2018 14th IEEE International Conference on Solid-State and Integrated Circuit Technology (ICSICT), Qingdao, China, 31 October–3 November 2018.
21. Robnik-Šikonja, M.; Kononenko, I. Theoretical and Empirical Analysis of ReliefF and RReliefF. *Mach. Learn.* **2003**, *53*, 23–69. [CrossRef]
22. Zhou, J.; Huang, S.; Zhou, T.; Armaghani, D.J.; Qiu, Y. Employing a genetic algorithm and grey wolf optimizer for optimizing RF models to evaluate soil liquefaction potential. *Artif. Intell. Rev.* **2022**, *55*, 5673–5705. [CrossRef]
23. Zuo, X.; Lu, H.; Zhang, Y.; Cheng, J.; Guo, Y. Graph convolutional network method for small-sample classification of hyperspectral imagery. *Acta Geod. Cartogr. Sin.* **2021**, *50*, 1358–1369.
24. Hong, D.; Gao, L.; Yao, J.; Zhang, B.; Plaza, A.; Chanussot, J. Graph Convolutional Networks for Hyperspectral Image Classification. *IEEE Trans. Geosci. Remote Sens.* **2021**, *59*, 5966–5978. [CrossRef]
25. Zhu, C.; Li, D.; Shao, Z.; Yin, J.; Wang, J. Evaluation method for segmentation accuracy of high-resolution remote sensing imagery based on object-oriented approach. *High Power Laser Part. Beams* **2015**, *27*, 43–49.
26. Chen, Y.; Mo, W.; Mo, J.; Wang, J.; Zhong, S. Extraction of rice planting areas in southern China based on object-oriented classification. *Remote Sens. Technol. Appl.* **2011**, *26*, 163–168.
27. Zhang, K.; Su, H.; Dou, Y. A novel method for evaluating classification accuracy of multi-class tasks based on confusion matrix. *Comput. Eng. Sci.* **2021**, *43*, 1910–1919.
28. Chen, X.; Tang, L.; Wang, Z. Research on the current situation and countermeasures of characteristic agricultural development in Huoshan County. *Shanxi Agric. Econ.* **2022**, *22*, 147–149.
29. Sun, W.; Yang, G.; Chen, C.; Chang, M.; Huang, K.; Meng, X.; Liu, L. Current status and literature analysis of Earth observation remote sensing satellite development in China. *J. Remote Sens.* **2020**, *24*, 479–510.
30. Shan, J.; Wang, Y.; Liu, J.; Zhao, Y.; Li, Y. Monitoring rice planting areas based on multi-temporal Gaofen-1 satellite imagery. *Jiangsu Agric. Sci.* **2017**, *45*, 229–232.
31. Cao, M.; Shi, Z.; Shen, Q. Study on optimal band selection of ALOS imagery in land cover classification. *Bull. Surv. Mapp.* **2008**, *12*, 16–18+27.
32. Chen, Z.; Zhang, Q.; Chi, T.; Fu, W.; Li, C. Gaofen-3 satellite system and its applications. *Satell. Appl.* **2024**, *2*, 20–26.
33. Wang, Z.; Wang, M.; Huang, Y.; Li, Y. A review on remote sensing image classification using support vector machine. *Comput. Sci.* **2016**, *43*, 11–17+31.
34. Ma, Y.; Jiang, Q.; Meng, Z.; Li, Y.; Wang, D.; Liu, H. Land use classification of agricultural areas based on random forest algorithm. *Trans. Chin. Soc. Agric. Mach.* **2016**, *47*, 297–303.
35. He, H.; Yang, G.; Li, H.; Feng, H.; Xu, B.; Wang, L. Water body extraction from high-resolution remote sensing imagery based on improved U-Net network. *Geo-Inf. Sci.* **2020**, *22*, 2010–2022.
36. Guan, X.; Huang, C.; Liu, H.; Meng, X.; Liu, Q. Mapping rice cropping systems in Vietnam using an NDVI-based time-series similarity measurement based on DTW distance. *Remote Sens.* **2016**, *8*, 19. [CrossRef]
37. Guan, S.; Fukami, K.; Matsunaka, H.; Okami, M.; Takahashi, K. Assessing correlation of high-resolution NDVI with fertilizer application level and yield of rice and wheat crops using small UAVs. *Remote Sens.* **2019**, *11*, 112. [CrossRef]
38. Akbari, E.; Boloorani, A.; Samany, N.N.; Hamzeh, S.; Pignatti, S. Crop mapping using random forest and particle swarm optimization based on multi-temporal Sentinel-2. *Remote Sens.* **2020**, *12*, 1449. [CrossRef]
39. Sun, C.; Zhang, H.; Ge, J.; Wang, C.; Li, L.; Xu, L. Rice mapping in a subtropical hilly region based on Sentinel-1 time series feature analysis and the dual branch BiLSTM model. *Remote Sens.* **2022**, *14*, 3213. [CrossRef]
40. Zhu, W.; Pan, Y.; Hu, Z.; Jin, Y.; Bai, Y. Decline in planting areas of double-season rice by half in southern China over the last two decades. *Remote Sens.* **2024**, *16*, 440. [CrossRef]

41. Gui, B.; Song, C.; Li, Y.; Wang, J.; Shang, S. SAGRNet: A novel object-based graph convolutional neural network for diverse vegetation cover classification in remotely-sensed imagery. *ISPRS J. Photogramm. Remote Sens.* **2025**, *227*, 99–124. [CrossRef]
42. Jiang, H.; Zhang, C.; Qiao, Y.; Zhang, Z.; Meng, W. CNN feature based graph convolutional network for weed and crop recognition in smart farming. *Comput. Electron. Agric.* **2020**, *174*, 105450. [CrossRef]
43. Gao, J.; Yuan, Q.; Li, J.; Zhang, H.; Su, X. Cloud removal with fusion of high resolution optical and SAR images using generative adversarial networks. *Remote Sens.* **2020**, *12*, 191. [CrossRef]
44. Tian, X.; Zhou, Q.; Wang, L.; Feng, H.; Yang, P. Crop classification in mountainous areas using object-oriented methods and multi-source data: A case study of Xishui County, China. *Agronomy* **2023**, *13*, 3037. [CrossRef]

remote sensing

MDPI

Article

High-Throughput Identification and Prediction of Early Stress Markers in Soybean Under Progressive Water Regimes via Hyperspectral Spectroscopy and Machine Learning

Caio Almeida de Oliveira [1], Nicole Ghinzelli Vedana [1], Weslei Augusto Mendonça [1], João Vitor Ferreira Gonçalves [1], Dheynne Heyre Silva de Matos [1], Renato Herrig Furlanetto [2], Luis Guilherme Teixeira Crusiol [3], Amanda Silveira Reis [1], Werner Camargos Antunes [4], Roney Berti de Oliveira [1], Marcelo Luiz Chicati [1], José Alexandre M. Demattê [5], Marcos Rafael Nanni [1] and Renan Falcioni [1,4,]*

[1] Graduate Program in Agronomy, State University of Maringá, Av. Colombo, 5790, Maringá 87020-900, Paraná, Brazil; pg55482@uem.br (C.A.d.O.); pg405864@uem.br (N.G.V.); pg405896@uem.br (W.A.M.); pg403805@uem.br (J.V.F.G.); ra143742@uem.br (D.H.S.d.M.); asreis@uem.br (A.S.R.); rboliveira@uem.br (R.B.d.O.); mlchicati@uem.br (M.L.C.); mrnanni@uem.br (M.R.N.)

[2] Gulf Coast Research and Education Center, University of Florida, Wimauma, FL 33598, USA; re.herrigfurlane@ufl.edu

[3] Embrapa Soja (National Soybean Research Center—Brazilian Agricultural Research Corporation), Rodovia Carlos João Strass, s/n°, Distrito de Warta, Londrina 86001-970, Paraná, Brazil; luis.crusiol@colaborador.embrapa.br

[4] Department of Biology, State University of Maringá, Av. Colombo, 5790, Maringá 87020-900, Paraná, Brazil; wcantunes@uem.br

[5] Department of Soil Science, Luiz de Queiroz College of Agriculture, University of São Paulo, Av. Pádua Dias, 11, Piracicaba 13418-260, São Paulo, Brazil; jamdemat@usp.br

* Correspondence: renanfalcioni@gmail.com or rfalcioni2@uem.br; Tel.: +55-443-0118-940

Highlights

What are the main findings?

- Hyperspectral spectroscopy combined with machine learning enables high-accuracy, nondestructive prediction of early stress markers (pigments, osmolytes, antioxidants, cell wall compounds, and water status) in soybean under progressive drought via remote sensing and machine learning models.
- Tree-based ensemble and neural network models (e.g., random forest, MLP) achieved >95% accuracy in classifying drought severity, outperformed distance- and probability-based classifiers, and effectively distinguished eleven water regimes across the full or range UV–VIS–NIR–SWIR spectrum.

What is the implication of the main finding?

- The integration of hyperspectral sensors and machine learning provides a rapid, field-deployable solution for early drought detection and precision irrigation management in soybean, potentially reducing the reliance on time-consuming laboratory assays via remote sensing tools.
- Selecting minimal and informative spectral bands paves the way for simplified, cost-effective proximal or UAV-mounted sensors for large-scale drought phenotyping and smart agriculture applications.

Abstract

The soybean *Glycine max* (L.) Merrill is a key crop in Brazil's agricultural sector and is essential for both domestic food security and international trade. However, water stress severely impacts its productivity. In this study, we examined the physiological and biochemical responses of soybean plants to various water regimes via hyperspectral reflectance (350–2500 nm) and machine learning (ML) models. The plants were subjected to eleven distinct water regimes, ranging from 100% to 0% field capacity, over 14 days. Seventeen key physiological parameters, including chlorophyll, carotenoids, flavonoids, proline, stress markers and water content, and hyperspectral data were measured to capture changes induced by water deficit. Principal component analysis (PCA) revealed significant spectral differences between the water treatments, with the first two principal components explaining 88% of the variance. Hyperspectral indices and reflectance patterns in the visible (VIS), near-infrared (NIR), and shortwave-infrared (SWIR) regions are linked to specific stress markers, such as pigment degradation and osmotic adjustment. Machine learning classifiers, including random forest and gradient boosting, achieved over 95% accuracy in predicting drought-induced stress. Notably, a minimal set of 12 spectral bands (including red-edge and SWIR features) was used to predict both stress levels and biochemical changes with comparable accuracy to traditional laboratory assays. These findings demonstrate that spectroscopy by hyperspectral sensors, when combined with ML techniques, provides a nondestructive, field-deployable solution for early drought detection and precision irrigation in soybean cultivation.

Keywords: agricultural management; biochemical prediction; computational intelligence; phenotyping in plants; precision agriculture; spectroscopy in plants; UV–VIS–NIR–SWIR sensors

1. Introduction

Soybean (*Glycine max* (L.) Merrill) plays a central role in Brazilian agriculture, underpinning both domestic food security and international trade [1,2]. Over the past decade, Brazil has solidified its position as the world's leading soybean producer, with output reaching a record 169.0 million tons in the 2024/25 marketing year, cultivated on some 47.4 million ha (USDA, 2024) [3]. This oilseed accounts for nearly 60% of global soybean exports, representing a cornerstone of the national economy and a critical source of foreign exchange [3]. Consequently, such prominence renders soybean both a driver of agronomic innovation and a sensitive indicator of environmental stressors.

In this context, drought is defined as a sustained soil water deficit that impairs plant physiological function and poses one of the greatest challenges to soybean productivity [4]. Water limitation disrupts carbon assimilation through stomatal closure, reduces the chlorophyll content, and accelerates the generation of reactive oxygen species (ROS), precipitating oxidative damage to the photosynthetic apparatus and membrane lipids [5,6]. Moreover, the accumulation of osmolytes such as proline and antioxidative compounds (e.g., flavonoids and phenolics) reflects a plant's protective response, linking water status to both photochemical efficiency and oxidative homeostasis [7].

Hyperspectral proximal sensing, which spans ultraviolet (UV), visible (VIS), near-infrared (NIR), and shortwave-infrared (SWIR) regions (350–2500 nm), offers a nondestructive window into these physiological and biochemical changes [8–10]. Narrowband features in the VIS region capture pigment absorption peaks (chlorophylls, carotenoids), whereas SWIR bands correspond to overtones of water and cell wall constituents (e.g., cellulose, lignin, and other compounds) [11]. Moreover, recent studies have demonstrated that spectroscopy via hyperspectral indices and continuum-removed reflectance metrics can reliably predict leaf water content and stress markers, with accuracies exceeding those of broadband sensors [12,13].

Narrowband reflectance vegetation indices (VIs) synthesize key wavelengths into simple ratios or differences, such as the NDVI, PRI and WBI, that increase sensitivity to chlorophyll content, photosynthetic efficiency and water status [14,15]. By condensing thousands of bands into targeted indices, VIs reduce noise and highlight physiological shifts, serving both as interpretable proxies of plant health and as robust input features for downstream models [16–18]. When integrated within machine learning frameworks, these indices improve prediction and characterization by emphasizing the most stress-responsive spectral signals while limiting redundant information [19].

Complementing spectral approaches, machine learning (ML) and deep learning (DL) techniques enable the classification and quantification of stress responses from high-dimensional reflectance data [20,21]. Algorithms such as random forest, support vector machines, and convolutional neural networks have achieved classification accuracies above 95% in distinguishing drought-stressed vs. well-watered soybeans by recognizing subtle spectral shifts associated with pigment degradation and osmotic adjustment [2,11,22]. These models not only automate stress detection but also provide real-time decision support in precision agriculture. For example, partial least squares or CNN-based classifiers can rapidly flag drought conditions with minimal false alarms, allowing timely irrigation or management interventions. The integration of ML thus adds robustness and scalability to hyperspectral drought diagnostics [23–25].

Critical to both spectral and ML-based methods is the targeted selection of informative wavelengths [26]. Meta-analyses have identified discrete bands at approximately 550–750 nm (the VIS to red-edge region) and in the SWIR near 1450 nm and 1900 nm as especially predictive of plant water status and structural changes [27]. The reflectance in the VIS to red-edge transition zone is highly sensitive to chlorophyll content and stress-induced pigment changes, whereas the reflectances at ~1450 nm and ~1900 nm correspond to major water absorption features where the LWC strongly influences the signal [2]. Leveraging these "spectral pin-points" can drastically reduce the data volume while preserving predictive power, enabling the development of simplified handheld sensors and UAV-mounted instruments tailored for drought monitoring. By focusing on a minimal set of diagnostic bands, one can design field-deployable multispectral systems that approximate the performance of full hyperspectral instruments in detecting water stress [28].

A comprehensive characterization of the drought response thus benefits from integrating multiple analytical layers, spectral indices, machine-learning classification, and biochemical assays to capture both macroscopic reflectance patterns and underlying molecular dynamics [29,30]. Such a multilayer approach can detect stress onset and quantify key metabolites before visible symptoms appear. Recent work has shown that hyperspectral models can predict drought-induced changes in leaf metabolites (e.g., proline, abscisic acid, and electrolyte leakage) with good accuracy, indicating that optical data contain biochemical information about stress [11,31,32]. By combining optical and biochemical perspectives, one can achieve robust early warning of water deficit and a mechanistic understanding of the stress response [33]. This framework promises more precise estimation of plant health and stress status, advancing both fundamental knowledge and practical management of drought in soybean [2,3,34].

In this study, we evaluated the efficacy of combined hyperspectral and ML methods for the nondestructive prediction of key physiological parameters (leaf water status and biochemical and metabolic parameters in cells) and oxidative stress biomarkers (flavonoids, proline, and phenolics) in soybean (*Glycine max* (L.) Merrill) plants under a gradient of water regimes. We hypothesize that a minimal suite of ~12 strategically selected wavelengths, coupled with ensemble learning models, can predict both drought severity and biochemical composition with accuracy comparable to that of traditional laboratory assays. Validating this hypothesis would demonstrate a rapid, field-deployable approach to soybean drought phenotyping, enabling early stress detection and informing water-management decisions.

2. Material and Methods

2.1. Plant Materials

Soybean seeds (*Glycine max* (L.) Merrill) were initially germinated under controlled laboratory conditions. Vigorous, morphologically uniform seedlings were selected and transplanted into 1 L plastic pots containing sterilized substrate. The plants were grown in a controlled-environment chamber, where the day/night temperatures were maintained at 26 °C/23 °C, the relative humidity was 70%, and a 16/8 h photoperiod was used. The photosynthetically active radiation was set at 500 μmol m^{-2} s^{-1} and was calibrated via a LI-COR 1800 quantum sensor (LI-COR Inc., Lincoln, NE, USA).

After transplanting, the seedlings were irrigated to full substrate capacity for seven days to ensure acclimation. The plants were subsequently subjected to eleven distinct water regimes, defined as 100% (W100), 90% (W90), 80% (W80), 70% (W70), 60% (W60), 50% (W50), 40% (W40), 30% (W30), 20% (W20), 10% (W10), and 0% (W0) of the substrate field capacity. Watering volumes were determined gravimetrically, simulating a gradient of water restriction. Hoagland's solution was applied every two days to ensure an adequate nutrient supply.

The experiment was carried out in a completely randomized design, with eleven water regimes, eight plants per treatment, and three technical replicates per plant, resulting in a total of 264 analysed samples. Pots were randomly distributed within the chamber. The water regimes were maintained for 14 days, with the environmental and irrigation parameters continuously monitored to ensure experimental consistency.

2.2. Hyperspectral Reflectance Data

Hyperspectral reflectance spectra were obtained from the adaxial surface of fully expanded leaves via a FieldSpec 3 spectroradiometer (ASD Inc., Boulder, CO, USA) equipped with a PlantProbe® leaf clip. The measurements covered the 350–2500 nm spectral range (UV–VIS–NIR–SWIR). Prior to each measurement session, the instrument was calibrated with both a Spectralon® white reference and a dark reference to ensure accuracy. For each leaf, 50 consecutive scans were averaged to reduce noise and improve signal quality. All reflectance data were processed via ViewSpec Pro® software version 5 (ASD Inc., Boulder, CO, USA), following standard protocols for baseline correction and interpolation. Only reflectance values were considered for subsequent statistical and multivariate analyses [35]. All the measurements were conducted under controlled ambient light conditions.

2.3. Chlorophylls and Carotenoids Extraction

Simultaneous quantification of total chlorophyll (Chl), total carotenoids (Car), and flavonoids (Flv) was performed as described by Gitelson and Solovchenko (2018) [36] and Falcioni et al. (2022) [37], with adaptations for soybean. Leaf segments (0.5 cm^2) were homogenized in 1.5 mL microtubes containing chloroform–methanol solution (2:1, v/v) in the presence of $CaCO_3$. After complete pigment extraction, distilled water (20% of the total extract volume) was added to promote phase separation. The samples were subsequently centrifuged at 15,000 rpm for 5 min to ensure clear separation of the polar and apolar phases. The absorbance was measured in a 96-well microplate using a Biochrom Asys UVM-340 microplate reader (Biochrom Ltd., Cambridge, UK) with ScanPlus VisibleWell® software version 1.0.2 (Biochrome Ltd., Milton Road, Cambridge, UK).

2.3.1. Quantification of Chlorophylls and Carotenoids

Chlorophyll a (Chl a), chlorophyll b (Chl b), total chlorophyll (Chl a + b), and carotenoids (Car; carotenes + xanthophylls) were quantified from the acetone phase by adding 200 μL of extract to each well. The absorbance was measured at 470, 652, and

665 nm, and 100% methanol was used as the blank. The base area (mg m^{-2}) and mass (mg g^{-1}) concentrations were calculated via the following equations:

$$\text{Chl a} = 16.72 \times \text{Abs}_{665} - 9.16 \times \text{Abs}_{652}$$

$$\text{Chl b} = 34.09 \times \text{Abs}_{652} - 15.28 \times \text{Abs}_{665}$$

$$\text{Chl } a + b = \text{Chl a} + \text{Chl b}$$

$$\text{Car} = \frac{1000 \times \text{Abs}_{470} - 1.63 \times \text{Chl a} - 104.96 \times \text{Chl b}}{221}$$

2.3.2. Quantification of Flavonoids

Flavonoids (Flv) were quantified in the polar methanol extract. The upper phase, containing extrachloroplastic pigments, was used for total flavonoid content determination by absorbance at 358 nm (ε_{358} = 25 mM^{-1} cm^{-1}; Gitelson & Solovchenko, 2018) [36].

2.4. Quantification of Proline

The proline content in the leaf samples was determined following the methods of Falcioni et al. (2025) [11], with adaptations for microplate analysis. Fresh leaf segments (100 mg) were homogenized in 2 mL of 3% (*w/v*) sulfosalicylic acid and centrifuged at 15,000 rpm for 10 min. A 100 µL aliquot of the supernatant was transferred to a 96-well microplate, to which 100 µL of acid ninhydrin solution (prepared by mixing 1.25 g of ninhydrin in 30 mL of glacial acetic acid and 20 mL of 6 M phosphoric acid) and 100 µL of glacial acetic acid were added. The plate was sealed and incubated at 95 °C for 1 h. After cooling to room temperature, 200 µL of toluene was added to each well, and the mixture was agitated for 30 s. The absorbance of the chromophore-containing toluene phase was measured at 520 nm via a microplate reader. The proline concentration was calculated from a standard curve constructed with L-proline and expressed as µmol proline per gram of fresh mass (µmol g^{-1} FM). All analyses were performed in triplicate.

2.5. Quantification of Soluble Phenolic Compounds (Phe)

The soluble phenolic compounds (Phe) in the leaf samples were quantified via an adapted Folin–Ciocalteu method (Ragaee, 2006) [38]. The methanolic extracts (150 µL) were mixed with 70 µL of Folin–Ciocalteu reagent (1 M), 140 µL of sodium carbonate (Na_2CO_3, 3.56 M), and 850 µL of deionized water in 2 mL tubes. The reaction mixture was incubated in the dark for 50 min at room temperature and then centrifuged at 15,000 rpm for 2 min. The absorbance of the supernatant was measured at 725 nm via a spectrophotometer. The phenolic content was calculated against a gallic acid standard curve ($\hat{Y}$ = 87.651x + 1.6515; R^2 = 0.993) and expressed as gallic acid equivalents per sample. All measurements were performed in triplicate.

2.6. Preparation of Protein-Free Cell Wall Fraction (PFCW) and Quantification of Lignin and Cellulose

Protein-free cell wall fractions (PFCWs) were prepared from dried, powdered leaf tissue. Aliquots of 150 mg were weighed into 2 mL microtubes. The samples were sequentially washed five times with 50 mM potassium phosphate buffer (pH 7.0), five times with Triton X-100 (pH 7.0), four times with 1 M NaCl (pH 7.0), four times with distilled water, and three times with acetone. After each wash, the samples were centrifuged at 15,000 rpm for 3 min. The final pellets were oven-dried at 60 °C for 24 h and used as the PFCW fraction, which was free of both water-soluble polar and apolar compounds.

2.6.1. Lignin Content Determination

The lignin content in the PFCW fraction was quantified via the acetyl bromide method. For each sample, 20 mg of PFCW was transferred to a new microtube and mixed with 130 µL of freshly prepared acetyl bromide solution (25% *v*/*v* in glacial acetic acid). The samples were incubated at 70 °C for 30 min and then cooled rapidly on ice. Subsequently, 0.24 mL of 2 M NaOH, 0.02 mL of 5 M hydroxylamine-HCl, and 1.6 mL of glacial acetic acid were added for complete solubilization of the lignin extract. The samples were subsequently centrifuged at 1400× *g* for 5 min. The lignin content was determined spectrophotometrically at 280 nm (ε = 22.1 g L^{-1} cm^{-1}) via a FlexStation 3 plate spectrophotometer (Molecular Devices LLC., San Jose, CA, USA) and SoftMax® Pro Software version 5 (Molecular Devices LLC., San Jose, CA, USA). The results are expressed as mg lignin per g of PFCW. All analyses were performed in triplicate.

2.6.2. Cellulose Content Determination

The cellulose was quantified according to standard protocols with adaptations for leaf tissue. The dried tissue samples were incubated at 70 °C for 1 h. The ethanol was then replaced with acetic/nitric acid solution for extraction, and the mixture was subsequently discarded. The samples were washed with distilled water and then treated with freshly prepared anthrone in sulfuric acid. Quantification was performed at 620 nm using a Biochrom Asys UVM-340 microplate reader (Biochrom Ltd., Cambridge, UK). The cellulose content was expressed as glucose equivalents (µmol glucose per g dry mass) and was calculated according to a glucose standard curve. All analyses were performed in triplicate.

2.7. Antioxidant Activity (RSA%)

Antioxidant activity, expressed as relative scavenging activity (RSA%), was determined via the DPPH (2,2-diphenyl-1-picrylhydrazyl) free radical assay following Falcioni et al. (2025) [11] with adaptations for soybean leaves. Methanolic leaf extracts (50 µL) were added to 200 µL of 1 mM DPPH solution in each well of a quartz 96-well microplate. The mixture was agitated and incubated in the dark at room temperature for 60 min. The absorbance was then measured at 515 nm via a microplate spectrophotometer (Biochrom Asys UVM-340). The RSA (%) was calculated as follows:

$$\text{RSA }(\%) = \left[1 - \left(\frac{A_{\text{sample}}}{A_{\text{blank}}}\right)\right] \times 100$$

where A_sample is the absorbance of the reaction mixture and A_{blank} is the absorbance of the DPPH solution without extraction. Higher RSA values indicate greater antioxidant activity. All measurements were performed in triplicate.

2.8. Electrolyte Leakage (ELK%)

Electrolyte leakage (ELK%) was determined to assess membrane integrity, following standard protocols with adaptations for soybean leaves. Fresh leaf discs (0.5 cm diameter) were rinsed thoroughly with deionized water to remove surface-adhered electrolytes. The discs were placed in test tubes containing 10 mL of deionized water and incubated at room temperature for 24 h. The initial conductivity (C_1) of the bath solution was measured via a conductivity meter (model, manufacturer). The samples were then autoclaved at 121 °C for 20 min to ensure complete membrane rupture and allowed to cool to room temperature, after which the final conductivity (C_2) was recorded. Electrolyte leakage was calculated as the percentage ratio of initial conductivity to total conductivity:

$$\text{ELK }(\%) = \left(\frac{C_1}{C_2}\right) \times 100$$

All measurements were performed in triplicate. The results are expressed as the percentage of total electrolytes released, reflecting the degree of cell membrane damage.

2.9. Relative Water Content (RWC%)

The relative water content (RWC) was determined to evaluate the leaf water status. Fresh leaf discs were collected and immediately weighed to obtain the fresh mass (FM). The discs were then floated on distilled water in Petri dishes for 24 h at 4 °C in the dark to reach full turgidity. After gentle blotting to remove surface water, the turgid mass (TM) was recorded. The samples were subsequently dried at 70 °C for 48 h and weighed to determine the dry mass (DM). The RWC was calculated as follows:

$$\text{RWC (\%)} = \left(\frac{\text{FM} - \text{DM}}{\text{TM} - \text{DM}}\right) \times 100$$

All measurements were performed in triplicate. The results are presented as the percentage of water content relative to full turgor.

2.10. Statistical Analyses

2.10.1. Analysis of Variance and Descriptive Statistics

Descriptive statistical analysis was conducted for all agronomic, physiological, biochemical, and spectral datasets, including the calculation of the mean, standard error of the mean (SEM), minimum and maximum values, and coefficient of variation (CV%). Treatment differences were evaluated by analysis of variance (ANOVA), with statistical significance considered at $p < 0.05$. Multiple mean comparisons were performed via Duncan's test at the same significance level. Relationships between physiological variables, vegetation indices, and growth parameters were investigated via Pearson's correlation test. All univariate analyses and the generation of summary tables were performed via custom Statistica 10® (StatSoft Inc., Tulsa, OK, USA), SigmaPlot 10.0® (Systat Inc., Santa Clara, CA, USA), CorelDraw 2020® (Corel Corp., Ottawa, ON, Canada) and Python scripts (version 3.11) [39].

2.10.2. Principal Component Analyses (PCA)

Principal component analysis (PCA) was applied to physiological and spectral data to reduce dimensionality and identify major groupings and response patterns to water treatments. The optimal number of principal components was determined according to the cumulative variance explained. All PCA analyses were conducted in Python, adopting a significance level of $p < 0.05$. In addition to scores and clustering, the proportion of explained variance (scree plot), spectral loadings, and regression coefficients for each principal component were extracted to determine the most informative wavelength regions. These outputs supported subsequent feature selection and modelling steps.

2.10.3. Vegetation Indices (VIs)

A comprehensive set of vegetation indices was calculated to evaluate their sensitivity in detecting differences among water regimes, foliar pigmentation, photochemical efficiency, water status, pigment content, and structural–physiological alterations in soybean leaves. The indices analysed were, in order, NDVI (normalized difference vegetation index), GNDVI (green normalized difference vegetation index), EVI (enhanced vegetation index), SAVI (soil-adjusted vegetation index), OSAVI (optimized soil-adjusted vegetation index), MSAVI2 (modified soil-adjusted vegetation index 2), SIPI (structure insensitive pigment index), PSSRc (pigment specific simple ratio—carotenoids), RARS (red-edge anthocyanin reflectance signal), WBI (water band index), MSI (moisture stress index), NDII (normalized difference infrared index), NDMI (normalized difference moisture index), NDDI (normalized difference drought index), NMDI (normalized multi-band drought index), NDWI1640 (normalized difference water index with 1640 nm band), NDWI2130 (normalized difference water index with 2130 nm), ARI1 (anthocyanin reflectance 1), ARI2 (anthocyanin reflectance 2), CRI1 (carotenoid reflectance 1), CRI2 (carotenoid reflectance 2), VOG1 (Vogelmann red edge 1), VOG2 (Vogelmann red edge 2), NPQI (normalized phaeophytinization index), and PRI (photochemical reflectance index).

The mean values for each index were compared across treatments, and their relationships with physiological parameters were assessed via Pearson's correlation coefficient. This integrated approach enabled robust discrimination of plant responses to water deficit and supported the identification of informative spectral markers for physiological and structural changes in the leaves.

2.10.4. Hierarchical and Cluster Analysis

Pairwise Euclidean distances were calculated between the mean spectral profiles of each water regime. A hierarchical clustering algorithm was applied to these distances to assess the similarity and grouping of treatments. The resulting Euclidean distance matrix was visualized as a heatmap, and a corresponding dendrogram was constructed to illustrate hierarchical relationships among treatments. All calculations and visualizations were performed in Python.

2.10.5. Machine Learning Models

Predictive classification models were developed to discriminate among water regimes via hyperspectral reflectance data. The algorithms employed were support vector machine (SVM), random forest (RF), k-nearest neighbor (KNN), naive Bayes (NB), decision tree (DT), logistic regression (LR), gradient boosting (GBoost), and multilayer perceptron (MLP classifier). All the data were split such that 60% of the samples were used for model training (calibration), 40% were used for internal validation, and an independent external set was reserved for model testing. The training and test groups were defined on the basis of both random sampling and hierarchical clustering outcomes to ensure representativeness across treatments.

Model performance was evaluated via the external test set, with efficiency defined as models achieving training accuracy above 95%. The performance metrics included the confusion matrix, accuracy, precision, recall (sensitivity), and F1 score for each class (i.e., water regime). The confusion matrices of all the models are presented for direct visual comparison of classification performance and error patterns across all the treatments.

2.10.6. Correlation and Heatmap Analyses

Pearson's correlation coefficients ($p < 0.001$) were calculated for all the physiological, biochemical, structural, and spectral variables. For each trait, a correlation matrix was generated with all individual wavelengths, and the results were visualized as heatmaps, with color scales representing the strength and direction of the correlation (from -1 to $+1$). Additionally, a second heatmap was produced to visualize the correlation matrix among all measured variables, allowing for an integrated assessment of relationships between biochemical, physiological, and structural traits.

2.10.7. Selection of Responsive Spectral Bands

To identify the most responsive wavelengths for each physiological and biochemical parameter, multiple variable selection algorithms were applied via custom Python scripts. The following methods were implemented: partial least squares regression (PLSR), variable importance in projection (VIP), interval partial least squares (iPLS-VIP), the genetic algorithm (GA), random forest (RF), competitive adaptive repeated sampling (CARS), Boruta, Lasso, mutual information, recursive feature elimination (RFE), and linear discriminant analysis (LDA). For each variable, the top 50 wavelengths were selected on the basis of the importance ranking or selection criterion of each algorithm. The convergence and specificity of selected bands across algorithms are visualized as dot plots, where wavelength positions (x-axis) are grouped by algorithm (y-axis). This integrative approach enabled robust identification of the spectral regions most informative for each target trait.

2.10.8. Partial Least Squares Regression (PLSR)

Partial least squares regression (PLSR) was used to predict physiological, biochemical, and agronomic parameters from the hyperspectral reflectance data. For each variable, the spectral and target data were synchronized and standardized via z score normalization prior to model fitting. The datasets were randomly split into calibration (training) and validation (testing) sets, typically using 60% of the samples for model calibration and 40% for validation; the exact proportion was user-defined.

The optimal number of PLSR components was selected by the user on the basis of data dimensionality and prediction stability. The PLSR models were fitted via the NIPALS algorithm. For each target variable, model performance was evaluated in both the training and testing sets by calculating the coefficient of determination (R^2), root mean square error (RMSE), mean absolute error (MAE), and prediction bias. Additionally, linear regression analysis between the observed and predicted values was performed to estimate the slope and intercept of the regression line. All observed and predicted values, together with the minimum and maximum ranges, were recorded.

Model predictions and metrics were visualized in scatter plots of observed versus predicted values for each variable, highlighting the 1:1 line, fitted regression, and color-coded gradients by the observed variable.

2.10.9. Analysis of Optimized Hyperspectral Vegetation Indices

To identify the most informative spectral regions for each physiological and biochemical parameter, all possible combinations of two wavelengths within the hyperspectral reflectance range were evaluated via a generalized normalized difference vegetation index (NDVI) formula:

$$\mathrm{HVI}(\lambda_1, \lambda_2) = \frac{R(\lambda_1) - R(\lambda_2)}{R(\lambda_1) + R(\lambda_2)}$$

where $R(\lambda)$ is the standardized reflectance at wavelength λ.

For each pairwise combination of wavelengths, the resulting HVI was calculated across all samples and correlated with the target variable via the coefficient of determination (R^2). This generated a two-dimensional spectral map (contour map) of R^2 values, highlighting spectral regions where the NDVI-like index was most predictive of the trait. All calculations were performed via custom Python scripts. The resulting R^2 maps were visualized for each variable, with the colour scale representing the strength of the correlation and the axes corresponding to λ_1 and λ_2. This approach enabled the identification of hyperspectral index combinations most responsive to variations in physiological, biochemical, and structural leaf parameters (Figure 1 and Table 1).

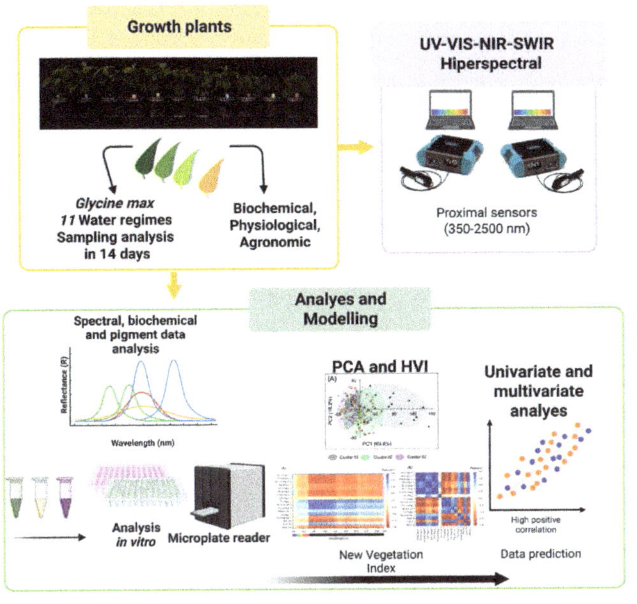

Figure 1. Experimental workflow for the assessment of the biochemical, physiological, and spectral responses of *Glycine max* under eleven water regimes. The diagram illustrates the experimental pipeline, including the cultivation of soybean plants under eleven distinct water regimes with sampling every 14 days. Biochemical, physiological, and agronomic parameters were analysed in parallel with hyperspectral data acquisition (UV–VIS–NIR–SWIR range, 350–2500 nm) via proximal sensors. Spectral, biochemical, and pigment analyses were performed both in vitro and through multivariate modelling approaches, including principal component analysis (PCA), hyperspectral vegetation indices (HVI), selected most responsive wavelengths and prediction models. Univariate and multivariate analyses were applied to predict and interpret the relationships between variables and to discriminate among treatments.

Table 1. Narrowband vegetation indices calculated from the leaf spectral reflectance.

Vegetation Index	Formula	Reference
NDVI	$NDVI = \frac{R_{NIR} - R_{Red}}{R_{NIR} + R_{Red}}$	[40]
GNDVI	$GNDVI = \frac{R_{NIR} - R_{Green}}{R_{NIR} + R_{Green}}$	[41]
EVI	$EVI = 2.5 \times \frac{R_{NIR} - R_{Red}}{R_{NIR} + 6\,R_{Red} - 7.5\,R_{Blue} + 1}$	[42]
SAVI	$SAVI = (1 + L)\,\frac{R_{NIR} - R_{Red}}{R_{NIR} + R_{Red} + L}\ (L = 0.5)$	[42]
OSAVI	$OSAVI = \frac{R_{NIR} - R_{Red}}{R_{NIR} + R_{Red} + 0.16}$	[42]
MSAVI2	$MSAVI2 = \frac{2\,R_{NIR} + 1 - \sqrt{(2\,R_{NIR} + 1)^2 - 8\,(R_{NIR} - R_{Red})}}{2}$	[43]
SIPI	$SIPI = \frac{R_{800} - R_{445}}{R_{800} - R_{680}}$	[44]
PSSRc	$PSSRc = \frac{R_{800}}{R_{470}}$	[44,45]
RARS	$RARS = \frac{R_{675}}{R_{700}}$	[44]
WBI	$WBI = \frac{R_{900}}{R_{970}}$	[46]
MSI	$MSI = \frac{R_{1600}}{R_{820}}$	[47]
NDII	$NDII = \frac{R_{819} - R_{1649}}{R_{819} + R_{1649}}$	[46]
NDMI	$NMDI = \frac{R_{860} - (R_{1640} - R_{2130})}{R_{860} + (R_{1640} - R_{2130})}$	[48]

Table 1. *Cont.*

Vegetation Index	Formula	Reference
NDDI	$NDDI = \frac{\text{NDVI}-\text{NDWI}}{\text{NDVI}+\text{NDWI}}$	[49]
NMDI	$NMDI = \frac{R_{860}-(R_{1640}-R_{2130})}{R_{860}+(R_{1640}-R_{2130})}$	[50]
NDWI1640	$NDWI_{1640} = \frac{R_{858}-R_{1640}}{R_{858}+R_{1640}}$	[50]
NDWI2130	$NDWI_{2130} = \frac{R_{858}-R_{2130}}{R_{858}+R_{2130}}$	[50]
ARI1	$ARI1 = \frac{1}{R_{550}} - \frac{1}{R_{700}}$	[51]
ARI2	$ARI2 = R_{800} \times \left(\frac{1}{R_{550}} - \frac{1}{R_{700}}\right)$	[51]
CRI1	$CRI1 = \frac{1}{R_{510}} - \frac{1}{R_{550}}$	[50]
CRI2	$CRI2 = \frac{1}{R_{510}} - \frac{1}{R_{700}}$	[50]
VOG1	$VOG1 = \frac{R_{740}}{R_{720}}$	[52]
VOG2	$VOG2 = \frac{R_{734}-R_{747}}{R_{715}+R_{726}}$	[52]
NPQI	$NPQI = \frac{R_{415}-R_{435}}{R_{415}+R_{435}}$	[53]
PRI	$\text{PRI} = \frac{\text{R}_{531}-\text{R}_{570}}{\text{R}_{531}+\text{R}_{570}}$	[54]

3. Results

3.1. Photosynthetic and Protective Pigments, Stress Markers, and Leaf Biochemical Parameters

The mean concentrations of photosynthetic pigments expressed per unit area were 391.51 mg m^{-2} for chlorophyll a, 187.01 mg m^{-2} for chlorophyll b, 578.52 mg m^{-2} for total chlorophyll (a + b), and 69.10 mg m^{-2} for carotenoids. The coefficients of variation ranged from 42.29% to 67.05%. When expressed per unit mass, chlorophyll a had a mean of 21.18 mg g^{-1}, chlorophyll b 9.49 mg g^{-1}, total chlorophyll (a + b) 30.66 mg g^{-1}, and carotenoids 3.86 mg g^{-1}, with coefficients of variation ranging from 28.35% to 55.00%.

For protective compounds, the mean values observed were 42.26 mg g^{-1} for flavonoids (mass basis), 67.91 nmol cm^{-2} for flavonoids (area basis), 23.38 µmol g^{-1} for proline, and 135.88 mL cm^{-2} for phenolic compounds. The coefficients of variation for these compounds ranged from 22.02% to 44.59%.

The analysis of the stress markers revealed a mean lignin concentration of 27.53 mg g^{-1}, cellulose content of 103.92 nmol mg^{-1}, antioxidant activity (RSA) of 64.75%, electrolyte leakage (ELK) of 39.70%, and relative water content (RWC) of 70.46%. The coefficients of variation for these variables varied between 16.54% and 37.98%. The distribution of values (minimum, median, and maximum) for each parameter is detailed in Table 2.

3.2. Spectral Reflectance Profiles Under Water Regimes

The mean reflectance spectra (350–2500 nm) of fully expanded *Glycine max* leaves exhibited distinct patterns across the eleven water regime treatments (W100 to W0). In the VIS region (350–700 nm), all the treatments presented low reflectance values, with a clear separation between the regimes: leaves under relatively high water availability (W100, W90, W80) consistently presented relatively low reflectance, whereas those under relatively severe water restriction (notably W30, W20, W10, and especially W0) presented a progressive increase in reflectance, particularly near the red edge (approximately 700 nm).

In the NIR region (700–1350 nm), all the treatments demonstrated a pronounced increase in reflectance. Compared with those under greater water deficit, the treatments with greater water availability (W100 to W60) maintained higher NIR reflectance values, with W0 consistently presenting the lowest NIR reflectance values.

Table 2. Descriptive statistics for physiological and biochemical parameters assessed in leaf samples. The table presents the mean, median, minimum, maximum, and coefficient of variation (CV, %) for each parameter measured across 264 samples. Photosynthetic pigments are expressed both per area (mg m^{-2} for chlorophyll a, chlorophyll b, total chlorophyll (a + b), and carotenoids) and per dry mass (mg g^{-1}). Protective compounds include flavonoids (mg g^{-1} and nmol cm^{-2}), proline (μmol g^{-1}), and phenolic compounds (mL cm^{-2}). The stress markers included lignin (mg g^{-1}), cellulose (nmol mg^{-1}), radical scavenging activity (RSA, %), electrolyte leakage (ELK, %), and relative water content (RWC, %). The CV indicates the degree of variability for each parameter within the sample set.

Physiological Groups	Parameters	Count (n)	Mean	Median	Min	Max	CV (%)
Photosynthetic pigments (area)	Chl a (mg m^{-2})	264	391.51	428.75	75.72	680.70	42.29
	Chl b (mg m^{-2})	264	187.01	187.92	3.77	484.47	67.05
	Chl a + b (mg m^{-2})	264	578.52	662.99	84.90	1145.01	47.20
	Car (mg m^{-2})	264	69.10	60.53	13.91	180.98	55.78
Photosynthetic pigments (mass)	Chl a (mg g^{-1})	264	21.18	21.21	5.45	36.48	28.35
	Chl b (mg g^{-1})	264	9.49	10.11	0.29	22.96	55.00
	Chl a + b (mg g^{-1})	264	30.66	33.06	6.11	58.18	31.49
	Car (mg g^{-1})	264	3.86	3.56	0.60	9.89	48.95
Protective compounds	Flv (mg g^{-1})	264	42.26	36.54	15.14	105.20	44.59
	Flv (nmol cm^{-2})	264	67.91	67.55	37.43	109.67	22.02
	Pro (umol g^{-1})	264	23.38	24.16	4.93	43.32	37.98
	Phe (mL cm^{-2})	264	135.88	130.45	68.71	238.30	27.93
Stress markers	Lig (mg g^{-1})	264	27.53	27.11	10.55	49.20	26.63
	Cel (nmol mg^{-1})	264	103.92	103.98	64.17	144.28	16.54
	RSA (%)	264	64.75	66.38	37.62	83.48	17.39
	ELK (%)	264	39.70	41.55	21.82	53.12	20.62
	RWC (%)	264	70.46	68.69	38.20	108.52	21.68

In the SWIR region (1350–2500 nm), two marked absorption features were observed (SWIR1 and SWIR2), with all the treatments resulting in a reduction in reflectance at these wavelengths. The separation among water regimes was most apparent in this domain: the W0 treatment presented the highest reflectance values across SWIR1 and SWIR2, whereas intermediate treatments followed a gradation corresponding to the severity of water deficit but did not progressively change.

Statistically significant differences in reflectance were detected among all the treatments throughout the entire spectral range ($F = 26.97$, $p < 0.001$; $n = 24$), as presented in Figure 2.

3.3. Principal Component Analysis of Leaf Reflectance

Principal component analysis (PCA) of the leaf reflectance spectra of *Glycine max* under different water regimes revealed clear separation among the treatments. The first two principal components explained 69.8% (PC1) and 18.2% (PC2) of the total variance, respectively. In the biplot (Figure 3A), three main clusters were identified, each corresponding to specific groupings of the water regime treatments.

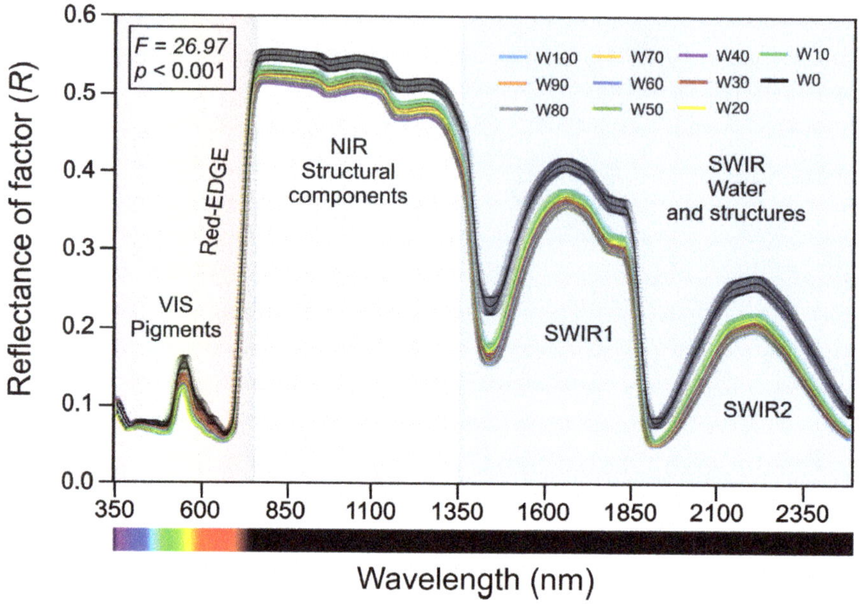

Figure 2. UV–VIS–NIR–SWIR reflectance profiles (350–2500 nm) of fully expanded *Glycine max* leaves under distinct water regimes. Reflectance spectra are shown for all the treatments, ranging from W100 to W0. The spectral domains are segmented as follows: VIS (350–700 nm; pigment absorption), NIR (700–1350 nm; structural leaf properties), and SWIR (1350–2500 nm; water and additional structural components). Significant differences between treatments were detected via one-way ANOVA ($F = 26.97$, $p < 0.001$). (Mean ± SE). (n = 24).

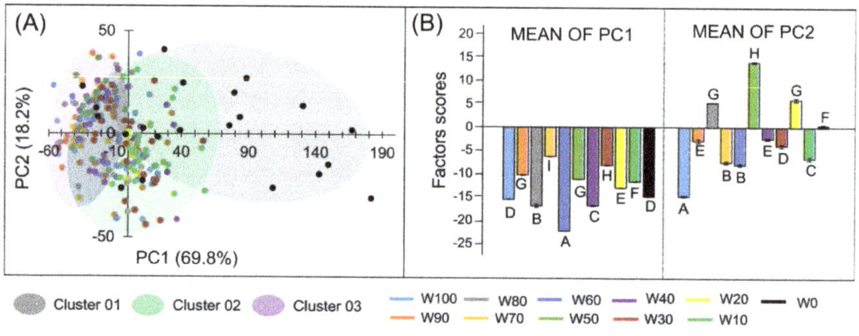

Figure 3. Principal component analysis (PCA) of leaf reflectance data from *Glycine max* under contrasting water regimes. (**A**) Biplot of the first two principal components (PC1 and PC2), explaining 69.8% and 18.2% of the total variance, respectively. Individual samples are coloured according to dot treatments, and ellipses highlight the main clusters identified by the analysis. (**B**) Mean factor scores (±SE) for each treatment along PC1 and PC2. Different uppercase letters indicate significant differences among treatments ($p < 0.05$). The treatments ranged from well-watered (W100) to severe water deficit (W0) (n = 24). The colours for each treatment are represented in the legend with the corresponding line styles.

The distribution of individual samples within the PCA space revealed that well-watered treatments (W100, W90, W80) were predominantly grouped within Cluster 1. Intermediate water regimes (W70, W60, W50, W40, W30, W20) were associated mainly with Cluster 2, whereas the most severe deficit treatments (W10, W0) were clearly separated into Cluster 03.

The mean factor scores for each treatment along PC1 and PC2 (Figure 3B) further supported these groupings. W100 and W80 presented the most negative mean values on PC1, whereas W0 and W10 presented the highest positive scores.

Principal component analysis of the hyperspectral reflectance data revealed that the first three principal components accounted for the majority of the spectral variability in the soybean leaves. PC1, PC2, and PC3 explained 69.8%, 18.2%, and 5.2% of the total variance, respectively, with the cumulative variance reaching over 93% by the third component (Figure 4A).

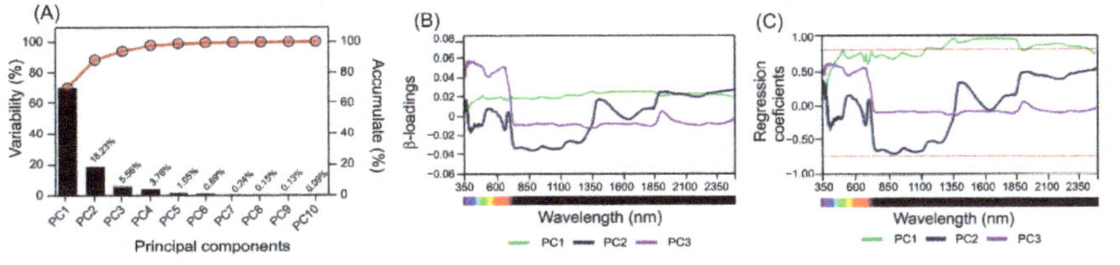

Figure 4. Principal component analysis (PCA) of hyperspectral reflectance data from soybean leaves. (**A**) Percentage of variance explained by the first ten principal components, with the cumulative variance indicated by the red line. (**B**) Standardized β-loadings for the first three principal components (PC1, PC2, PC3) as a function of wavelength, highlighting the main spectral regions contributing to variability. (**C**) Regression coefficients for the first three principal components across the spectral range (350–2500 nm), showing the wavelengths most strongly associated with each PC. The red dotted lines represent the limits of −0.75 and +0.75 limited in regression coefficients.

The standardized β-loadings for PC1, PC2, and PC3 (Figure 4B) revealed that the major spectral regions contributing to variability were distributed across the VIS, NIR, and SWIR domains. PC1 was associated primarily with broad features spanning the entire spectrum, whereas PC2 and PC3 exhibited more distinct peaks and troughs, especially in regions at approximately 600 nm, 1400 nm, and 1900 nm.

The regression coefficients for the first three principal components (Figure 4C) highlighted the wavelengths most strongly linked to each PC. Notably, the PC1 coefficients remained high across the VIS and NIR regions, whereas the PC2 and PC3 coefficients varied more markedly in the SWIR region, identifying specific wavelengths that contributed to the separation of water regimes in the PCA. All analyses were conducted using reflectance data in the 350–2500 nm range.

3.4. Variable Importance of Vegetation Indices for Leaf Trait Prediction

The evaluation of 25 spectral vegetation indices revealed substantial variation in their relative importance for predicting physiological and biochemical responses in soybean leaves across water regimes. Among all indices, CRI2 exhibited the highest variable importance, reaching approximately 35%. ARI2 and ARI1 followed, with relative importance values of approximately 16% and 12%, respectively. Other indices, such as CRI1 (8%), VOG1 (5%), and PSSRc (2%), also contributed to the predictive modelling but with a lower impact.

Conversely, the classic indices related to greenness, moisture, and general vegetation status (e.g., the NDVI, GNDVI, EVI, SAVI, OSAVI, MSI, NDMI, and PRI) all presented relative importance values below 2%. The majority of indices were below this threshold, indicating limited relevance for predicting physiological and biochemical variation under the tested water regimes (Figure 5).

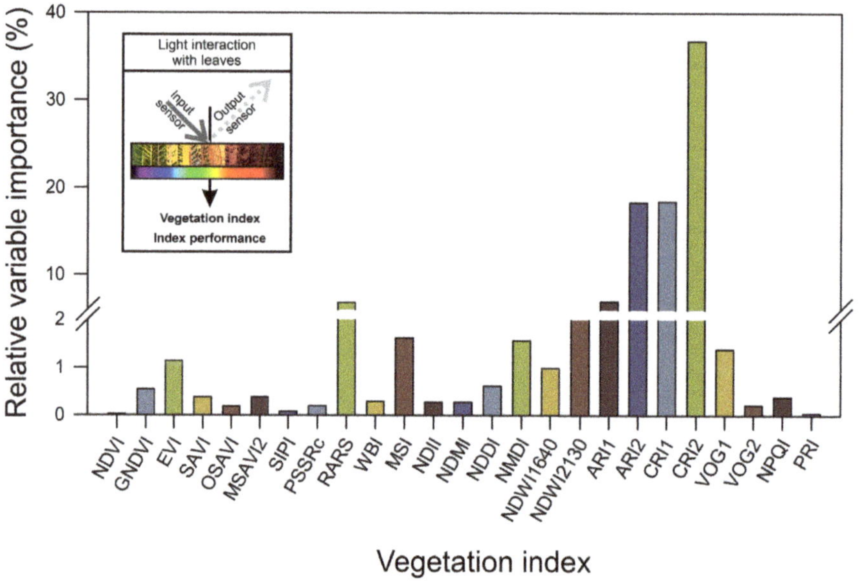

Figure 5. Relative importance of spectral vegetation indices for predicting physiological and biochemical responses in soybean leaves under different water regimes. The bar chart displays the relative variable importance (%) of 25 vegetation indices derived from hyperspectral data. The inset illustrates the principle of light interaction with leaves and the vegetation index calculation. Index abbreviations: NDVI, normalized difference vegetation index; GNDVI, green normalized difference vegetation index; EVI, enhanced vegetation index; SAVI, soil-adjusted vegetation index; OSAVI, optimized soil-adjusted vegetation index; MSAVI2, modified soil-adjusted vegetation index 2; SIPI, structure insensitive pigment index; PSSRc, pigment specific simple ratio—carotenoids; RARS, red-edge anthocyanin reflectance signal; WBI, water band index; MSI, moisture stress index; NDII, normalized difference infrared index; NDMI, normalized difference moisture index; NDDI, normalized difference drought index; NMDI, normalized multi-band drought index; NDWI1640, normalized difference water index with 1640 nm band; NDWI2130, normalized difference water index with 2130 nm; ARI1, anthocyanin reflectance 1; ARI2, anthocyanin reflectance 2; CRI1, carotenoid reflectance 1; CRI2, carotenoid reflectance 2; VOG1, Vogelmann red edge 1; VOG2, Vogelmann red edge 2; NPQI, normalized phaeophytinization index; PRI, photochemical reflectance index, shows greater contribution to the predictive modelling of leaf responses.

These results demonstrate that carotenoid and anthocyanin indices are the most informative spectral metrics for the prediction of leaf responses to water deficit in soybean (Figure 5).

3.5. Hierarchical Clustering of Spectral Profiles Under Water Regimes

Hierarchical clustering analysis based on spectral data revealed clear separation among the *Glycine max* treatments according to water availability. The Euclidean distance matrix (Figure 6A) revealed low dissimilarity between the well-watered treatments (W100, W90, W80, W70, W60), with pairwise distances ranging from 0.01–0.34. The moderate water deficit treatments (W50, W40, W30) also clustered closely together (distances < 0.5) but presented increasing divergence from both the well-watered and extreme deficit groups.

The most pronounced spectral dissimilarity was observed between W0 and all the other treatments, with Euclidean distances ranging from 1.4–2.0, and between W0 and W10, with a distance of 0.66. The dendrogram (Figure 6B) highlighted the formation of distinct clusters, with W0 forming a separate branch and W10 and W20 grouping together, clearly separated from the well-watered (W100–W60) and moderate-deficit (W50–W30) treatments. This structure reflects the spectral divergence resulting from the gradient of water restriction, with severe deficit and extreme restriction regimes forming distinct clusters.

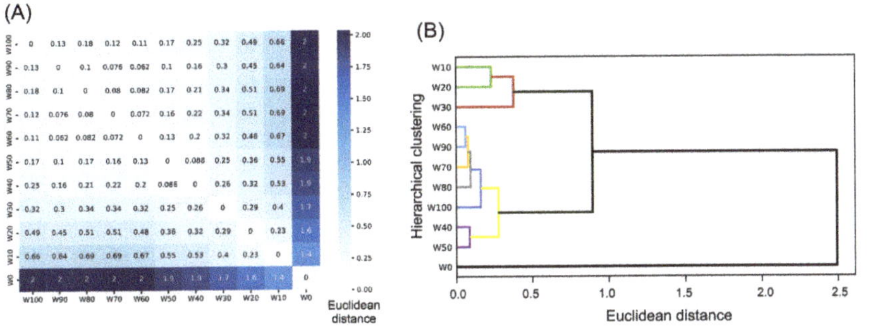

Figure 6. Hierarchical clustering of *Glycine max* treatments under different water regimes on the basis of spectral data. (**A**) Euclidean distance matrix showing pairwise distances among all water regimes (W100 to W0), with the intensity of blue shading indicating increasing dissimilarity. (**B**) Dendrogram from hierarchical clustering analysis, highlighting the formation of distinct groups among treatments. The separation reflects the spectral divergence imposed by the water availability gradient, with severe deficit (W0) and extreme restriction (W10, W20) forming distinct clusters from the well-watered and moderate-deficit groups according to the Euclidean distance. The colours in the hierarchical clustering indicate the proximity among similar treatments.

The confusion matrices in Figure 7 summarize the classification results of the eleven water regimes via eight machine learning models on the basis of hyperspectral reflectance data. The random forest, decision tree, gradient boosting, and MLP classifiers produced perfect classifications, with all samples correctly assigned to their original classes (values of 1.00 along the diagonal and zeros elsewhere). Logistic regression also achieved high accuracy, with a few misclassifications, mostly between intermediate treatments such as W80, W70, and W60.

In contrast, SVM, KNN, and naive Bayes showed increased rates of confusion, particularly among adjacent or similar water regimes. SVM and KNN presented off-diagonal values up to 0.22, indicating some overlap in prediction between regimes such as W10, W20, and W0. Naive Bayes resulted in the highest number of misclassifications, with proportions up to 0.22 for certain deficit treatments, reflecting reduced discriminatory power in more challenging scenarios.

Overall, tree-based ensemble models and neural network approaches demonstrated the highest predictive precision for distinguishing water regimes in soybean leaves, whereas distance- and probability-based models struggled to separate spectrally similar classes.

3.6. Correlation Between Spectral Data and Leaf Biochemical/Biophysical Traits

The Pearson correlation heatmap (Figure 8A) demonstrated distinct association patterns between hyperspectral reflectance (350–2500 nm) and key biochemical and biophysical leaf traits in *Glycine max*. Photosynthetic pigments, both per area and mass (Chl a, Chl b, Chl a + b, Car), exhibited strong positive correlations with reflectance in the near-infrared region (NIR; >750 nm), with r values approaching +1. Conversely, negative correlations were observed in the visible region (VIS; 350–700 nm), particularly between 600–700 nm (r values near −1). Flavonoids, proline, and total phenolics displayed the opposite trend, being positively correlated with VIS reflectance and negatively correlated with the NIR bands.

Lignin, cellulose, and some stress markers (RSA, ELK, RWC) were generally negatively correlated with reflectance across much of the spectrum. The most pronounced negative associations were found in the shortwave infrared (SWIR; >1400 nm) band, where the r values reached below −0.5, especially for RSA, ELK, and RWC, indicating a strong inverse relationship between reflectance and these parameters under varying water regimes.

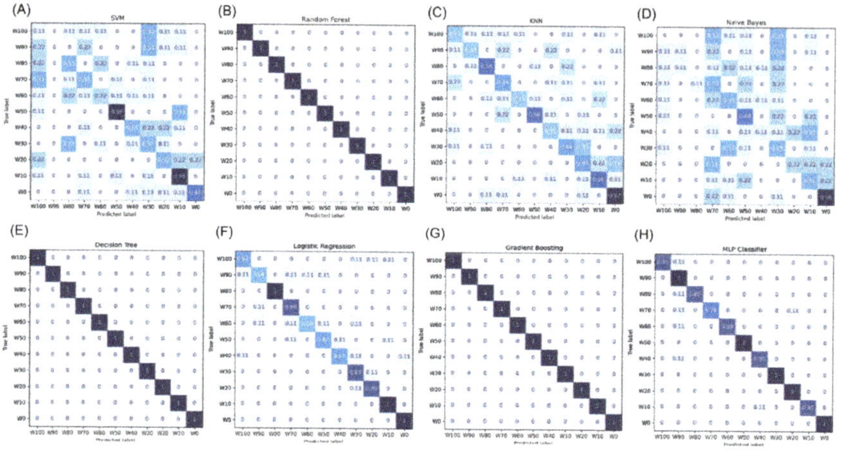

Figure 7. Confusion matrices for eight machine learning classifiers for predicting water regimes in soybean plants on the basis of hyperspectral reflectance data. The classification performance of the following models is shown: (**A**) support vector machine (SVM), (**B**) random forest, (**C**) k-nearest neighbors (KNN), (**D**) naive Bayes, (**E**) decision tree, (**F**) logistic regression, (**G**) gradient boosting, and (**H**) multilayer perceptron (MLP classifier). The predicted classes (x-axis) and true classes (y-axis) correspond to the eleven water regimes (W100 to W0). Each cell shows the proportion of correctly or incorrectly classified samples per class in increasing light to dark blue. The values in the boxes indicate correct classifications, whereas the off-diagonal entries denote misclassifications.

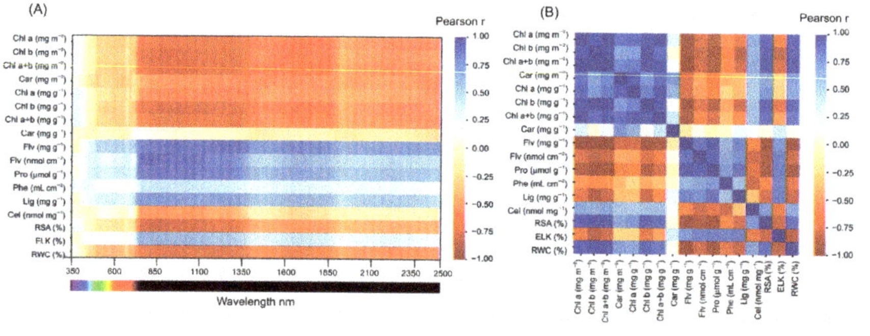

Figure 8. Pearson correlation heatmaps between leaf spectral and biochemical/biophysical traits in *Glycine max*. (**A**) Correlation matrix showing the Pearson correlation coefficient (r) between hyperspectral reflectance (350–2500 nm) and selected biochemical and biophysical variables: chlorophyll a (Chl a), chlorophyll b (Chl b), total chlorophyll (Chl a + b), carotenoids (Car), flavonoids (Flv), proline (Pro), total phenolics (Phe), lignin (Lig), cellulose (Cel), radical scavenging activity (RSA), electrolyte leakage (ELK), and relative water content (RWC), all expressed per area or mass as indicated. (**B**) Pearson correlation matrix among the same leaf traits. The color scale ranges from −1 (strong negative correlation, blue) to +1 (strong positive correlation, red), with yellow indicating near-zero correlations.

The pairwise trait correlation matrix (Figure 8B) revealed clear groupings. Photosynthetic pigments strongly positively correlated with each other (r > 0.8). Flavonoids, proline, and phenolics were also positively correlated (r > 0.5) but strongly negatively correlated with pigments (r < −0.5). Lignin and cellulose were positively correlated (r ≈ 0.6) and negatively associated with RWC (r ≈ -0.6). RSA and ELK were positively correlated, but both were negatively related to RWC.

These results highlight that the VIS and NIR spectral bands are decisive for distinguishing major biochemical components, whereas the SWIR bands are more strongly associated with stress-related traits that align structures and cell compounds. This highlights the value of full-spectrum reflectance profiling for the functional and physiological characterization of soybean leaves under contrasting water availability conditions.

3.7. Wavelength Selection for the Prediction and Classification of Leaf Traits

The selection of informative wavelengths by multiple multivariate algorithms revealed both common and variable-specific patterns across the 350–2500 nm range for all foliar traits in *Glycine max* (Figure 9). For all the parameters evaluated, the majority of the algorithms consistently identified wavelengths within the visible (VIS, 350–700 nm) and near-infrared (NIR, 700–1350 nm) regions as the most relevant for predictive modelling.

For pigments expressed per area (Chl a, Chl b, Chl a + b, Car; Figure 9A–D), the most frequently selected wavelengths were concentrated between 550 and 750 nm, spanning the red-edge transition. Pigments expressed per mass (Figure 9E–H) similarly presented selection peaks at the VIS–NIR boundary, but several algorithms also highlighted informative bands in the SWIR region (1400–1900 nm).

For secondary metabolites (Flv, Pro, Phe; Figure 9I–L) and cell wall components (Lig, Cel; Figure 9M,N), the algorithms predominantly selected wavelengths in both the VIS and SWIR regions. Notably, the SWIR region (between 1450–1900 nm and >2100 nm) was highly represented in the selection for phenolics, lignin, and cellulose.

For the stress and physiological indices (RSA, ELK, RWC; Figure 9O–Q), informative wavelengths were distributed across the spectrum, with mutual importance placed on the SWIR region (particularly around water absorption features at 1450 nm, 1940 nm, and 2200 nm). The random forest, Boruta, and genetic algorithm approaches often select broader ranges, whereas the LDA and VIP methods consistently highlight more discrete bands.

The repeated selection of red-edge (between 700–740 nm) and SWIR bands (especially 1450, 1940, and 2200 nm) by distinct algorithms reinforces their mechanistic link to pigment absorption and water status, providing strong candidates for compact multispectral sensor design and trait-specific remote phenotyping (Figure 9A–Q).

3.8. Predictive Modelling Using Hyperspectral Reflectance

The partial least squares regression (PLSR) models demonstrated robust predictive performance for multiple foliar traits in *Glycine max* on the basis of the hyperspectral reflectance data (Figures 10 and 11).

For the prediction of foliar pigment concentrations (Figure 10), high coefficients of determination (R^2) were observed for both the training and test sets. The models for chlorophyll a, chlorophyll b, total chlorophyll, and carotenoids (base area) achieved R^2 values ranging from 0.74–0.88, with low bias and a root mean square error (RMSE) typically less than 100. Pigments expressed per mass also showed satisfactory predictive capacity, with R^2 values between 0.44 and 0.67 for chlorophyll a, chlorophyll b, total chlorophyll, and carotenoids. The regression slopes for all pigment models were close to unity, and the MAE values remained low across the range of observed concentrations.

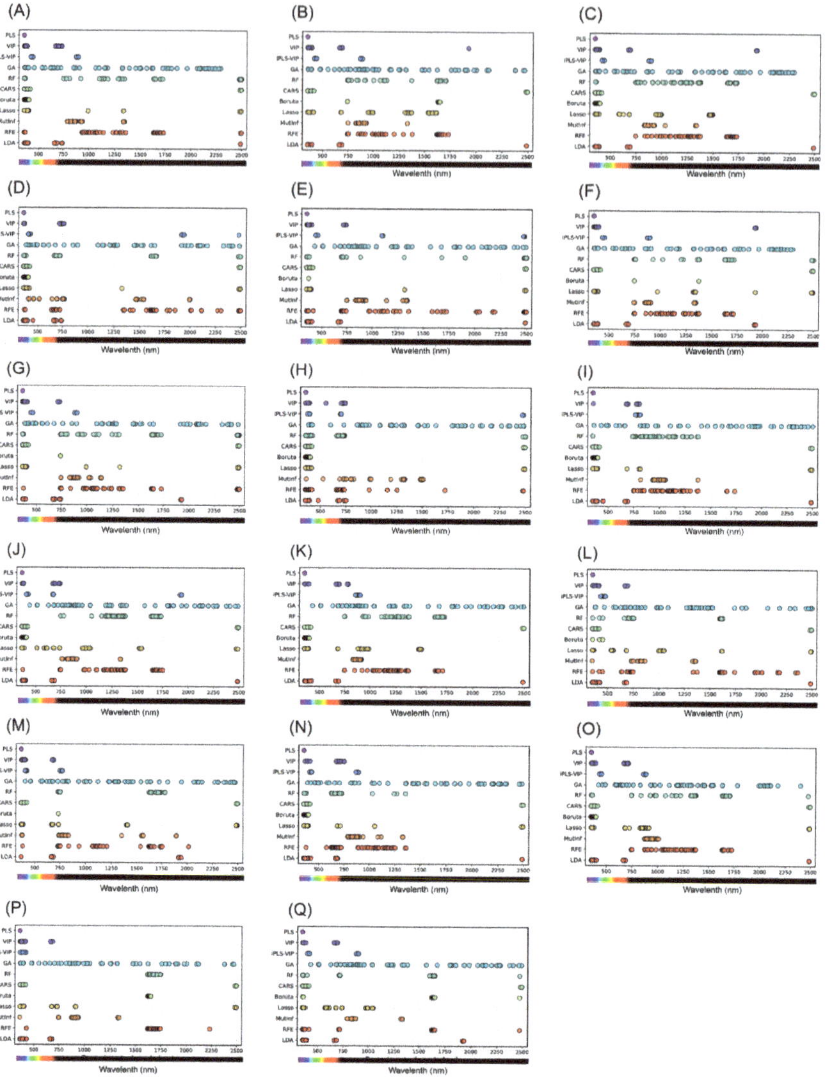

Figure 9. Wavelengths selected by different algorithms for the prediction and classification of foliar traits in *Glycine max*. (**A**–**Q**) represent the most informative wavelengths identified for each variable (in order): (**A**) chlorophyll a (Chl a, mg m^{-2}), (**B**) chlorophyll b (Chl b, mg m^{-2}), (**C**) total chlorophyll (Chl a + b, mg m^{-2}), (**D**) carotenoids (Car, mg m^{-2}), (**E**) chlorophyll a (mg g^{-1}), (**F**) chlorophyll b (mg g^{-1}), (**G**) total chlorophyll (mg g^{-1}), (**H**) carotenoids (mg g^{-1}), (**I**) flavonoids (Flv, mg g^{-1}), (**J**) flavonoids (nmol cm^{-2}), (**K**) proline (Pro, µmol g^{-1}), (**L**) phenolics (Phe, mL cm^{-2}), (**M**) lignin (Lig, mg g^{-1}), (**N**) cellulose (Cel, nmol mg^{-1}), (**O**) radical scavenging activity (RSA, %), (**P**) electrolyte leakage (ELK, %), (**Q**) relative water content (RWC, %). Each dot color indicates a selected wavelength for the respective algorithm. The y-axis lists the algorithms used by each color: PLS (Partial Least Squares), VIP (Variable Importance in Projection), iPLS-VIP (Interval Partial Least Squares), GA (Genetic Algorithm), RF (Random Forest), CARS (Competitive Adaptive Repeated Sampling), Boruta, Lasso, MutInf (Mutual Information), RFE (Recursive Feature Elimination), LDA (Linear Discriminant Analysis).

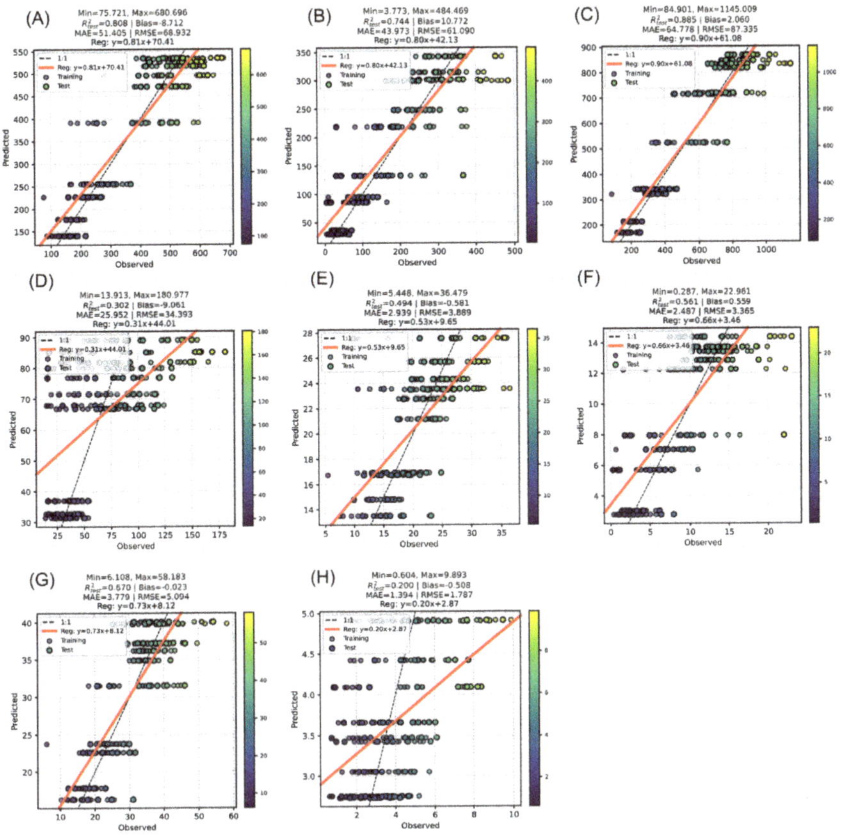

Figure 10. Performance of PLSR models for predicting foliar pigment concentrations in *Glycine max* using hyperspectral reflectance data. Scatter plots of observed versus predicted values for (**A**) chlorophyll a (Chl a, mg m^{-2}), (**B**) chlorophyll b (Chl b, mg m^{-2}), (**C**) total chlorophyll (Chl a + b, mg m^{-2}), (**D**) carotenoids (Car, mg m^{-2}), (**E**) chlorophyll a (mg g^{-1}), (**F**) chlorophyll b (mg g^{-1}), (**G**) total chlorophyll (mg g^{-1}), and (**H**) carotenoids (mg g^{-1}). The training and test sets are indicated, with the red line representing the 1:1 relationship. Each image shows the regression equation, coefficient of determination (R^2), bias, mean absolute error (MAE), and root mean square error (RMSE) for both sets. The color scale denotes the sample density within each observed range.

For the biochemical and physiological parameters (Figure 11), the PLSR models yielded moderate to high R^2 values for most variables. The prediction of flavonoids (mg g^{-1}) reached an R^2 of 0.72, with a low bias and RMSE of 4.52. Flavonoids (nmol cm^{-2}) and proline (μmol g^{-1}) achieved R^2 values of 0.36 and 0.15, respectively, reflecting a lower predictive strength for these variables. The models for phenolics, lignin, and cellulose presented R^2 values between 0.31 and 0.96, with the best performance obtained for cellulose (R^2 = 0.96, RMSE = 4.83). The radical scavenging activity (RSA) was also well predicted, with R^2 = 0.81 and RMSE = 3.01.

For all the traits, scatter plots of the observed versus predicted values indicated close clustering along the 1:1 line in the best-performing models, particularly for pigments and cellulose. The color density scale confirmed good model generalizability across the observed trait ranges in both the training and test sets.

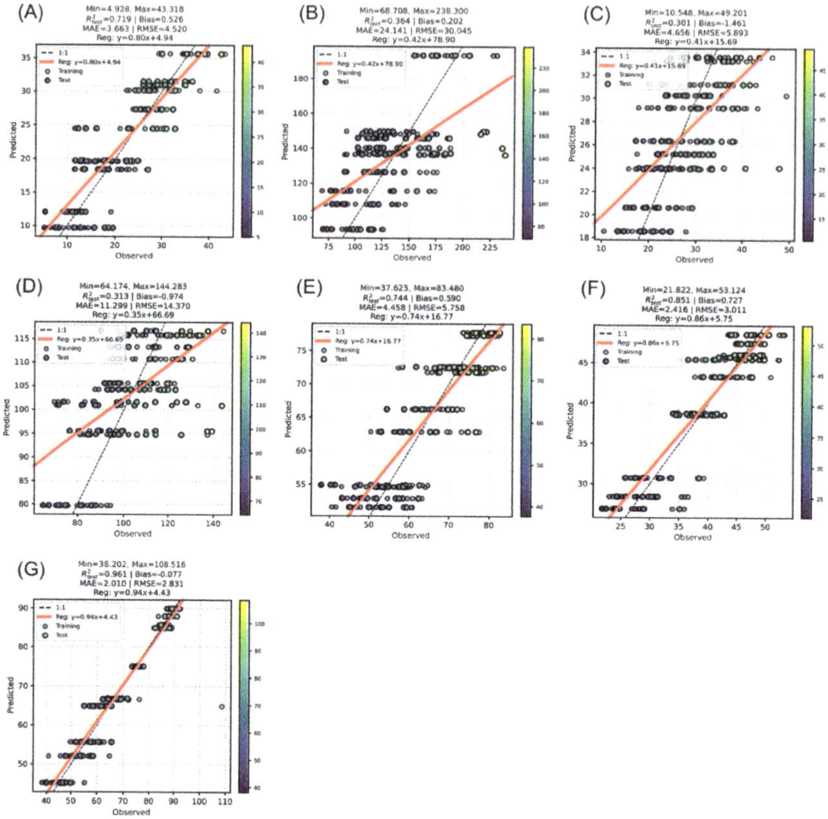

Figure 11. Predictive performance of PLSR models for biochemical and physiological traits in *Glycine max* leaves using hyperspectral reflectance data. Scatter plots of observed versus predicted values for (**A**) flavonoids (Flv, mg g^{-1}), (**B**) flavonoids (nmol cm^{-2}), (**C**) proline (Pro, μmol g^{-1}), (**D**) total phenolics (Phe, mL cm^{-2}), (**E**) lignin (Lig, mg g^{-1}), (**F**) cellulose (Cel, nmol mg^{-1}), and (**G**) radical scavenging activity (RSA, %). The training and test sets are indicated, with the red line representing the 1:1 relationship. For each trait, the regression equation, coefficient of determination (R^2), bias, mean absolute error (MAE), and root mean square error (RMSE) are reported. The color bar represents the sample density across the observed range.

3.9. Identification of the Most Responsive Wavelength Pairs via Spectral Correlation Analysis

The spectral correlation (R^2) maps generated via the hyperspectral vegetation index (HVI) method revealed the wavelength combinations most strongly associated with the prediction of each foliar biochemical and physiological trait in *Glycine max* (Figure 12).

Across all the traits, the majority of the wavelength pairs produced low R^2 values (dark blue), indicating weak or negligible predictive power for most combinations within the 350–2500 nm range. However, each map displays discrete clusters of yellow to red points, representing pairs of wavelengths with notably high predictive capacity (R^2 approaching or exceeding 0.75).

For chlorophyll-related traits (Figure 12A–C,E–G), the highest R^2 values were observed in combinations involving the red-edge (approximately 700–750 nm) and NIR bands (750–900 nm), often in association with SWIR wavelengths near 1400 nm or 1900 nm. These hotspots indicate regions where the synergy between pigment absorption features and water/structural absorption results in optimal sensitivity for model prediction.

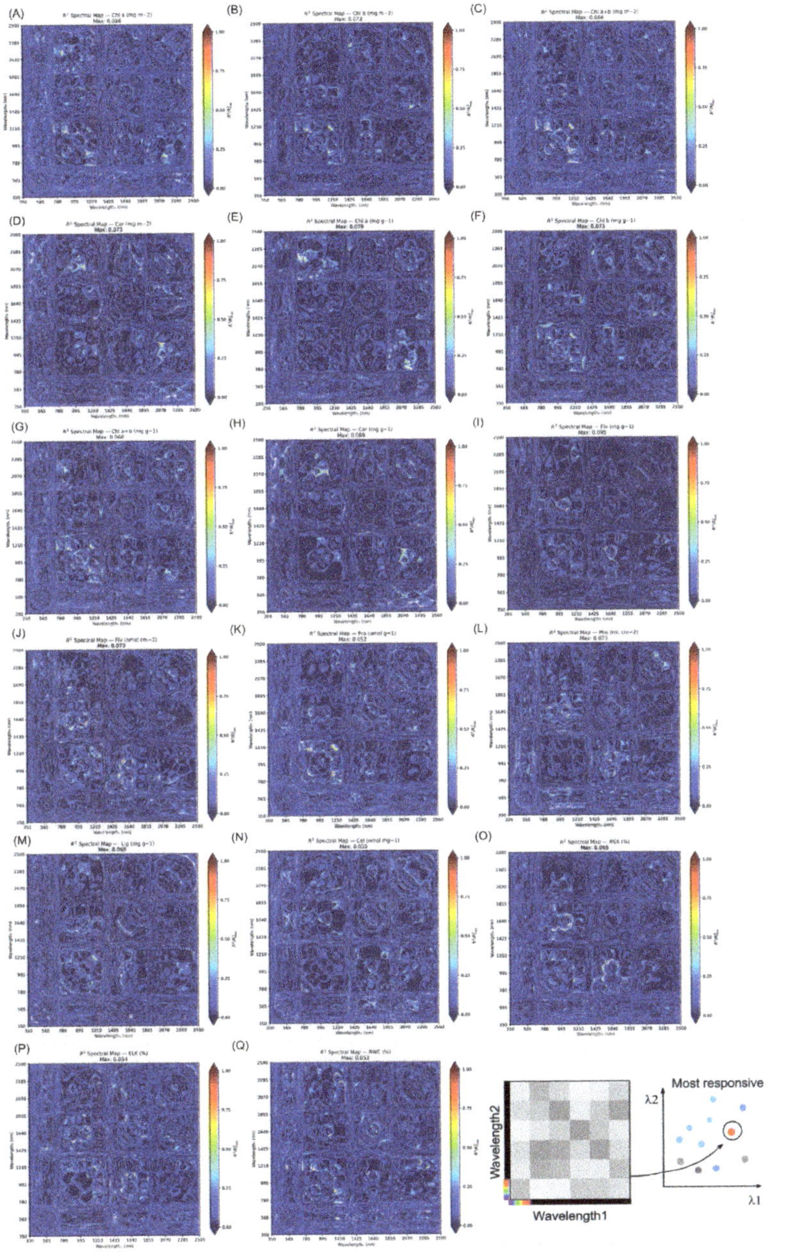

Figure 12. Spectral correlation (R^2) maps for all pairwise wavelength combinations via the hyperspectral vegetation index (HVI) method for the prediction of foliar biochemical and physiological traits in *Glycine max*. (**A–Q**) R^2 maps for (**A**) chlorophyll a (Chl a, mg m^{-2}), (**B**) chlorophyll b (Chl b, mg m^{-2}), (**C**) total chlorophyll (Chl a + b, mg m^{-2}), (**D**) carotenoids (Car, mg m^{-2}), (**E**) chlorophyll a (mg g^{-1}), (**F**) chlorophyll b (mg g^{-1}), (**G**) total chlorophyll (mg g^{-1}), (**H**) carotenoids (mg g^{-1}), (**I**) flavonoids (Flv, mg g^{-1}), (**J**) flavonoids (nmol cm^{-2}), (**K**) proline (Pro, µmol g^{-1}), (**L**) phenolics (Phe, mL cm^{-2}), (**M**) lignin (Lig, mg g^{-1}), (**N**) cellulose (Cel, nmol mg^{-1}), (**O**) radical scavenging activity (RSA, %), (**P**) electrolyte leakage (ELK, %), (**Q**) relative water content (RWC, %). The bottom right schematic illustrates the approach used to identify the most responsive wavelength pairs via the HVI algorithm (red points).

Carotenoids (Figure 12D,H) and secondary metabolites such as flavonoids, proline, and phenolics (Figure 12I–L) also presented the most responsive points between the VIS–NIR boundary and specific SWIR intervals (especially at approximately 1450, 1940, and 2200 nm), with the highest R^2 clusters scattered but generally located along these axes.

For cell wall components (lignin and cellulose; Figure 12M,N) and physiological stress markers (RSA, ELK, RWC; Figure 12O–Q), the strongest correlations were detected between pairs of SWIR wavelengths. Particularly those spanning the water absorption regions near 1450, 1940, and 2200 nm.

These responsive points, where R^2 approaches 1.0, represent the wavelength pairs with the greatest sensitivity for the nondestructive prediction of each respective trait. The schematic in the lower right of Figure 12 illustrates the HVI approach, highlighting how only a minority of wavelength combinations yield maximal prediction power, reflected as isolated yellow/red points within the correlation landscape (Figure 12).

In summary HVI, shows the R^2 correlation maps not only confirm the known associations between certain spectral regions and foliar biochemistry but also visually reveal the unique, trait-specific wavelength pairs driving predictive performance. This approach highlights the potential of hyperspectral analysis not just for nondestructive trait estimation, but also for advancing our understanding of the optical signatures underlying plant physiological status. Such findings pave the way for the development of optimized indices and the rational design of next-generation remote sensing tools tailored to key functional traits in soybean and other crops.

4. Discussion

4.1. Overview of Key Findings

This study provides compelling evidence of the significant impacts of various water regimes on soybean (*Glycine max*) leaf physiology, biochemical composition, and hyperspectral reflectance. As anticipated, water stress led to a progressive decrease in photosynthetic pigments, specifically chlorophyll (Chl a and Chl b) and carotenoids, and an increase in protective compounds such as flavonoids (Flv) and proline (Pro), which are commonly associated with stress tolerance mechanisms [8,55]. Furthermore, spectral reflectance data revealed distinct patterns associated with different water treatments, with notable shifts in the visible (VIS), near-infrared (NIR), and shortwave infrared (SWIR) regions, reflecting physiological changes in response to water availability [8,30,56].

Recent studies corroborate our findings, highlighting that drought stress significantly reduces the contents of photosynthetic pigments such as chlorophyll and carotenoids in soybean plants. For example, a study by Nehra et al., (2025) [57] reported a decrease in chlorophyll content under drought stress, attributed to increased reactive oxygen species (ROS) production leading to oxidative damage and chlorophyll degradation. Similarly, carotenoids, which play crucial roles in photoprotection and antioxidant defense, are also affected by water stress. Their reduction under stress limits nonphotochemical quenching and impairs their ability to scavenge ROS [58,59]. In our study, the reduction in chlorophyll and carotenoid contents under water stress conditions aligns with these reports, indicating a compromised photosynthetic capacity and increased susceptibility to oxidative damage [60,61].

The accumulation of protective compounds such as flavonoids and proline under water stress is a well-documented response in plants [62,63]. Flavonoids, known for their antioxidant properties, help mitigate oxidative stress by scavenging free radicals. Proline acts as an osmoprotectant, stabilizing cellular structures and maintaining cellular functions under stress conditions [64,65]. The elevated levels of these compounds under water stress coincide with SWIR reflectance changes, highlighting their mechanistic role in osmotic adjustment and antioxidative defense, which together increase stress tolerance in soybean (Figures 1–12).

Hyperspectral reflectance analysis has emerged as a powerful tool for monitoring plant stress responses [32]. In our study, distinct shifts in reflectance patterns were observed across the VIS, NIR, and SWIR regions under different water regimes. These spectral changes are indicative of physiological alterations in plants, such as changes in pigment content and water status caused by changes in the thylakoid membrane. For example, a study by Almakas et al., (2025) [66] revealed that cold stress (similarly water or drought stress) in soybean led to significant changes in spectral reflectance, reflecting alterations in chlorophyll fluorescence and pigment composition. Similarly, our findings of spectral shifts under water stress conditions suggest that hyperspectral reflectance can serve as a noninvasive and efficient method for the early detection of stress in soybean plants [67].

The integration of physiological, biochemical, and hyperspectral data provides a comprehensive approach for understanding plant responses to water stress [2,66]. By combining these datasets, we can develop predictive models to assess plant health and stress levels more accurately. For example, machine learning algorithms can be trained on hyperspectral data to classify stress levels on the basis of spectral features, enabling real-time monitoring of crop health [20,21]. Moreover, the identification of specific spectral bands associated with stress indicators can aid in the development of targeted remote sensing tools for precision agriculture.

Therefore, spectral reflectance shifts are a direct consequence of pigment degradation and metabolic adjustment under water stress, highlighting how hyperspectral sensors capture underlying physiological mechanisms noninvasively [8,68,69]. This mechanistic understanding supports both breeding for drought tolerance and improved management of water-limited systems.

4.2. Principal Component Analysis and Model Performance

Principal component analysis (PCA) of the hyperspectral reflectance data effectively revealed spectral variability across eleven water treatments in soybean (*Glycine max*). The first two principal components accounted for 88% of the variance, clearly distinguishing treatments on the basis of water availability. This aligns with findings by Furlanetto et al. (2024) [70], who utilized PCA to classify potassium deficiency in soybean plants, achieving 100% variance across various developmental stages and seasons.

The separation of the well-watered treatments from those under intermediate and severe water deficit conditions underscores the utility of PCA in capturing water stress-induced spectral changes. This approach has been corroborated by studies [28,47], which applied PCA to hyperspectral data to monitor plant stress responses, demonstrating its efficacy in classifying varying stress levels.

With respect to model performance, machine learning algorithms such as random forest (RF), gradient boosting, and multilayer perceptron (MLP) have achieved perfect classification accuracy in predicting water regimes on the basis of hyperspectral data. These ensemble methods effectively handle complex, high-dimensional data, as supported by research by Furlanetto et al. (2020) [70], who reported high classification accuracies when RF was used to identify potassium deficiency in soybean.

Conversely, simpler models such as support vector machines (SVMs) and k-nearest neighbors (KNNs) struggled with misclassifications, particularly between regimes with similar spectral signatures. This highlights a fundamental issue in the use of traditional models. For example, while they are computationally efficient, they lack the capacity to capture the complex interactions between multiple spectral features. These models are limited by their inherent assumption of linearity or proximity-based classification, which often fails when the spectral data are highly variable and multidimensional [71–74].

Therefore, while PCA serves as a robust tool for dimensionality reduction and visualization of spectral data, the integration of advanced machine learning algorithms enhances the predictive accuracy for water stress classification in soybean. The combination of PCA and ensemble learning models offers a powerful framework for the nondestructive assessment of plant physiological responses to varying water regimes, facilitating informed decision-making in precision agriculture to reduce water stress in plants [68,75,76].

4.3. Correlation Analysis and Implications for Spectral Indices

The selection of the most responsive wavelengths identified specific wavelengths within the VIS–NIR and SWIR regions as pivotal for predicting soybean leaf physiological traits under varying water regimes. Notably, wavelengths of approximately 1450 nm, 1940 nm, and 2200 nm have emerged as critical for assessing water status and stress markers. This aligns with findings of Wijewardana et al. (2019) [77], as SWIR reflectance specifically tracks changes in leaf water content and cell wall plasticity, reflecting the initial molecular events that precede irreversible tissue damage.

The convergence of multiple variable selection algorithms, partial least squares regression (PLSR), random forest (RF), and variable importance in projection (VIP), on these spectral regions underscores their robustness for the nondestructive assessment of soybean leaf physiology [20,21]. For example, PLSR has been effectively utilized to model leaf physiological responses to water stress, as demonstrated in studies by Wijewardana et al. (2019) [77].

Interestingly, our study also revealed correlations between spectral reflectance and biochemical markers of stress, such as proline and flavonoids [78]. The accumulation of these compounds, known for their roles in osmotic regulation and antioxidative defense, respectively, was reflected in spectral changes, suggesting that hyperspectral imaging can serve as a noninvasive tool to monitor plant physiological responses to water stress.

Furthermore, the identification of optimal spectral bands for predicting key physiological traits can enhance the development of spectral indices tailored for water stress detection. For example, indices incorporating wavelengths of approximately 1450 nm, 1940 nm, and 2200 nm could be particularly effective in monitoring the water status of soybean. This approach aligns with the findings of Wong et al. (2023) [79], who emphasized the importance of specific spectral bands in developing effective vegetation indices for drought monitoring.

Collectively, these results validate that hyperspectral reflectance, combined with robust chemometric modelling, enables reliable and simultaneous prediction of a wide array of physiological and biochemical traits in soybean leaves. However, they also highlight the importance of variable selection and model calibration for traits with weaker spectral signatures or higher biological noise, emphasizing the need for continuous refinement and validation across diverse genotypes and environments.

4.4. Wavelength Selection for Predictive Modelling

Hyperspectral datasets comprising thousands of contiguous bands offer detailed plant physiological insights, yet their practical application hinges on identifying minimal but informative spectral subsets [80]. The hybrid feature-selection strategy, which integrates partial least squares regression (PLSR) with the ensemble (random forest), kernel-based (support vector machine) and Boruta algorithms, demonstrates that a strategic dozen wavelengths capture over 95 % of the predictive variance for key leaf traits. Figure 9A–Q map algorithm- specific band importance scores, consistently showing chlorophyll a and b absorption peaks (≈550–750 nm) as primary predictors of photosynthetic capacity and stress onset, corroborating established physiological absorption profiles.

Shortwave-infrared (SWIR) absorption windows at ~1450 nm and ~1940 nm are crucial for estimating water content and turgor, with independent validation yielding R^2 values > 0.74 (Figure 10A–C). These bands coincide with first and second water vibrational overtones, underscoring their mechanistic link to leaf hydration. Furthermore, mid-infrared overtone bands at ~1700 nm and ~2230 nm correlate strongly (R^2 test > 0.65; Figure 12) with cell-wall constituents such cellulose and lignin, reflecting their absorption by C–H and O–H bond vibrations. This spectral pinpointing aligns with molecular absorption theory and provides direct insight into cell- wall maturity and structural rigidity [81,82].

The transition from full-spectrum analyses to a targeted twelve-band model reduces the data volume and processing time by >80 %, while maintaining the model error within ±5%. This efficiency gain is critical for real-time, in-field monitoring, such UAV platforms and autonomous greenhouse systems can incorporate tunable laser-line filters at these wavelengths [68,69], enabling continuous mapping of hydration, nutrient status and biomechanical properties. Decision-support algorithms can then trigger precision interventions, dynamic irrigation adjustments, targeted nutrient delivery or the application of growth regulators well before conventional stress symptoms manifest.

Re-evaluation of conventional vegetation indices (e.g., NDVI and PRI) via the existing full-spectrum models against trait-tuned hyperspectral indices ("Cell Wall Index" blending red-edge and ~2230 nm; "Turgor Stress Index" combining red-edge and ~1450 nm) demonstrated improvements of approximately 10–15 % in R^2 during both calibration and independent validation (Figure 10D–F). These indices were derived from the same dataset and model parameters, confirming that strategic wavelength selection alone enhances predictive performance without requiring retraining of new models [41,69,83]. Trait-tuned HVIs thus distil multi-band complexity into single metrics directly correlated with agronomic endpoints [10,22], facilitating seamless integration into precision-agriculture dashboards for both water-limited systems and biomass-focused bioenergy crops.

The spectral correlation matrices show patchwork patterns of high-correlation hotspots interspersed with low-information regions, highlighting the unequal distribution of explanatory power across wavelengths, similar to [84,85]. This nonuniformity validates that a tiered modelling approach by PLSR offers rapid calibration and interpretable loadings for operational sensor deployment; random forest excels at detecting spectral anomalies in time-series data; and Boruta maintains ongoing marker discovery, adapting to novel genotypic and environmental conditions.

In summary, while most algorithms converge on the red-edge region (particularly ~700 nm) and classical water absorption bands (notably 1450, 1940, and 2200 nm) [41,69,83], some models highlight additional diagnostic bands (e.g., 675, 1200, and >2100 nm) depending on the target trait (Figures 1–12). This convergence across multiple feature selection methods underscores the physiological relevance and robustness of these specific wavelengths as predictors of drought-related responses in soybean [10,22],. The identification of these consistent "spectral hot spots" suggests that future proximal or UAV-based sensors can be engineered to target a minimal set of key wavelengths, drastically reducing sensor complexity and data redundancy without compromising prediction accuracy.

Expanding upon these findings, future work should explore the robustness of the twelve-band model under varying illumination angles, leaf orientations and canopy structures, as well as its transferability across species and developmental stages. Integrating these discrete wavelengths into miniaturized spectrometers or hyperspectral snapshot cameras could revolutionize plant phenotyping workflows, enabling high-throughput screening in breeding programmes and real-time crop monitoring at the field scale.

5. Conclusions

This study demonstrates that a minimal set of 12 strategically selected wavelengths—primarily in the red-edge (550–750 nm) and shortwave infrared (1450, 1940, 2200 nm) regions—enables highly accurate, nondestructive predictions of physiological and biochemical drought responses in soybean under controlled conditions. The ensemble machine learning models (random forest, gradient boosting, and multilayer perceptron) achieved over 95% accuracy in classifying eleven water regimes, and the PLSR models predicted key leaf traits with R^2 values of 0.96 (cellulose) and 0.88 (chlorophyll a). This compact-band approach reduced the data dimensionality by more than 80% with minimal loss of predictive power, providing a cost-efficient basis for real-time drought phenotyping and sensor development.

Our hybrid feature selection and modelling framework is novel compared with previous studies that relied on full-spectrum data or generic indices, offering practical advances for precision agriculture. Future studies should validate these findings under diverse field conditions and further integrate hyperspectral and other sensing modalities for robust crop monitoring.

Author Contributions: Conceptualization, C.A.d.O., N.G.V., W.C.A., M.R.N. and R.F.; Data curation, C.A.d.O., J.A.M.D., M.R.N. and R.F.; Formal analysis, C.A.d.O., N.G.V., W.A.M., D.H.S.d.M., R.H.F., L.G.T.C., A.S.R., R.B.d.O., J.A.M.D., M.R.N. and R.F.; Funding acquisition, R.B.d.O., M.L.C., J.A.M.D., M.R.N. and R.F.; Investigation, C.A.d.O., N.G.V., W.A.M., J.V.F.G., D.H.S.d.M., R.H.F., L.G.T.C., R.B.d.O., M.L.C., J.A.M.D., M.R.N. and R.F.; Methodology, C.A.d.O., N.G.V., W.A.M., J.V.F.G., R.H.F., L.G.T.C., A.S.R., W.C.A., R.B.d.O., M.L.C., J.A.M.D., M.R.N. and R.F.; Project administration, M.R.N. and R.F.; Resources, M.R.N. and R.F.; Software, C.A.d.O., N.G.V., W.A.M., J.V.F.G., R.H.F., L.G.T.C., A.S.R., W.C.A., R.B.d.O., M.L.C., J.A.M.D., M.R.N. and R.F.; Supervision, M.R.N. and R.F.; Validation, C.A.d.O., W.A.M., D.H.S.d.M., R.H.F., L.G.T.C., A.S.R., R.B.d.O., M.L.C., J.A.M.D., M.R.N. and R.F.; Visualization, M.R.N. and R.F.; Writing—original draft, C.A.d.O., N.G.V., W.A.M., J.V.F.G., D.H.S.d.M., R.H.F., L.G.T.C., A.S.R., W.C.A., R.B.d.O., M.L.C., J.A.M.D., M.R.N. and R.F.; Writing—review & editing, C.A.d.O., N.G.V., W.A.M., D.H.S.d.M., R.H.F., L.G.T.C., W.C.A., R.B.d.O., M.L.C., J.A.M.D., M.R.N. and R.F. All authors have read and agreed to the published version of the manuscript.

Funding: Programa de Apoio à Fixação de Jovens Doutores no Brasil (CNPq 168180/2022-7), Fundação Araucária (CP 19/2022—Jovens Doutores), Coordenação de Aperfeiçoamento de Pessoal de Nível Superior—Brasil (CAPES)—Finance Code 001.

Institutional Review Board Statement: Not applicable.

Informed Consent Statement: Not applicable.

Data Availability Statement: The original contributions presented in this study are included in the article. Further inquiries can be directed to the corresponding author.

Acknowledgments: We would like to thank the editors and reviewers for their comments and suggestions to improve our work and the State University of Maringá for supporting the analyses.

Conflicts of Interest: The authors declare that they have no conflicts of interest. The funders had no role in the design of the study; in the collection, analyses, or interpretation of data; in the writing of the manuscript; or in the decision to publish the results.

References

1. dos Santos, G.L.A.A.; Reis, A.S.; Besen, M.R.; Furlanetto, R.H.; Rodrigues, M.; Crusiol, L.G.T.; de Oliveira, K.M.; Falcioni, R.; de Oliveira, R.B.; Batista, M.A.; et al. Spectral Method for Macro and Micronutrient Prediction in Soybean Leaves Using Interval Partial Least Squares Regression. *Eur. J. Agron.* **2023**, *143*, 126717. [CrossRef]
2. Crusiol, L.G.T.; Nanni, M.R.; Furlanetto, R.H.; Sibaldelli, R.N.R.; Sun, L.; Gonçalves, S.L.; Foloni, J.S.S.; Mertz-Henning, L.M.; Nepomuceno, A.L.; Neumaier, N.; et al. Assessing the Sensitive Spectral Bands for Soybean Water Status Monitoring and Soil Moisture Prediction Using Leaf-Based Hyperspectral Reflectance. *Agric. Water Manag.* **2023**, *277*, 108089. [CrossRef]

3. Vargas-Almendra, A.; Ruiz-Medrano, R.; Núñez-Muñoz, L.A.; Ramírez-Pool, J.A.; Calderón-Pérez, B.; Xoconostle-Cázares, B. Advances in Soybean Genetic Improvement. *Plants* **2024**, *13*, 3073. [CrossRef]
4. Yoosefzadeh-Najafabadi, M.; Tulpan, D.; Eskandari, M. Using Hybrid Artificial Intelligence and Evolutionary Optimization Algorithms for Estimating Soybean Yield and Fresh Biomass Using Hyperspectral Vegetation Indices. *Remote Sens.* **2021**, *13*, 2555. [CrossRef]
5. Flexas, J.; Medrano, H. Drought-Inhibition of Photosynthesis in C3plants: Stomatal and Non-Stomatal Limitations Revisited. *Ann. Bot.* **2002**, *89*, 183–189. [CrossRef]
6. Chaudhry, S.; Sidhu, G.P.S. Climate Change Regulated Abiotic Stress Mechanisms in Plants: A Comprehensive Review. *Plant Cell Rep.* **2022**, *41*, 1–31. [CrossRef]
7. Wang, Q.; Zuo, Z.; Wang, X.; Gu, L.; Yoshizumi, T.; Yang, Z.; Yang, L.; Liu, Q.; Liu, W.; Han, Y.-J.; et al. Photoactivation and Inactivation of in Arabidopsis Cryptochrome 2. *Science* **2016**, *354*, 343–347. [CrossRef] [PubMed]
8. Shurygin, B.; Chivkunova, O.; Solovchenko, O.; Solovchenko, A.; Dorokhov, A.; Smirnov, I.; Astashev, M.E.; Khort, D. Comparison of the Non-Invasive Monitoring of Fresh-Cut Lettuce Condition with Imaging Reflectance Hyperspectrometer and Imaging PAM-Fluorimeter. *Photonics* **2021**, *8*, 425. [CrossRef]
9. Nievola, C.C.; Carvalho, C.P.; Carvalho, V.; Rodrigues, E. Rapid Responses of Plants to Temperature Changes. *Temperature* **2017**, *4*, 371–405. [CrossRef]
10. Falcioni, R.; Gonçalves, J.V.F.; de Oliveira, K.M.; de Oliveira, C.A.; Reis, A.S.; Crusiol, L.G.T.; Furlanetto, R.H.; Antunes, W.C.; Cezar, E.; de Oliveira, R.B.; et al. Chemometric Analysis for the Prediction of Biochemical Compounds in Leaves Using UV-VIS-NIR-SWIR Hyperspectroscopy. *Plants* **2023**, *12*, 3424. [CrossRef]
11. Falcioni, R.; de Oliveira, C.A.; Vedana, N.G.; Mendonça, W.A.; Gonçalves, J.V.F.; da Silva Haubert, D.D.F.; de Matos, D.H.S.; Reis, A.S.; Antunes, W.C.; Crusiol, L.G.T.; et al. Progressive Water Deficit Impairs Soybean Growth, Alters Metabolic Profiles, and Decreases Photosynthetic Efficiency. *Plants* **2025**, *14*, 2615. [CrossRef] [PubMed]
12. Falcioni, R.; de Oliveira, R.B.; Chicati, M.L.; Antunes, W.C.; Demattê, J.A.M.; Nanni, M.R. Fluorescence and Hyperspectral Sensors for Nondestructive Analysis and Prediction of Biophysical Compounds in the Green and Purple Leaves of Tradescantia Plants. *Sensors* **2024**, *24*, 6490. [CrossRef] [PubMed]
13. Hassanzadeh, A.; Murphy, S.P.; Pethybridge, S.J.; van Aardt, J. Growth Stage Classification and Harvest Scheduling of Snap Bean Using Hyperspectral Sensing: A Greenhouse Study. *Remote Sens.* **2020**, *12*, 3809. [CrossRef]
14. Buchhorn, M.; Raynolds, M.K.; Walker, D.A. Influence of BRDF on NDVI and Biomass Estimations of Alaska Arctic Tundra. *Environ. Res. Lett.* **2016**, *11*, 1–13. [CrossRef]
15. Chaves, M.E.D.; De Carvalho Alves, M.; De Oliveira, M.S.; Sáfadi, T. A Geostatistical Approach for Modeling Soybean Crop Area and Yield Based on Census and Remote Sensing Data. *Remote Sens.* **2018**, *10*, 680. [CrossRef]
16. Sims, D.A.; Gamon, J.A. Relationships between Leaf Pigment Content and Spectral Reflectance across a Wide Range of Species, Leaf Structures and Developmental Stages. *Remote Sens. Environ.* **2002**, *81*, 337–354. [CrossRef]
17. Jia, M.; Li, D.; Colombo, R.; Wang, Y.; Wang, X.; Cheng, T.; Zhu, Y.; Yao, X.; Xu, C.; Ouer, G.; et al. Quantifying Chlorophyll Fluorescence Parameters from Hyperspectral Reflectance at the Leaf Scale under Various Nitrogen Treatment Regimes in Winter Wheat. *Remote Sens.* **2019**, *11*, 2838.
18. Ryu, J.H.; Jeong, H.; Cho, J. Performances of Vegetation Indices on Paddy Rice at Elevated Air Temperature, Heat Stress, and Herbicide Damage. *Remote Sens.* **2020**, *12*, 2654. [CrossRef]
19. Li, K.; Wang, C.; Rong, G.; Wei, S.; Liu, C.; Yang, Y.; Sudu, B.; Guo, Y.; Sun, Q.; Zhang, J. Dynamic Evaluation of Agricultural Drought Hazard in Northeast China Based on Coupled Multi-Source Data. *Remote Sens.* **2023**, *15*, 57. [CrossRef]
20. Sobejano-Paz, V.; Mikkelsen, T.N.; Baum, A.; Mo, X.; Liu, S.; Köppl, C.J.; Johnson, M.S.; Gulyas, L.; García, M. Hyperspectral and Thermal Sensing of Stomatal Conductance, Transpiration, and Photosynthesis for Soybean and Maize under Drought. *Remote Sens.* **2020**, *12*, 3182. [CrossRef]
21. Osman, S.O.M.; Saad, A.S.I.; Tadano, S.; Takeda, Y.; Konaka, T.; Yamasaki, Y.; Tahir, I.S.A.; Tsujimoto, H.; Akashi, K. Chemical Fingerprinting of Heat Stress Responses in the Leaves of Common Wheat by Fourier Transform Infrared Spectroscopy. *Int. J. Mol. Sci.* **2022**, *23*, 2842. [CrossRef]
22. Crusiol, L.G.T.; Sun, L.; Sun, Z.; Chen, R.; Wu, Y.; Ma, J.; Song, C. In-Season Monitoring of Maize Leaf Water Content Using Ground-Based and UAV-Based Hyperspectral Data. *Sustainability* **2022**, *14*, 9039. [CrossRef]
23. da Silva Junior, C.A.; Nanni, M.R.; Shakir, M.; Teodoro, P.E.; de Oliveira-Júnior, J.F.; Cezar, E.; de Gois, G.; Lima, M.; Wojciechowski, J.C.; Shiratsuchi, L.S. Soybean Varieties Discrimination Using Non-Imaging Hyperspectral Sensor. *Infrared Phys. Technol.* **2018**, *89*, 338–350. [CrossRef]
24. Baio, F.H.R.; Santana, D.C.; Teodoro, L.P.R.; de Oliveira, I.C.; Gava, R.; de Oliveira, J.L.G.; da Silva Junior, C.A.; Teodoro, P.E.; Shiratsuchi, L.S. Maize Yield Prediction with Machine Learning, Spectral Variables and Irrigation Management. *Remote Sens.* **2023**, *15*, 79.

25. Vian, C.E.D.F.; Andrade Júnior, A.M.; Baricelo, L.G.; Da Silva, R.P. Origens, Evolução e Tendências Da Indústria de Máquinas Agrícolas. *Rev. Econ. Sociol. Rural* **2013**, *51*, 719–744.
26. Sonobe, R.; Wang, Q. Hyperspectral Indices for Quantifying Leaf Chlorophyll Concentrations Performed Differently with Different Leaf Types in Deciduous Forests. *Ecol. Inform.* **2017**, *37*, 1–9.
27. Braga, P.; Crusiol, L.G.T.; Nanni, M.R.; Caranhato, A.L.H.; Fuhrmann, M.B.; Nepomuceno, A.L.; Neumaier, N.; Farias, J.R.B.; Koltun, A.; Gonçalves, L.S.A.; et al. Vegetation Indices and NIR-SWIR Spectral Bands as a Phenotyping Tool for Water Status Determination in Soybean. *Precis. Agric.* **2021**, *22*, 249–266.
28. Mondal, S.; Karmakar, S.; Panda, D.; Pramanik, K.; Bose, B.; Singhal, R.K. Crucial Plant Processes under Heat Stress and Tolerance through Heat Shock Proteins. *Plant Stress* **2023**, *10*, 100227. [CrossRef]
29. Galvão, L.S.; Formaggio, A.R.; Tisot, D.A. Discrimination of Sugarcane Varieties in Southeastern Brazil with EO-1 Hyperion Data. *Remote Sens. Environ.* **2005**, *94*, 523–534. [CrossRef]
30. Kováč, D.; Veselovská, P.; Klem, K.; Večeřová, K.; Ač, A.; Peñuelas, J.; Urban, O. Potential of Photochemical Reflectance Index for Indicating Photochemistry and Light Use Efficiency in Leaves of European Beech and Norway Spruce Trees. *Remote Sens.* **2018**, *10*, 1202.
31. de Oliveira, K.M.; Ferreira Gonçalves, J.V.; Falcioni, R.; Almeida de Oliveira, C.; de Fatima da Silva Haubert, D.; Mendonça, W.A.; Teixeira Crusiol, L.G.; Berti de Oliveira, R.; Reis, A.S.; Cezar, E.; et al. Classification of Soil Horizons Based on VisNIR and SWIR Hyperespectral Images and Machine Learning Models. *Remote Sens. Appl. Soc. Environ.* **2024**, *36*, 101362. [CrossRef]
32. Sexton, T.; Sankaran, S.; Cousins, A.B. Predicting Photosynthetic Capacity in Tobacco Using Shortwave Infrared Spectral Reflectance. *J. Exp. Bot.* **2021**, *72*, 4373–4383. [CrossRef] [PubMed]
33. Falcioni, R.; de Oliveira, R.B.; Chicati, M.L.; Antunes, W.C.; Demattê, J.A.M.; Nanni, M.R. Estimation of Biochemical Compounds in Tradescantia Leaves Using VIS-NIR-SWIR Hyperspectral and Chlorophyll a Fluorescence Sensors. *Remote Sens.* **2024**, *16*, 1910.
34. Crusiol, L.G.T.; Sun, L.; Sibaldelli, R.N.R.; Junior, V.F.; Furlaneti, W.X.; Chen, R.; Sun, Z.; Wuyun, D.; Chen, Z.; Nanni, M.R.; et al. Strategies for Monitoring Within-Field Soybean Yield Using Sentinel-2 Vis-NIR-SWIR Spectral Bands and Machine Learning Regression Methods. *Precis. Agric.* **2022**, *23*, 1093–1123.
35. Falcioni, R.; Gonçalves, J.V.F.; de Oliveira, K.M.; Antunes, W.C.; Nanni, M.R. VIS-NIR-SWIR Hyperspectroscopy Combined with Data Mining and Machine Learning for Classification of Predicted Chemometrics of Green Lettuce. *Remote Sens.* **2022**, *14*, 6330. [CrossRef]
36. Gitelson, A.; Solovchenko, A. Non-Invasive Quantification of Foliar Pigments: Possibilities and Limitations of Reflectance- and Absorbance-Based Approaches. *J. Photochem. Photobiol. B Biol.* **2018**, *178*, 537–544.
37. Falcioni, R.; Moriwaki, T.; Gibin, M.S.; Vollmann, A.; Pattaro, M.C.; Giacomelli, M.E.; Sato, F.; Nanni, M.R.; Antunes, W.C. Classification and Prediction by Pigment Content in Lettuce (*Lactuca sativa* L.) Varieties Using Machine Learning and ATR-FTIR Spectroscopy. *Plants* **2022**, *11*, 3413. [CrossRef]
38. Ragaee, S. Antioxidant Activity and Nutrient Composition of Selected Cereals for Food Use. *Food Chem.* **2006**, *98*, 32–38. [CrossRef]
39. Izenman, A.J. *Modern Multivariate Statistical Techniques*, 1st ed.; Springer: Berlin/Heidelberg, Germany, 2008.
40. Carlson, T.N.; Ripley, D.A. On the Relation between NDVI, Fractional Vegetation Cover, and Leaf Area Index. *Remote Sens. Environ.* **1997**, *62*, 241–252. [CrossRef]
41. Gitelson, A.A.; Kaufman, Y.J.; Merzlyak, M.N. Use of a Green Channel in Remote Sensing of Global Vegetation from EOS-MODIS. *Remote Sens. Environ.* **1996**, *58*, 289–298.
42. Ahamed, T.; Tian, L.; Zhang, Y.; Ting, K.C. A Review of Remote Sensing Methods for Biomass Feedstock Production. *Biomass and Bioenergy* **2011**, *35*, 2455–2469. [CrossRef]
43. Bannari, A.; Morin, D.; Bonn, F.; Huete, A.R. A Review of Vegetation Indices. *Remote Sens. Rev.* **1995**, *13*, 95–120. [CrossRef]
44. Blackburn, G.A. Spectral Indices for Estimating Photosynthetic Pigment Concentrations: A Test Using Senescent Tree Leaves. *Int. J. Remote Sens.* **1998**, *19*, 657–675. [CrossRef]
45. Falcioni, R.; Antunes, W.C.; Demattê, J.A.M.; Nanni, M.R. Biophysical, Biochemical, and Photochemical Analyses Using Reflectance Hyperspectroscopy and Chlorophyll a Fluorescence Kinetics in Variegated Leaves. *Biology* **2023**, *12*, 704. [CrossRef]
46. Peñuelas, J.; Gamon, J.A.; Fredeen, A.L.; Merino, J.; Field, C.B. Reflectance Indices Associated with Physiological Changes in Nitrogen- and Water-Limited Sunflower Leaves. *Remote Sens. Environ.* **1994**, *48*, 135–146. [CrossRef]
47. Hunt, E.R.; Rock, B.N. Detection of Changes in Leaf Water Content Using Near- and Middle-Infrared Reflectances. *Remote Sens. Environ.* **1989**, *30*, 43–54.
48. Cibula, W.G.; Zetka, E.F.; Rickman, D.L. Response of Thematic Mapper Bands to Plant Water Stress. *Int. J. Remote Sens.* **1992**, *13*, 1869–1880. [CrossRef]
49. Patil, P.P.; Jagtap, M.P.; Khatri, N.; Madan, H.; Vadduri, A.A.; Patodia, T. Exploration and Advancement of NDDI Leveraging NDVI and NDWI in Indian Semi-Arid Regions: A Remote Sensing-Based Study. *Case Stud. Chem. Environ. Eng.* **2024**, *9*, 100573.

50. Staszel, J.; Lupa, M.; Adamek, K.; Wilkosz, M.; Marcinkowska-Ochtyra, A.; Ochtyra, A. Spatial Insights into Drought Severity: Multi-Index Assessment in Małopolska, Poland, via Satellite Observations. *Remote Sens.* **2024**, *16*, 836. [CrossRef]
51. Gitelson, A.A.; Zur, Y.; Chivkunova, O.B.; Merzlyak, M.N. Assessing Carotenoid Content in Plant Leaves with Reflectance Spectroscopy. *Photochem. Photobiol.* **2002**, *75*, 272. [CrossRef]
52. Vogelmann, J.E.; Rock, B.N.; Moss, D.M. Red Edge Spectral Measurements from Sugar Maple Leaves. *Int. J. Remote Sens.* **1993**, *14*, 1563–1575. [CrossRef]
53. Peñuelas, J.; Filella, I. Visible and Near-Infrared Reflectance Techniques for Diagnosing Plant Physiological Status. *Trends Plant Sci.* **1998**, *3*, 151–156. [CrossRef]
54. Peñuelas, J.; Filella, I.; Gamon, J.A. Assessment of Photosynthetic Radiation-Use Efficiency with Spectral Reflectance. *New Phytol.* **1995**, *131*, 291–296. [CrossRef]
55. Zheng, W.; Lu, X.; Li, Y.; Li, S.; Zhang, Y. Hyperspectral Identification of Chlorophyll Fluorescence Parameters of Suaeda Salsa in Coastal Wetlands. *Remote Sens.* **2021**, *13*, 2066. [CrossRef]
56. Barnes, M.L.; Breshears, D.D.; Law, D.J.; van Leeuwen, W.J.D.; Monson, R.K.; Fojtik, A.C.; Barron-Gafford, G.A.; Moore, D.J.P. Beyond Greenness: Detecting Temporal Changes in Photosynthetic Capacity with Hyperspectral Reflectance Data. *PLoS ONE* **2017**, *12*, e0189539. [CrossRef]
57. Nehra, A.; Kalwan, G.; Taneja, D.; Jangra, R.; Joshi, K.; Kumar, A.; Jain, P.K.; Nehra, K.; Ansari, M.W.; Singh, K.; et al. Comprehensive Structural, Evolutionary and Functional Analysis of Superoxide Dismutase Gene Family Revealed Critical Role in Salinity and Drought Stress Responses in Chickpea (*Cicer arietinum* L.). *Plant Physiol. Biochem.* **2025**, *226*, 110042. [CrossRef] [PubMed]
58. Valentini, R.; Epron, D.; De Angelis, P.; Matteucci, G.; Dreyer, E. In Situ Estimation of Net CO2 Assimilation, Photosynthetic Electron Flow and Photorespiration in Turkey Oak (*Q. cerris* L.) Leaves: Diurnal Cycles under Different Levels of Water Supply. *Plant. Cell Environ.* **1995**, *18*, 631–640. [CrossRef]
59. Gill, S.S.; Tuteja, N. Reactive Oxygen Species and Antioxidant Machinery in Abiotic Stress Tolerance in Crop Plants. *Plant Physiol. Biochem.* **2010**, *48*, 909–930. [CrossRef]
60. Zhou, Y.H.; Zhang, Y.Y.; Zhao, X.; Yu, H.J.; Shi, K.; Yu, J.Q. Impact of Light Variation on Development of Photoprotection, Antioxidants, and Nutritional Value in *Lactuca sativa* L. *J. Agric. Food Chem.* **2009**, *57*, 5494–5500. [CrossRef]
61. Pinnola, A.; Bassi, R. Molecular Mechanisms Involved in Plant Photoprotection. *Biochem. Soc. Trans.* **2018**, *46*, 467–482. [CrossRef]
62. Gitelson, A.; Chivkunova, O.; Zhigalova, T.; Solovchenko, A. In Situ Optical Properties of Foliar Flavonoids: Implication for Non-Destructive Estimation of Flavonoid Content. *J. Plant Physiol.* **2017**, *218*, 258–264. [CrossRef]
63. Hichri, I.; Barrieu, F.; Bogs, J.; Kappel, C.; Delrot, S.; Lauvergeat, V. Recent Advances in the Transcriptional Regulation of the Flavonoid Biosynthetic Pathway. *J. Exp. Bot.* **2011**, *62*, 2465–2483. [CrossRef]
64. Zhou, Z.; Gao, H.; Ming, J.; Ding, Z.; Lin, X.; Zhan, R. Combined Transcriptome and Metabolome Analysis of Pitaya Fruit Unveiled the Mechanisms Underlying Peel and Pulp Color Formation. *BMC Genom.* **2020**, *21*, 734. [CrossRef]
65. Baranović, G.; Šegota, S. Infrared Spectroscopy of Flavones and Flavonols. Reexamination of the Hydroxyl and Carbonyl Vibrations in Relation to the Interactions of Flavonoids with Membrane Lipids. *Spectrochim. Acta Part A Mol. Biomol. Spectrosc.* **2018**, *192*, 473–486. [CrossRef]
66. Almakas, A.; Elrys, A.S.; Desoky, E.-S.M.; Al-Shuraym, L.A.; Alhag, S.K.; Alshaharni, M.O.; Alnadari, F.; NanNan, Z.; Farooq, Z.; El-Tarabily, K.A.; et al. Enhancing Soybean Germination and Vigor under Water Stress: The Efficacy of Bio-Priming with Sodium Carboxymethyl Cellulose and Gum Arabic. *Front. Plant Sci.* **2025**, *15*, 1475148. [CrossRef] [PubMed]
67. Boshkovski, B.; Doupis, G.; Zapolska, A.; Kalaitzidis, C.; Koubouris, G. Hyperspectral Imagery Detects Water Deficit and Salinity Effects on Photosynthesis and Antioxidant Enzyme Activity of Three Greek Olive Varieties. *Sustainability* **2022**, *14*, 1432. [CrossRef]
68. Wang, D.; Cao, W.; Zhang, F.; Li, Z.; Xu, S.; Wu, X. A Review of Deep Learning in Multiscale Agricultural Sensing. *Remote Sens.* **2022**, *14*, 559.
69. Zhou, Q.; Yu, L.; Zhang, X.; Liu, Y.; Zhan, Z.; Ren, L.; Luo, Y. Fusion of UAV Hyperspectral Imaging and LiDAR for the Early Detection of EAB Stress in Ash and a New EAB Detection Index NDVI(776,678). *Remote Sens.* **2022**, *14*, 2428. [CrossRef]
70. Furlanetto, R.H.; Moriwaki, T.; Falcioni, R.; Pattaro, M.; Vollmann, A.; Sturion Junior, A.C.; Antunes, W.C.; Nanni, M.R. Hyperspectral Reflectance Imaging to Classify Lettuce Varieties by Optimum Selected Wavelengths and Linear Discriminant Analysis. *Remote Sens. Appl. Soc. Environ.* **2020**, *20*, 100400. [CrossRef]
71. Ge, Y.; Atefi, A.; Zhang, H.; Miao, C.; Ramamurthy, R.K.; Sigmon, B.; Yang, J.; Schnable, J.C. High-Throughput Analysis of Leaf Physiological and Chemical Traits with VIS–NIR–SWIR Spectroscopy: A Case Study with a Maize Diversity Panel. *Plant Methods* **2019**, *15*, 66.
72. Wang, L.; Chang, Q.; Li, F.; Yan, L.; Huang, Y.; Wang, Q.; Luo, L. Effects of Growth Stage Development on Paddy Rice Leaf Area Index Prediction Models. *Remote Sens.* **2019**, *11*, 361.

73. Hu, Y.; Wang, Z.; Li, X.; Li, L.; Wang, X.; Wei, Y. Nondestructive Classification of Maize Moldy Seeds by Hyperspectral Imaging and Optimal Machine Learning Algorithms. *Sensors* **2022**, *22*, 6064. [CrossRef]
74. Fu, P.; Meacham-Hensold, K.; Guan, K.; Bernacchi, C.J. Hyperspectral Leaf Reflectance as Proxy for Photosynthetic Capacities: An Ensemble Approach Based on Multiple Machine Learning Algorithms. *Front. Plant Sci.* **2019**, *10*, 730. [CrossRef] [PubMed]
75. Cotrozzi, L.; Lorenzini, G.; Nali, C.; Pellegrini, E.; Saponaro, V.; Hoshika, Y.; Arab, L.; Rennenberg, H.; Paoletti, E. Hyperspectral Reflectance of Light-Adapted Leaves Can Predict Both Dark- and Light-Adapted Chl Fluorescence Parameters, and the Effects of Chronic Ozone Exposure on Date Palm (*Phoenix dactylifera*). *Int. J. Mol. Sci.* **2020**, *21*, 6441. [PubMed]
76. Nalepa, J. Recent Advances in Multi- and Hyperspectral Image Analysis. *Sensors* **2021**, *21*, 6002. [CrossRef]
77. Wijewardana, C.; Reddy, K.R.; Krutz, L.J.; Gao, W.; Bellaloui, N. Drought Stress Has Transgenerational Effects on Soybean Seed Germination and Seedling Vigor. *PLoS ONE* **2019**, *14*, e0214977. [CrossRef]
78. Gururani, M.A.; Venkatesh, J.; Ganesan, M.; Strasser, R.J.; Han, Y.; Kim, J.-I.; Lee, H.Y.; Song, P.S. In Vivo Assessment of Cold Tolerance through Chlorophyll-a Fluorescence in Transgenic Zoysiagrass Expressing Mutant Phytochrome A. *PLoS ONE* **2015**, *10*, e0127200.
79. Wong, C.Y.S.; Gilbert, M.E.; Pierce, M.A.; Parker, T.A.; Palkovic, A.; Gepts, P.; Magney, T.S.; Buckley, T.N. Hyperspectral Remote Sensing for Phenotyping the Physiological Drought Response of Common and Tepary Bean. *Plant Phenomics* **2023**, *5*, 21. [CrossRef]
80. Sun, Y.; Liu, B.; Yu, X.; Yu, A.; Gao, K.; Ding, L. From Video to Hyperspectral: Hyperspectral Image-Level Feature Extraction with Transfer Learning. *Remote Sens.* **2022**, *14*, 5118.
81. Nogales-Bueno, J.; Baca-Bocanegra, B.; Rooney, A.; Miguel Hernández-Hierro, J.; José Heredia, F.; Byrne, H.J. Linking ATR-FTIR and Raman Features to Phenolic Extractability and Other Attributes in Grape Skin. *Talanta* **2017**, *167*, 44–50. [CrossRef]
82. Prats-Mateu, B.; Felhofer, M.; de Juan, A.; Gierlinger, N. Multivariate Unmixing Approaches on Raman Images of Plant Cell Walls: New Insights or Overinterpretation of Results? *Plant Methods* **2018**, *14*, 52. [CrossRef] [PubMed]
83. Zhang, Y.; Li, X.; Wang, C.; Zhang, R.; Jin, L.; He, Z.; Tian, S.; Wu, K.; Wang, F. PROSPECT-PMP+: Simultaneous Retrievals of Chlorophyll a and b, Carotenoids and Anthocyanins in the Leaf Optical Properties Model. *Sensors* **2022**, *22*, 3025. [CrossRef]
84. Chen, J.; de Hoogh, K.; Gulliver, J.; Hoffmann, B.; Hertel, O.; Ketzel, M.; Bauwelinck, M.; van Donkelaar, A.; Hvidtfeldt, U.A.; Katsouyanni, K.; et al. A Comparison of Linear Regression, Regularization, and Machine Learning Algorithms to Develop Europe-Wide Spatial Models of Fine Particles and Nitrogen Dioxide. *Environ. Int.* **2019**, *130*, 104934. [CrossRef] [PubMed]
85. Zhu, H.; Chu, B.; Zhang, C.; Liu, F.; Jiang, L.; He, Y. Hyperspectral Imaging for Presymptomatic Detection of Tobacco Disease with Successive Projections Algorithm and Machine-Learning Classifiers. *Sci. Rep.* **2017**, *7*, 4125. [CrossRef] [PubMed]

remote sensing

MDPI

Article

Physics-Driven Machine-Learning Retrieval and Uncertainty Quantification of Crop Leaf Area Index

Wei Liu [1,2], Xiaohua Zhu [1,*], Suyi Yang [3] and Zhihai Gao [4]

[1] Aerospace Information Research Institute, Chinese Academy of Sciences, Beijing 100094, China
[2] School of Electronic, Electrical and Communication Engineering, University of Chinese Academy of Sciences, Beijing 100049, China
[3] School of Geosciences and Info-Physics, Central South University, Changsha 410083, China
[4] Institute of Forest Resource Information Techniques, Chinese Academy of Forestry, Beijing 100091, China
* Correspondence: zhuxh@aircas.ac.cn

Highlights

What are the main findings?

- A PROSAIL-driven, GA-optimised MLP (NN–GA) reliably retrieves crop LAI from Sentinel-2B at 10 m, achieving RMSE/R^2 = 0.44/0.73 (Minqin) and 0.40/0.56 (Zhangye), outperforming the SNAP/SL2P benchmark.
- A staged uncertainty quantification (UQ) workflow separates physical-driver and machine-learning contributions and synthesises them to report retrieval relative uncertainties (Minqin 21.37%, Zhangye 17.31%).

What is the implication of the main finding?

- The framework improves 10 m LAI retrieval accuracy and delivers a reproducible, end-to-end uncertainty decomposition to support confidence-aware agronomic applications.
- The results prioritise reductions in machine-learning stage stochasticity and recommend including uncertainty as a routine product layer to increase LAI product reliability.

Abstract

Leaf Area Index (LAI) is a key biophysical descriptor of crop canopies and is essential for growth monitoring and yield estimation. We present a physics-driven machine-learning framework for operational LAI retrieval and end-to-end uncertainty quantification that couples the PROSAIL radiative transfer model with a genetic-algorithm-optimised multilayer perceptron (NN–GA). PROSAIL is sampled across plausible parameter priors and spectra are convolved with Sentinel-2B spectral response functions to build a 30,000-sample training library; a GA is used to globally optimise network weights and biases. Total retrieval uncertainty is decomposed into a simulation component (PROSAIL parameter variability) and a training component (variability across repeated NN–GA trainings) and combined via the law of propagation of uncertainty. The model was developed in Minqin (modelling/testing area; entirely maize) and transferred to Zhangye (transfer/validation area; predominantly maize, with one sunflower plot). Sentinel-2B validation results were RMSE/R^2 = 0.44/0.73 (Minqin) and 0.40/0.56 (Zhangye), indicating reasonable cross-site generalisation. The uncertainty split indicates physical-driven contributions of 11.42% and 11.48% and machine-learning contributions of 18.06% and 12.96%, respectively. The framework improves 10 m LAI retrieval accuracy and supplies a reproducible, per-pixel uncertainty budget to guide product use and refinement.

Keywords: leaf area index; neural network; genetic algorithm; uncertainty quantification; PROSAIL radiative transfer model

1. Introduction

Leaf Area Index (LAI), defined as one half of the total one-sided leaf area per unit ground area, is a central biophysical descriptor of crop canopies that controls light interception and photosynthetic productivity [1]. In agricultural applications, LAI underpins crop-growth modelling, yield forecasting, irrigation and fertilisation scheduling, and phenological and pest/disease monitoring; hence accurate, spatially explicit LAI estimates are of direct practical value for agronomic decision support and precision farming. Because in situ LAI measurements (e.g., destructive sampling or canopy analysers) provide only sparse point observations and are labour-intensive to acquire at scale, satellite and airborne remote sensing have become the primary means to generate continuous, wall-to-wall LAI products over operational extents [2,3].

Remote retrieval approaches for LAI broadly fall into empirical statistical methods and physics-based radiative transfer model (RTM) methods [4]. Empirical regressions or vegetation-index approaches are simple to implement and often perform well within the calibration domain, but their transferability across sensors, scene types, and phenological stages is limited [5]. Physics-based RTMs—most notably, the coupled PROSPECT (leaf optics) and SAIL (canopy scattering) system, PROSAIL—simulate canopy reflectance as a function of biophysical variables and observation geometry, providing mechanistic constraints that improve interpretability and enable the generation of large synthetic training libraries for controlled testing [6]. Since forward RTMs do not directly yield parameter estimates, practical retrievals typically invert the models using lookup tables, optimisation schemes, or data-driven regressors (e.g., neural networks, random forests, support vector regression), combining physical realism with the flexibility of machine learning to fit complex, nonlinear mappings [7,8].

Despite their strengths, coupling radiative transfer models (RTMs) with data-driven learners still faces fundamental challenges. RTMs typically rely on simplified parameterizations that cannot fully capture canopy and physiological heterogeneity; as a result, inversion problems can exhibit parameter equifinality and ill-posedness, while data-driven approaches remain highly sensitive to the representativeness and distribution of the training data, limiting out-of-domain generalisation. Together, these factors reduce the reliability and transportability of retrievals across differing crop conditions and sensor observations [9–11]. To mitigate these issues, global optimisation and ensemble-based strategies have been explored. In particular, genetic algorithms (GAs) provide a population-based global search that helps explore wide, multimodal parameter landscapes [7,12,13]. This capability can (i) reveal equifinality by locating distinct parameter sets that produce similar spectral responses, (ii) support multi-objective or regularised fitness formulations that penalise physically implausible parameter combinations (thereby suppressing practically irrelevant equifinal solutions), and (iii) facilitate ensemble generation (via multiple independent GA runs or archives of near-optimal solutions) whose spread offers an empirical characterisation of solution non-uniqueness and model uncertainty. When combined with a feedforward neural network (NN–GA), GAs' global search complements the NN's expressive power: the hybrid improves training stability and reproducibility, enables direct optimisation of bespoke fitness functions (for example, validation-weighted loss), and supplies practical diagnostics and ensemble outputs that inform uncertainty quantification and robustness analyses.

A further, often overlooked challenge is rigorous end-to-end uncertainty quantification (UQ). Existing studies on remote-sensing uncertainty predominantly focus on upstream data quality (for example, radiance, surface reflectance, or atmospheric-correction residuals), while a traceable, quantitative decomposition of errors across the entire retrieval chain—from observations and RTM forward simulation to inversion and downstream LAI estimates—remains limited [4]. Operational products such as global MODIS LAI include quality flags and per-pixel standard deviation layers (e.g., STD_LAI), and many validation studies report aggregate accuracy metrics; however, these provisions do not substitute for a systematic partitioning and quantitative propagation of uncertainties arising from prior parameter uncertainty, model structural error, and inversion-stage algorithmic variability [14]. In practice, retrieval products are typically summarised by RMSE or R^2 against reference data, but are seldom accompanied by confidence intervals or a source-resolved error budget—information that is essential for uncertainty-aware model assimilation and operational decision support [15].

Some recent efforts (for example, product reprocessing and noise reduction for MODIS LAI) have improved temporal consistency and reduced spurious variability, yet they mainly address time-series quality control rather than providing a principled accounting of how specific error sources propagate through each stage of the retrieval pipeline [16]. Concurrently, the rapid adoption of machine-learning methods introduces additional uncertainty modes—such as calibration uncertainty, probabilistic predictive outputs, and model confidence—which require dedicated UQ treatments. Although methodological research on neural-network uncertainty estimation is advancing, its practical application to operational biophysical-parameter retrievals remains limited [17].

Motivated by these gaps, this study develops a physics-driven machine-learning framework that couples the PROSAIL radiative transfer model with a genetic-algorithm-optimised multilayer perceptron (NN–GA) inversion and an explicit, staged UQ protocol. Our objectives are to deliver a broadly applicable LAI retrieval model and a reproducible UQ workflow to support robust product generation and to provide methodological guidance for uncertainty estimation of other land-surface variables. Concretely, we (1) construct a PROSAIL-based forward simulation library matched to Sentinel-2 spectral response functions (SRFs); (2) integrate a feedforward multilayer perceptron (MLP) with a GA for global optimisation of network parameters; (3) evaluate the approach across contrasting agricultural sites to test transferability; and (4) decompose the LAI retrieval uncertainty-propagation chain into constituent stages, estimate the uncertainty contribution of each stage, and synthesise these estimates to produce a comprehensive quantification and assessment of the overall LAI retrieval uncertainty.

2. Materials and Methods

2.1. Overview of the Research Area and Data Collection

The study was carried out at two agricultural sites in northwest China. Site 1 (Minqin County, Wuwei City, Gansu Province; 102°56′–103°07′E, 38°28′–38°37′N) served as the model-development and testing area for PROSAIL simulations, NN–GA training, and preliminary parameter tuning. Site 2 (Ganzhou District, Zhangye City; 100°16′–100°27′E, 38°45′–38°53′N) was used as a transfer and demonstration/validation area to evaluate model generalisation and transfer performance. Field campaigns synchronous (or near-synchronous) with the Sentinel-2 acquisitions were conducted on 31 August–1 September 2024 (maize, maturity stage) and 3–9 July 2025 (maize and sunflower; maize at tasseling). The locations and the spatial distributions of field sampling points are shown in Figure 1.

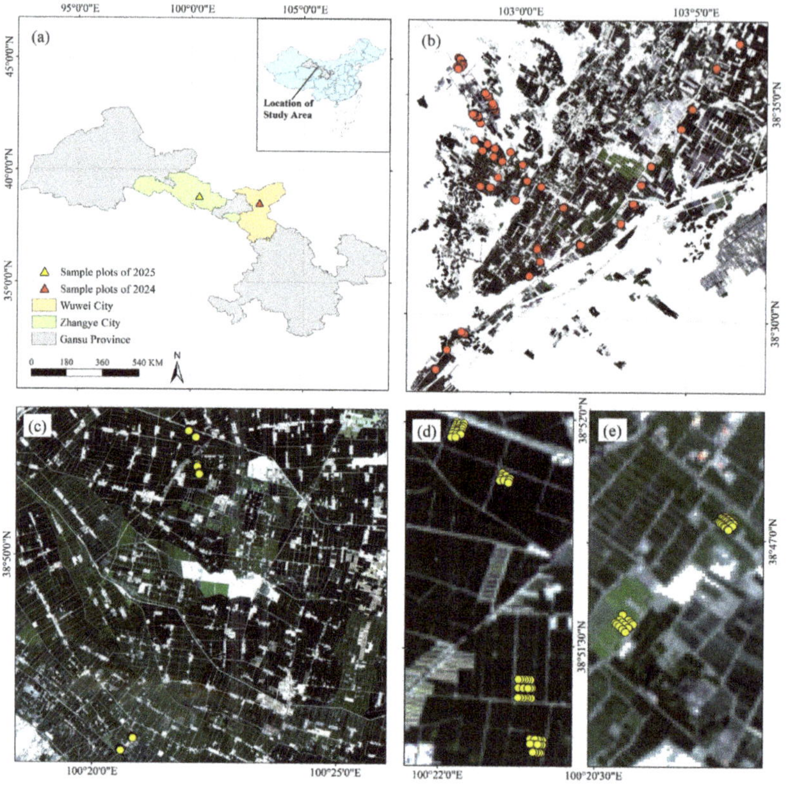

Figure 1. Study area and sampling design. (**a**) Geographic locations of the two study sites in Gansu Province, China: Minqin County and Ganzhou District, Zhangye; (**b**) Spatial distribution of field sampling points acquired in 2024 (red markers). Base map: Sentinel-2B Level-2A true-colour composite (B4/B3/B2), ESA/Copernicus (acquisition: 29 August 2024); (**c**) Spatial distribution of field sampling points acquired in 2025 (yellow markers). Base map: Sentinel-2B Level-2A true-colour composite (B4/B3/B2), ESA/Copernicus (acquisition: 8 July 2025); (**d**) Detailed view of the four intensive sampling subplots in the northern sector of panel (**c**). Base map: same as panel (**c**); (**e**) Detailed view of the two intensive sampling subplots in the southern sector of panel (**c**). Base map: the same as panel (**c**). All base maps showing bottom-of-atmosphere (Level-2A) reflectance were resampled to a common 10 m grid for display, and are presented in WGS84 (EPSG:4326). Data source: Copernicus Sentinel-2 MSI (© Copernicus Sentinel data 2024/2025).

Both sites lie in the mid-latitude zone of northwest China and are characterised by a warm-temperate to temperate continental arid–semi-arid climate. Minqin, located near the desert margin, exhibits pronounced aridity and desertification, whereas Ganzhou (Zhangye) is influenced by the summer monsoon and contains irrigated oasis agriculture with more concentrated precipitation. The landscape is dominated by plains and oasis farmland at low to moderate elevations; summers are hot by day and cool at night, annual precipitation is low, and rainfall is strongly seasonal. Major crops include winter wheat, spring/summer maize, and potato; the maize growing season typically extends from late June to mid–late September.

To balance regional representativeness with accurate pixel-scale matching to satellite data, two complementary ground-sampling strategies were adopted. In Minqin (modelling/testing site), 101 sparsely distributed sample points were established (inter-point distance $\geq$ 300 m) to capture landscape-scale heterogeneity in topography, soil, and management practices, and to support regional generalisation tests. In Zhangye (transfer/validation site), we selected three representative 500 × 500 m blocks and deployed dense sampling grids within each block: every block contained two 48 × 48 m sampling plots in which points were laid out along three transects at 2 m intervals (≈81 points per plot) to provide robust, pixel-scale LAI estimates. All field sampling locations were sited at least 10 m from field or canopy edges to minimise the probability of boundary/mixed-pixel contamination.

Canopy LAI was measured in situ with a LAI-2200 Canopy Analyzer (LI-COR, Lincoln, NE, USA) following an A–B–B–B–B sequence (one above-canopy reading "A" alternated with four below-canopy readings "B") using the 270° view cap. For each sampling location, the four below-canopy B readings were taken to capture ridge–row variability: on the ridge, at 1/4 of the inter-row distance, at the inter-row centre, and at 3/4 of the inter-row distance. LAI values were derived using the standard LI-COR processing workflow: the single above-canopy (A) reading was paired with each below-canopy (B) reading and the LAI-2200 software (gap-fraction algorithm) produced four instantaneous PAI/LAI estimates; these four estimates were then averaged to yield the final LAI assigned to that sampling position. No additional clumping correction was applied because all field measurements were acquired within diffuse-sky conditions (uniform overcast or thick, spatially uniform cloud cover) in which directional illumination effects are minimal and the uncorrected gap-fraction retrieval is appropriate (see the LI-COR user manual). For plot- or pixel-level validation, individual point LAI measurements within a sampling plot were averaged to obtain a single representative LAI for comparison with the 10 m Sentinel-2 retrieval.

2.2. Satellite Imagery Acquisition and Processing

Sentinel-2 is an optical Earth-observation mission within the Copernicus programme, comprising two twin satellites (Sentinel-2A, launched June 2015, and Sentinel-2B, launched March 2017). Its MultiSpectral Instrument (MSI) acquires 13 spectral bands spanning the visible, red-edge, near-infrared, and shortwave-infrared portions of the spectrum, with nominal spatial resolutions of 10 m, 20 m, and 60 m. For this study, we used Level-2A bottom-of-atmosphere (BOA) surface reflectance products obtained from the ESA data platform (Dataspace); these products are radiometrically calibrated and atmospherically corrected and are therefore appropriate for quantitative vegetation retrieval.

Two Sentinel-2B scenes (29 August 2024 and 8 July 2025) were used. All reflectance bands were resampled to a common 10 m grid to ensure tight spatial co-registration for model input: continuous reflectance layers were resampled with bilinear or cubic-convolution interpolation to preserve spectral smoothness, while the Scene Classification Layer (SCL) was resampled with nearest-neighbour interpolation to avoid fractional/mixed class labels produced by continuous interpolants. Whole-scene cloud cover was 13.71% for the Minqin image and 13.57% for the Zhangye image; clouds were largely confined to desert areas, so we cropped each scene to the sampling-area centre before analysis. An SCL-derived quality mask was then applied to exclude clouds, cloud shadows, and other invalid pixels. To focus the retrieval on substantive vegetation, we applied a fixed NDVI mask: pixels with NDVI $\leq$ 0.20 were treated as non-vegetated and excluded from further analysis. The fixed-threshold approach defines a single lower NDVI bound below which vegetation is considered absent regardless of location or local conditions; NDVI values > 0.20 are interpreted as indicating increasing canopy cover (though the NDVI–cover relationship is not strictly linear) [18]. This conservative, widely used threshold helps remove bare soil and very sparse vegetation prior to LAI retrieval.

The final analysis dataset comprises spatially co-registered, quality-controlled 10 m multispectral reflectance layers used for LAI retrieval and uncertainty analysis: B2 (blue), B3 (green), B4 (red), B7 (red-edge, resampled from 20 m to 10 m), and B8 (near-infrared). Band centre wavelengths and native spatial resolutions are summarised in Table 1. The choice of this five-band set balances spectral sensitivity to canopy structure and the practical considerations of native spatial resolution and noise: red-edge and NIR channels (B7, B8, B8A) are known to carry information relevant to LAI and canopy structure (ESA Sentinel-2 MSI Technical Guide), but B8A (20 m) overlaps substantially with B8 while requiring upsampling that can introduce scale-mismatch noise and amplify saturation effects at high LAI; therefore, B8 (10 m) was preferred to maintain native spatial consistency.

Table 1. Sentinel-2B bands used in this study.

Band	Description	Centre Wavelength	Resolution
B2	Blue	492.3 nm	10 m
B3	Green	558.9 nm	10 m
B4	Red	664.9 nm	10 m
B7	Red-edge	779.7 nm	20 m (resampled to 10 m)
B8	Near-infrared	832.9 nm	10 m

2.3. *PROSAIL Radiative Transfer Modelling*

The PROSAIL model couples the PROSPECT leaf-optical model with the SAIL canopy scattering model to provide a physics-based link between leaf optical properties and canopy-scale directional reflectance. In PROSPECT, a leaf is represented as a multilayer homogeneous plate; leaf absorption and scattering across the 400–2500 nm spectral domain are computed from biophysical and biochemical inputs including the leaf structure parameter (N), chlorophyll content (Cab), dry-matter content (Cm), and equivalent water thickness (Cw). The SAIL component treats the canopy as a homogeneous continuous medium and solves the radiative-transfer problem at canopy scale, accounting for canopy structural parameters (e.g., LAI and leaf angle distribution, parameterized here by average leaf angle, ALA) together with solar–sensor geometry to simulate bidirectional reflectance.

2.3.1. Sensitivity Analysis

To identify the PROSAIL inputs that most strongly influence simulated reflectance in the 450–950 nm interval, we adopted a two-step hierarchical sensitivity-analysis protocol. First, the Morris screening method was applied as a computationally efficient global filter to rank candidate parameters by their elementary effects [19]. Second, the subset of parameters selected by Morris was subjected to Sobol variance-based decomposition to obtain quantitative first-order (S_1) and total-effect (S_T) sensitivity indices [20].

For a model $Y = f(\mathbf{X})$, the Morris elementary effect (EE) for the i-th input at a sampling point **x** with perturbation Δ_i is defined as

$$EE_i(\mathbf{x}) = \frac{f(x_1, \ldots, x_i + \Delta_i, \ldots, x_d) - f(x_1, \ldots, x_i, \ldots, x_d)}{\Delta_i}. \quad (1)$$

Repeating the procedure over multiple, independent base points yields a sample of EE_i values for each parameter. The commonly reported statistics are the mean absolute elementary effect $\mu_i^* = \text{mean}(| EE_i |)$, which measures overall parameter importance, and the standard deviation $\sigma_i = \text{std}(EE_i)$, which indicates nonlinearity or interaction [21].

Morris screening was performed on nine candidate parameters (N, Cab, Car, Cw, Cm, LIDFa, LAI, hspot, and psoil) using 40 repeats and a perturbation amplitude equal to 10%

of each parameter's range. Parameters were ranked by μ_i^* and the top seven were retained for subsequent Sobol analysis.

Sobol variance decomposition was then applied to the selected seven-parameter subset. Using Monte Carlo sampling with a base sample size $N = 1000$, we estimated first-order indices S_1 and total-effect indices S_T across the 450–950 nm spectral interval. For summary reporting, wavelength-resolved S_1 and S_T curves were averaged over 450–950 nm to yield scalar importance measures. The Sobol runs required a total of 9000 PROSAIL model evaluations. This hierarchical scheme reduces the dimensionality of the Sobol calculation while retaining the ability to quantify main effects and interactions reliably.

This apparent discrepancy highlights the different emphases of the two methods: Morris μ^* measures the absolute magnitude of local responses to finite perturbations and is therefore sensitive to parameters that induce strong spectral derivatives in localised wavelength bands [21]; Sobol indices quantify each parameter's contribution to the overall output variance across the full input space [22]. Combining Morris screening and Sobol decomposition therefore provides an efficient strategy to (i) identify parameters that produce strong local spectral responses and (ii) select those that drive population-scale variance—a useful basis for reducing inversion dimensionality and for designing representative forward simulation libraries.

2.3.2. Synthetic Sample Generation

Guided by the sensitivity analysis, seven influential parameters were selected for sampling: N, Cab, Cm, Cw, LAI, psoil, and ALA. Each parameter was sampled independently from a uniform distribution over a physically plausible range to generate a synthetic ensemble of 30,000 parameter combinations (Table 2).

Table 2. PROSAIL (version D) input parameters and sampling ranges used to generate the simulated dataset.

Radiative Transfer Model	Parameter	Symbol	Unit	Ranges
PROSPECT-D	Leaf structure index	N	unitless	1.0–2.0
	Chlorophyll a + b content	Cab	μg cm^{-2}	30–70
	Leaf mass per area	Cm	g cm^{-2}	0.004–0.007
	Equivalent water thickness	Cw	g cm^{-2}	0.005–0.030
SAIL	Leaf area index	LAI	m^2 m^{-2}	0.2–8.0
	Soil reflectance	psoil	unitless	0–1.0
	Average leaf inclination angle	ALA	degrees	10–70
	Solar zenith angle	tts	degrees	Fixed *
	Sensor zenith angle	tto	degrees	Fixed *
	Relative azimuth angle	psi	degrees	Fixed *

* tts, tto, and psi were set to the Sentinel-2B observation geometry for each scene; parameters not included in the sampling set were held at empirically based constant values. For uncertainty-propagation experiments, we additionally applied a small relative perturbation ($\pm$1%) to these fixed geometry parameters to represent minor sensor–scene geometry jitter.

For each sampled parameter vector, PROSAIL produced a continuous reflectance spectrum $\rho(\lambda)$. These spectra were convolved with the Sentinel-2B SRFs (Figure 2) to synthesise band-level bottom-of-atmosphere reflectances corresponding to bands B2, B3, B4, B7, and B8:

$$\rho_{band} = \frac{\int_{\lambda_1}^{\lambda_2} SRF(\lambda) \cdot \rho(\lambda) d\lambda}{\int_{\lambda_1}^{\lambda_2} SRF(\lambda) d\lambda}. \tag{2}$$

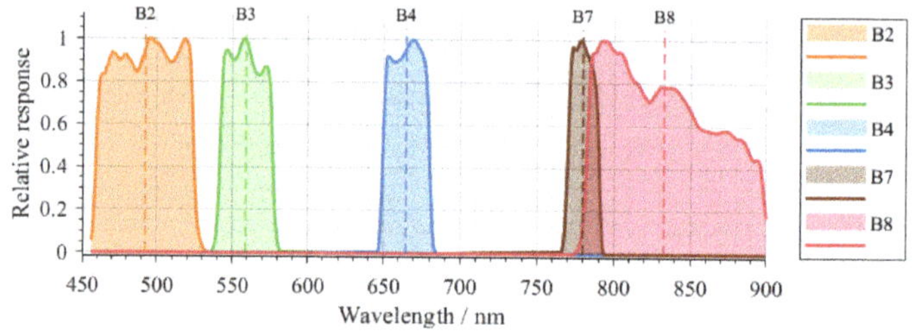

Figure 2. Sentinel-2B SRFs for bands B2, B3, B4, B7 and B8.

These band-aggregated reflectances, together with the associated LAI values, formed a 30,000-sample simulated training library used for subsequent NN–GA model development and uncertainty quantification.

2.4. NN–GA Coupled Inversion Framework

A fully connected MLP with two hidden layers was adopted to map Sentinel-2B multispectral reflectances to LAI. The network accepts a 5-dimensional input vector (band reflectances B2, B3, B4, B7, and B8) and outputs a single scalar (LAI). The overall inversion architecture is illustrated in Figure 3.

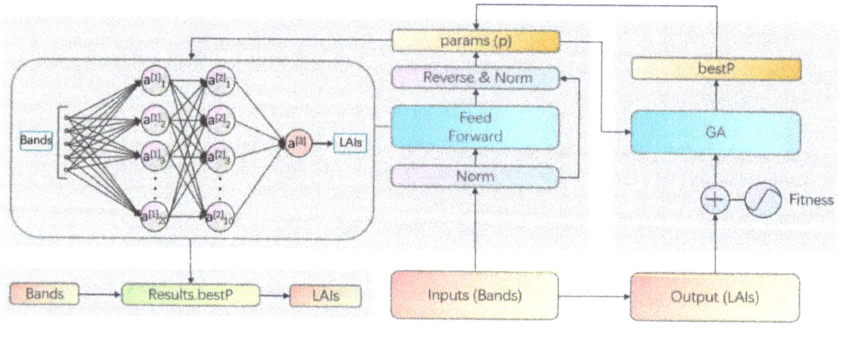

Figure 3. NN–GA coupled inversion framework for LAI retrieval.

To promote global optimisation and explicitly favour generalisation, a population-based genetic algorithm [23–25] was used to optimise all network weights and biases. The PROSAIL-derived simulated dataset (30,000 samples) was randomly partitioned into training, validation, and test subsets in a 60:20:20 ratio, where the training portion supplies sufficient examples to fit the multi-parameter MLP, the validation portion is reserved to guide model selection and to guard against overfitting during optimisation, and the held-out test portion provides an unbiased estimate of final generalisation performance. All network parameters (weights and biases) were encoded as real-valued chromosomes and evolved by the GA. The GA fitness function was defined as a weighted sum of training and validation root-mean-square errors (RMSEs):

$$\mathrm{Fit}(p) = \mathrm{RMSE}_{\mathrm{train}}(p) + 2 \times \mathrm{RMSE}_{\mathrm{val}}(p), \tag{3}$$

$$\mathrm{RMSE}_{\mathrm{train}}(p) = \sqrt{\frac{1}{N_{\mathrm{t}}}\sum_{i\in\mathcal{T}}(y_i - \hat{y}_i(p))^2},\ \mathrm{RMSE}_{\mathrm{val}}(p) = \sqrt{\frac{1}{N_{\mathrm{v}}}\sum_{i\in\mathcal{V}}(y_i - \hat{y}_i(p))^2}, \tag{4}$$

where p denotes a GA individual encoding the network parameters, $\mathcal{T}$ and $\mathcal{V}$ are the index sets of training and validation samples with sizes N_{t} and N_{v}, respectively, y_i is the reference (normalised) LAI of sample i, and $\hat{y}_i(p)$ is the MLP prediction under parameters p. The validation error is weighted more heavily to explicitly promote generalisation during evolution and to mitigate overfitting.

Chromosome values were constrained to the interval $[-5, 5]$. The GA optimisation was executed independently 30 times and the individual with the best fitness was selected as the final network. We emphasise that the GA is not employed because the MLP is non-differentiable (the MLP and its activation functions are differentiable), but because evolutionary, population-based search offers practical advantages for this application. In particular, a GA enables the direct optimisation of bespoke, potentially non-differentiable fitness formulations (e.g., our weighted training/validation RMSE), explores highly non-convex and multimodal parameter landscapes to reduce sensitivity to initial weights and local minima, and naturally handles mixed or constrained parameter encodings and custom feasibility rules used during evolution [12,13]. These properties have motivated prior uses of evolutionary algorithms for neural-network training and are valuable when robustness, reproducibility, and custom fitness definitions are required. The NN–GA hybrid therefore marries the expressive power of the MLP with the GA's global search capability to improve stability and generalisation when decoding simulated spectra and applying the trained model to Sentinel-2B imagery.

2.5. Accuracy Assessment

Model performance was assessed using the coefficient of determination (R^2) and the root-mean-square error (RMSE). R^2 is defined as

$$R^2 = 1 - \frac{\sum_{i=1}^{n} (y_i - \hat{y}_i)^2}{\sum_{i=1}^{n} (y_i - \overline{y})^2}, \tag{5}$$

where y_i and $\hat{y}_i$ are the observed and predicted LAI values, respectively, and $\overline{y}$ is the mean of the observations. RMSE is defined as

$$\text{RMSE} = \sqrt{\frac{1}{n}\sum_{i=1}^{n} (y_i - \hat{y}_i)^2}. \tag{6}$$

Higher R^2 values (approaching 1) and lower RMSE values (approaching 0) indicate better agreement between predictions and observations. These metrics were computed on the independent test sets and on the spatially co-located field observations used for Sentinel-2 validation.

2.6. Uncertainty Quantification (UQ) Method

Uncertainty quantification follows the measurement-model paradigm of the Guide to the Expression of Uncertainty in Measurement (GUM) [26]: the quantity of interest is treated as a function of input quantities, and the propagation of their uncertainties through that function yields an uncertainty estimate for the output. In the LAI retrieval context, the measurement model can be written as

$$y = f(x_1, x_2, \ldots, x_i, \ldots), \tag{7}$$

where y denotes retrieved LAI and each input x_i has an estimable central value $\overline{x}_i$ and standard uncertainty u_{x_i}.

Because LAI retrieval typically involves highly nonlinear, numerically driven algorithms, we use two complementary UQ approaches: the law of propagation of uncertainty (LPU), based on local linearization, and the Monte Carlo method (MCM), based on sampling [26–28]. The LPU employs a first-order Taylor expansion to linearize the measurement function locally and propagates input uncertainties via analytical or numerical derivatives. In contrast, MCM draws large numbers of samples from the input error (or prior) distributions, propagates each sample through the full (possibly nonlinear, black-box) retrieval pipeline, and empirically approximates the resulting output distribution [29].

2.6.1. Physics-Driven-Stage Uncertainty

In this study, the physics-driven-stage is defined as the propagation of uncertainty arising from variability in non-LAI PROSAIL inputs (e.g., Cab, Cm, Cw, N, psoil, ALA) through the forward model and the Sentinel-2 spectral response convolution into band reflectances. For the physics-driven component, and for each fixed LAI level, non-LAI PROSAIL inputs x_i are modelled as truncated normal perturbations $x_i \sim \mathcal{N}(\overline{x}_i, \sigma^2_{x_i})$. The nominal value $\overline{x}_i$ is set to the midpoint of the prescribed range and σ_{x_i} is set to one-sixth of the range so that $\pm 3\sigma$ approximately covers the interval. For a given LAI level, N Monte-Carlo draws $\left\{\mathbf{x}^{(j)}\right\}_{j=1}^{N}$ are drawn from the joint (marginal, initially independent) distributions and each sample is propagated through PROSAIL and the Sentinel-2 SRFs to produce band reflectances $y^{(j)} = f(\mathbf{x}^{(j)})$. From the output sample set $\left\{y^{(j)}\right\}$, the sample mean $\overline{y}$ and the relative standard uncertainty associated with perturbations of the inputs can be computed. In particular, the relative uncertainty induced by the input ensemble is estimated as the coefficient of variation in the output:

$$u(x_i) = \sqrt{\frac{\frac{1}{N-1}\sum_{j=1}^{N}\left(y^{(j)} - \overline{y}\right)^2}{\overline{y}^2}}. \quad (8)$$

To account for statistical dependence between PROSAIL inputs, pairwise associations between input-induced output responses are estimated and propagated. For two distinct inputs x_p and x_q, the corresponding output response vectors $\mathbf{Y}_p = (y_p^{(1)}, \ldots, y_p^{(N)})$ and $\mathbf{Y}_q = (y_q^{(1)}, \ldots, y_q^{(N)})$ are formed (each obtained by perturbing one parameter while sampling others). The Pearson correlation coefficient between these two response series is computed as

$$r(x_p, x_q) = \frac{\sum_{j=1}^{N}\left(y_p^{(j)} - \overline{y}_p\right)\left(y_q^{(j)} - \overline{y}_q\right)}{\sqrt{\sum_{j=1}^{N}\left(y_p^{(j)} - \overline{y}_p\right)^2}\sqrt{\sum_{j=1}^{N}\left(y_q^{(j)} - \overline{y}_q\right)^2}}, \quad (9)$$

where $\overline{y}_p$ and $\overline{y}_q$ are the sample means of $\mathbf{Y}_p$ and $\mathbf{Y}_q$, respectively. The corresponding error covariance is then approximated by

$$u(x_p, x_q) = r(x_p, x_q)\, u(x_p)\, u(x_q). \quad (10)$$

With per-input variances $u^2(x_i)$ and covariances $u(x_i, x_j)$ thus estimated, the law of propagation of uncertainty (LPU) is applied to obtain the band-level output variance for the forward model $y = f(\mathbf{x})$:

$$u^2_{\text{overall}}(y) = \sum_{i=1}^{n}\left(\frac{\partial f}{\partial x_i}\right)^2 u^2(x_i) + 2\sum_{i=1}^{n-1}\sum_{j=i+1}^{n}\left(\frac{\partial f}{\partial x_i}\right)\left(\frac{\partial f}{\partial x_j}\right)u(x_i, x_j). \quad (11)$$

The sensitivity coefficients $\partial f/\partial x_i$ are evaluated numerically by central finite differences (appropriate when no closed-form derivative is available) [30]:

$$\frac{\partial f}{\partial x_i} \approx \frac{f(\ldots, x_i + h, \ldots) - f(\ldots, x_i - h, \ldots)}{2h}, \quad (12)$$

With step h set to 1% of the corresponding input parameter range. In addition, to represent small sensor–scene geometry jitter during inversion, we independently perturb the observation geometry (solar zenith tts, sensor zenith tto, and relative azimuth psi) by ±1% when propagating uncertainties. For a fixed LAI level, this procedure yields band-level relative uncertainties $u_{\text{overall},b}$ for each Sentinel-2 band b. The bandwise uncertainties are then combined into a single physics-driven-stage uncertainty for that LAI level by a weighted-variance sum:

$$u_{\text{sim}}(\text{LAI}) = \sqrt{\sum_{b=1}^{k} WL_b u_{\text{overall},b}^2}, \quad (13)$$

where WL_b denotes the weight assigned to band b. It should be noted that band-width weighting is a pragmatic and computationally efficient approximation to account for differing spectral information content and to partially accommodate inter-band correlation.

Finally, the representative physics-driven-stage uncertainty reported in the paper is obtained by RMS aggregation over a discrete set of L LAI levels:

$$u_{\text{sim}} = \sqrt{\frac{1}{L}\sum_{i=1}^{L} u_{\text{sim}}^2(\text{LAI}_i)}. \quad (14)$$

This workflow preserves the sample-based strengths of Monte-Carlo propagation while introducing a straightforward, empirically grounded estimate of input parameter dependence. The approach balances improved realism (through covariance terms) with computational tractability for the multi-parameter PROSAIL forward model.

2.6.2. Machine-Learning-Stage Uncertainty

To quantify uncertainty arising from the stochasticity of the NN–GA training, the full training pipeline was repeated R independent times to produce R trained models. Using an NDVI-guided sampling strategy, M representative pixels were drawn from the target scene. Let $y_{p,j}$ denote the LAI predicted for pixel p by model j. For each pixel, we compute the model mean and the absolute standard uncertainty:

$$\overline{y}_p = \frac{1}{R}\sum_{j=1}^{R} y_{p,j}, \; u_{\overline{y}_p} = \text{std}\{y_{p,j}\}, \quad (15)$$

And the pixelwise relative training uncertainty is $u_p^{(\text{train})} = u_{\overline{y}_p}/\overline{y}_p$. The summary machine-learning-stage relative uncertainty is the RMS over the M sampled pixels:

$$u_{\text{train}} = \sqrt{\frac{1}{M}\sum_{p=1}^{M} \left(u_p^{(\text{train})}\right)^2}. \quad (16)$$

2.6.3. Uncertainty Combination

The total uncertainty of a single LAI retrieval is decomposed into two stage-wise components (Figure 4). Building on the pixelwise machine-learning-stage uncertainty estimates, we compute the corresponding pixelwise physics-driven-stage uncertainties for the selected sample pixels and then evaluate the statistical dependence between the two components. The total variance is therefore expressed as

$$u_{\text{total}}^2 = u_{\text{sim}}^2 + u_{\text{train}}^2 + 2\,\text{cov}(u_{sim}, u_{train}), \quad (17)$$

where u_{sim} denotes the physics-driven-stage relative standard uncertainty and u_{train} denotes the machine-learning-stage relative standard uncertainty. The term $\text{cov}(u_{\text{sim}}, u_{\text{train}})$ is the covariance between the two stage-wise relative uncertainties (evaluated across the representative pixel set). In practice, we estimate this covariance empirically as the sample covariance over M representative pixels:

$$\text{cov}(u_{\text{sim}}, u_{\text{train}}) = \frac{1}{M-1}\sum_{p=1}^{M}(u_{\text{sim},p} - \overline{u}_{\text{sim}})(u_{\text{train},p} - \overline{u}_{\text{train}}), \tag{18}$$

where $\overline{u}_{\text{sim}}$ and $\overline{u}_{\text{train}}$ are the sample means. If the empirical covariance is negligible, Equation (17) reduces to the usual LPU (quadrature) combination $u_{\text{total}} = \sqrt{u_{\text{sim}}^2 + u_{\text{train}}^2}$; otherwise, the covariance term is retained to account for statistical dependence between the two stages.

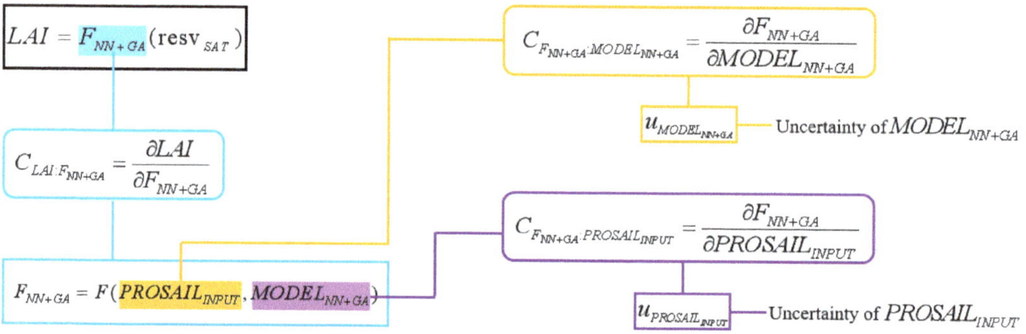

Figure 4. Uncertainty decomposition and computational workflow for LAI retrieval.

2.7. Workflow Summary

The technical workflow of the proposed physics-driven machine-learning framework is summarised in Figure 5 and comprises three sequential steps. In Step 1, a hierarchical sensitivity analysis (Morris screening followed by Sobol decomposition over 450–950 nm) identified seven influential PROSAIL inputs; PROSAIL was sampled across prior ranges and simulated spectra were convolved with Sentinel-2B SRFs to synthesise a 30,000-sample five-band BOA reflectance library (B2, B3, B4, B7, B8). In Step 2, a two-hidden-layer MLP was trained on the simulated library with all weights and biases globally optimised by a GA (NN–GA); the optimal model was then applied to Sentinel-2B reflectance imagery to produce spatial LAI maps. In Step 3, the total relative uncertainty of the LAI product was decomposed into the physics-driven-stage component (uncertainty propagated from non-LAI PROSAIL input variability to band reflectances, quantified by MCM) and the machine-learning-stage component (prediction variability across repeated NN–GA trainings); each component was quantified independently and the two standard uncertainties were combined under the LPU to produce per-retrieval relative uncertainty estimates for the LAI maps.

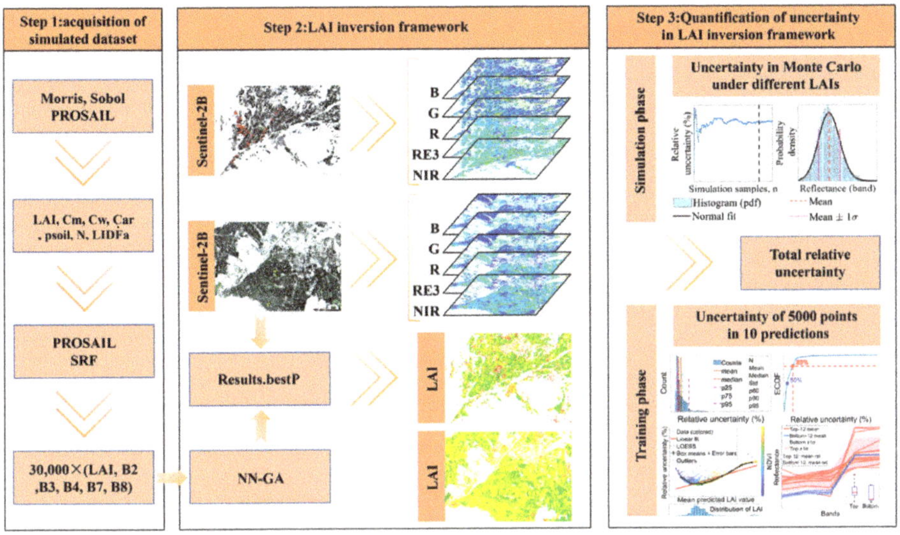

Figure 5. Workflow diagram illustrating computational steps for coupled PROSAIL–NN–GA LAI retrieval and uncertainty quantification.

3. Results

3.1. Parameter Sensitivity Analysis and Sample Generation Based on the PROSAIL Model

A hierarchical sensitivity-analysis protocol combining Morris screening and Sobol variance decomposition was applied to quantify the sensitivity of PROSAIL-simulated reflectance across 450–950 nm. The wavelength-resolved outcomes are summarised in Figure 6.

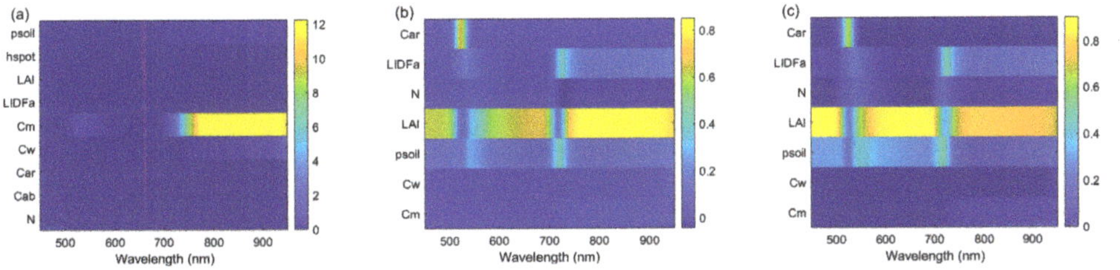

Figure 6. Sensitivity analysis results for PROSAIL-simulated reflectance (450–950 nm). (**a**) Morris μ_i^* heatmap showing the mean absolute elementary effects of model inputs across wavelength; (**b**) Sobol first-order index $S_1(\lambda)$ heatmap indicating the fraction of output variance attributable to each input; (**c**) Sobol total-effect index $S_T(\lambda)$ heatmap representing each input's overall contribution including interactions.

Morris screening (averaged across 450–950 nm) ranked the top seven parameters by the mean absolute elementary effect μ^* as: C_m (μ^* = 5.1220), C_w (μ^* = 0.1244), p_{soil} (μ^* = 0.0670), LAI (μ^* = 0.0254), N (μ^* = 0.0241), LIDFa (μ^* = 0.0012), and C_{ar} (μ^* = 0.0005). The Morris results indicate that C_m produces substantially larger local responses than the other parameters in the 700–950 nm region; i.e., the model exhibits high local sensitivity to C_m in the NIR.

Sobol variance decomposition applied to the Morris-selected subset provided complementary, population-level information. Averaged over 450–950 nm, LAI dominates the simulated reflectance variance: the mean first-order index for LAI is ≈0.64 (64%) and the mean total-effect index is ≈0.75 (75%), indicating that LAI accounts for the largest share of output variance with only partial entanglement in interactions. By contrast, although C_m ranks highest in Morris, its mean Sobol first-order contribution is small (mean $S_1 \approx 0.0115$), reflecting that C_m's strong influence is concentrated in a limited spectral/parameter region and therefore contributes little to the global variance under the prescribed input distributions.

Guided by the sensitivity analysis, seven influential parameters (listed above) were sampled uniformly within physically plausible priors to generate the synthetic dataset. PROSAIL spectra were convolved with the Sentinel-2B SRFs to produce band-level (B2, B3, B4, B7, B8) bottom-of-atmosphere reflectances paired with LAI. The resulting simulated data distributions for the site are shown as histograms with kernel density estimates in Figure 7.

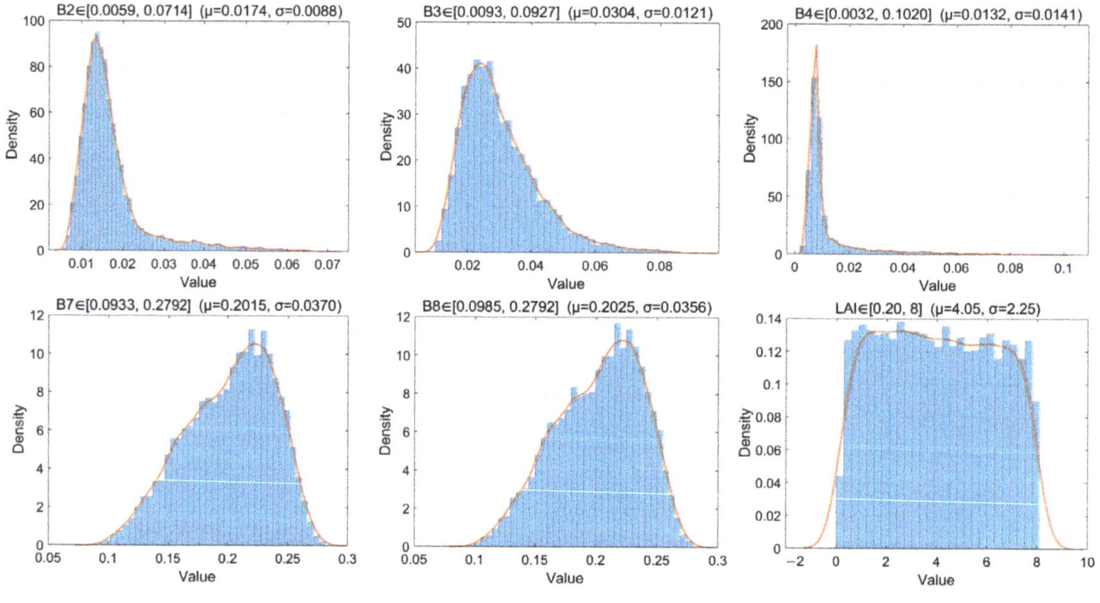

Figure 7. Distributions (histogram + kernel density estimate) of the simulated Sentinel-2B five-band reflectances (B2, B3, B4, B7, B8) and associated LAI values used for model training and uncertainty analysis (30,000 samples).

3.2. Crop LAI Retrieval and Accuracy Analysis Based on NN–GA

3.2.1. Crop LAI Retrieval

The convergence, stability, and generalisation behaviour of the NN–GA training were examined via the diagnostic metrics summarised in Figure 8. The diagnostics include: generation-wise best and mean fitness, a smoothed view of the best fitness trend (moving average and median), a per-generation stall indicator, and the cumulative relative improvement referenced to generation 1.

At initialization, the best individual fitness was 1.56; the algorithm reduced this value rapidly during early evolution, reaching the run minimum of 0.5386 at generation 89 and finishing with a best-of-run value of 0.58 at the final generation (i.e., the final best is slightly above the observed run minimum). The population mean fitness fell from 7.62 at initialization to ≈0.87 at termination, indicating consistent improvement across the population as a whole. Most of the aggregate fitness reduction occurred in the early-to-mid phase of the run (roughly generations 10–40), corresponding to broad global exploration; later generations produced smaller, incremental gains consistent with local exploitation and fine-tuning. Calculating relative improvement from the initial best to the run minimum gives $(1.56 - 0.5386)/1.56 \approx 66\%$. Performance on the simulated test data indicates that the NN–GA framework locates stable, high-quality solutions: the training-set RMSE was 0.73 and the independent test-set RMSE was 0.85, demonstrating reasonable generalisation while leaving scope for further improvements (e.g., cross-validation or local fine-tuning). Overall, the GA-based global search efficiently identified promising regions of the parameter space and produced reproducible network solutions suitable for application to Sentinel-2B imagery.

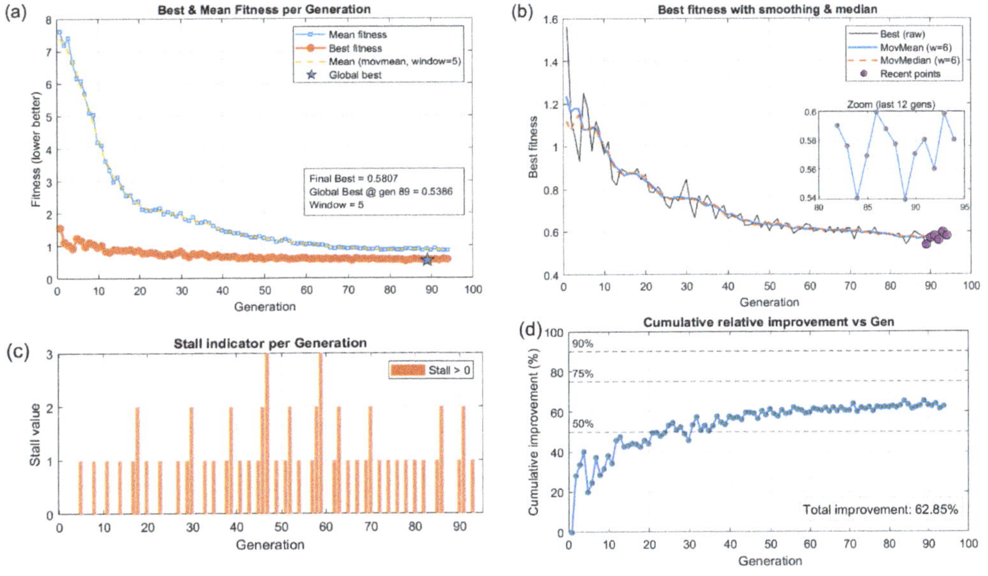

Figure 8. Training optimisation diagnostics for the NN–GA inversion framework. (**a**) Best and mean fitness per generation; (**b**) smoothed best fitness trend (moving average and median); (**c**) per-generation stall indicator; (**d**) cumulative relative improvement versus generation 1 (percentage).

3.2.2. Accuracy Assessment of LAI Retrieval

To assess the practical performance and transferability of the trained NN–GA mapping, we applied the model to near-synchronous Sentinel-2B scenes and validated the pixelwise retrievals against LI-COR LAI-2200 field measurements. The spatial distributions of predictions and validation samples for Minqin (modelling/testing area) and Zhangye (transfer/validation area) are shown in Figure 9.

Pointwise comparison (Figures 10 and 11) returns the following validation statistics: Minqin (101 measured points)—RMSE = 0.44, R^2 = 0.73; Zhangye (112 measured points)—RMSE = 0.40, R^2 = 0.56. The Minqin results show tighter scatter about the 1:1 line and higher explained variance, whereas the Zhangye results exhibit greater dispersion and unexplained variability despite a comparable RMSE.

Residual diagnostics for Zhangye (Figure 12) indicate a small overall bias (0.015) and a weak pattern of overestimation at low LAI and underestimation at high LAI, with near-zero bias in the midrange (≈2.75–4.25).

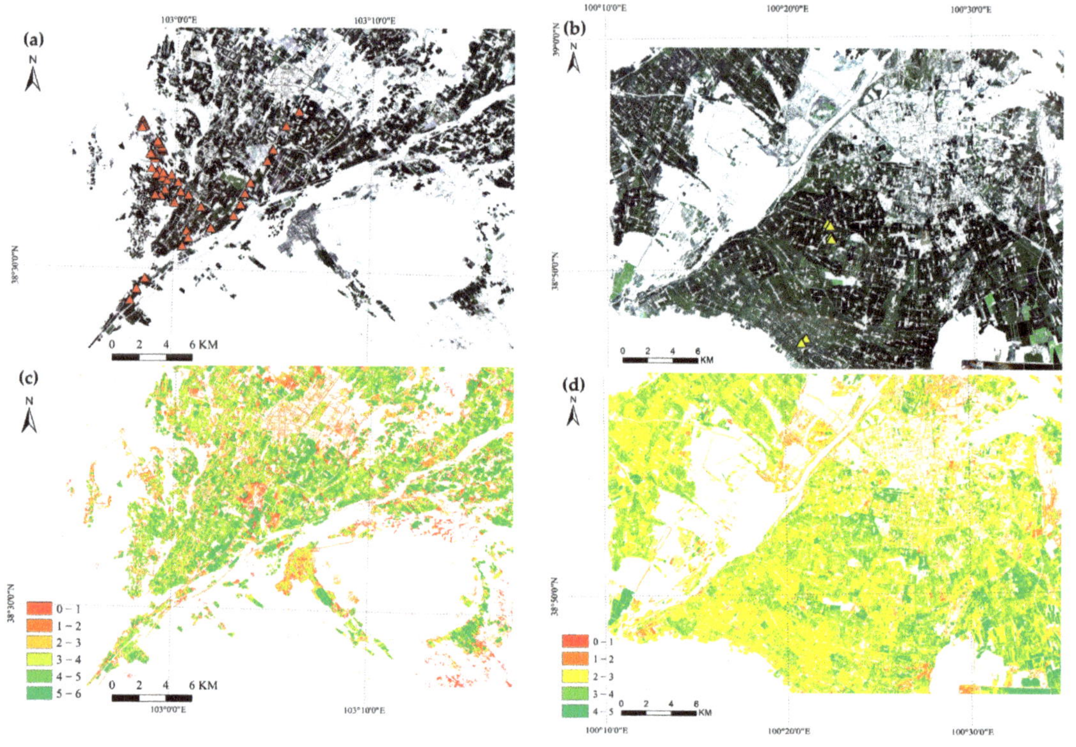

Figure 9. Spatial distributions of predicted LAI and ground-sample locations. (**a**) True-colour composite of the Minqin site; red triangles denote LAI-2200 measurement locations. (**b**) True-colour composite of the Zhangye site; yellow triangles denote LAI-2200 measurement locations. (**c**) LAI retrieval map for Minqin. (**d**) LAI retrieval map for Zhangye.

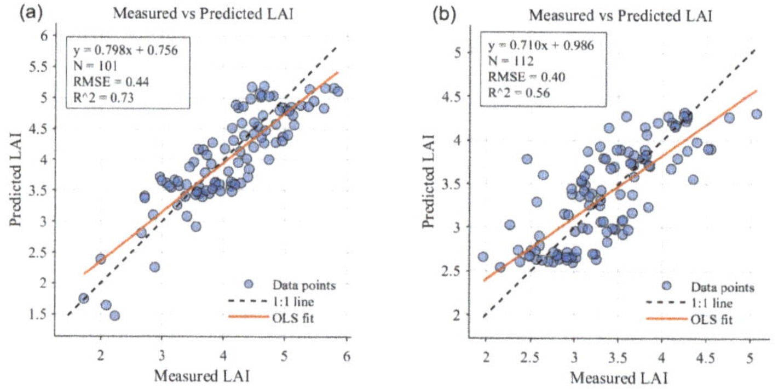

Figure 10. Observed versus predicted LAI (LAI-2200 vs. NN–GA). (**a**) Minqin (101 measured points). (**b**) Zhangye (112 measured points). Dashed line: 1:1 reference; solid line: least-squares regression.

Bland–Altman analysis produces 95% limits of agreement of approximately [−0.77, 0.80], implying that individual prediction errors can reach on the order of ±0.8 LAI units. Binned error statistics reveal a U-shaped dependence on observed LAI: both RMSE and MAE increase at low (<2.5) and high (>4.5) LAI, with several extreme residuals located in the distribution tails.

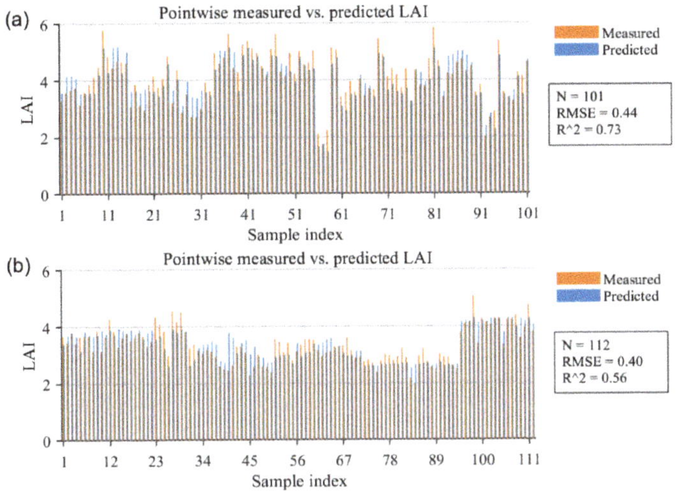

Figure 11. Observed versus predicted LAI (LAI-2200 vs. NN–GA) shown as point-by-point bar charts. (**a**) Minqin (101 measured points). (**b**) Zhangye (112 measured points). Orange and blue bars indicate measured and predicted values, respectively.

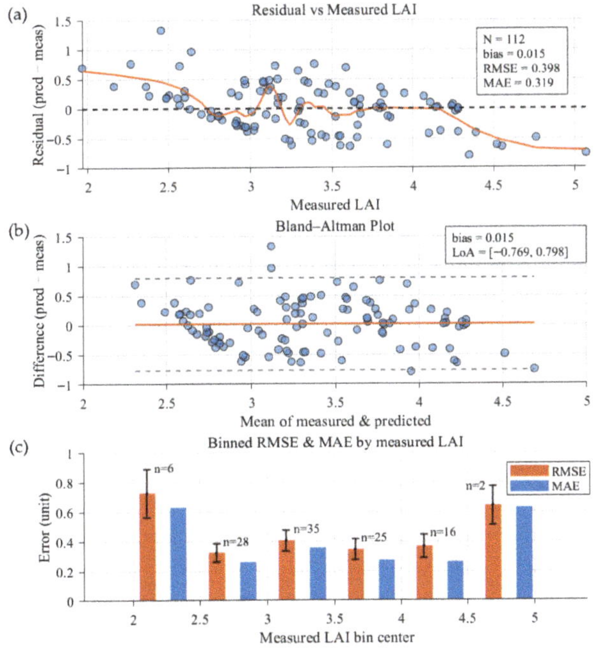

Figure 12. Residual analysis for Zhangye (Site 2; 112 measured points). (**a**) Residuals vs. measured LAI with a LOESS smoothed trend line; (**b**) Bland–Altman plot showing the mean bias (0.015, solid line) and 95% limits of agreement (dashed lines); (**c**) Binned error statistics grouped by measured LAI. Orange and blue bars denote RMSE and MAE, respectively; error bars represent the standard error.

In summary, the NN–GA framework provides satisfactory LAI estimates at both sites (Zhangye: RMSE = 0.40, MAE = 0.32, R^2 = 0.56), with stronger agreement at the model-development site. The error patterns—poorer performance at very low and very high LAI and the presence of outliers—are plausibly attributable to factors such as crop structural anomalies during the satellite overpass (e.g., exposed ears/inflorescences and altered leaf-angle distributions associated with tasseling) that increase optical heterogeneity [31], and to optical saturation and background-mixing effects that disproportionately affect extreme LAI values [32]. These observations motivate the following section's systematic uncertainty decomposition of the retrieval chain, which aims to identify dominant error sources and guide targeted improvements.

3.3. Quantification of LAI Retrieval Uncertainty via a Coupled Physics-Driven and Machine-Learning Approach

We decompose and quantify two stochastic uncertainty sources in the retrieval chain: the physics-driven-stage uncertainty arising from variability of non-LAI PROSAIL inputs propagated to Sentinel-2 band reflectances, and the machine-learning-stage uncertainty arising from variability across repeated NN–GA trainings. Below, we summarise the physics-driven-stage analysis and present band-wise and aggregated uncertainty results; Machine-Learning-Stage diagnostics follow.

To determine a stable Monte-Carlo sample size for the physics-driven-stage propagation, we examined the convergence of the estimated output uncertainty for a representative case with LAI fixed at 3. As shown in Figure 13, the estimated standard relative uncertainty decreases and stabilises as the number of Monte-Carlo draws N increases; beyond $N \approx 5000$, the uncertainty curve is effectively converged and fluctuations are substantially reduced. Balancing statistical stability and computational cost, we therefore adopted $N = 6000$ for all subsequent Monte-Carlo runs.

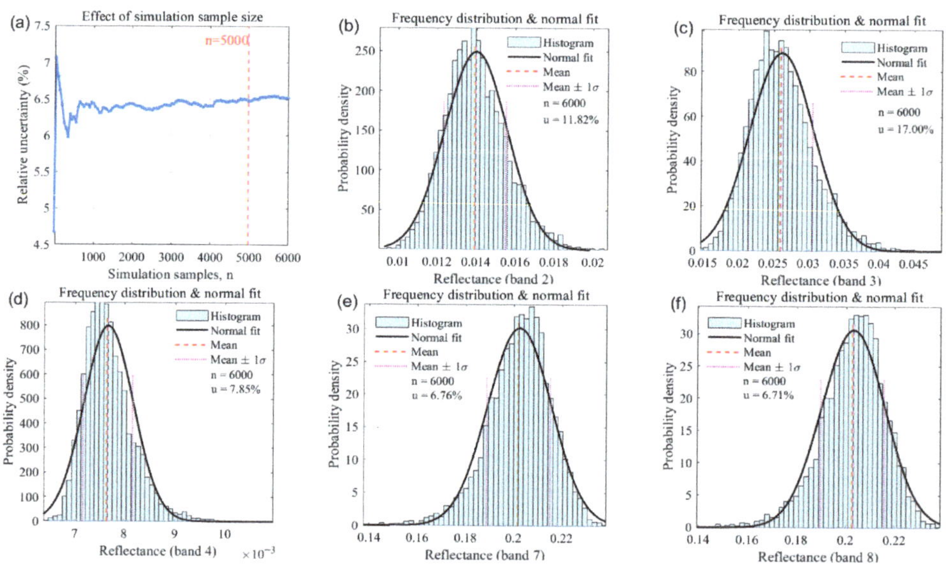

Figure 13. Physics-driven-stage uncertainty diagnostics and bandwise reflectance perturbation distributions (LAI = 3). (**a**) Monte Carlo convergence of the estimated standard relative uncertainty as a function of sample size (diagnostic used to select $N = 6000$); (**b**–**f**) show the bandwise reflectance distributions under parameter perturbations for Sentinel-2B bands B2, B3, B4, B7, and B8 (MinQin).

Applying the same Monte-Carlo propagation across the sampled LAI range, Tables 3 and 4 summarise the band-wise relative uncertainties and the aggregated (combined) standard relative uncertainty for Minqin (modelling/testing area) and Zhangye (transfer/validation area), respectively. When LAI is fixed and other PROSAIL inputs are perturbed within the prescribed prior ranges, the aggregated physics-driven-stage standard relative uncertainties are 11.42% for Minqin and 11.48% for Zhangye. The slightly larger value for Zhangye primarily reflects differences in the prescribed prior ranges (notably the soil-background factor psoil) used in the two site-specific simulations. Band- and LAI-resolved patterns in Tables 3 and 4 indicate that relative uncertainty depends on both LAI and wavelength: certain bands (e.g., B3 and B4 in several LAI bins) exhibit higher relative uncertainty, and the aggregated uncertainty varies across LAI levels where spectral sensitivity to perturbed inputs is larger. These estimates form the basis for combination with machine-learning-stage uncertainty and for producing uncertainty maps of the final LAI product.

Table 3. Band-wise and aggregated physics-driven-stage uncertainties for Minqin (modelling/testing area), listed by LAI bin.

LAI	B2 (%)	B3 (%)	B4 (%)	B7 (%)	B8 (%)	Input-Parameter Uncertainty (%)	Combined Standard Uncertainty (%)
1	15.17	15.26	20.78	10.96	10.96	14.35	11.42
2	9.97	15.22	9.75	8.05	8.00	10.09	
3	11.82	17.00	7.85	6.76	6.71	10.39	
4	12.91	17.68	11.42	5.99	5.95	11.00	
5	13.31	17.91	11.69	5.51	5.49	11.20	
6	13.46	18.00	12.03	5.23	5.22	11.25	
7	13.53	18.03	12.16	5.06	5.05	11.27	
8	13.57	18.06	12.21	4.97	4.96	11.28	

Table 4. Band-wise and aggregated physics-driven-stage uncertainties for Zhangye (transfer/validation area), listed by LAI bin.

LAI	B2 (%)	B3 (%)	B4 (%)	B7 (%)	B8 (%)	Input-Parameter Uncertainty (%)	Combined Standard Uncertainty (%)
1	15.33	15.33	21.37	11.15	11.13	14.50	11.48
2	9.93	15.19	10.14	8.30	8.24	10.21	
3	11.74	17.02	7.41	7.01	6.96	10.41	
4	12.91	17.76	10.27	6.22	6.18	11.03	
5	13.34	18.01	11.47	5.72	5.69	11.25	
6	13.49	18.09	11.86	5.40	5.39	11.30	
7	13.55	18.12	11.98	5.22	5.21	11.30	
8	13.57	18.13	12.01	5.11	5.10	11.29	

For the machine-learning-stage, the complete NN–GA training pipeline was repeated R = 10 independent times to produce R trained models. Using NDVI-guided sampling of the target scenes, M = 5000 representative pixels were drawn and each pixel was predicted by the R models. The aggregated machine-learning-stage standard relative uncertainty u_{train} was then computed as the root-mean-square (RMS) of pixelwise relative standard deviations. The resulting values are u_{train} = 18.06% for Minqin and u_{train} = 12.96% for Zhangye, indicating that stochastic factors in NN–GA training (e.g., random initialization and evolutionary search variability) substantially affect output stability.

Figures 14 and 15 present summary diagnostics of per-pixel machine-learning-stage relative uncertainty for Minqin and Zhangye, respectively. For Minqin (Figure 14), 75% of the sampled pixels have relative uncertainty $\leq$ 11.9% and 80% $\leq$ 14.1%; the empirical cumulative distribution shows a slow rise across the lower 80% and a steeper tail across the upper 20%, indicating a minority of pixels with markedly higher training uncertainty. A LOESS-smoothed scatter of mean predicted LAI versus relative uncertainty reveals a

U-shaped dependence: uncertainty is elevated at low (≤2) and high (≥6) LAI and minimal in the mid-range (3–6). Spectral inspection shows that many high-uncertainty pixels are mixed-boundary pixels (vegetation–soil/road), supporting the hypothesis that mixed pixels and limited training coverage in distribution tails drive elevated training uncertainty. Zhangye (Figure 15) displays broadly similar behaviour, indicating spatial transferability of the NN–GA uncertainty patterns.

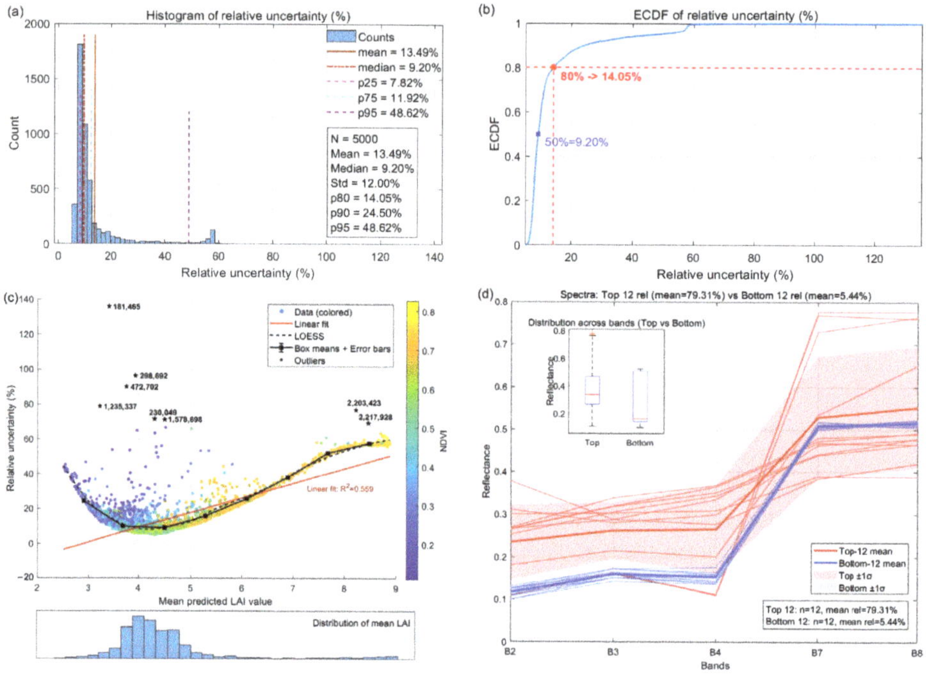

Figure 14. Comprehensive analysis of per-pixel machine-learning-stage relative uncertainty for Minqin (M = 5000). (**a**) Histogram; (**b**) ECDF; (**c**) mean predicted LAI vs. relative uncertainty with LOESS smoothing (point colour indicates NDVI); (**d**) mean spectra for Top-12 and Bottom-12 pixels.

We evaluated the statistical dependence between the physics-driven and machine-learning stage uncertainties and quantified its effect on the combined retrieval uncertainty. The results are reported for two study areas (N = 5000 sample pixels each).

In Minqin, the sample means of the stage-wise relative standard uncertainties are $u_{\text{sim}} = 0.1103$ and $u_{\text{train}} = 0.1350$. The paired-sample Pearson correlation between these per-pixel uncertainties is $r = 0.14$ (two-sided $p = 5.36 \times 10^{-24}$); a permutation test (2000 permutations) confirms that this correlation is highly unlikely under the null hypothesis of independence (permutation $p < 10^{-3}$). Bootstrap estimation (B = 2000) yields $\text{cov}(u_{\text{sim}}, u_{\text{train}}) \approx 1.4426 \times 10^{-4}$ with 95% CI = $[5.5880 \times 10^{-4}, 2.2389 \times 10^{-4}]$. Substituting the empirically estimated covariance into Equation (17), the combined standard uncertainty increases from ≈0.1743 (covariance neglected) to ≈0.1752 (with covariance), corresponding to an absolute increase of ≈8.26×10^{-4} and a relative change of about +0.47%. In Zhangye, we find $u_{\text{sim}} = 0.1044$, $u_{\text{train}} = 0.1060$, Pearson $r = 0.17$ (two-sided $p = 1.88 \times 10^{-34}$), and bootstrap covariance $\approx 4.2018 \times 10^{-5}$ (95% CI = $[3.5340 \times 10^{-5}, 4.8561 \times 10^{-5}]$). Including this covariance increases the combined standard uncertainty from ≈0.1488 to ≈0.1491, an absolute change of $\approx 2.82 \times 10^{-4}$ or +0.19% relative.

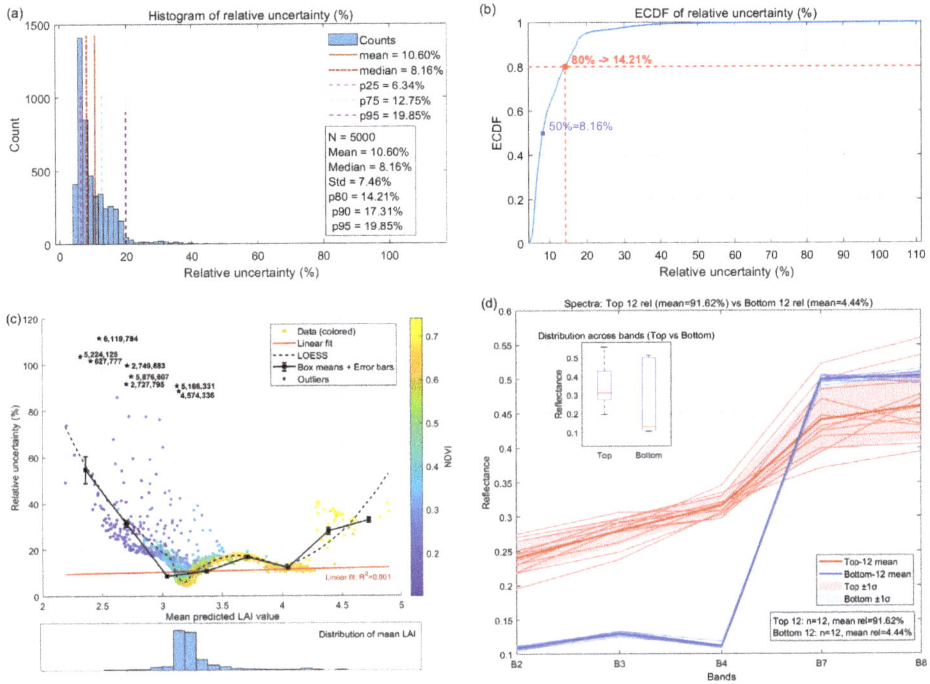

Figure 15. The same diagnostics as Figure 14, but for Zhangye.

Overall, although a weak but statistically significant positive correlation between the two stage-wise uncertainties is observed in both regions, the covariance term has a negligible numerical impact on the final combined uncertainty (well below 1% relative). The physics-driven-stage standard uncertainty u_{sim} and the machine-learning-stage standard uncertainty u_{train} were combined according to the law of propagation of uncertainty (root-sum-square of independent standard uncertainties) to yield the final per-retrieval relative uncertainty. Table 5 lists the combined results. After combination, the single-retrieval relative uncertainties are 21.37% for Minqin and 17.31% for Zhangye. Both values exceed the GCOS reference target of 15% for certain biophysical products, indicating that further methodological improvements—for example, refined sampling strategies, explicit handling of mixed pixels, inclusion of ancillary indices such as NDVI, or multisource data fusion—will be required to reach climate-grade uncertainty levels.

Table 5. Final relative uncertainties for single LAI retrievals after combination of physics-driven-stage and machine-learning-stage components.

Study Area	Component	Relative Uncertainty (%)	Combined Standard Relative Uncertainty (%)
Minqin	u_{sim}	11.42	21.37
	u_{train}	18.06	
Zhangye	u_{sim}	11.48	17.31
	u_{train}	12.96	

In summary, the staged UQ procedure identifies dominant contributors to retrieval uncertainty and points to concrete mitigation strategies (e.g., targeted augmentation of training samples in extreme LAI regimes, mixed-pixel treatment, and reduction in training stochasticity) that can directly reduce the overall LAI product uncertainty.

4. Discussion

4.1. Method Performance and Comparison

SL2P (Simplified Level-2 Product Processor) is a canopy biophysical retrieval algorithm developed by the European Space Agency for Sentinel-2 data [33]. The method uses feed-forward neural networks trained on a radiative-transfer model simulation library to combine physical realism with the computational speed of empirical retrievals [34]. SL2P is capable of efficiently estimating key biophysical variables from Sentinel-2 imagery, including leaf area index (LAI), fraction of absorbed photosynthetically active radiation (fAPAR), and fractional vegetation cover (FVC). Because SL2P delivers consistent wall-to-wall estimates at 10 m resolution and has been widely adopted as a baseline algorithm, we selected it as the primary retrieval method for this study [35].

In this work, we ran the SL2P implementation available in SNAP's S2_10m Biophysical Processor (SNAP v10.0.0). The processor was supplied with Sentinel-2 Level-2A bottom-of-atmosphere (BOA) reflectance as input and was executed with its default configuration: no user-tunable lookup tables (LUTs), priors, or inversion parameters were modified. Output biophysical products were quality-filtered using the Scene Classification Layer (SCL) to remove clouds, cloud shadow, and other invalid pixels. Because the S2_10m configuration uses only the 10 m bands, the resulting LAI, fAPAR, and FVC products are provided at 10 m spatial resolution.

The NN–GA retrieval was validated against site-level LI-COR LAI-2200 measurements. Pointwise statistics are: Minqin (modelling/testing area; 101 measured points)—RMSE = 0.44, R^2 = 0.73; Zhangye (transfer/validation area; 112 measured points)—RMSE = 0.40, R^2 = 0.56. For benchmarking, we compared these results with operational Sentinel-2 LAI estimates from the SNAP/SL2P processor evaluated on the same validation sets: Minqin (SL2P)—RMSE = 0.83, R^2 = 0.04; Zhangye (SL2P)—RMSE = 0.61, $R^2 = -0.03$. Under the present data and parameter settings, the NN–GA framework substantially reduces absolute error (ΔRMSE $\approx$ 0.39, $\sim$47% relative improvement for Minqin; ΔRMSE $\approx$ 0.21, $\sim$35% for Zhangye) and markedly increases explained variance versus the SL2P baseline.

Observed differences between the two study areas are largely attributable to sampling design and sample representativeness. Minqin was sampled sparsely across the landscape (101 points, inter-point distance $\geq$ 300 m) to capture regional heterogeneity, whereas Zhangye used dense, within-field sampling ($\approx$81 points per 48 $\times$ 48 m plot) to represent pixel-scale variability. Sparse designs tend to enlarge the reference variance and—when a model captures broad spatial contrasts—may yield higher R^2, but they also increase the risk of pixel–plot mismatch and measurement noise, which can inflate RMSE. Conversely, dense sampling typically reduces pixel–plot mismatch and RMSE, but may lower R^2 because of reduced sample variance. These sampling effects also help explain differences in SL2P performance across sites. Accordingly, algorithm comparisons and validation statements should explicitly account for sampling strategy and scale; we recommend reporting stratified or weighted validation statistics and clearly documenting sampling layouts to enable fair, reproducible assessments.

Overall, the NN–GA framework shows improved absolute accuracy and explanatory power relative to the SNAP/SL2P benchmark in our experiments, indicating stronger fit and transfer potential under the tested conditions. Nonetheless, the evaluation underscores that sampling design and measurement repeatability remain critical determinants of apparent performance and should be integral to future method comparisons and operational assessments.

4.2. Analysis of Uncertainty Sources

The two study areas exhibited broadly consistent uncertainty patterns (distributional shapes, LOESS-indicated nonlinearity, and the spectral/spatial characteristics of high-uncertainty pixels), suggesting that the proposed NN–GA retrieval and the staged UQ workflow generalise across the regions and phenological stages examined. Quantitatively, the machine-learning-stage uncertainty was comparable to or larger than the physics-driven-stage uncertainty: u_{train} = 18.06%/12.96% versus u_{sim} = 11.42%/11.48% for Minqin/Zhangye, respectively. This finding indicates that stochasticity in model training (e.g., random initialisation, evolutionary search variability, and hyperparameter choices) is a major contributor to overall retrieval variance and therefore a primary target for improvement.

We also note key simplifying assumptions in the present UQ analysis that may bias the absolute numerical estimates: (i) non-LAI PROSAIL inputs were modelled as truncated normal perturbations with independently applied draws; (ii) input parameter correlations were neglected; and (iii) NDVI-guided pixel sampling was used as a proxy for scene variability. These choices were adopted for tractability and reproducibility but restrict the interpretation of the reported uncertainty as a baseline conditional on the stated assumptions. Future work should relax these assumptions by explicitly incorporating parameter covariances, enlarging the prior sample space and observational constraints, and by adopting fully probabilistic (e.g., Bayesian) UQ approaches that jointly estimate parameter posteriors and predictive uncertainty. Additional practical avenues to reduce total uncertainty—suggested by our diagnostics—include (a) reducing training stochasticity via ensemble/aggregation methods, systematic hyperparameter optimisation, or regularisation; (b) targeted augmentation of training samples in LAI distribution tails; (c) explicit treatment of mixed pixels (sub-pixel unmixing or masking); and (d) fusion with ancillary sources (e.g., higher-resolution UAV data, topographic data, or management metadata). Implementing these strategies is expected to lower both the machine-learning-stage and physics-driven-stage contributions and to move operational LAI retrievals closer to application-level uncertainty targets.

5. Conclusions

We developed and validated a physics-driven machine-learning framework for crop Leaf Area Index (LAI) retrieval that jointly addresses accuracy and end-to-end UQ. The approach couples PROSAIL forward simulations (convolved with Sentinel-2B SRFs) to construct a 30,000-sample training library, trains a two-hidden-layer multilayer perceptron whose weights and biases are globally optimised by a genetic algorithm (NN–GA), and implements a two-stage uncertainty workflow: Monte-Carlo propagation to characterise uncertainty arising in the physics-driven-stage, and repeated independent NN–GA trainings to quantify uncertainty originating in the machine-learning-stage. The two standard uncertainty components are combined according to the law of propagation of uncertainty to yield a single, reproducible estimate of retrieval uncertainty.

A hierarchical sensitivity analysis confirmed that LAI is the dominant driver of reflectance variance across the visible–near-infrared bands considered, justifying the chosen parameter reduction and guiding efficient sampling. Validation against LI-COR LAI-2200 measurements at two northwest China sites demonstrates robust retrieval performance and cross-site transferability: Minqin (modelling/testing site) achieved RMSE = 0.43 and R^2 = 0.73, while Zhangye (transfer/validation site) achieved RMSE = 0.40 and R^2 = 0.56.

Compared with the SNAP/SL2P baseline, the NN–GA framework substantially reduced absolute error and increased explained variance at the sample scale.

The staged uncertainty decomposition exposes the relative contributions of the retrieval chain: physics-driven-stage standard relative uncertainties were 11.42% (Minqin) and 11.48% (Zhangye), while machine-learning-stage standard relative uncertainties were 18.06% (Minqin) and 12.96% (Zhangye). After variance synthesis, the single-run relative uncertainties are 21.37% (Minqin) and 17.31% (Zhangye). These results indicate that, under the current implementation, stochasticity associated with model training (initialization, hyperparameter search, and optimisation variability) is a major contributor to total uncertainty; reducing training variability (for example, via ensemble strategies, more systematic hyperparameter optimisation, or regularisation) should therefore be a priority to lower final product uncertainty.

In conclusion, the proposed NN–GA retrieval pipeline advances 10 m Sentinel-2 crop LAI estimation by combining improved accuracy with a transparent, reproducible uncertainty-quantification workflow. The framework supplies actionable uncertainty diagnostics that can guide targeted algorithmic improvements and supports the production of confidence-aware LAI products for agricultural monitoring and downstream applications. Future work should focus on (i) reducing machine-learning-stage stochasticity, (ii) incorporating parameter correlations and Bayesian UQ to relax current assumptions, and (iii) exploring multi-source data fusion and mixed-pixel handling to further tighten uncertainty and broaden operational applicability.

Author Contributions: Conceptualization, W.L. and X.Z.; methodology, W.L. and S.Y.; software, W.L. and S.Y.; validation, X.Z. and Z.G.; formal analysis, W.L. and S.Y.; investigation, W.L. and X.Z.; writing—original draft preparation, W.L. and S.Y.; writing—review and editing, X.Z. and Z.G.; supervision, X.Z. and Z.G. All authors have read and agreed to the published version of the manuscript.

Funding: This work was funded by the National Key Research and Development Program of China (2023YFB3905804; 2022YFB3903501), Key R&D Program of Shandong Province, China (2025CXGC010113); and the Future-Star program of the Aerospace Information Research Institute (E2Z106010F).

Data Availability Statement: The data presented in this study are available on request from the corresponding author, the data are not publicly available due to privacy.

Acknowledgments: The authors acknowledge the use of the Sentinel-2 Level-2A (MSI) product from the Copernicus Programme/European Space Agency (ESA), accessed via the Google Earth Engine platform. We thank the teams at ESA/Copernicus for their support and for providing the data that have been essential to our study.

Conflicts of Interest: The authors declare no conflicts of interest.

References

1. Chen, J.M.; Black, T.A. Defining leaf area index for non-flat leaves. *Plant Cell Environ.* **1992**, *15*, 421–429. [CrossRef]
2. Belward, A.; Bourassa, M.; Dowell, M.; Briggs, S.; Dolman, H.A.J.; Holmlund, K.; Husband, R.; Quegan, S.; Simmons, A.; Sloyan, B.; et al. *The Global Observing System for Climate: Implementation Needs*; WMO: Geneva, Switzerland, 2016.
3. Wei, S.; Yin, T.; Dissegna, M.A.; Whittle, A.J.; Ow, G.L.F.; Yusof, M.L.M.; Lauret, N.; Gastellu-Etchegorry, J. An assessment study of three indirect methods for estimating leaf area density and leaf area index of individual trees. *Agric. For. Meteorol.* **2020**, *292*, 108101. [CrossRef]
4. Yin, G.; Verger, A.; Qu, Y.; Zhao, W.; Xu, B.; Zeng, Y.; Liu, K.; Li, J.; Liu, Q. Retrieval of High Spatiotemporal Resolution Leaf Area Index with Gaussian Processes, Wireless Sensor Network, and Satellite Data Fusion. *Remote Sens.* **2019**, *11*, 244. [CrossRef]
5. Jacquemoud, S.; Verhoef, W.; Baret, F.; Bacour, C.; Zarco-Tejada, P.J.; Asner, G.P.; François, C.; Ustin, S.L. PROSPECT + SAIL models: A review of use for vegetation characterization. *Remote Sens. Environ.* **2009**, *113*, S56–S66. [CrossRef]

6. Colombo, R.; Bellingeri, D.; Fasolini, D.; Marino, C.M. Retrieval of leaf area index in different vegetation types using high resolution satellite data. *Remote Sens. Environ.* **2003**, *86*, 120–131. [CrossRef]
7. Fang, H.; Liang, S.; Kuusk, A. Retrieving leaf area index using a genetic algorithm with a canopy radiative transfer model. *Remote Sens. Environ.* **2003**, *85*, 257–270. [CrossRef]
8. Campos-Taberner, M.; Moreno-Martínez, Á.; García-Haro, F.J.; Camps-Valls, G.; Robinson, N.P.; Kattge, J.; Running, S.W. Global Estimation of Biophysical Variables from Google Earth Engine Platform. *Remote Sens.* **2018**, *10*, 1167. [CrossRef]
9. Fernández-Guisuraga, J.M.; Verrelst, J.; Calvo, L.; Suárez-Seoane, S. Hybrid inversion of radiative transfer models based on high spatial resolution satellite reflectance data improves fractional vegetation cover retrieval in heterogeneous ecological systems after fire. *Remote Sens. Environ.* **2021**, *255*, 112304. [CrossRef]
10. Berger, K.; Atzberger, C.; Danner, M.; D Urso, G.; Mauser, W.; Vuolo, F.; Hank, T. Evaluation of the PROSAIL Model Capabilities for Future Hyperspectral Model Environments: A Review Study. *Remote Sens.* **2018**, *10*, 85. [CrossRef]
11. Cao, Z.; Jiang, L.; Yue, P.; Gong, J.; Hu, X.; Liu, S.; Tan, H.; Liu, C.; Shangguan, B.; Yu, D. A large scale training sample database system for intelligent interpretation of remote sensing imagery. *Geo-Spat. Inf. Sci.* **2024**, *27*, 1489–1508. [CrossRef]
12. Rocha, M.; Cortez, P.; Neves, J. Evolutionary Neural Network Learning. In *Progress in Artificial Intelligence*; Pires, F.M., Abreu, S., Eds.; Springer: Berlin/Heidelberg, Germany, 2003; pp. 24–28.
13. Fusco, P.; Amato, A.; Venticinque, S. Cloud-based analysis of aerial imagery for unveiling ancient archaeological patterns. *J. Cloud Comput.* **2025**, *14*, 22. [CrossRef]
14. Myneni, R.B.; Park, Y. *MODIS Collection 6 (C6) LAI/FPAR Product User's Guide*; NASA EOSDIS Land Processes DAAC (LP DAAC); USGS EROS Center: Sioux Falls, SD, USA, 2020.
15. Fang, H.; Wei, S.; Liang, S. Validation of MODIS and CYCLOPES LAI products using global field measurement data. *Remote Sens. Environ.* **2012**, *119*, 43–54. [CrossRef]
16. Yan, K.; Wang, J.; Peng, R.; Yang, K.; Chen, X.; Yin, G.; Dong, J.; Weiss, M.; Pu, J.; Myneni, R.B. HiQ-LAI: A high-quality reprocessed MODIS leaf area index dataset with better spatiotemporal consistency from 2000 to 2022. *Earth Syst. Sci. Data.* **2024**, *16*, 1601–1622. [CrossRef]
17. Gawlikowski, J.; Tassi, C.R.N.; Ali, M.; Lee, J.; Humt, M.; Feng, J.; Kruspe, A.M.; Triebel, R.; Jung, P.; Roscher, R.; et al. A survey of uncertainty in deep neural networks. *Artif. Intell. Rev.* **2021**, *56*, 1513–1589. [CrossRef]
18. Yan, P.; Zhou, P.; Chen, H.; Lei, S.; Tan, Z.; Huang, J.; Guo, Y. Process-Based Remote Sensing Analysis of Vegetation–Soil Differentiation and Ecological Degradation Mechanisms in the Red-Bed Region of the Nanxiong Basin, South China. *Remote Sens.* **2025**, *17*, 3462. [CrossRef]
19. Morris, M.D. Factorial sampling plans for preliminary computational experiments. *Technometrics* **1991**, *33*, 161–174. [CrossRef]
20. Sobol, I.M. Global sensitivity indices for nonlinear mathematical models and their Monte Carlo estimates. *Math. Comput. Simul.* **2001**, *55*, 271–280. [CrossRef]
21. Campolongo, F.; Cariboni, J.; Saltelli, A. An effective screening design for sensitivity analysis of large models. *Environ. Modell. Softw.* **2007**, *22*, 1509–1518. [CrossRef]
22. Saltelli, A.; Annoni, P.; Azzini, I.; Campolongo, F.; Ratto, M.; Tarantola, S. Variance based sensitivity analysis of model output. Design and estimator for the total sensitivity index. *Comput. Phys. Commun.* **2010**, *181*, 259–270. [CrossRef]
23. Goldberg, D.E. *Genetic Algorithms in Search, Optimization, and Machine Learning*; Addison-Wesley: Boston, MA, USA, 1989.
24. Conn, A.R.; Gould, N.I.; Toint, P. A globally convergent augmented Lagrangian algorithm for optimization with general constraints and simple bounds. *Siam J. Numer. Anal.* **1991**, *28*, 545–572. [CrossRef]
25. Conn, A.; Gould, N.; Toint, P. A globally convergent Lagrangian barrier algorithm for optimization with general inequality constraints and simple bounds. *Math. Comput.* **1997**, *66*, 261–288. [CrossRef]
26. Joint, C.F.G.I. *Evaluation of Measurement Data—Guide to the Expression of Uncertainty in Measurement*; Bureau International des Poids et Mesures: Sèvres, France, 2008.
27. Cox, M.; Siebert, B. The use of a Monte Carlo method for evaluating uncertainty and expanded uncertainty. *Metrologia* **2006**, *43*, S178. [CrossRef]
28. Gorroño, J.; Guanter, L.; Graf, L.V.; Gascon, F. A Framework for the Estimation of Uncertainties and Spectral Error Correlation in Sentinel-2 Level-2A Data Products. *IEEE Trans. Geosci. Remote Sens.* **2024**, *62*, 5634613. [CrossRef]
29. Woolliams, E.; Hueni, A.; Gorroño, J. *Intermediate Uncertainty Analysis for Earth Observation (Instrument Calibration Module): Training Course Textbook*; National Physical Laboratory: Teddington, UK, 2015; pp. 14–28. Available online: https://www.meteoc.org/wp-content/uploads/sites/35/2017/11/uaeo-int-trg-course-v2.pdf (accessed on 3 January 2025).
30. Leveque, R. *Finite Difference Methods for Ordinary and Partial Differential Equations: Steady-State and Time-Dependent Problems*; Society for Industrial and Applied Mathematics: Philadelphia, PA, USA, 2007.
31. Jiang, Y.; Cheng, Z.; Yang, G.; Zhao, D.; Zhang, C.; Xu, B.; Feng, H.; Feng, Z.; Ren, L.; Zhang, Y.; et al. Fine-Scale Quantification of the Effect of Maize Tassel on Canopy Reflectance with 3D Radiative Transfer Modeling. *Remote Sens.* **2024**, *16*, 2721. [CrossRef]

32. Yan, L.; Liu, X.; Jing, X.; Geng, L.; Che, T.; Liu, L. Enhancing Leaf Area Index Estimation for Maize with Tower-Based Multi-Angular Spectral Observations. *Sensors* **2023**, *23*, 9121. [CrossRef]
33. Weiss, M.; Baret, F.; Jay, S. *S2ToolBox Level 2 products: LAI, FAPAR, FCOVER*; Version 2.1; INRAE: Avignon, France, 2020; p. 59.
34. Najib, D.; Fernandes, R.; Sun, L.; Canisius, F.; Hong, G. Python Version of Simplified Level 2 Prototype Processor for Retrieving Canopy Biophysical Variables from Sentinel 2 Multispectral Instrument Data. 2024. Available online: https://zenodo.org/records/10654520 (accessed on 5 December 2024).
35. Djamai, N.; Zhong, D.; Fernandes, R.; Zhou, F. Evaluation of Vegetation Biophysical Variables Time Series Derived from Synthetic Sentinel-2 Images. *Remote Sens.* **2019**, *11*, 1547. [CrossRef]

remote sensing

MDPI

Article

Corn Plant Detection Using YOLOv9 Across Different Soil Background Colors, Growth Stages, and UAV Flight Heights

Thiago O. C. Barboza [1], Adão Felipe dos Santos [1], Emily K. Bedwell [2], George Vellidis [3,4] and Lorena N. Lacerda [3,5,*]

1 Agriculture Department (DAG), Lavras School of Agricultural Sciences, Federal University of Lavras (UFLA), Lavras 37200-900, Brazil; thiago.barboza1@estudante.ufla.br (T.O.C.B.); adao.felipe@ufla.br (A.F.d.S.)

2 Department of Soil and Water Systems, Kimberly R&E Center, University of Idaho, Kimberly, ID 83341, USA; ebedwell@uidaho.edu

3 Institute for Integrative Precision Agriculture, University of Georgia, Athens, GA 30602, USA; yiorgos@uga.edu

4 Department of Crop and Soil Sciences, University of Georgia, Tifton, GA 31794, USA

5 Department of Crop and Soil Sciences, University of Georgia, Athens, GA 30602, USA

* Correspondence: llacerda@uga.edu

Highlights

What are the main findings?

- Gray/red-brown soils and a no-tilled system presented the most challenging soil backgrounds in the detection of corn plants in different corn growth stages at both flight heights (30 and 70 m).
- The V3 and V5 were the best predicted corn growth stages for 70 m flight height using the YOLOv9-small version.

What is the implication of the main findings?

- Gray/red-brown and no-tilled soil backgrounds need more powerful models as well as hyperparameter optimization for lightweight models.
- Corn stand count and prediction can be performed at 70 m flight heights, increasing the scalability of plant count models to map larger fields.

Abstract

Accurate stand count and growth stage detection are essential for crop monitoring, since traditional methods often overlook field variability, leading to poor management decisions. This study evaluated the performance of the YOLOv9-small model for detecting and counting corn plants under real field conditions. The model was tested across three soil background types, two flight heights (30 and 70 m), and four corn growth stages (V2, V3, V5, and V6). Unmanned aerial vehicle (UAV) imagery was collected from three distinct fields and cropped into 640 × 640 pixels. Datasets were split into training (70%), validation (20%), and testing (10%) datasets. Model performance was assessed using precision, recall, classification loss, and mean average precision of 50% and 50–90%. The results showed that the V3 and V5 stages yielded the highest detection accuracy, with mAP50 values exceeding 85% in conventional tillage fields and slightly lower performance in gray/red-brown conditions due to background interference. Increasing flight height to 70 m reduced accuracy by 8–12%, though precision remained high, particularly at V5, and performance was poorest for V2 and V6. In conclusion, YOLOv9-small is effective for early-stage corn detection, particularly at V3 and V5, with 30 m providing optimal results. However, 70 m may be acceptable at V5 to optimize mapping time.

Keywords: computer vision; deep learning; UAV image; soil background; smart farming

1. Introduction

Corn is an important crop for both human consumption and animal feed. Its production is heavily concentrated in the United States, China, and Brazil, which together account for approximately 32%, 23%, and 10% of global corn output, respectively [1]. Despite its significance, corn production faces several limitations, particularly those related to stand quality, which directly impacts yield potential and economic returns [2–4]. Plant density also plays a critical role and is influenced by factors such as hybrid selection, crop maturity, weather and soil conditions, length of the growing season, management practices, and planting date [4,5]. Modifying plant density has long been employed as a strategy to increase yields [6–8]. However, many of these approaches do not take into account field-level spatial variability, which can lead to suboptimal outcomes, as microenvironmental differences may restrict the crop's yield potential.

Early detection of corn growth stages and accurate stand counting are essential for supporting precision agriculture practices. These tasks enable farmers to make informed decisions regarding post-emergence herbicide applications, replanting needs, fertilizer management, and assessment of spatial field variability relative to yield maps [4,9]. However, identifying corn stages and performing stand counts remain challenging due to the limitations of conventional methods. Traditionally, stand counting involves walking through the field and sampling specific areas by measuring row spacing and manually counting plants within small plots [3,4]. This manual process is labor-intensive, time-consuming, and fails to capture variability across the entire field. In most cases, assessments are limited to field borders or isolated sample points, which may not accurately represent the true stand distribution [4].

Advancements in computer science and precision agriculture have significantly contributed to the mapping of field variability and corn stand count. Various studies have addressed this task using approaches such as Mask Region-Based Convolutional Neural Network [10], U-Net [4,11], adapted You Only Look Once (YOLO) models [12], small ground robots [13], and the integration of multispectral imagery with large-scale deep learning models [14]. However, these studies predominantly rely on complex two-stage deep learning architectures, which require extensive hyperparameter tuning and long processing times. Moreover, most experiments were conducted at very low altitudes, typically between 10 and 25 m, limiting their applicability to small, experimental plots rather than large-scale agricultural fields. Additionally, they often overlook key environmental factors such as variations in soil background color, especially under no-tillage systems, which introduce substantial challenges for accurate corn detection. These gaps highlight the need for more scalable, efficient, and robust solutions capable of operating under realistic field conditions.

The YOLO family has been widely applied to agricultural tasks for a long time. Introduced by Joseph Redmon in 2016 [15], YOLOv1 represents the first version of this family of one-stage object detectors. Subsequent versions, culminating in YOLOv3, introduced several key improvements, including multi-scale training, dimensional clustering, spatial pyramid pooling (SPP) blocks, and the Darknet-53 backbone. In 2020, Alexey Bochkovskiy proposed YOLOv4 with the CSPDarknet-53 backbone, followed by YOLOv5 developed by Glenn Joche, which incorporated anchor-free detection, a Swish-based activation function, and a PANet-style neck [16]. The Meituan Vision AI Department released YOLOv6, incorporating self-attention mechanisms to improve accuracy. Xu et al. (2022) introduced

https://doi.org/10.3390/rs18010014

YOLOv7 with transformer components and E-ELAN re-parameterization, and Ultralytics presented YOLOv8 in 2023, leveraging Generative Adversarial Network techniques to enhance the model's ability to handle different scenarios [16]. More recently, Wang and Liao [17] developed YOLOv9 with the PGI and GELAN modules, Wang et al. [17] introduced YOLOv10 as a non-maximum suppression-free (NMS-free) detector, Ultralytics released YOLOv11 with C3k2 and C2PSA modules in the backbone and neck, and Tian et al. [18] proposed YOLOv12 as the first attention-centric YOLO model, incorporating the R-ELAN mechanism [19].

One-stage models, such as those in the YOLO family, have significantly advanced the field of computer vision and play a fundamental role in agricultural applications. These models perform object classification and bounding box prediction simultaneously, making them well-suited for real-time tasks due to their speed and efficiency [20,21]. Recent versions, such as YOLOv9, introduce innovations like Programmable Gradient Information (PGI) and the Generalized Efficient Layer Aggregation Network (GELAN), which improve model performance by enhancing gradient flow and reducing information loss in deeper layers [20]. These innovations increase the performance and decrease the number of flops and hyperparameters, particularly in the MS COCO dataset and a lightweight version, such as the "small" model. They also contribute to high precision in heterogeneous backgrounds with multiple targets, emphasizing the use for different background and environment conditions. YOLOv9 is also available in multiple configurations: nano, small, medium, large, and X-large, allowing users to balance computational cost and detection accuracy based on the complexity of the task. As a result, YOLOv9 has seen widespread use in agriculture for diverse object detection tasks, including growth stage classification and identification in corn and disease detection [3,22,23].

Despite recent advances, applying object detection models in complex agricultural environments remains challenging, particularly in the presence of high background noise variability caused by soil color variations and management practices such as no-tillage. Furthermore, integrating these models into large-scale unmanned aerial vehicle (UAV) mapping operations at higher flight altitudes continues to be a limitation, as most studies focus on ultra-low altitudes in controlled experimental plots. Both flight altitude and background complexity are known to affect model performance, and their integration into training data can improve the model's ability to generalize across diverse field conditions [13,24]. Therefore, a highly variable dataset is needed to create robust models that can be generalized to accurately count corn plants under different soil conditions. Additionally, creating robust datasets by labeling images is a time-consuming and labor-intensive process that can contribute to poor performance for small models, since they require larger datasets to learn the pattern. Similarly, robust models such as transformers and Fast-RCNN seem to be an alternative to improve the model's performance. However, these models demand powerful hardware, which can contribute to limited applications depending on the task.

This study aims to evaluate the performance of the YOLOv9-small model for detecting and counting corn plants at higher flight altitudes and varying soil backgrounds, mimicking conditions of large cropping areas. Specifically, the model was tested across three distinct soil background colors, two flight heights suitable for large-scale mapping (30 and 70 m), and four corn growth stages. This evaluation was guided by two formulated hypotheses. Hypothesis I: Corn plants with three or more fully developed leaves provide more favorable conditions for accurate object detection, regardless of background, whereas earlier stages are associated with higher misclassification due to greater visual similarity with the background. Hypothesis II: Higher UAV flight heights (70 m) can be used for corn plant detection across different growth stages with moderate accuracy.

https://doi.org/10.3390/rs18010014

2. Materials and Methods

2.1. Study Location and Characterization

The data collection occurred at three different fields located across Georgia, USA (Figure 1). Field one was located at the Stripling Irrigation Research Park (SIRP) in Mitchell County, Georgia. Fields two and three were located at the Iron Horse Plant Sciences Research Farm, called Iron Horse I (IHI) and Iron Horse II (IHII), located in Greene County, Georgia. The fields presented different soil types and textures. The IHII presents the characteristics of Chewacla (ChA) silt loam, 0 to 2 percent slopes, frequently flooded, and Wickham sandy loam (WkB), 2 to 6 percent slopes, rarely flooded. On the other hand, the IHI has the soil characteristics of Wickham sandy loam (WkB), 2 to 6 percent slopes, rarely flooded, and the SIRP field has Lucy loamy sandy (LmB) soils, 0 to 5 percent slopes, according to the Natural Resources Conservation Services (NRCS) Web Soil Survey (https://websoilsurvey.nrcs.usda.gov/app/) (accessed on 05 August 2025).

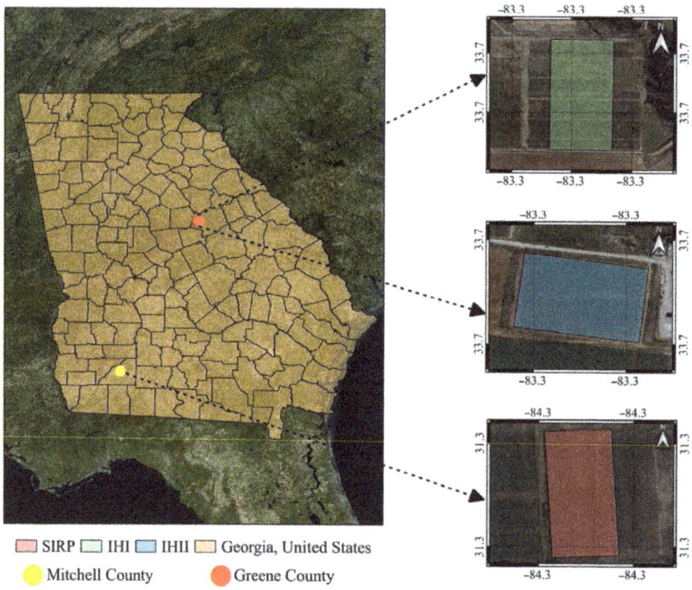

Figure 1. Location map showing the Stripling Irrigation Research Park (SIRP) field site in Mitchell County, GA (red polygon), and Iron Horse I (IHI) (blue polygon) and Iron Horse II (IHII) (green polygon) sites in Greene County, GA, United States. The yellow and red points in the Georgia map (orange) represent the location of the farms in Mitchell and Greene Counties, respectively.

All three fields were planted with corn in 2024. At the SIRP location, corn was planted on 8 April 2024, followed by IHI on 2 May 2024, and IHII on 15 May 2024. The tillage systems varied by field. A no-tillage system was used at SIRP (Figure 2A), while IHI and IHII used a conventional tillage system. However, the IHI presents red-brown soil color (Figure 2B), while IHII has a gray/red-brown soil color (Figure 2C). The soil background was one of the key features used to train the deep learning model. Elements like straw and stones in the background can oftentimes present similar shapes and colors as the crop canopy and can affect the model's performance. These visual similarities can either improve the model's capabilities by making the training process more complex by introducing challenges in distinguishing between plants and backgrounds.

https://doi.org/10.3390/rs18010014

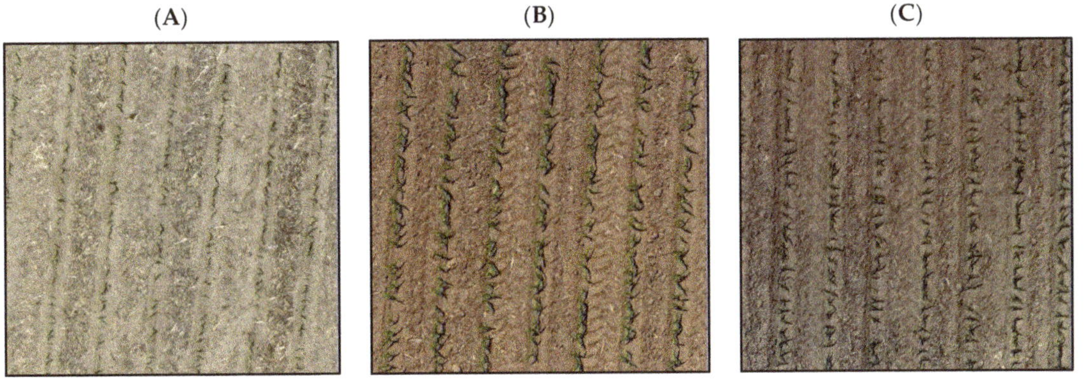

Figure 2. Field images taken at 30 m flight height at the V3 corn growth stage showing the no-tillage field located in Stripling Irrigation Research Park (SIRP) in Mitchell County, GA (**A**), the red-brown field located in Iron Horse I (IHI) at Greene County, GA (**B**), and the gray/red-brown soil for Iron Horse II (IHII) located in Greene County, GA (**C**).

The corn hybrid used for SIRP and Iron Horse I was DKC68-78SS, while at Iron Horse II was the Integra 6641 Bayer SmartStax. The plant population adopted for SIRP and Iron Horse I was 74,000 plants, while at the Iron Horse field, the plant population was 75,000 plants per hectare. Row spacing of 0.75 m was used for Iron Horse I and Iron Horse II, and 0.90 m for SIRP.

2.2. Corn Stages

Determining the corn phenology stages depends on the weather, soil, and management conditions. Thus, the present study used the crop phenology scale developed by Ritchie and Hannway [25] and split the stages for data collection in V2, V3, V5, and V6. Each of these labels represents the vegetative stage, followed by the number of completely open leaves. For a leaf to be considered "completely open", the leaf must present a visible collar [25,26]. Since the field conditions and sowing dates of each field were different, growth stages were monitored, and fully open leaves were manually counted to ensure that imagery of each field was collected at the same crop growth stage.

2.3. UAV Flights and Parameters

The image collection started at V2, followed by V3, V5, and V6 for every field. The V4 stage was skipped due to adverse weather conditions, which prevented image collection during this period. The UAV used was the DJI® Mavic 3M (Shenzhen, Guangdong, China), which contains five different sensors, including a 20-MP RGB sensor with a maximum resolution of 5280 × 3956 pixels, and a multispectral sensor subdivided into four sensors, Red (650 nm), Green (560 nm), Red-edge (730 nm) and near-infrared (860 nm) with 5-MP and maximum resolution of 2592 × 1944 pixels. Despite the DJI Mavic 3M having a multispectral sensor, the RGB images were selected for this study due to their higher spatial resolution. In addition, RGB sensors require simpler processing and are more affordable, which increases the developed model's applications for growers.

The flight heights used for all fields were 30 and 70 m, and the overlap during the collection was 80% for both front and side overlap. The flight heights were selected to evaluate the performance of the corn plant-counting model while balancing detection accuracy with the speed and efficiency required for large-scale data collection [4,13,27]. All flights were performed within 2 h of solar noon between 11:00 and 15:00, only on completely sunny conditions, without the presence of clouds over the target fields.

https://doi.org/10.3390/rs18010014

2.4. Dataset Preparation

The dataset preparation process included labeling and data augmentation. Initially, ten random images were selected, then cropped to a size of 640 × 640 pixels. From these, a total of 1920 images were generated for each corn growth stage (V2, V3, V5, and V6) at both 30 m and 70 m flight height. The dataset was organized by field location, flight height, and corn stage. Using the Roboflow® tool (Des Moines, IA, USA), the images were annotated with bounding boxes, marking 50 plants per image, resulting in a total of 96,000 annotated instances across all corn phenology stages. The dataset was then split into training (70%), testing (20%), and validation (10%) datasets. For the training set, data augmentation techniques such as brightness adjustment, rotation, and contrast variation were applied to increase image variability. Finally, the dataset was exported into the data.yaml format for use in model training.

2.5. You Only Look Once (YOLO) Model

Deep learning models have been frequently applied for object detection tasks. These models present different characteristics, performance, and architecture that can contribute to the object detection task. You Only Look Once models (Figure 3) represent one family of one-shot models with high speed and accuracy in detecting objects [16,22]. The one-shot model, such as YOLO, works by dividing the image into small grids and making predictions in each grid [16,22]. This process is different from conventional convolutional neural networks (CNNs) that present heavy training and test steps. Thus, lightweight models such as YOLO can be applied to camera sensors and smartphones, highlighting their efficiency and enhancing the model's capability for real-time applications on limited devices [22,28,29].

The YOLO model has different variations from the original model. In the present study, the YOLOv9-small was selected and used to train each dataset. This model presents the PGI and GELAN functions. Both techniques increase model performance and avoid the information bottlenecks, where features can be lost as data propagate through deep network layers [21,30]. The implementation of YOLO v9 follows the Roboflow® tool (Des Moines, IA, USA) API model. The number of epochs used to train the model was 500 epochs. However, the patience parameter was set to 70 epochs, meaning that if the model did not detect any changes in the loss values in the training step after 70 epochs, it would be stopped. The image size input used was 640 × 640 pixels, the batch size was 8 images, and the number of data workers was 8. The optimizer SGD (Stochastic Gradient Descendent) was used to train the model with a learning rate starting at 0.001 and reaching the best at 0.0168 and a momentum of 0.5. The parameters were selected following the limitations imposed for the small video plate used in the present study. To adapt YOLOv9-small to the specific characteristics of the UAV imagery, a hyperparameter tuning was performed using the Ultralytics Tuner v8.3.239 (Ultralytics, Frederick, MD, USA). This procedure implements an evolutionary search over optimization and data-augmentation hyperparameters, including initial learning rate, learning-rate decay, loss weights, and mosaic/scale/color jitter probabilities. In each iteration, a candidate set of hyperparameters is obtained by mutating the best configuration found. A YOLOv9-small model is trained on the dataset for a reduced number of epochs, and a composite fitness score based on the validation dataset computes the precision, recall, and mAP. The process is repeated for multiple iterations, and the configuration with the highest fitness is stored in a best_hyperparameters.yaml file, which is then used to train the final YOLOv9-small model.

https://doi.org/10.3390/rs18010014

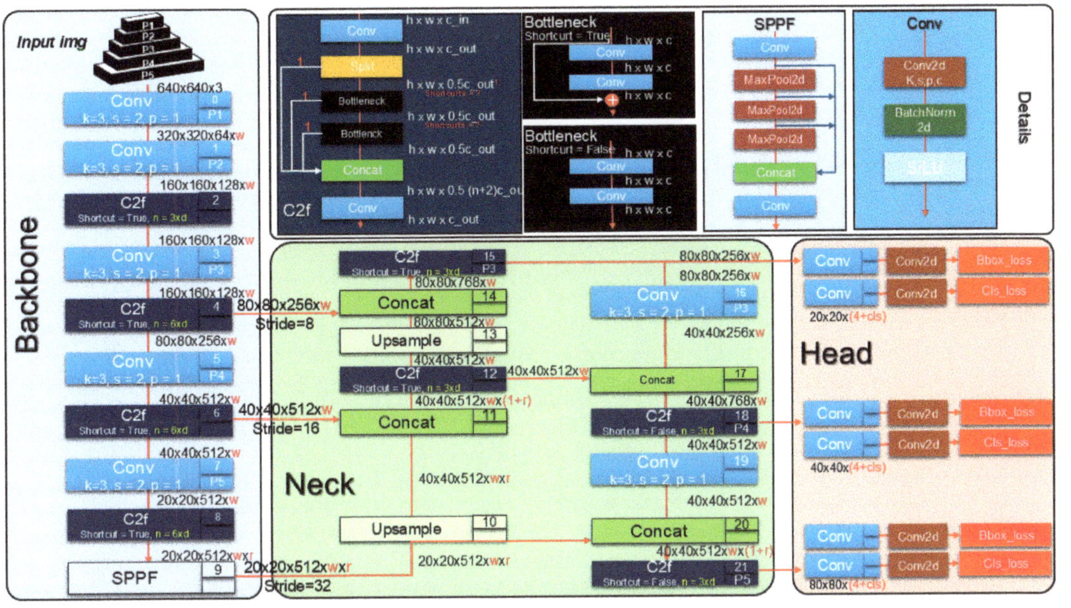

Figure 3. You Only Look Once (Yolo) version 9 architecture model used to train, validate, and test the different soil background colors. The network consists of three main components: Backbone (blue), Neck (green), and Head (red). The Backbone extracts hierarchical features at multiple scales (P1–P5) through successive Conv and C2f blocks. The Neck implements a Feature Pyramid Network (FPN) combined with a Path Aggregation Network (PAN) to fuse multi-scale features via Upsample and Concat operations. The Head performs bounding box regression (Bbox_loss) and classification (Cls_loss) at three detection scales: P3 (80 × 80, stride 8), P4 (40 × 40, stride 16), and P5 (20 × 20, stride 32), enabling detection of small, medium, and large objects, respectively. Conv: Convolution block comprising Conv2d, BatchNorm2d, and SiLU activation; Conv2d: two-dimensional convolution operation parameterized by k (kernel size), s (stride), p (padding), and c (number of output channels); C2f: Cross Stage Partial block with two convolutions and n Bottleneck modules. Bottleneck: residual block with optional shortcut connection. SPPF: Spatial Pyramid Pooling Fast module for multi-scale receptive field aggregation. MaxPool2d: max pooling operation with 5 × 5 kernel applied sequentially within SPPF to capture contextual information at multiple scales without increasing computational cost. Concat: channel-wise concatenation operation that merges feature maps from different layers along the channel dimension to combine semantic and spatial information. Upsample: bilinear interpolation for spatial upsampling. For YOLOv8-small, the depth multiplier is 0.33, and the width multiplier is 0.50. Notation: h × w × c denotes height × width × channels. w: indicates the width scaling factor applied to the number of channels. d: indicates the depth scaling factor determining the number of repeated blocks. r: denotes the channel expansion ratio used in the Neck to adjust feature map dimensions after concatenation operations. The shortcuts at C2f represent the bottleneck layers (black box), with the option to use true or false in the shortcut to regulate the function.

2.6. Performance Metrics

The metrics used to assess the model's performance in each dataset were the mean average precision of 50% (mAP50) (Equation (1)) and 50–95% (mAP50–95) (Equation (2)), followed by Precision (Equation (3)) and Recall (Equation (4)). The results from each model were also compared using the confusion matrix, comparing the true positive, negative, and false positive and negative.

$$\text{mAP50} = \frac{1}{N}\sum_{i=l}^{N} AP_i \quad (1)$$

https://doi.org/10.3390/rs18010014

mAP50 represents the mean average precision of intersection over union 50% (IoU 50%). N refers to the number of samples, and AP represents the average precision values.

$$\text{mAP50} - 95 = \frac{1}{N}\sum_{i=l}^{N} AP_i \tag{2}$$

mAP50-95 represents the mean average precision of intersection over union between 50% to 95% representing a IoU 75%. N refers to the number of samples, and AP represents the average precision values.

$$\text{Precision} = \frac{TP}{TP + FP} \tag{3}$$

Precision refers to the proportion of correct detections made by the model over all detections made by the model. TP represents the true positives; FP represents the false positives.

$$\text{Recall} = \frac{TP}{TP + FN} \tag{4}$$

Recall represents the proportion of correct objects detected by the model over the total of correct objects in the image or dataset. The FN refers to the false negative samples that were calculated. Nevertheless, the models were evaluated using the Train and Validation Classification Loss (Val_cls_loss and Train_cls_loss). Both metrics represent the model's performance on validation and training datasets across the epochs used to train the models. After training and validation, the model for each corn growth stage, flight height, and field was implemented in one image to compare the results.

2.7. Video Plate and Software

Large models such as YOLOv9 have a wide number of hyperparameters (7.2 million) that need to be trained. These parameters represent convolutional filters and layers and use many mathematical operations that have a significant time cost. To address this cost, Python 3.12 (Python Software Foundation, Wilmington, DE, USA) associated with Pytorch (Meta AI, Menlo Park, CA, USA) and CUDA 11.4 were installed before running the model on an NVIDIA GTX 1660 GPU (6 GB) (NVIDIA Corporation, Santa Clara, CA, USA). All the models were analyzed using Python 3.12 and the graphs plotted using the software RStudio 4.3 (Boston, MA, USA).

3. Results

Figure 4 summarizes the training and validation metrics across epochs using box plots. For each metric, the central horizontal line inside the box represents the median value, while the bottom and top edges of the box correspond to the 25th (Q1) and 75th (Q3) percentiles, respectively. The whiskers extend to the minimum and maximum values, and points plotted beyond the whiskers (black dots) are considered outliers. Values closer to the top of the box and whiskers indicate better performance for the corresponding metric (precision, recall, mAP50 and mAP50-95). This representation provides a clear way to visualize the variability of each metric over the training epochs and directly compares the distributions between training and validation sets. The highest variation was observed in the Val_cls_loss and Train_cls_loss metrics. The training and validation classification losses (Val_cls_loss and Train_cls_loss) represent the difference between the true values and those predicted by the model, with lower values indicating better model performance. During the initial stages of training, these values are typically high, but as the model adjusts its weights, the losses decrease. Consequently, higher values can be seen early in training, especially at the V2 and V6 growth stages, which exhibited greater variability compared to

https://doi.org/10.3390/rs18010014

V3 and V5, except for val_cls_loss at 70 m for IHII and II. This suggests that, regardless of flight height, the model struggles more to identify plants at the V2 and V6 stages, likely due to difficulty in distinguishing corn plants from the background at V2 and misidentifying individual plants at V6 due to increased leaf overlap between plants. Conversely, the model demonstrated a more consistent performance for Precision and Recall at V3 and V5, regardless of background conditions. This indicates that for these two stages, both 30- and 70 m flight heights yielded similar results in terms of Precision and Recall. However, when analyzing mAP50 and mAP50–95, the model showed significantly higher reliability at the 30 m flight height, highlighting its superior performance.

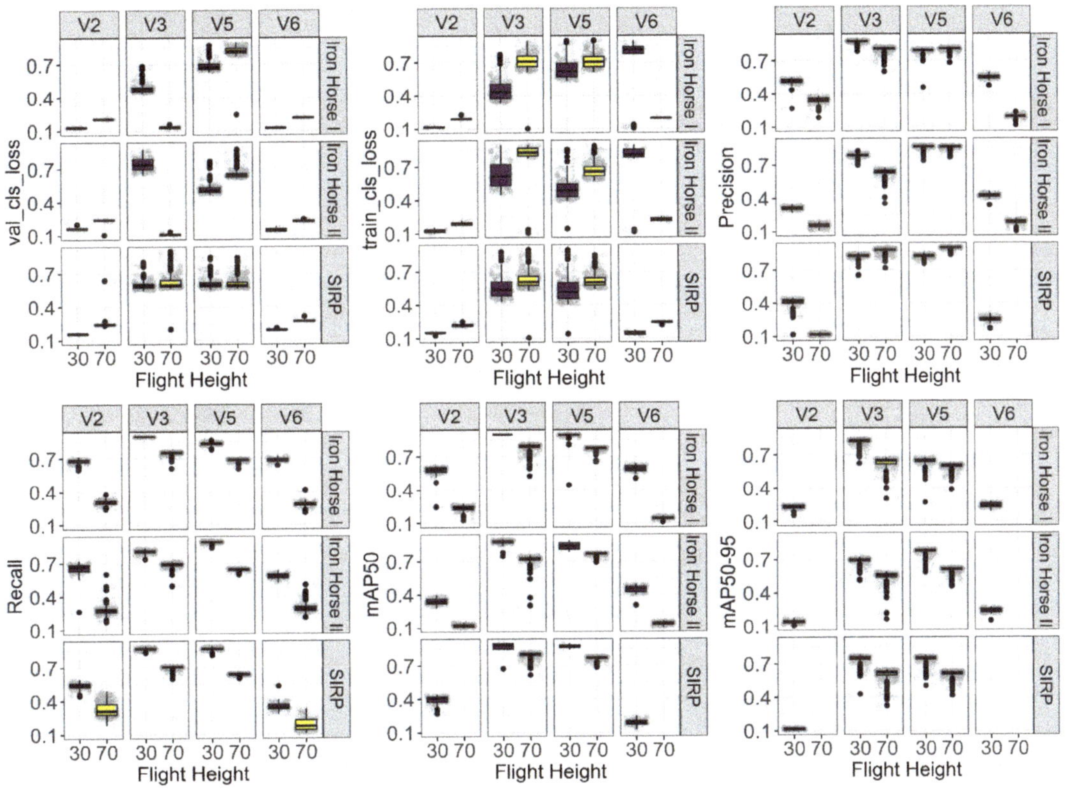

Figure 4. Boxplot analysis comparing the training and validation model metrics variability for classification losses (Train_cls_loss and Val_cls_loss), mean average of 50 and 50–95 (mAP50 and mAP 50–95), Precision, and Recall (primary y-axis) at 30- and 70 m flight heights (primary x-axis) for the Stripling Irrigation Research Park (SIRP), Iron Horse I (IHI) and Iron Horse II (IHII) fields (secondary y-axis) at different growth stages (secondary x-axis).

The initial results revealed the mean values obtained for each corn growth stage at different flight altitudes and across various fields (Figure 5). In all cases, the 30 m flight height outperformed the 70 m height. Notably, the V3 and V5 growth stages consistently showed the highest detection performance, even when the background varied between fields. This trend was also reflected in the Precision, Recall, mAP50, and mAP50–95 metrics. At 70 m, the mAP50 values for the V2 and V6 stages were the lowest. However, at 30 m, these values nearly doubled, increasing from 0.25 to 0.50. Although mAP50 values improved with the lower flight height, the mAP50–95 values for V2 and V6 decreased to approximately 0.2, indicating limited model effectiveness in detecting these stages.

https://doi.org/10.3390/rs18010014

Furthermore, while Precision and Recall were highest for the V3 and V5 stages, performance still varied by location, with IHI achieving the best results and the SIRP field showing the lowest.

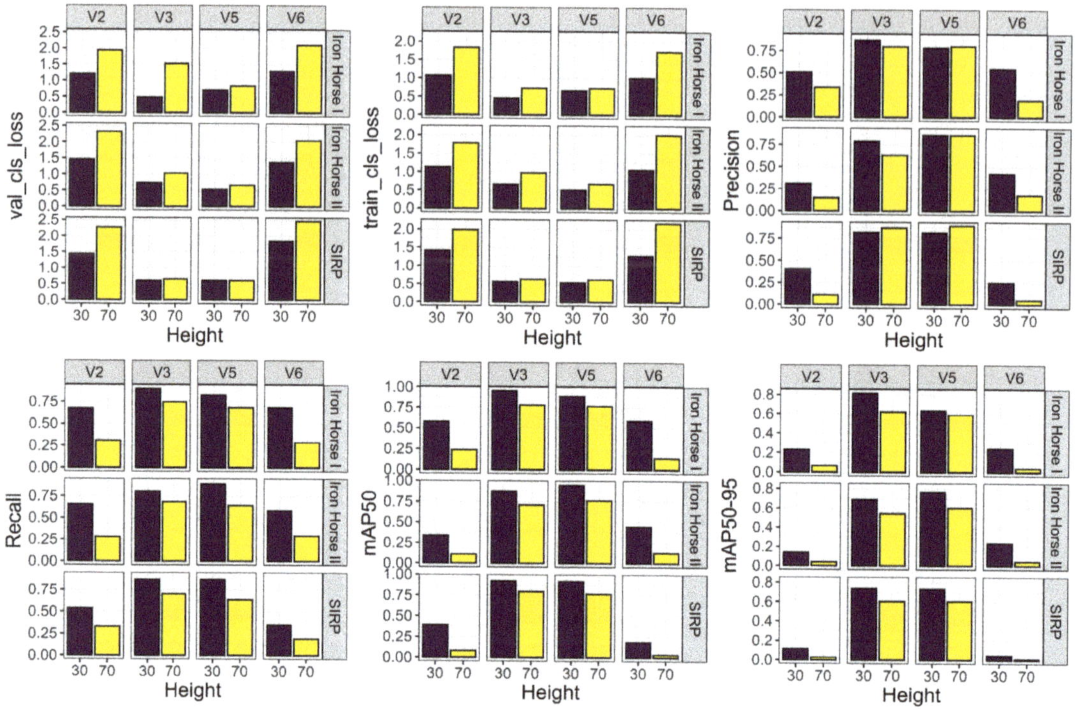

Figure 5. Bar graphics results comparing the models' mean values for the metrics training and validation classification losses (Train_cls_loss and Val_cls_loss), mean average of 50 and 50–95 (mAP50 and mAP 50–95), Precision and Recall (primary y-axis) at 30- and 70 m flight heights represented by the purple and yellow color bars for the Stripling Irrigation Research Park (SIRP), Iron Horse I (IHI) and Iron Horse II (IHII) fields (secondary y-axis) at different growth stages (secondary x-axis).

Figure 6 supports the results obtained using YOLOv9-small for detecting corn plants across different growth stages, flight heights, and backgrounds. The most detectable stages, V3 and V5, showed variability in the number of epochs required to train the model. At the lowest flight height (30 m), the model converged more quickly, as indicated by lower initial loss values and a reduced number of training epochs. In contrast, the SIRP field required nearly 300 epochs for model convergence at 70 m, and this setting also resulted in the highest Val_cls_loss values. Correspondingly, the 70 m flight height yielded the lowest performance in mAP50, mAP50–95, and Recall across all growth stages. With the exception of SIRP (V3 and V5), IHI (V5), and IHII (V5), the 70 m flight height generally resulted in higher Precision values. Precision measures the accuracy of model predictions, indicating how many predicted objects are correct. Therefore, at V3 and V5, in complex backgrounds like at the SIRP field or at IHI and IHII fields, the 70 m flight may offer more accurate predictions. However, when analyzing the Recall metric, the 30 m flight height consistently produced better results. This indicates that although the model detected objects at 70 m, many were likely false positives, as evidenced by the lower Recall scores. Recall reflects how well the model detects all objects in the image. Thus, high Recall means fewer missed detections, which is expected from images with higher spatial resolution.

https://doi.org/10.3390/rs18010014

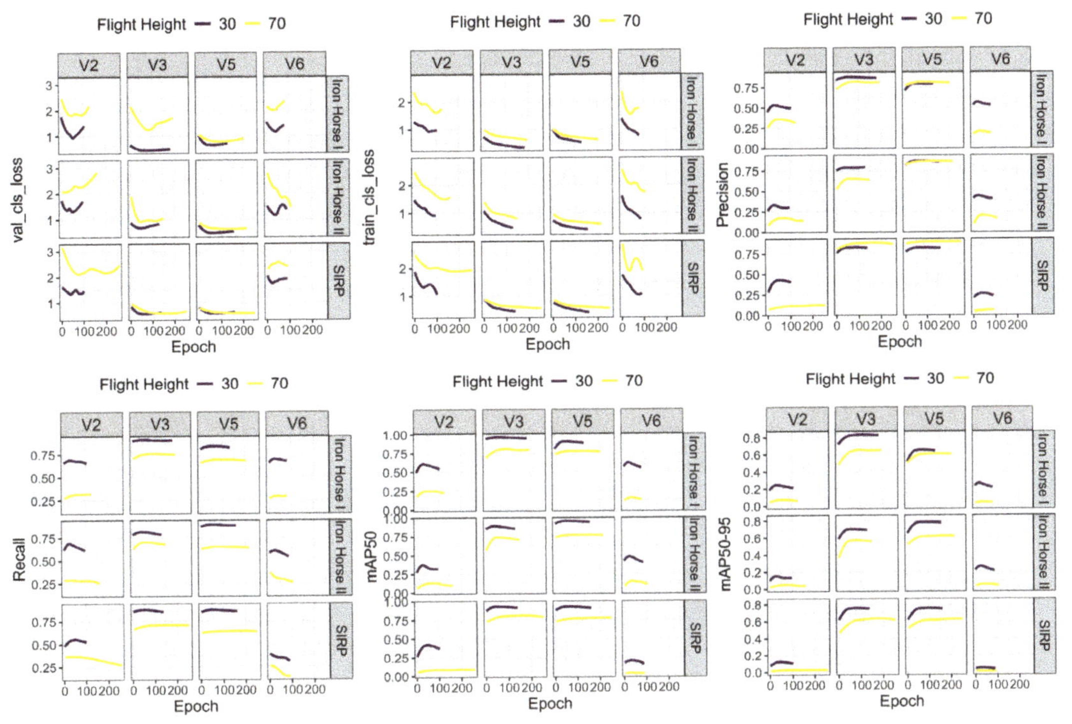

Figure 6. Smooth line graphics results comparing the metrics for training and validation classification losses (Train_cls_loss and Val_cls_loss), mean average of 50 and 50–95 (mAP50 and mAP 50–95), Precision and Recall (primary y-axis) at 30- and 70 m flight heights and the number of epochs (primary x-axis) for the Stripling Irrigation Research Park (SIRP), Iron Horse I (IHI), and Iron Horse II (IHII) fields (secondary y-axis) at different growth stages (secondary x-axis).

The interaction between background, corn growth stages, and flight height was further analyzed using the confusion matrix (Figure 7). The confusion matrix was used to understand the misclassification of the corn growth stages and flight height. The X-axis represents the true label, i.e., the real label created manually, while the Y-axis represents the predicted label, i.e., the label created by the model. Thus, the true negative (top-left cell) is the background, correctly classified as background, while the true positive (bottom-right cell) is the plants correctly classified as plants. The false positive (top-right cell) represents the true background (x-axis) predicted as a plant (Y-axis), while the false negative (bottom-left cell) represents the corn plants the model failed to detect.

At the V2 growth stage with a 30 m flight height, 67% of background pixels were misclassified as corn at V2. This misclassification rate was also observed at the field level, with 44%, 51%, and 51% of predictions being incorrectly identified as corn for SIRP, IHI, and IHII, respectively. However, when using the 70 m flight height at the same V2 stage, the misclassification significantly increased, especially at the SIRP and IHII fields, where 95% and 51% V2 corn plants were misclassified as background, respectively.

The confusion matrix also highlighted the highest model performance at 30 m for the following combinations: SIRP field at the V5 stage, IHI at V3, and IHII at V5. These results indicate the V3 and V5 stages as more favorable for accurate plant count. Nonetheless, even at the most suitable stages, increasing flight height can lead to a decrease in model performance. The SIRP field, for example, which features a no-tillage background, an increase in flight height to 70 m resulted in a 26% decrease in model accuracy for detecting the V5 growth stage. Other results showed that flying the UAV at 70 m can reduce accuracy by approximately 24% to 36% depending on background conditions.

Figure 7. Confusion Matrix used to evaluate the model's performance at different corn growth stages (V2, V3, V5, and V6) and flight height (30 and 70 m). The primary y-axis represents the predicted values by the model, and the x-axis represents the observed values (True), while the secondary x-axis represents the flight height (30 and 70 m).

Model validation was also performed on 20 images collected from each field (Figure 8). The 70 m flight height presented no successful detections for the V2 growth stage at IHI and for V6 at the SIRP and IHII fields. Additionally, the best performance on the validation dataset was observed for V5 at both 30 and 70 m, reaching R^2 values of 0.73 and 0.79 with errors varying between 18 plants for 30 m to 168 plants for 70 m at the SIRP field. IHI presented the best performance for V5 at 30 m with R^2 0.41 and an error of 75 and 68 plants for RMSE and MAE, respectively. IHII showed the worst performance with a maximum R^2 of 0.27 found at the V5 corn stage for 30 m and an error of 21 and 15 plants for RMSE and MAE.

The models' performance can also be observed in Figure 9, in which the number of counted plants for V3 and V5 stages, represented by the red dots, is shown for 30- and 70 m flight heights. The comparison results across all stages and field conditions validate that soil contrast represents one of the most important factors influencing model accuracy, especially at a flight height of 70 m. At this height, if the soil–plant contrast, caused by the soil color background, was high, the model accurately counted the corn plants, as observed for IHII, V5 stage, where 818 corn plants were identified, and for SIRP, V3 stage, in which 722 corn plants were identified. Soil color is another important factor. Gray soils or dark-hue conditions led to better detection performance for YOLOv9-small model, even for higher flights, such as at 70 m, highlighting the model's feasibility for larger-scale field mapping.

https://doi.org/10.3390/rs18010014

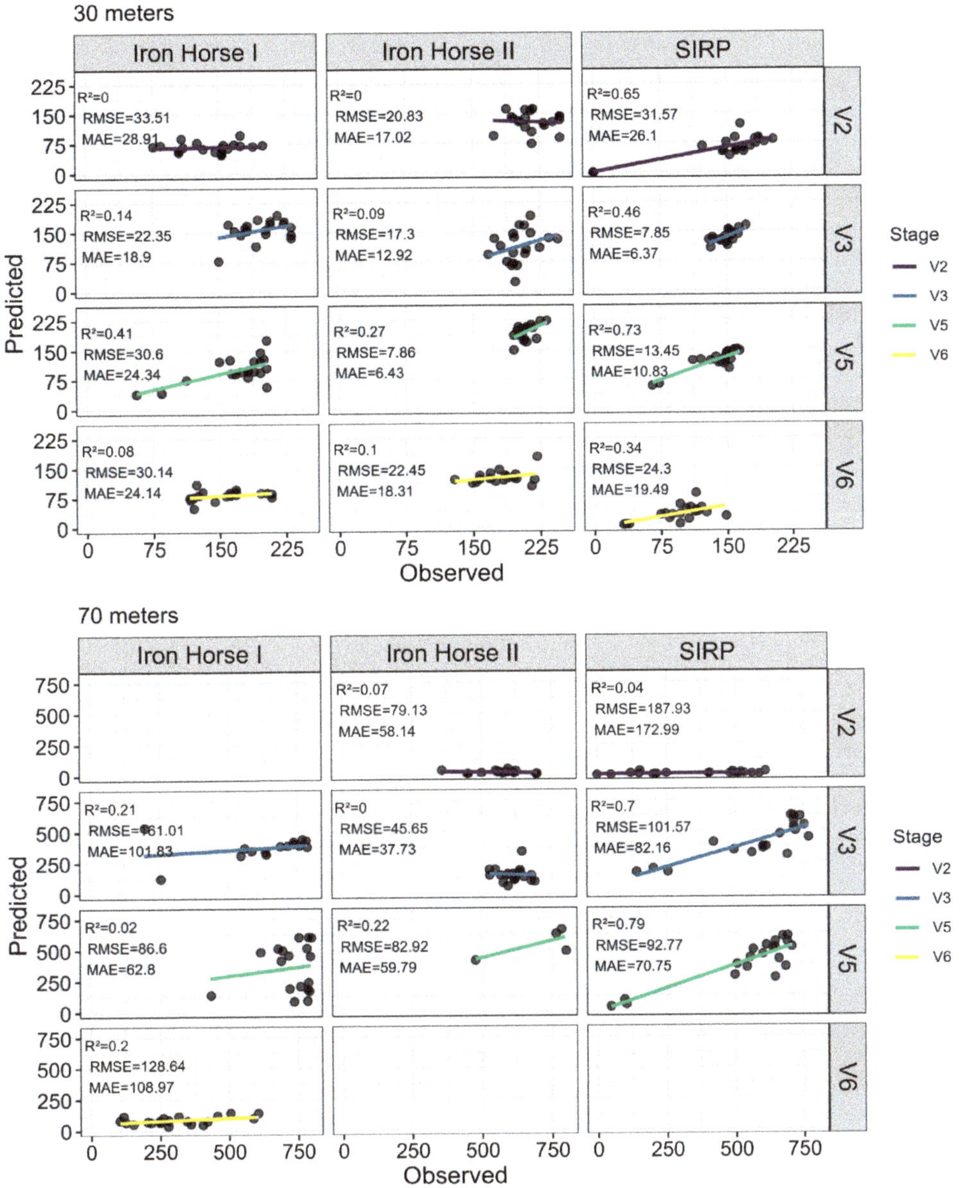

Figure 8. Scatter plots comparing observed (manual counting) (primary x-axis) versus predicted (YOLOv9s model) corn plant counts (primary y-axis) across three experimental fields: Iron Horse I, Iron Horse II, and Stripling Irrigation Research Park (SIRP) (secondary x-axis). Results are presented for two unmanned aerial vehicle (UAV) flight altitudes (30 and 70 m) and four corn growth stages (V2, V3, V5, and V6) (secondary y-axis). Each panel displays the coefficient of determination (R^2), root mean square error (RMSE), and mean absolute error (MAE). Colored lines represent linear regression fits for each growth stage. Missing panels at 70 m indicate combinations where data acquisition was not performed or model predictions were not feasible due to image resolution constraints.

https://doi.org/10.3390/rs18010014

Figure 9. Prediction results using the training model for each flight height, field, and corn growth stage. H—height (meters); C—corn plant count; F—field location. SIRP—Stripling Irrigation Research Park; IHI—Iron Horse I; IHII—Iron Horse II.

4. Discussion

The first hypothesis states that corn plants with three or more fully developed leaves provide more favorable conditions for accurate object detection, regardless of background, whereas earlier stages with fewer than three fully developed leaves are associated with higher misclassification due to greater visual similarity with the background. This pattern can be explained by the observed interaction between flight height and background conditions. Corn development begins at emergence (VE) and progresses through vegetative (Vn)

https://doi.org/10.3390/rs18010014

and reproductive (Rn) stages. The V1 stage, characterized by one fully open leaf, represents one of the earliest growth stages, where the cotyledonary leaf is still predominant. With each accumulation of approximately 66 growing degree days, the corn plant develops a new leaf [26]. By the time the plant reaches V3, its lower leaves are larger and more structurally defined, making them easier for detection models to identify from higher flight altitudes. At the V2 stage, the smaller leaf size and denser background, particularly in no-tillage systems, reduce detection accuracy, even at lower flight heights such as 30 m. In no-tillage systems, such as the SIRP field, the background interference is a major challenge for early-stage detection. However, contrasting soil characteristics can influence model performance. In fields like IHII, where light to dark-gray soil colors enhance visual contrast between plants and background, detection accuracy is high, even at the 70 m flight height. This suggests that soil color and texture play a role in optimizing detection outcomes. Overall, the transition stages between V3–V4 and V5–V6 are the most effective for corn stand counting and detection, regardless of background or flight height [31,32]. In a study evaluating different backgrounds, flight heights, and model types for corn density estimation, Jia et al. [3] found that a flight height of 20 m yielded the best performance. However, such low altitudes have trade-offs, including increased data acquisition time and larger data volumes for processing.

Following the second hypothesis, lower UAV flight altitudes (30 m) resulted in higher corn plant detection accuracy across growth stages, particularly V3 and V5, by reducing background influence and enhancing model convergence. It was observed that the higher resolution increases the model performance in detecting the corn plants, mainly at the V3 and V5 stages, while also decreasing the time for convergence. This means that a lower flight height does not necessarily need powerful models with high hyperparameters and complex layers, i.e., the high resolution contributed to models' performance with the highest mAP50, mAP50–95, Precision, and Recall [3,10,31]. Conversely, increased flight heights (70 m) allow for faster field mapping, but more powerful models with higher parameters are required to detect subtle differences in the background. These findings align with those observed by Pang et al. [33] that at a flight height of 30 m, the Mask-CNN model exhibited 87% accuracy, and when changed to heights of 40, 50, and 60 m, the average precision dropped to 79.4%, 75.5%, and 68.1%, respectively. Although lower flight heights improve model training efficiency and performance, allowing the use of small models, the applications for large-scale fields remain a limitation. Higher flight heights, as presented in this study, facilitate faster decision-making by the farmers, with reduced processing time and storage, unlike other studies [3,10,31], which, despite achieving higher performance, have limited applicability in large-scale fields.

YOLO models have significantly transformed one-stage object detection tasks, offering a balance between speed and accuracy that contrasts with more complex models such as Mask R-CNN and Faster R-CNN. While these two-stage detectors often achieve higher accuracy, their implementation is limited due to the large number of hyperparameters and high computational demand. In this study, YOLOv9-small was evaluated for its effectiveness in corn detection tasks, introducing two innovative architectural components: PGI and the GELAN. PGI helps preserve gradient flow across layers, minimizing information loss during training, while GELAN efficiently aggregates features using multiple computational blocks, including CSPBlocks, ResBlocks, and DarkBlocks, ensuring effective multiscale feature extraction and representation [20]. This efficiency has been demonstrated in other applications, with YOLOv9 achieving up to 94% mAP50 in coffee plant detection using various sensors [34,35]. Previous research with Mask R-CNN on corn at the V4–V5 stage and 30 m flight height also reported high performance (mAP50 = 94%), while segmentation models such as U-Net achieved Recall and Precision values of 0.95 and 0.96 at ultra-low

https://doi.org/10.3390/rs18010014

flight heights of 10 m [4]. Similarly, Lu et al. [34] reported mAP50 values of 82% at the V3 stage and 86.3% at V7 using YOLOv5 at 12 m. Under conditions of high weed infestation, improved YOLOv5 models reached mAP50 scores of 93% and 89% at 15 and 30 m, respectively [33]. However, what distinguishes the present study is the ability to achieve high performance metrics (85% mAP50, Precision, and Recall) even at a flight height of 70 m, using a lightweight model like YOLOv9-small, as observed for the fast convergence in initial stages (Figure 6). This is a significant advancement, as most previous studies relied on either more complex models or lower flight altitudes to obtain similar results. Moreover, the use of RGB cameras in this study underscores the practical and cost-effective nature of the approach. Unlike multispectral or modified sensors, which are expensive and require specialized expertise, RGB cameras are affordable, user-friendly, and easily integrated into UAV systems. This makes the proposed method highly suitable for scalable, real-world agricultural applications.

YOLOv9 achieved high accuracy at both corn growth stages (V3 and V5), and its performance in this study is consistent with results reported in the literature for agricultural applications. Wang et al. [32] reported high precision and mAP scores (precision = 89.30%, mAP0.5 = 94.60%, mAP0.5–0.95 = 64.60%) when comparing YOLOv5, YOLOv7, and YOLOv8 for coffee detection and counting. Additionally, Sharma et al. [36] showed that YOLOv9 outperforms YOLOv11, YOLOv10, YOLOv8, and Faster R-CNN in terms of mAP0.5, mAP0.5–0.95, precision, and recall, although at the cost of longer inference time. Similarly, in an orchard scenario using ground-based cameras for apple fruit detection and counting, Sapkota et al. [16] reported that YOLOv9-GELAN-E achieved higher mAP0.5 than YOLOv10, YOLOv8, YOLOv11, and YOLOv12 and obtained precision (0.903 versus 0.908) and recall (0.899 versus 0.900) values comparable to YOLOv10x and YOLOv12l, while YOLOv12 was approximately three times faster in terms of inference speed. Overall, these studies indicate that agricultural tasks require fast, real-time, and robust object detection, demanding models that balance accuracy and speed, as is characteristic of the YOLO family [37]. At the same time, agricultural environments involve substantial field variability and therefore require large, diverse datasets and careful fine-tuning. This was also observed in the present study, despite the high precision and recall achieved by YOLOv9-small. Its higher latency and lower inference speed may limit its applicability in strict real-time conditions, especially for autonomous robotics or UAV-based imagery. Thus, modified versions using improved functions [38–40], different optimizers [27], and adding transfer learning [15,41] or pre-processing steps [15] could be a solution to creating powerful models with a balance between precision, accuracy and hardware requirements.

Limitations and Future Research

This study investigated the influence of flight height and field background on the accuracy of counting corn plants at different growth stages using a YOLO-based object detection model. While this study tested a higher flight height to increase the scalability of plant stand count to larger fields, limitations remain when applying these models to very large commercial crop fields at early growth stages. Results showed a lower plant stand count accuracy in the very early stages of corn development (V2). Acquiring additional training data, focused specifically on early corn growth stages, combined with lower flight altitudes, can enhance YOLO performance and support early field decisions. Nonetheless, UAV operational constraints, such as limited battery life, flight time, and legal restrictions, create a trade-off between flight height, ground sampling distance, and field coverage. Another limitation is related to the cost of hardware and software needed to perform this task. This work focused on using compact models, such as YOLOv9-small, which allow large numbers of image tiles to be processed with moderate computational resources.

https://doi.org/10.3390/rs18010014

Future work exploring edge computing, local farm servers, and model compression strategies to reduce inference time and hardware requirements could further improve plant count accuracy.

Environmental factors such as wind, dust, crop residue, and heterogeneous soil moisture can also degrade image quality and lead to missed or false detections, reinforcing the need for multi-temporal acquisitions, UAV image quality control, and fine-tuning with field-specific data. The data used in this study focused on comparing the model's accuracy at different soil background colors. However, the soil background color was used only as a visual reference. Future studies considering the soil color histogram or classification under controlled conditions, while also measuring the residues and brightness, might add valuable information for the model's performance.

It is worth noting that the light-hue soil colors can challenge the model in recognizing patterns and counting plants, sometimes reducing confidence values to less than 0.1 [42,43]. Another approach would be to increase the dataset with images reflecting field variability, especially in no-till and gray/red-brown colors, which can contribute to model optimization. A third approach would be to apply different augmentation strategies or generative approaches, such as Generative Adversarial Networks (GANs), to expand the dataset and simulate diverse environments considering soil background colors. Additionally, large datasets that include soil moisture, different illumination levels, brightness, soil background color, and a no-tillage system could support the training of highly robust and accurate models.

Models that account for the overlap between the leaves and the dense canopy at the V6 corn growth stage, combining YOLO-based and segmentation models, could be a feasible application to enhance the model's performance at high flight heights. Although heavier models often achieve higher accuracy, their deployment using affordable devices remains challenging due to their large size, increasing the cost of developing the hardware and software and limiting their use in large-scale field applications. Thus, creating efficient models and hardware can help support extension professionals, growers, consultants, and other stakeholders to make confident, data-driven management decisions by monitoring the field variability and crop development.

5. Conclusions

This study demonstrates that both background conditions and flight height affect object detection performance in corn. The highest model accuracy was observed at stages V3 and V5, with mAP50 = 0.85, mAP50–95 = 0.75, and Precision and Recall near 0.80, highlighting the strong potential for early-stage detection. Performance was low at the V2 stage due to small plant size and background interference, and at V6 due to leaf overlap. Complex backgrounds in no-tillage fields reduced the model's accuracy, while red-brown soils enhanced contrast and improved model performance. Lower flight heights enabled effective detection even with lightweight models, whereas higher altitudes (70 m) reduced detection at V2 and V6, indicating the need for more advanced models under these conditions.

From an operational perspective, these results highlight a clear trade-off between spatial resolution, model complexity, and field coverage. Low-altitude flights are more preferable when precise per-plant counts are required in sample plots, whereas higher altitudes are more suitable for large-scale monitoring of intermediate growth stages, where timely decisions and reduced data volume are priorities. The study also has limitations that suggest directions for future research. Soil background was treated as a visual reference in RGB imagery, and detailed soil properties (standardized color scales, residue cover, structure, and moisture) were not measured systematically. Future work should combine

https://doi.org/10.3390/rs18010014

UAV imagery with in situ measurements of soil and residue properties, expand the dataset to include a wider range of soil backgrounds, illumination conditions, and no-tillage systems, and explore advanced augmentation or generative approaches (such as GANs) to better represent challenging backgrounds and early growth stages.

Author Contributions: Conceptualization, L.N.L. and A.F.d.S.; methodology, T.O.C.B., L.N.L. and A.F.d.S.; software, T.O.C.B.; validation, E.K.B., L.N.L. and A.F.d.S.; formal analysis, T.O.C.B.; investigation, T.O.C.B. and L.N.L.; resources, L.N.L., A.F.d.S. and G.V.; data curation, T.O.C.B., E.K.B., L.N.L. and A.F.d.S.; writing—original draft preparation, T.O.C.B. and L.N.L.; writing—review and editing, L.N.L., A.F.d.S., E.K.B. and G.V.; visualization, T.O.C.B. and L.N.L.; supervision, L.N.L.; project administration, A.F.d.S. and L.N.L.; funding acquisition, L.N.L. All authors have read and agreed to the published version of the manuscript.

Funding: This research was funded by the Georgia Commodity Commission for Corn grant number [CR2409] and by the USDA Foreign Agricultural Service grant number [6000028574].

Data Availability Statement: Data available upon request.

Acknowledgments: The author would like to thank the following Brazilian agencies: National Council for Scientific and Technological Development (CNPq), Brazilian Federal Agency for Support and Evaluation of Graduate Education (CAPES), Fundação de Amparo a Pesquisa de Minas Gerais (FAPEMIG). Also, the authors acknowledge and thank the University of Georgia for its support during the project development and the Lacerda Research Group (LRG) for its support during the data collection and processing. The authors extend the acknowledgments to the Extension and Research Group in Digital Agriculture (GEPAD) undergraduate students from Federal University of Lavras (UFLA) for the labor-intensive label creation process.

Conflicts of Interest: The authors declare no conflicts of interest.

Abbreviations

The following abbreviations are used in this manuscript:

IHI	Iron Horse I
IHII	Iron Horse II
SIRP	Stripling Irrigation Research Park
YOLO	You Only Look Once

References

1. United States Department of Agriculture. *Corn Explorer*; USDA: Washington, DC, USA, 2025.
2. Tang, B.; Zhou, J.; Zhao, C.; Pan, Y.; Lu, Y.; Liu, C.; Ma, K.; Sun, X.; Zhang, R.; Gu, X. Using UAV-based multispectral images and CGS-YOLO algorithm to distinguish maize seeding from weed. *Artif. Intell. Agric.* **2025**, *15*, 162–181. [CrossRef]
3. Jia, Y.; Fu, K.; Lan, H.; Wang, X.; Su, Z. Maize tassel detection with CA-YOLO for UAV images in complex field environments. *Comput. Electron. Agric.* **2024**, *217*, 108562. [CrossRef]
4. Vong, C.H.; Conway, L.S.; Zhou, J.; Kitchen, N.R.; Sudduth, K.A. Early corn stand count of different cropping systems using UAV-imagery and deep learning. *Comput. Electron. Agric.* **2021**, *186*, 106214. [CrossRef]
5. Sangoi, L. Understanding plant density effects on maize growth and development: An important issue to maximize grain yield. *Ciência Rural* **2001**, *31*, 159–168. [CrossRef]
6. Coulter, J.A.; Nafziger, E.D.; Abendroth, L.J.; Thomison, P.R.; Elmore, R.W.; Zarnstorff, M.E. Agronomic responses of corn to stand reduction at vegetative growth stages. *Agron. J.* **2011**, *103*, 577–583. [CrossRef]
7. Stanger, T.F.; Lauer, J.G. Optimum plant population of Bt and non-Bt corn in Wisconsin. *Agron. J.* **2006**, *98*, 914–921. [CrossRef]
8. Roekel, R.J.V.; Coulter, J.A. Agronomic Responses of Corn to Planting Date and Plant Density. *Agron. J.* **2011**, *103*, 1464–1565. [CrossRef]
9. da Silva, E.E.; Baio, F.H.R.; Kolling, D.F.; Júnior, R.S.; Zanin, A.R.A.; Neves, D.C.; Fontoura, J.V.P.F.; Teodoro, P.E. Variable-rate in corn sowing for maximizing grain yield. *Sci. Rep.* **2021**, *11*, 12711. [CrossRef]
10. Kitano, B.T.; Mendes, C.T.; Geus, A.R.; Oliveira, H.C.; Souza, J.R. Corn Plant Counting Using Deep Learning and UAV Images. *IEEE Geosci. Remote Sens. Lett.* **2019**, 1–5. [CrossRef]

https://doi.org/10.3390/rs18010014

11. Wang, L.; Xiang, L.; Tang, L.; Jiang, H. A Convolutional Neural Network-Based Method for Corn Stand Counting in the Field. *Sensors* **2021**, *21*, 507. [CrossRef] [PubMed]
12. Zhang, Z.; Kayacan, E.; Thompson, B.; Chowdhary, G. High precision control and deep learning-based corn stand counting algorithms for agricultural robots. *Auton. Robot.* **2020**, *44*, 1289–1302. [CrossRef]
13. Shen, J.; Wang, Q.; Zhao, M.; Hu, J.; Wang, J.; Shu, M.; Liu, Y.; Guo, W.; Qiao, H.; Niu, Q.; et al. Mapping Maize Planting Densities Using Unmanned Aerial Vehicles, Multispectral Remote Sensing, and Deep Learning Technology. *Drones* **2024**, *8*, 140. [CrossRef]
14. Yaseen, M. What Is YOLOV9: An In-Depth Exploration of the Internal Features of the Next-Generation Object Detector. In *Computer Science, Computer Vision and Pattern Recognizition*; Cornell University: Ithaca, NY, USA, 2024.
15. Akdogan, C.; Ozer, T.; Oguz, Y. PP-YOLO: Deep learning based detection model to detect apple and cherry trees in orchard based on Histogram and Wavelet preprocessing techniques. *Comput. Electron. Agric.* **2025**, *232*, 110052. [CrossRef]
16. Sapkota, R.; Meng, Z.; Churuvija, M.; Du, X.; Mab, Z.; Karkee, M. Comprehensive Performance Evaluation of YOLOv12, YOLO11, YOLOv10, YOLOv9 and YOLOv8 on Detecting and Counting Fruitlet in Complex Orchard Environments. *arXiv* **2025**, arXiv:2407.12040.
17. Wang, C.; Yeh, I.; Liao, H.M. YOLOv9: Learning What You Want to Learn Using Programmable Gradient Information. *arXiv* **2024**, arXiv:2402.13616.
18. Tian, Y.; Ye, Q.; Doermann, D. YOLOv12: Attention-Centric Real-Time Object Detectors. *arXiv* **2025**, arXiv:2502.12524.
19. Alif, M.A.R.; Hussain, M. Yolov1 to Yolov10: A Comprehensive Review of Yolo Variants and Their Application in the Agricultural Domain. *arXiv* **2024**, arXiv:2406.10139. [CrossRef]
20. Vijayakumar, A.; Vairavasundaram, S. YOLO-based Object Detection Models: A Review and its Applications. *Multimed Tools Appl.* **2024**, *83*, 83535–83574. [CrossRef]
21. Yang, J.; Bagavathiannan, M.; Wang, Y.; Chen, Y.; Yu, J. A comparative evaluation of convolutional neural networks, training image sizes, and deep learning optimizers for weed detection in alfalfa. *Weed Technol.* **2023**, *36*, 512–522. [CrossRef]
22. Wang, B.; Zhou, J.; Costa, M.; Kaeppler, S.M.; Zhang, Z. Plot-Level Maize Early Stage Stand Counting and Spacing Detection Using Advanced Deep Learning Algorithms Based on UAV Imagery. *Agronomy* **2023**, *13*, 1728. [CrossRef]
23. Ritchie, S.W.; Hanway, J.J. *How a Corn Plant Develops*; Iowa State University of Science and Technology, Cooperative Extension Service: Ames, IA, USA, 1982; Special Report No. 48.
24. Fu, X.; Li, A.; Meng, Z.; Yin, X.; Zhang, C.; Zhang, W.; Qi, L. A Dynamic Detection Method for Phenotyping Pods in a Soybean Population Based on an Improved YOLO-v5 Network. *Agronomy* **2022**, *12*, 3209. [CrossRef]
25. Nleya, T.; Chungu, C.; Kleinjan, J. Chapter 5: Corn growth and development. In *SDSU Extension Corn: Best Management Practices*; South Dakota State University: Brookings, SD, USA, 2019.
26. Yang, T.; Zhu, S.; Zhang, W.; Zhao, Y.; Song, X.; Yang, G.; Yao, Z.; Wu, W.; Liu, T.; Sun, C.; et al. Unmanned Aerial Vehicle-Scale Weed Segmentation Method Based on Image Analysis Technology for Enhanced Accuracy of Maize Seedling Counting. *Agriculture* **2024**, *14*, 175. [CrossRef]
27. Zhao, Z.Q.; Zheng, P.; Xu, S.T.; Wu, X. Object Detection with Deep Learning: A Review. *IEEE Trans. Neural Netw. Learn. Syst.* **2019**, *30*, 11. [CrossRef]
28. Soviany, P.; Ionescu, R.T. Optimizing the Trade-Off between Single-Stage and Two-Stage Deep Object Detectors using Image Difficulty Prediction. In Proceedings of the 20th International Symposium on Symbolic and Numeric Algorithms for Scientific Computing (SYNASC), Timisoara, Romania, 20–23 September 2018.
29. Ali, M.L.; Zhang, Z. The YOLO Framework: A Comprehensive Review of Evolution, Applications, and Benchmarks in Object Detection. *Computers* **2024**, *13*, 336. [CrossRef]
30. Liu, M.; Su, W.H.; Wang, X.Q. Quantitative Evaluation of Maize Emergence Using UAV Imagery and Deep Learning. *Remote Sens.* **2023**, *15*, 1979. [CrossRef]
31. Mota-Delfin, C.; López-Canteñs, G.d.J.; López-Cruz, I.L.; Romantchik-Kriuchkova, E.; Olguín-Rojas, J.C. Detection and Counting of Corn Plants in the Presence of Weeds with Convolutional Neural Networks. *Remote Sens.* **2021**, *14*, 4892. [CrossRef]
32. Wang, X.; Zhang, C.; Qiang, Z.; Liu, C.; Wei, X.; Cheng, F. A Coffee Plant Counting Method Based on Dual-Channel NMS and YOLOv9 Leveraging UAV Multispectral Imaging. *Remote Sens.* **2024**, *16*, 3810. [CrossRef]
33. Pang, Y.; Shi, Y.; Gao, S.; Jiang, F.; Veeranampalayam-sivakumar, A.; Thompson, L.; Luck, J.; Liu, C. Improved crop row detection with deep neural network for early-season maize stand count in UAV imagery. *Comput. Electron. Agric.* **2020**, *178*, 105766. [CrossRef]
34. Lu, C.; Nnadozie, E.; Camenzind, M.P.; Hu, Y.; Yu, K. Maize plant detection using UAV-based RGB imaging and YOLOv5. *Front. Plant Sci.* **2024**, *14*, 1274813. [CrossRef] [PubMed]
35. Feng, A.; Zhou, J.; Vories, E.; Sudduth, K.A. Evaluation of Cotton Emergence Using UAV-Based Imagery and Deep Learning. *Comput. Electron. Agric.* **2020**, *177*, 105711. [CrossRef]
36. Sharma, A.; Kumar, V.; Longchamps, L. Comparative performance of YOLOv8, YOLOv9, YOLOv10, YOLOv11 and Faster R-CNN models for detection of multiple weed species. *Smart Agric. Technol.* **2024**, *9*, 100648. [CrossRef]

https://doi.org/10.3390/rs18010014

37. Badgujar, C.H.; Poulose, A.; Gan, H. Agricultural object detection with You Only Look Once (YOLO) Algorithm: A bibliometric and systematic literature review. *Comput. Electron. Agric.* **2024**, *223*, 109090. [CrossRef]
38. Bai, Y.; Nie, C.; Wang, H.; Cheng, M.; Liu, S.; Yu, X.; Shao, M.; Wang, Z.; Wang, S.; Tuohuti, N.; et al. A fast and robust method for plant count in sunflower and maize at different seedling stages using high-resolution UAV RGB imagery. *Precis. Agric.* **2022**, *23*, 1720–1742. [CrossRef]
39. Wei, P.; Yan, X.; Yan, W.; Sun, L.; Xu, J.; Yuan, H. Precise extraction of targeted apple tree canopy with YOLO-Fi model for advanced UAV spraying plans. *Comput. Electron. Agric.* **2024**, *226*, 109425. [CrossRef]
40. Tseng, H.H.; Yang, M.D.; Saminathan, R.; Hsu, Y.C.; Yang, C.Y.; Wu, D.H. Rice Seedling Detection in UAV Images Using Transfer Learning and Machine Learning. *Remote Sens.* **2022**, *14*, 2837. [CrossRef]
41. Coulibaly, S.; Foguem, B.K.; Kmissoko, D.; Traore, D. Deep neural networks with transfer learning in millet crop images. *Comput. Ind.* **2019**, *108*, 115–120. [CrossRef]
42. Zhang, F.; Dong, D.; Jia, X.; Guo, J.; Yu, X. Sugarcane-YOLO: An Improved YOLOv8 Model for Accurate Identification of Sugarcane Seed Sprouts. *Agronomy* **2024**, *14*, 2412. [CrossRef]
43. Sun, D.; Zhang, K.; Zhong, H.; Xie, J.; Xue, X.; Yan, M.; Wu, W.; Li, J. Efficient Tobacco Pest Detection in Complex Environments Using an Enhanced YOLOv8 Model. *Agriculture* **2024**, *14*, 353. [CrossRef]

https://doi.org/10.3390/rs18010014

MDPI AG
Grosspeteranlage 5
4052 Basel
Switzerland
Tel.: +41 61 683 77 34

Remote Sensing Editorial Office
E-mail: remotesensing@mdpi.com
www.mdpi.com/journal/remotesensing

MDPI

www.ingramcontent.com/pod-product-compliance
Lightning Source LLC
Chambersburg PA
CBHW061809210926
51850CB00006B/654
* 9 7 8 3 7 2 5 8 6 6 1 0 6 *